Arctic Ecosystems in a Changing Climate

Physiological Ecology
A Series of Monographs, Texts, and Treatises

List continues at the end of the volume.

Arctic Ecosystems in a Changing Climate

An Ecophysiological Perspective

Edited by

F. Stuart Chapin III
Department of Integrative Biology
University of California, Berkeley
Berkeley, California

Robert L. Jefferies
Department of Botany
University of Toronto
Toronto, Ontario
Canada

James F. Reynolds
Department of Botany
Duke University
Durham, North Carolina

Gaius R. Shaver
Ecosystems Center
Marine Biological Laboratory
Woods Hole, Massachusetts

Josef Svoboda
Department of Botany
University of Toronto
Toronto, Ontario
Canada

Ellen W. Chu
Developmental Editor

1992

Academic Press, Inc.
Harcourt Brace Jovanovich, Publishers
San Diego   New York   Boston   London   Sydney   Tokyo   Toronto

Front cover photo: Aerial view of a river drainage system with hills of dwarf shrub tundra. For details see Chapter 4, p. 67 and Color Plate 3.

Academic Press, Inc.
San Diego, California 92101

United Kingdom Edition published by
Academic Press Limited
24–28 Oval Road, London NW1 7DX

Library of Congress Cataloging-in-Publication Data

Arctic ecosystems in a changing climate : an ecophysiological
    perspective / F. Stuart Chapin III ... [et al.].
        p.   cm. -- (Physiological ecology series)
    Includes index.
    ISBN 0-12-168250-1
    1. Plant physiological ecology--Arctic regions.   2. Vegetation and
climate--Arctic regions.   3. Vegetation dynamics--Arctic regions.
    4. Global warming.   I. Chapin, F. Stuart (Francis Stuart), iii.
    II. Series.
QK474.A48   1991
581.5'2621--dc20   9120355
                                                           CIP

PRINTED IN THE UNITED STATES OF AMERICA
91 92 93 94   9 8 7 6 5 4 3 2 1

William Dwight Billings has provided inspiration, advice, example, argument, and encouragement to generations of students of ecology. He has played a leading role in the development of physiological ecology over the past forty years, and his research continues to lead the way to important advances in our understanding of the Arctic. Virtually every chapter in this book has been influenced, directly or indirectly, by Dwight's contributions. It is thus with respect and affection that we dedicate this book to William Dwight Billings.

Contents

Introduction

1. Arctic Plant Physiological Ecology: A Challenge for the Future
F. S. Chapin III, R. L. Jefferies, J. F. Reynolds, G. R. Shaver, and J. Svoboda

Part I
The Arctic System

2. Arctic Climate: Potential for Change under Global Warming
Barrie Maxwell

3. Arctic Hydrology and Climate Change
Douglas L. Kane, Larry D. Hinzman, Ming-ko Woo, and Kaye R. Everett

Part II
Carbon Balance

8. Photosynthesis, Respiration, and Growth of Plants in the Soviet Arctic

O. A. Semikhatova, T. V. Gerasimenko, and T. I. Ivanova

9. Phenology, Resource Allocation, and Growth of Arctic Vascular Plants

Gaius R. Shaver and Jochen Kummerow

10. The Ecosystem Role of Poikilohydric Tundra Plants

J. D. Tenhunen, O. L. Lange, S. Hahn, R. Siegwolf, and S. F. Oberbauer

11. Arctic Tree Line in a Changing Climate

Bjartmar Sveinbjörnsson

Part III
Water and Nutrient Balance

Part IV
Interactions

17. Response of Tundra Plant Populations to Climatic Change
James B. McGraw and Ned Fetcher

18. Controls over Secondary Metabolite Production by Arctic Woody Plants
John P. Bryant and Paul B. Reichardt

19. Tundra Grazing Systems and Climatic Change
R. L. Jefferies, J. Svoboda, G. Henry, M. Raillard, and R. Ruess

20. Modeling the Response of Arctic Plants to Changing Climate
James F. Reynolds and Paul W. Leadley

Summary

21. Arctic Plant Physiological Ecology in an Ecosystem Context

F. S. Chapin III, R. L. Jefferies, J. F. Reynolds, G. R. Shaver, and J. Svoboda

Contributors

Numbers in parentheses indicate the pages on which the authors' contributions begin.

Frank Berendse (337), Centre for Agrobiological Research, NL-6700 AA Wageningen, The Netherlands

W. D. Billings (91, 139), Department of Botany, Duke University, Durham, North Carolina 29634

Caroline S. Bledsoe (301), Department of Land, Air, and Water Resources, University of California, Davis, California 95616

L. C. Bliss (59, 111), Department of Botany, University of Washington, Seattle, Washington 98195

John P. Bryant (377), Institute of Arctic Biology, University of Alaska, Fairbanks, Alaska 99775

David M. Chapin (301), EA Engineering, Science, and Technology, Redmond, Washington 98052

F. Stuart Chapin III (3, 321, 441), Department of Integrative Biology, University of California, Berkeley, Berkeley, California 04720

Todd E. Dawson (259), Section of Ecology and Systematics, Cornell University, Ithaca, New York 14853

Kaye R. Everett (35), Byrd Polar Research Center, Ohio State University, Columbus, Ohio 43210

Ned Fetcher (359), Department of Biology, Faculty of Natural Sciences, University of Puerto Rico, Rio Piedras, Puerto Rico 00931

T. V. Gerasimenko (169), Laboratory of Ecology of Photosynthesis, Komarov Botanical Institute, Academy of Sciences of the U.S.S.R., Leningrad 197022, U.S.S.R.

Anne E. Giblin (281), The Ecosystems Center, Marine Biological Laboratory, Woods Hole, Massachusetts 02543

S. Hahn (213), Lehrstuhl für Botanik II, Universität Würzburg, D-8700 Würzburg, Germany

G. Henry (391), Department of Geography, University of Alberta, Edmonton, Alberta T6G 2E1, Canada

Larry D. Hinzman (35), Water Research Center, Institute of Northern Engineering, University of Alaska, Fairbanks, Alaska 99775

T. I. Ivanova (169), Laboratory of Ecology of Photosynthesis, Komarov Botanical Institute, Academy of Sciences of the U.S.S.R., Leningrad 197022, U.S.S.R.

R. L. Jefferies (3, 391, 441), Department of Botany, University of Toronto, Toronto, Ontario M5S 3B2

Sven Jonasson (337), Institute of Plant Ecology, Øster, Farimagsgade 2D, 1353 Copenhagen K, Denmark

Douglas L. Kane (35), Water Research Center, Institute of Northern Engineering, University of Alaska, Fairbanks, Alaska 99775

Knut Kielland (321), Institute of Arctic Biology, University of Alaska, Fairbanks, Alaska 99775

Jochen Kummerow (193), Department of Biology, San Diego State University, San Diego, California 92182

O. L. Lange (213), Lehrstuhl für Botanik II, Universität Würzburg, D-8700 Würzburg, Germany

Paul W. Leadley (413), Department of Botany, Duke University, Durham, North Carolina 27706

A. E. Linkins (281), Department of Biology, Clarkson University, Potsdam, New York 13676

N. V. Matveyeva (59), Geobotanical Division, Komarov Botanical Institute, of the Academy of Sciences of the U.S.S.R., Leningrad, 197022 U.S.S.R.

Barrie Maxwell (11), Canadian Climate Centre, Atmospheric Environment Service, Downsview, Ontario M3H 5T4, Canada

James B. McGraw (359), Department of Biology, West Virginia University, Morgantown, West Virginia 26506

Knute J. Nadelhoffer (281), The Ecosystems Center, Marine Biological Laboratory, Woods Hole, Massachusetts 02543

Steven F. Oberbauer (213, 259), Department of Biological Sciences, Florida International University, Miami, Florida 33199

Walter C. Oechel (139), Department of Biology, San Diego State University, San Diego, California 92182

K. M. Peterson (111), Department of Biology, University of Alaska, Anchorage, Alaska 99508

M. Raillard (391), Department of Botany, University of Toronto, Toronto, Ontario M5S 3B2, Canada

Paul B. Reichardt (377), Department of Chemistry, University of Alaska, Fairbanks, Alaska 99775

J. F. Reynolds (3, 413, 441), Department of Botany, Duke University, Durham, North Carolina 27706

R. Ruess (391), Institute of Arctic Biology, University of Alaska, Fairbanks, Alaska 99775

O. A. Semikhatova (169), Laboratory of Ecology of Photosynthesis, Komarov Botanical Institute, Academy of Sciences of the U.S.S.R., Leningrad 197022, U.S.S.R.

G. R. Shaver (3, 193, 281, 441), Ecosystem Center, Marine Biology Laboratory, Woods Hole, Massachusetts 02543

R. Siegwolf (213), Paul Scherrer Institut, Villigen (PSI), Switzerland

Bjartmar Sveinbjörnsson (239), Department of Biological Sciences, University of Alaska, Anchorage, Anchorage, Alaska 99508

J. Svoboda (3, 391, 441), Department of Botany, University of Toronto, Toronto, Ontario M5S 3B2, Canada

J. D. Tenhunen (213), Systems Ecology Research Group, San Diego State University, San Diego, California 92182

Ming-ko Woo (35), Geography Department, McMaster University, Hamilton, Ontario L8S 4L8, Canada

Preface

This book reviews the physiological ecology of arctic plants, suggests a new role for physiological ecology in studying biotic controls over community and ecosystem processes, and provides a physiological basis for predicting how arctic plant communities will respond to global climate change. The Arctic has always fascinated physiological ecologists because its rigorous environment challenges the physiological capabilities of organisms. For this reason, physiological ecologists have, in the past, focused on the effects the physical environment has on plants. In this book we review conclusions drawn from this work but then go on to ask how physiological traits affect community and ecosystem processes.

This approach is vital today, given the need to predict how the world's ecosystems will respond to human-induced climatic change. Moreover, the Arctic is expected to undergo the earliest and most pronounced effects of global warming. In this book we examine the complex paths by which climatic change might affect arctic plants and how these plant responses might in turn influence arctic ecosystems and their potential feedbacks to the globe. The book should thus prove useful to any practicing scientists, students of ecology, and policy makers who are interested in the Arctic, physiological ecology, ecosystem ecology, and global climate change.

The opening chapters describe the present and expected future environment of the Arctic, including climate, hydrology, and soils. The following chapters discuss the diversity of arctic vegetation, its history, and the controls over its limits at tree line. Chapters on individual physiological processes examine how each process responds to the environment, how it influences community and ecosystem processes, and how we might expect these controls to change in response to altered climate. In the last chapter, the major interactions and feedbacks between plants and their environment are considered as a whole in the light of a changing climate.

This book grew out of a conference funded by the National Science Foundation and the United States Department of Energy. It is the first in a series that will examine the responses of different ecosystems to global climate change.

Introduction

1

Arctic Plant Physiological Ecology: A Challenge for the Future

F. S. Chapin III, R. L. Jefferies, J. F. Reynolds, G. R. Shaver, and J. Svoboda

I. Introduction

Many of the most exciting advances in knowledge occur when previously independent disciplines merge. By integrating physiology and ecology, physiological ecology has contributed substantially to our understanding of the environmental controls over physiological processes and of the capacity of organisms to tolerate different environments.

Although ecophysiological studies have been important from an evolutionary perspective and offer insights into the mechanisms by which plants and microbes cope with their environment, we suggest that the field has reached a plateau and that most current research simply refines our understanding of various physiological processes. In the absence of a broader context, physiological ecology can become a sterile series of "just-so" stories— organisms live where they do because they have the physiological traits enabling them to do so (Gould and Lewontin, 1979). Because the physiological activities of plants and other organisms control many community and ecosystem processes, community and ecosystem ecology can provide the needed context for interpreting the importance of ecophysiological traits beyond the level of the individual. Thus, studying the role of organisms in controlling community and ecosystem processes should be a major research direction in physiological ecology. Such an interdisciplinary approach to ecosystem processes is imperative if we are to understand how impending changes in climate will affect our globe.

II. Physiological Ecology and Ecosystem Studies

Physiological ecology has been used to predict ecosystem response to environmental change in three primary ways, each of which has been criticized. For physiological ecology to play a constructive role in ecosystem studies, the limitations of the three approaches must be recognized and addressed:

1. Ecosystem response to environmental change is often predicted from simple physiological responses of organisms (e.g., estimation of plant production from temperature and light responses of photosynthesis). In the long term, however, feedbacks among processes often govern performance under natural circumstances more strongly than do short-term, kinetic responses of individual processes. For example, even though photosynthesis always responds to CO_2 concentration in the short term, compensatory changes in photosynthetic potential when plants are grown under different CO_2 concentrations may counteract these short-term effects (Tissue and Oechel, 1987; Chap. 7). Consequently, environmental factors that exert strong short-term effects may not be influential over longer time scales. To predict ecosystem response, we must study feedbacks as well as direct environmental effects on plants.
2. Simulation modeling based on physiological studies can incorporate many of the feedbacks that operate in natural ecosystems. Experience suggests, however, that modeling predictions cannot be usefully extrapolated beyond two levels of organization (biochemistry, cell, leaf, canopy, population, community, landscape, globe; Reynolds and Acock, 1985). For this reason, the types of controls studied at one level of organization may not be fully relevant at other levels. Most past ecophysiological work has emphasized the direct effect of environment on physiology. Yet we need to know the nature, strength, and timing of the feedbacks and time lags that control resource supply and, indirectly, the growth of organisms. This will require an integrated, whole-plant approach to physiological ecology. Ecosystem and community ecology may provide the criteria for deciding which feedbacks are important at these higher levels of organization.
3. Ecosystem response to environment has been predicted from comparisons of current performance in two different environments (e.g., latitudinal comparisons). A comparison of two ecosystems under equilibrium conditions says nothing about the trajectory an ecosystem might follow in going from one equilibrium state to another. Moreover, a new environment will be occupied by different combinations of species than those co-occurring

at present (Davis, 1981; COHMAP Members, 1988). If the dynamics of the present community are strongly influenced by specific competitive interactions, we may have difficulty predicting how new combinations of species will interact. Forecasting such interactions is particularly difficult because there are no present analogs of future climate, such as increased atmospheric CO_2. Latitudinal gradients in temperature, which might provide some insight into vegetation response to a changing thermal regime, are complicated by differences in day length and soils that will not be mimicked by climate change. Simulation modeling and whole-ecosystem experiments may enable us to test novel combinations of factors.

III. Physiological Ecology in the Arctic

The Arctic, the region north of the latitudinal treeline, is a logical place to examine the potential of physiological ecology to contribute to an understanding of ecosystem processes. Early in the history of their field, physiological ecologists were attracted to the Arctic because the adaptation of organisms to their physical environment is most evident in extreme habitats (e.g., Russell, 1940; Bliss, 1956; Mooney and Billings, 1961; Warren-Wilson, 1966). The basic patterns of physiological adaptation to Arctic life were delineated in several excellent early reviews (Bliss, 1962, 1971; Warren-Wilson, 1966; Billings and Mooney, 1968), which have been infrequently updated in the last 25 years (Tieszen, 1978; Chapin and Shaver, 1985).

The link between physiological ecology and ecosystem processes in the Arctic was actively developed from 1966 to 1974 during the International Biological Programme (Wielgolaski, 1975; Bliss, 1977; Brown *et al.*, 1980; Bliss *et al.*, 1981); this focus has continued in recent research programs (e.g., Miller *et al.*, 1984; Oechel, 1989). One reason for the effectiveness of this link in the Arctic is that the low stature and fine-grained spatial heterogeneity of arctic vegetation make it practical to manipulate the vegetation so that both physiological and ecosystem responses can be measured at similar scales. Furthermore, impending climate change is likely to be felt most strongly in the Arctic (Chap. 2), and we must therefore understand how arctic ecosystems will respond. Physiological ecology is likely to be a key discipline in achieving that understanding.

IV. Climate Change: A Theme for Arctic Physiological Ecology

Atmospheric CO_2 concentration is steadily rising (Keeling *et al.*, 1982; Trabalka, 1985), and consensus is growing that this will lead to substantial global warming. How will this affect the Arctic, and how will an altered arctic ecosystem affect the rest of the globe (Fig. 1)? If we wish to understand these arctic–global feedbacks, we should focus primarily on the global processes

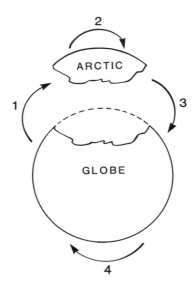

Figure 1 The role of the Arctic in global processes. Global climate affects the Arctic (arrow 1). The response of arctic biota to this altered climate changes arctic processes (arrow 2) which can feed back to alter global climate (arrow 3). The responses of nonarctic ecosystems to climate change (arrow 4) are beyond the scope of this book.

affecting the Arctic (Fig. 1: arrow 1) and the arctic processes that will significantly influence global climate (arrow 3). Global processes and interactions not involving the Arctic (arrow 4) are beyond the scope of this book, and those processes whose major effect is restricted to the Arctic (arrow 2), although important, are of less concern except insofar as they introduce time lags and feedbacks that influence global climate.

The increase in atmospheric CO_2 will probably warm the globe most strongly at the poles, perhaps by as much as 5–10° C (Solomon *et al.*, 1985; Chap. 2), and this will have widespread effects on other components of the arctic climate, such as evaporation, cloudiness, and precipitation. The climate in turn determines the thermal and hydrologic regimes of soils, and vegetation determines the degree of coupling between the climate and soil (Chap. 3). Already, the permafrost may be warming in the Arctic (Lachenbruch and Marshall, 1986).

About 80% of the organic carbon in the world's terrestrial ecosystems is in soil, and of this, 27% is stored in tundra and taiga ecosystems (Billings, 1987: Chap. 5). We must predict whether climatic warming will release substantial proportions of this soil carbon back to the atmosphere. The fate of this soil carbon will be determined by the biotic system and its response to climate change. A study of these interactions is the realm of physiological ecology. In addition, tundra and boreal wetlands are important sources of methane, a potent greenhouse gas that contributes substantially to global warming (Whalen and Reeburgh, 1990). The extent of methane oxidation versus release to the atmosphere is largely governed by soil moisture and vegetation type. Thus, processes that determine community composition strongly influence methane release.

In summary, global processes will cause changes in the Arctic primarily through changes in climate. The extent to which these climatic changes will affect arctic ecosystems depends on the physiological responses of arctic organisms to the altered climate and especially on the strength and nature of feedbacks coupling plant and soil processes. Our challenge is to demonstrate that physiological ecology can contribute substantially to predicting the nature and strengths of feedbacks between the Arctic and the rest of the globe.

References

Billings, W. D. (1987). Carbon balance of Alaskan tundra and taiga ecosystems: Past, present, and future. *Quat. Sci. Rev.* **6**, 165–177.

Billings, W. D., and Mooney, H. A. (1968). The ecology of arctic and alpine plants. *Biol. Rev. Camb. Philos. Soc.* **43**, 481–529.

Bliss, L. C. (1956). A comparison of plant development in microenvironments of arctic and alpine tundra. *Ecol. Monogr.* **26**, 303–337.

Bliss, L. C. (1962). Adaptations of arctic and alpine plants to environmental conditions. *Arctic* **15**, 117–144.

Bliss, L. C. (1971). Arctic and alpine plant life cycles. *Annu. Rev. Ecol. Syst.* **2**, 405–438.

Bliss, L. C., ed. (1977). "Truelove Lowland, Devon Island, Canada: A High Arctic Ecosystem." Univ. of Alberta Press, Edmonton.

Bliss, L. C., Heal, O. W., and Moore, J. J. (1981). "Tundra Ecosystems: A Comparative Analysis." Cambridge Univ. Press, Cambridge.

Brown, J., Miller, P. C., Tieszen, L. L., and Bunnell, F. L. (1980). "An Arctic Ecosystem: The Coastal Tundra at Barrow, Alaska." Dowden, Hutchinson & Ross, Stroudsburg, Pennsylvania.

Chapin, F. S., III, and Shaver, G. R. (1985). Arctic. *In* "Physiological Ecology of North American Plant Communities" (B. F. Chabot and H. A. Mooney, eds.), pp. 16–40. Chapman and Hall, London.

COHMAP Members. (1988). Climatic changes of the last 18,000 years: Observations and model simulations. *Science* **241**, 1043–1051.

Davis, M. B. (1981). Quaternary history and the stability of forest communities. *In* "Forest Succession" (D. C. West, H. H. Shugart, and D. B. Botkin, eds.), pp. 132–153. Springer-Verlag, New York.

Gould, S. J., and Lewontin, R. C. (1979). The spandrels of San Marco and the Panglosian paradigm: A critique of the adaptationist programme. *Proc. R. Soc. Lond. B* **205**, 581–598.

Keeling, C. D., Bacastow, R. B., and Whorf, T. P. (1982). Measurements of the concentration of carbon dioxide at Mauna Loa Observatory, Hawaii. *In* "Carbon Dioxide Review: 1982" (W. C. Clark, ed.), pp. 377–385. Oxford Univ. Press, New York.

Lachenbruch, A. H., and Marshall, B. V. (1986). Changing climate: Geothermal evidence from permafrost in the Alaskan Arctic. *Science* **234**, 689–696.

Miller, P. C., Miller, P. M., Blake-Jacobson, M., Chapin, F. S., III, Everett, K. R., Hilbert, D. W., Kummerow, J., Linkins, A. E., Marion, G. M., Oechel, W. C., Roberts, S. W., and Stuart, L. (1984). Plant–soil processes in *Eriophorum vaginatum* tussock tundra in Alaska: A systems modeling approach. *Ecol. Monogr.* **54**, 361–405.

Mooney, H. A., and Billings, W. D. (1961). Comparative physiological ecology of arctic and alpine populations of *Oxyria digyna*. *Ecol. Monogr.* **31**, 1–29.

Oechel, W. C. (1989). Nutrient and water flux in a small arctic watershed: An overview. *Holarct. Ecol.* **12**, 229–237.

Reynolds, J. F., and Acock, B. (1985). Predicting the response of plants to increasing carbon dioxide: A critique of plant growth models. *Ecol. Modell.* **29**, 107–129.

Russell, R. S. (1940). Physiological and ecological studies on an arctic vegetation. III. Observations on carbon assimilation, carbohydrate storage, and stomatal movement in relation to the growth of plants on Jan Mayen Island. *J. Ecol.* **28**, 289–309.

Solomon, A. M., Trabalka, J. R., Reichle, D. E., and Voorhees, L. D. (1985). The global cycle of carbon. *In* "Atmospheric Carbon Dioxide and the Global Carbon Cycle" (J. R. Trabalka, ed.), pp. 1–13. DOE/ER-0239. Natl. Tech. Info. Serv., Springfield, Virginia.

Tieszen, L. L., ed. (1978). "Vegetation and Production Ecology of an Alaskan Arctic Tundra." Springer-Verlag, New York.

Tissue, D. T., and Oechel, W. C. (1987). Response of *Eriophorum vaginatum* to elevated CO_2 and temperature in the Alaskan tussock tundra. *Ecology* **68**, 401–410.

Trabalka, J. R., ed. (1985). "Atmospheric Carbon Dioxide and the Global Carbon Cycle." DOE/ER-0239. Natl. Tech. Info. Serv., Springfield, Virginia.

Warren-Wilson, J. (1966). An analysis of plant growth and its control in arctic environments. *Ann. Bot.* **30**, 383–402.

Whalen, S. C., and Reeburgh, W. S. (1990). A methane flux transect along the trans-Alaska pipeline haul road. *Tellus* **42B**, 237–249.

Wielgolaski, F. E., ed. (1975). "Fennoscandian Tundra Ecosystems." Part I, Plants and Microorganisms. Springer-Verlag, Berlin.

Part I

The Arctic System

2

Arctic Climate: Potential for Change under Global Warming

Barrie Maxwell

I. Introduction

Climate is the main external factor controlling the functioning of global ecosystems. At present, there is a scientific consensus that the climate is likely to change within our lifetimes (Houghton *et al.*, 1990). Such change will take the form of a significant global warming, which is expected to be most pronounced at polar latitudes. Plant and animal species and communities have changed greatly in response to changing climates in the past, suggesting a substantial potential for ecosystem transformation following climate changes in the future (Warrick *et al.*, 1986). Arctic ecosystems, which have reacted significantly to past climatic changes (AES, 1985), are likely to be the most altered under future global warming (French, 1986; Maxwell and Barrie, 1989; Roots, 1989; Etkin, 1990a).

As a basis for assessing the implications of such warming, this chapter outlines prominent features of the present arctic climate and examines the anticipated "greenhouse effect," its causes, and the global climate models used to study it. Some of the implications of the greenhouse effect for characteristic features of arctic ecosystems are also discussed.

II. Present-Day Climate

The Arctic is characterized by continuous darkness during several of the winter months and continuous daylight in the summer, although these extremes are tempered in several ways. Complete darkness does not settle in until at least one month after the sun sets completely in the autumn, creating a prolonged arctic twilight. Even during the polar night, moonlight reflected from snow and ice surfaces noticeably lightens the darkness. In summer, because the days are so long, the total input of solar energy is approximately the same as at lower latitudes, but because arctic surfaces are highly reflective, a smaller percentage of the available energy actually remains to heat the earth's surface and the atmosphere.

Unequal solar heating of the globe, combined with the earth's rotation, drives general atmospheric circulation, which in turn directs large-scale storm movements. Regional circulation patterns are altered by mountain ranges and by differential solar heating from season to season over different surfaces such as forest, tundra, snow, ice, and water. Atmospheric circulation in the Arctic is dominated by a deep, cold, low-pressure area known as the circumpolar vortex, which extends through the middle and upper troposphere and lower stratosphere. This feature appears on the January 50-kPa mean height pattern as a pronounced area of low heights extending over the North Pole with three distinct troughs, the most intense of which is centred over eastern Canada near northern Baffin Island (Fig. la). Such a pattern steers weather systems from northwest to southeast across Canada in winter. In summer (Fig. lb), the vortex is less intense, with its deepest value now occurring over the pole and no longer solely identified with the eastern Canada trough. The trough has withdrawn northward, thus allowing weather systems to penetrate the Canadian Archipelago routinely.

An important climate control in the Arctic is the nature of the underlying surface. The Arctic Basin is an ice-covered ocean that plays an important role in determining the characteristics of large-scale arctic air masses, but the seas and channels separating the adjacent islands and land masses exert a dominant influence on the coasts and adjoining lands. For example, spring arrives considerably later in the Arctic than in more temperate areas. This phenomenon is due, first, to the melting of snow and ice, which requires considerable thermal energy and, second, to the high reflective properties of those surfaces. Snow and ice fields also affect the character and movement of low- and high-pressure areas and frontal systems. During summer and autumn, the climate is influenced by the stretches of cold, open water that develop. Thus, local onshore and offshore winds, low clouds, and fog occur frequently in summer, as do snow squalls in autumn. High land masses also

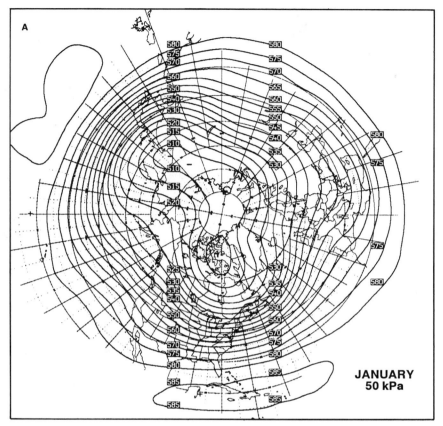

Figure 1   Circumpolar 50-kPa mean height pattern for January (A) and July (B) based on the 1949–1978 time period. Heights, given in decameters (dm) and contoured at 5-dm intervals, represent the elevation above the earth's surface at which a pressure of 50 kPa occurs. [Maps reprinted from Harley (1980).] (*Continues*)

affect local climate. In the eastern Canadian Arctic, for example, the mountain ranges of Ellesmere, Devon, and Baffin islands, together with Greenland, act as a barrier to mild, moist air that might otherwise penetrate from the North Atlantic. The leeward slopes also receive considerably more precipitation than adjacent inland areas, particularly on southeastern Baffin Island.

Figure 2 shows how these climate controls are reflected in the January and July surface mean air temperature patterns. A core of cold air over the Arctic Ocean extends into the Canadian Archipelago in winter. In summer, temperatures hover near 0° C over the permanent ice pack and are constrained to not much above 5° C over most of the adjacent channels and smaller islands. At certain inland locations on some of the larger islands, however, summer temperatures can be considerably warmer. An important winter

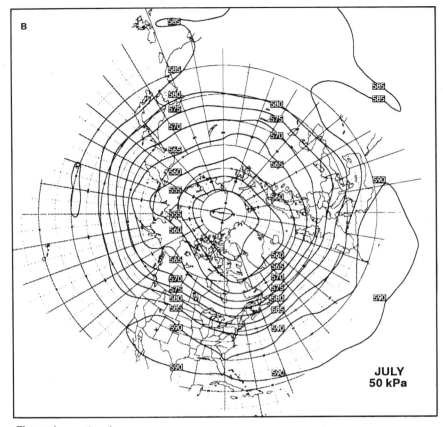

B

JULY
50 kPa

Figure 1   *continued.*

feature of the arctic surface temperature regime is the arctic inversion (temperature increases with height). At the top of the inversion, near 1500 m, typical temperatures are -20 to $-25°$ C—a temperature inversion of 10 to 20° C. In summer, the inversion is much shallower and usually restricted to areas with ice and cold water surfaces.

Figure 3 illustrates the two sides of the arctic precipitation regime: generally low totals (100–150 mm) over much of the area (the central arctic islands) punctuated by locally higher amounts resulting from topographic influences, nearness to major cyclonic trajectories, or both (e.g., the exposed coasts of the eastern arctic islands). Due to wind effects, actual precipitation may be double or more the amount recorded in standard precipitation gauges (Maxwell, 1980, 1982).

The growing season for plants is climatically defined by the total number of degree-days above 5° C (growing degree-days), that is, the cumulative total of the daily average temperature values that exceed 5° C. Fairly good approximations

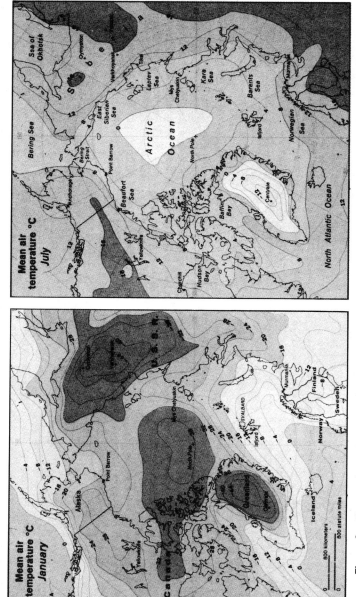

Figure 2 Circumpolar surface mean air temperature for January (left) and July (right). [Reprinted from CIA (1978).]

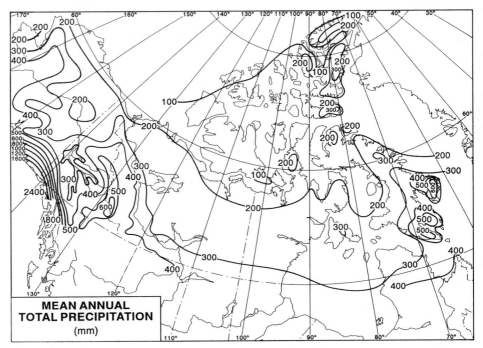

Figure 3 Mean annual precipitation (mm) over northern North America, assembled from AES (1986) and selected Alaskan station data.

of the boundaries of different ecosystems can be derived from contours of annual growing degree-day (GDD) totals, although the relationship may break down where precipitation limits growth (Zoltai, 1988). The boundaries between the Low and mid-Arctic and between the mid- and High Arctic (Fig. 4a) have been found to coincide roughly with the 6–8° C and 5–6° C mean July isotherms, respectively. Further zonation within the high-arctic area north of the Parry Channel is also possible on the basis of the 4 and 3° C isotherms. These relationships between arctic vegetation and growing degree-days or July mean temperatures are examples of analytical tools that may help in evaluating the impacts on ecosystems of changes in the current arctic climate.

III. The Greenhouse Effect

Incoming solar radiation (shortwave, or ultraviolet) heats the Earth–atmosphere system and is balanced by the emission of thermal radiation (long-wave, or infrared) from Earth to space. Equilibrium in the system would be maintained by an effective planetary radiating temperature of 255° K (−18° C), which would be the Earth's surface temperature were it not for the atmosphere.

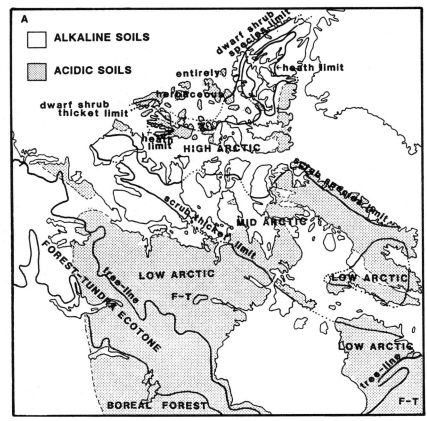

Figure 4 Pattern of vegetation boundaries in arctic Canada currently found (A), and projected under a scenario of a 2–3° C northward shift in summer isotherms and the matching of coincident vegetation patterns (B). [Reprinted from Edlund (1986).] (*continues*)

The atmosphere plays a moderating role because it contains certain gases that are transparent to incoming solar radiation while effectively absorbing outgoing long-wave radiation. Some of the absorbed longwave radiation is therefore radiated back to the Earth's surface, warming it further. As a result, the actual mean surface temperature of the Earth is about 15° C. This gas-induced increase in surface temperature has been termed the greenhouse effect.

Carbon dioxide is by far the most widely discussed of the greenhouse gases. Over time, its concentration in the atmosphere has varied substantially, from several thousand parts per million (ppm) more than one million years ago to the 200–300 ppm characteristic of the recent glacial–interglacial cycles (Fig. 5). From the end of the last ice age 10,000 years ago to the beginning of the nineteenth century, the concentration was relatively steady at about 260–290 ppm. Since then, measurable human influence, through fossil fuel

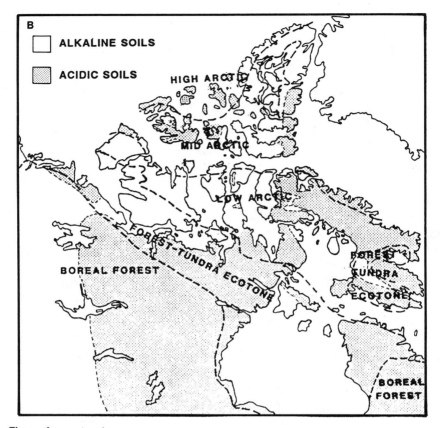

Figure 4   *continued.*

consumption and certain land-use practices, has caused atmospheric CO_2 concentrations to increase steadily to values approaching 350 ppm (Gammon *et al.*, 1985). It is this recent increase in concentration—of approximately 20% in less than two centuries at a rate that would double the total concentration from its present-day value before the middle of the next century—that has prompted the present global concern about future climate warming.

More and more, however, it is becoming apparent that the combined increases in other trace gases (methane, nitrous oxide, and chlorofluorocarbons [CFCs]) may play as big a role as the increase in carbon dioxide (Wang *et al.*, 1986; Ramanathan *et al.*, 1987; Hansen *et al.*, 1989; Lashof, 1989), even though the concentrations of these trace gases are low compared with CO_2 (Table I). The efficiency of a gas as a greenhouse gas is determined by its concentration, the wavelengths at which it absorbs radiation, its strength of absorption per molecule, and whether other gases absorb strongly at the same wavelengths (Mitchell, 1989). The trace gases generally absorb longwave radi-

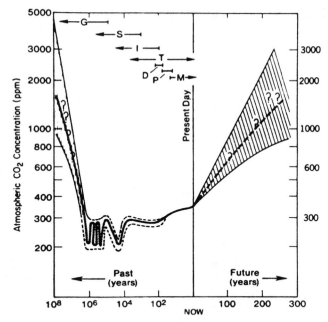

Figure 5 Atmospheric CO_2 concentrations from 10^8 years ago to the present day on a log–log scale. For comparison, possible CO_2 levels projected for the next few centuries are shown on a linear time scale. The analytical method appropriate to each time scale is indicated by the letters and arrows: G, geological carbon cycle models; S, ocean sediment cores; I, trapped air bubbles in ice cores; T, ^{13}C isotopic studies of tree rings; D, direct chemical measurements of the past century; P, spectroscopic plates from Smithsonian Solar Constant Program; M, Mauna Loa record and subsequent precise atmospheric CO_2 measurements by nondispersive infrared spectroscopy. The hatching and dashed lines indicate the general level of uncertainty about the estimated CO_2 concentration in each time interval. The ice-age cycling is only representative of the range of CO_2, not the specific number of glaciations. [Reprinted from Gammon *et al.* (1985).]

ation more efficiently than does CO_2. Moreover, because they remain in the atmosphere longer than CO_2, their role in greenhouse warming increases with time (Etkin, 1990b).

Water vapor may also act as a greenhouse gas through its feedbacks on climate. An increase in atmospheric temperature produces more water vapor, which in turn leads to further warming—a positive feedback process. Increased water vapor, however, also increases cloud water content and reflectivity, which reduces solar heating of the atmosphere and the surface, tending to reduce temperature. Indeed, the possible effects of greenhouse warming on cloudiness are uncertain enough that the magnitude of the greenhouse effect as a whole also remains uncertain (Mitchell, 1989).

Much of the work on the possible impact of greenhouse warming has been based on a doubling of present-day atmospheric CO_2. This concentration is expected in less than 50 years given the current rate of increase, a time

Table I   Past and Projected Greenhouse Gas Concentrations and Associated Changes in Greenhouse Heating (ΔQ)[a]

Gas	Assumed 1860 concentration (ppm)	ΔQ 1860–1985 (W m^{-2})	Estimated 2035 concentration (ppm)	Estimated ΔQ 1985–2035 (W m^{-2})
Carbon dioxide	275.0	1.3	475.0	1.8
Methane	1.1	0.4	2.8	0.5
Nitrous oxide	0.28	0.05	0.38	0.15
CFC11	0	0.06	1.6×10^{-3}	0.35
CFC12	0	0.12	2.8×10^{-3}	0.69
Total		1.9		3.5

[a] Reprinted from Mitchell (1989) with permission from the author.

interval within the planning range for land-use decisions, wildlife management, and industrial projects. It is important to remember, however, that because of greenhouse gases other than CO_2, warming equivalent to that caused by a doubling of CO_2 alone might occur much sooner than anticipated.

IV. Climate Models

One approach for predicting what might happen with future global warming is to look for analogs in past warm epochs (Palutikof, 1986). Another, more powerful approach is computer modeling, even though present model projections contain considerable uncertainties, particularly on regional and seasonal scales. A further drawback with this technique to date is that many of the experiments have simulated an instantaneous doubling of the atmospheric CO_2 concentration and then allowed the model to run to a steady-state situation. In reality, of course, CO_2 is increasing gradually, and this might result in a very different global response than if a sudden doubling were to occur. Moreover, the real atmosphere is not in equilibrium because of lags built into the system, so comparing a model equilibrium climate to the real climate is in itself open to question.

Early climate modeling efforts relied on one- and two-dimensional models that were both economical to run and relatively easy to analyze. Although such models are useful for specific applications, such as evaluating the sensitivity of more complex models to various feedback processes (Schlesinger and Mitchell, 1985), they do not shed light on the overall global pattern of climate change. Neither do they address regional patterns, which need to be known in order to evaluate impacts of a changing climate on society.

A. General Circulation Models

Three-dimensional general circulation models (GCMs) of the atmosphere do allow global and, to a limited extent, regional simulations. They are based on six equations: those of conservation of momentum, heat and moisture; that of mass continuity; the hydrostatic equation; and the atmospheric gas equation of state. The solution of these equations, using either finite difference or spectral methods, yields values for five main variables: temperature, humidity, surface pressure, and northerly and westerly winds. These values are determined for grids of approximately 5° latitude by 7° longitude. The models operate with 2 to 11 vertical levels (Mitchell, 1989).

In general, the large-scale processes embodied in these six equations are handled reasonably well by the GCMs. The functioning of the climate system, however, also depends on a number of important small-scale processes governing phenomena such as sea ice, cloud cover, precipitation, and surface friction, which are involved in feedback loops of particular significance to the Arctic. Indeed, the manner in which a particular model deals with these processes is perhaps the most important feature of that model (Gates, 1985).

More than a dozen GCMs are being used or developed worldwide (Cess *et al.*, 1989), but the doubled-CO_2 simulations most commonly referred to are those from four American models (the model from the Geophysical Fluid Dynamics Laboratory [GFDL; Wetherald and Manabe, 1986]; from the Goddard Institute for Space Sciences [GISS; Hansen *et al.*, 1984]; from the National Center for Atmospheric Research [NCAR; Washington and Meehl, 1984]; and that from Oregon State University [OSU; Schlesinger and Zhao, 1987]); a British model from the United Kingdom Meteorological Office (UKMO; Wilson and Mitchell, 1987); and, most recently, a Canadian model from the Canadian Climate Centre (CCC; Boer *et al.*, 1991). The characteristics of several of these models have been summarized by Schlesinger and Mitchell (1985, pp. 109–112); all have strengths and weaknesses. For example, representation of the ocean in these models has been fairly rudimentary, but such representation is potentially very important, particularly for the Arctic. Simulation in which the NCAR model was coupled with an ocean-circulation model (OCM; Washington and Meehl, 1989), for example, suggested average global surface warming and patterns of sea ice extent that differed from the projections of the NCAR model alone. The authors did note several errors in model outputs, especially in their simulations of present-day North Atlantic sea ice, and drift in transient runs of models involving a coupled GCM–OCM is a problem; further studies like these are thus needed.

One measure of the level of confidence in results generated by GCMs is the degree to which they reproduce current global climate. The models often agree well, with present climate and with one another, for seasonal or annual averages over large areas, but substantial disagreements appear as the scale is reduced below continental (Grotch, 1988). Thus we should not be overconfident of current GCM predictions for any region, including the Arctic (Table

II), and predictions of the effect of doubled CO_2 must remain rather qualitative, at least in the areas of biological and socioeconomic concern.

B. Feedback Mechanisms

Geophysical feedback mechanisms involve water vapor, cloud cover, snow, and ice (Hansen *et al.*; Dickinson, 1986). They tend to be accounted for in current climate GCMs with varying degrees of sophistication. Without their inclusion, model simulations of doubled CO_2 suggest a global warming of 1.2 to 1.3° C; with their inclusion, the range is 1.5 to 4.5° C or more (Lashof, 1989). In some instances, the issue is more complicated than just ensuring that a particular mechanism is realistically modeled. For example, a warmer climate might produce an altogether different circulation regime in the oceans, which could completely alter the pattern of regional climates over the globe (Manabe and Stouffer, 1988).

Biogeochemical feedbacks, in contrast to the geophysical, have been generally neglected so far. They include physical concomitants of warming (such as release of methane from hydrates and changes in ocean circulation and mixing), short-term biological responses to warming and to increased CO_2 (such as an increased flux of carbon dioxide and methane from soil organic matter

Table II   Comparison of Results from Three GCMs Simulating Present-Day Arctic Climate[a]

GISS	GFDL	NCAR
Correctly simulates low-level inversion in arctic winter	Simulates surface inversion in arctic winter	Correctly simulates low-level inversion and nearly isothermal structure of stratosphere in arctic winter
Largest temperature differences occur in stratosphere, with values 5–10° C colder than observed in winter and summer	Summer stratospheric temperatures 10° C warmer than observed	Absolute stratospheric temperature colder than observed by >10° C in lower region over pole
Largest differences in global maritime surface air temperatures found over arctic (and antarctic) sea ice; simulated values 5–10° C colder than observed	Winter temperature means 5° C colder than observed, summer means 5° C warmer; polar night minima 20° C colder than observed; Greenland 10–20° C warmer than observed in summer and winter	Maritime surface air temperature at high latitudes in winter generally colder than observed, thus overestimating equatorward extent of sea ice
Precipitation minima simulated as observed	Precipitation overestimated	Precipitation minima simulated as observed

[a] *Arctic* here refers to the area north of 60° N latitude; *summer* and *winter* refer to averaged results for June–August and December–February, respectively. Present-day climate data against which the GCM results are validated come from citations in Schlesinger and Mitchell (1985).

Table III CCC GCM Predictions of Mean Annual Changes from Doubled CO_2[a]

Variable	Globe (100%)	Land (29% of globe)	Ice-free ocean (63.5%)
Surface air temperature (° C)	3.5	4.4	2.7
Precipitation (%)	3.8	0.9	4.3
Evaporation (%)	3.8	3.8	3.3
Cloud cover (%)	− 2.2	− 1.9	− 3.3
Soil moisture (%)		− 6.6	
Sea ice mass (%)	− 66		

[a] Reprinted from Boer *et al.* (1990a).

to the atmosphere caused by higher rates of microbial activity), and effects produced by the reorganization of ecosystems (such as changes in surface albedo, terrestrial carbon storage, and biological pumping of carbon from the ocean surface to deeper waters). The total amplification of greenhouse effects from these feedbacks has been estimated to be perhaps one-fourth that for geophysical feedbacks (Lashof, 1989) and, if included in global models of doubled CO_2, would imply a mean global surface warming of from 1.9 to >10° C (although the high end of this range should not be taken seriously).

C. Model Projections

1. Temperature and Precipitation

GCM simulations indicate that a doubling of present atmospheric CO_2 would be likely to result in a mean annual global warming of 1.9 to 5.2° C. This warming would be accompanied by increased annual evaporation and precipitation, the latter by as much as 15% (Boer *et al.*, 1990). The CCC model results, which project a mean annual global warming of 3.5° C, are broken down for land and sea areas in Table III. Note the global decrease in cloud cover of approximately 2% and in soil moisture of between 6 and 7%.

For the Arctic, the GCMs predict a winter surface warming 2–2.4 times greater than the annual average warming for the globe; in summer, however, surface warming would be only 0.5–0.7 times the global average (Table IV; Jaeger, 1988). Several factors would be responsible for the enhanced winter warming. First, reduced snow and ice cover resulting from the greenhouse effect would reduce the earth's albedo, increasing the absorption of solar radiation and, therefore, terrestrial reradiation; this positive feedback process would further warm areas of snow and ice. Second, the stable surface (inversion) layer characteristic of the Arctic would confine the warming to a relatively shallow region instead of allowing it to spread throughout the entire depth of the troposphere, thus keeping the actual surface temperature higher than it would be otherwise. Third, more latent energy would be transported from the generally warmer, wetter atmosphere of lower latitudes

Table IV Regional Scenarios for Climate Change[a]

Region	Temperature change[b] Summer	Winter	Precipitation change
High latitudes (60–90°)	0.5–0.7 times	2.0–2.4 times	Enhanced in winter
Midlatitudes (30–60°)	0.8–1.0 times	1.2–1.4 times	Possibly reduced in summer
Low latitudes (0–30°)	0.9–0.7 times	0.9–0.7 times	Enhanced in places with heavy rainfall today

[a] Reprinted from Jaeger (1988).
[b] Expressed as a multiple of the annual global average.

to the Arctic, where it is released as latent heat of condensation (Manabe and Wetherald, 1975).

In summer, the decrease in surface albedo of the Arctic Ocean caused by disappearing sea ice and the puddling of remaining sea ice would again increase both the absorption of solar radiation and terrestrial reradiation. In this case, excess heating would not translate into increased surface temperature, but rather be absorbed in melting sea ice. The warming would spread through more layers of the atmosphere, since surface inversions are less frequent in summer, or it would be stored in the oceanic mixed layer, with its large thermal inertia. This stored heat would be released gradually later in the year, most likely delaying freeze-up as well as producing thinner ice cover, and this would allow more diffusion of heat from below, further contributing to winter warming (Manabe and Stouffer, 1980).

Any future warming in the Arctic will vary substantially on both an areal and temporal basis. The North American GCMs all predict generally higher temperatures throughout the North American Arctic in every season (Table V), with the greatest increases occurring in winter and the smallest in summer. Winter temperature increases range from about 7° C in the lower Mackenzie Valley and Keewatin areas to above 9° C over the arctic islands. Summer increases range from about 3° C over the arctic islands to 4° C in the lower Mackenzie Valley and Keewatin. The spring and autumn increases are generally intermediate to those of summer and winter—about 5 to 6° C for all regions except for the arctic islands in autumn, when the value is about 7.5° C. The individual results from the CCC GCM are closest to the average values for all regions.

For a variety of reasons, the models show less uniformity in their predictions for precipitation, and their projections should be treated with great caution. On the one hand, some of the models do not accurately simulate present-day precipitation patterns; on the other, some of the simulations may be more accurate than they seem, given the uncertainties in precipitation records. Most of the models project increasing precipitation, although a few decreases are

Table V Seasonal Comparison of Changes in Temperature (T) and Precipitation (P) Simulated by Four GCMs for Doubled CO_2 over the North American Arctic[a]

Region	Model	Spring T (°C)	Spring P (% 1 × CO₂)[b]	Summer T (°C)	Summer P (% 1 × CO₂)	Autumn T (°C)	Autumn P (% 1 × CO₂)	Winter T (°C)	Winter P (% 1 × CO₂)
Arctic islands	GISS	4–5	130–145	1–2.5	110–150	8–11	120–135	9–10	130–150
	GFDL	8–9	100–150	3.5–4.5	110–145	8–10	110–135	12–14	100–135
	OSU	3.5–4	110–140	2–3	95–110	2.5–3.5	105–120	5–8	110–135
	CCC	3–5	70–130	1–5	95–140	5–12	100–130	6–11	95–130
Average		5	120	3	120	7.5	120	9.5	125
Northern Alaska	GISS	4.5–5	115–130	1.5–3	120–135	6–10	110–130	8–9	110–130
	GFDL	8–8.5	130–140	4–6	130–160	7–9	125–135	11–15	115–140
	OSU	3	115–130	2.5–3	105–115	3–3.5	110–135	5.5–7	120–140
	CCC	4–5	100–110	2–5	120–135	5–6	110–120	3–9	95–110
Average		5	120	3.5	130	6	120	8.5	120
Lower Mackenzie	GISS	3.5–4.5	125–135	1.5–2.5	130–140	7–10	135–155	8–9	135–150
	GFDL	8–10	120–150	5.5–7	100–130	6–8	110–135	10–12	95–130
	OSU	3–3.5	115–125	3–3.5	95–110	3–3.5	100–140	4–5.5	105–115
	CCC	3.5–6	100–120	4–5	100–130	4.5–6	100–120	3–6	100–110
Average		5.5	125	4	115	6	125	7	120
Keewatin	GISS	3.5–5	120–130	2–3	130–170	6–8	120–145	8–9	130–140
	GFDL	9–10	120–150	4.5–6	85–130	6–7.5	110–130	8.5–11	100–125
	OSU	3.5–4.5	105–130	2.5–3.5	90–105	3–3.5	100–120	4–5	110–125
	CCC	3.5–5	100–130	4–5.5	100–140	3–6	105–130	5.5–8	90–130
Average		5.5	125	4	120	5.5	120	7	120
Eastern Canadian Arctic	GISS	4.5–6	120–150	2–3.5	100–130	5–10	100–130	9–11	130–145
	GFDL	9–11	100–140	4.5–5.5	110–140	5.5–8	110–130	8–13	90–135
	OSU	4–4.5	115–140	2–3	100–110	3–4	110–125	4.5–6	120–145
	CCC	3–4	90–120	2–5	100–140	2–5	100–130	6–11	90–125
Average		6	120	3.5	115	5.5	115	8.5	125

[a]Temperature and precipitation change values determined from copies of the gridded data from each model held in the Canadian Climate Centre.
[b]Values represent precipitation occurring under a scenario of doubled present-day atmospheric CO_2 expressed as a percentage (%) of present-day values (1 × CO_2).

also indicated (see Table V). There are no areas for which all models predict increases for all seasons, and only in autumn do all models predict increases for all areas. Nonetheless, the average of all the models' predictions for precipitation over northern North America does represent an increase over present-day values in all seasons—an increase amounting to about 20–25%.

The CCC GCM predicts higher temperatures throughout the circumpolar Arctic as well, including North America, Greenland, the Arctic Basin, the northern USSR, and northeastern Scandinavia (Table VI). Temperature increases in these last two areas are similar to those of northern North America, except for possibly slightly warmer winters over northeastern Siberia. The Arctic Basin shows the greatest winter increases (10–13° C) of any area for winter and the least for summer. The CCC model also predicts a net increase in precipitation throughout the circumpolar Arctic (see Table VI), although some small areas may receive less precipitation in certain seasons. Generally, the increased precipitation comes from increased storminess implied by the models' projection of decreased sea-level pressure throughout the Arctic both in winter and, except on the Eurasian side of the Arctic Basin, in summer.

2. Soil Moisture

All models predict increased winter soil moisture in the north, mainly because of increased precipitation and the melting of some snow throughout the season. The models disagree, however, in their predictions for summer. The GFDL, GISS, and UKMO models, for example, all show drying virtually throughout the area north of 60° N, whereas the NCAR model suggests increased soil moisture, particularly in Alaska. There are several possible explanations for this discrepancy. Increased precipitation might cause increased soil moisture. Alternatively, an earlier snowmelt could allow normal summer drying from evaporation to begin earlier and be stronger. Manabe *et al.* (1981) have suggested that this alternative is the dominant process at high latitudes. Washington and Meehl (1984) argue, however, that the NCAR model's soil moisture results are also related to earlier snowmelt but that a positive feedback among soil moisture, precipitation, and clouds is critical, leading to increased summer soil moisture.

A key factor in determining the sensitivity of soil moisture to climate change is the degree to which the models' present-day climate simulations correctly represent present-day soil moisture (Meehl and Washington, 1988). Determining how correct these representations are is difficult, however, because surface hydrology is complex and poorly understood, partly because appropriate observed data are lacking. An analysis of the NCAR and GFDL simulations by Meehl and Washington (1988) suggests that the NCAR model underestimates summer dryness from increased CO_2, whereas the GFDL model somewhat overestimates it. The CCC model, which has a more sophisticated treatment of hydrology (Boer *et al.*, 1991), also predicts modest drying over most of the Canadian mainland north of 60° N in summer but increased soil moisture on the arctic islands and in northern Alaska. On the Eurasian

Table VI   Comparison of Changes in Winter and Summer Temperature (T) and Precipitation
(P) Simulated by the CCC GCM for Doubled CO_2 over the Circumpolar Arctic

	Winter		Summer	
Region	T (°C)	P (% 1 × CO_2)[a]	T (°C)	P (% 1 × CO_2)
Arctic islands	6–11	95–130	1–5	95–140
Northern Alaska	3–9	95–110	2–5	120–135
Lower Mackenzie	3–6	100–110	4–5	100–130
Keewatin	5.5–8	90–130	4–5.5	100–140
Eastern Arctic	6–11	90–125	2–5	100–140
Greenland	5–10	100–120[b]	3–7	100–120[b]
Finland–European USSR	4–8	110–120[b]	3–5	95–110[b]
Northcentral USSR	6–8	95–120[b]	2–6	90–130[b]
Northeastern Siberia	8–10	95–110[b]	2–8	100–140[b]
Arctic Basin	10–13	95–120[b]	0–2	95–150[b]

[a] Values represent precipitation occurring under doubled present-day atmospheric CO_2 expressed as a percentage (%) of present-day values (1 × CO_2).
[b] Estimated.

side of the Arctic Basin, soil moisture in the coastal areas is generally shown to increase, whereas the areas farther inland show drying conditions. The contour for zero soil moisture change in the model generally coincides with the southern boundary of present-day continuous permafrost.

V. Implications for Snow, Permafrost, and Ice

Warmer winters tend to be more snowy at high latitudes because cyclonic activity is more frequent, and more moisture enters those areas in such winters. Yet, observational records indicate that the actual duration of snow cover may be only marginally affected by the increased snowfall. Significant future warming thus may well prolong the snow-free season (Barry, 1982). This assessment is reflected in the GCM protections, which generally agree on an earlier snowmelt in the spring. Although more snowfall is expected during the winter, it may well be more than counterbalanced by snowmelt right through the winter, plus an earlier start to the main melt season. Meehl and Washington (1988), for example, found that in the doubled-CO_2 simulations of both the GFDL and NCAR models, snowmelt was complete about one month earlier than at present (approximately the end of May as opposed to the end of June) for mainland North America north of 60° N. Earlier completion of spring snowmelt, combined with somewhat later dates of first frost, implies growing seasons lengthened by a month or more, depending on the specific region of the Arctic.

For most of the arctic tundra area of North America, evidence suggests earlier dates of snow disappearance in the last two decades than in the

1960s—a trend that seems to agree with trends in other climatic elements. If this trend continues, the likelihood of being able to attribute it to a specific cause, such as increased CO_2, would increase (Foster, 1989; but see Chap. 3).

With future warming, it is inevitable that large areas of permafrost, such as those in northern Canada and Siberia, will eventually disappear. During the next century, however, the greatest changes will probably occur in processes related to permafrost—indirect changes resulting from a deepening active layer (Chap. 3). These may include an increase in slope movement caused by mud flows, skin flows, and slumping; fewer and smaller ice wedges and pingos; and uncertain changes in mechanical soil properties, such as creep and relaxation.

To analyze changes in the southern limit of continuous permafrost in northern Canada that might occur under the GISS and GFDL CO_2-doubling simulations, Stuart (1986) applied a frost index based only on temperature and precipitation considerations (Nelson and Outcalt, 1983) to model the permafrost boundary. Both models suggest that in the western Canadian Arctic, the continuous permafrost boundary would shift northward by about 350 km. For the eastern Arctic, however, the models disagree over the magnitude of the northward movement, with the GFDL model suggesting about 200 km but the GISS model up to 800 km (Fig. 6).

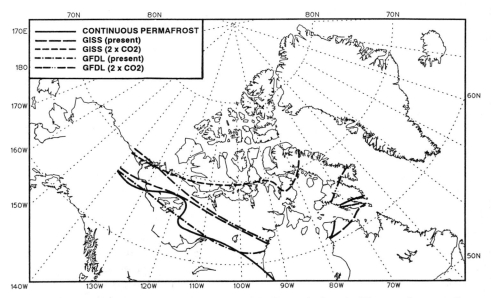

Figure 6   Boundaries of continuous permafrost in arctic Canada. The actual present-day boundary is shown in comparison with present and projected boundaries from the GISS and GFDL models. [Reprinted from Stuart (1986).]

The northward movement of permafrost could further increase the emission of greenhouse gases into the atmosphere through the release of CO_2 from soils that are highly organic, as in Siberia. If soils thaw more deeply, however, methane emissions from the tundra might also be reduced because of increased microbial oxidation of methane (Whalen and Reeburgh, 1990). Such biogeochemical feedbacks have not yet been incorporated in GCMs.

Changes to the arctic sea-ice regime under a warming scenario are likely to be concentrated in the areas between the boundaries of present-day sea ice at minimum (August–September) and maximum (March–April) extent. This means that summers are likely to be completely ice free around Svalbard, along the north Siberian coast, and among the Canadian arctic islands. In the latter area, as a result, multiyear ice would be much less of a concern. In winter, ice thicknesses would generally decrease (by about 35% from the 2.5 m current average in Canadian waters, for example), and the southern limits of sea ice would retreat northward substantially. The CCC model indicates that an instantaneous doubling of CO_2 would cause the ice to retreat from much of the southern half of Hudson Bay, the northern Labrador and southern Greenland seas, the entire Bering Sea, and a broad band of the Barents Sea stretching from Svalbard to Novaya Zemlya (G. Boer, pers. comm.).

For land ice, warming would increase summer ablation; stormier winters, however, would augment winter accumulation. Increased glacial calving, into Baffin Bay for example, is likely. On balance, the smaller mountain glaciers and ice caps would probably decrease in mass, resulting in a 10-to-30-cm rise in sea level (National Research Council, 1985). The impact on the large Greenland ice sheet is more uncertain. Although this ice sheet has the potential to increase sea level by about 6 m under complete melting or disintegration, its likely contribution over the next century (combined with that of the West Antarctica ice sheet) is estimated to fall between -10 and $+30$ cm. The sea-level increases for both small glaciers and polar ice sheets combined are roughly equivalent to sea-level rises expected from the thermal expansion of ocean water alone and lead to an overall anticipated global mean sea-level rise of <0.5–1 m (Frei *et al.*, 1988). This has important implications for those areas of the Arctic that are now subsiding (Egginton and Andrews, 1989), such as the western half of the North American arctic coastline.

VI. Implications for Ecosystems

On a global scale, the greatest ecological changes in response to rising levels of carbon dioxide are likely to take place in tundra, boreal forest, and polar desert zones (Tegart *et al.*, 1990). The impacts in the Arctic of the direct effects of increased CO_2, as well as of CO_2-induced alterations in climate, are likely to be both positive and negative.

For the Canadian Arctic, preliminary work suggests that if the temperature rises and precipitation remains the same (which is unlikely), vegetation zones might simply move northward (Edlund, 1986; Fig. 4b). The extent to which these changes occur will depend on soil suitability, the nature of the new moisture regime, and the restriction of species movement by geography. Even when conditions are favourable for migration, there will likely still be a lag of several hundred years. Such uncertainties also hold for projected northward migration of the boreal forest (Wheaton and Singh, 1989; Smith and Tirpak, 1990; Chap. 5) and for any improved agricultural potential (Eley and Findlay, 1977).

For arctic wildlife, which can move within and between ecosystems, alterations in habitat patterns, food availability, and predator-prey relationships may have a greater impact than the physiological effects of warming (Smith and Tirpak, 1990; Chaps. 18, 19), and changes in these factors may outweigh any advantage conferred by mobility. One of the most important consequences of a warming climate for wildlife will be the increasing isolation of individual arctic islands as open water becomes more widespread. With interisland migration no longer routinely possible, animal populations could be extinguished piecemeal by one or more unsuitable seasons (Harington, 1986).

Arctic land oases and polynyas (areas of open water surrounded by sea ice or land and sea ice) are critical to the survival of land mammals, breeding birds, freshwater fish, and marine mammals, yet these areas make up a small proportion of available habitats. A warmer and wetter climate might mean more areas with conditions similar to present-day oases, but it might also cause the disappearance of traditional oases or have other negative effects (e.g., Miller *et al.*, 1977). Polynyas are caused by a combination of currents, winds, tides, and upwellings, and their size and distribution are likely to change as the climate warms. In southern areas, they may expand or lose identity as open water supplants sea ice. In the north, new polynyas may develop and others, particularly those controlled by local winds, may shrink or disappear.

The complexity and uncertainty associated with predicting ecosystem response to climate change demand improvements in current study methods (Cohen, 1990), including more sophisticated model simulations (e.g., Dickinson *et al.*, 1989) and the further definition of precise relationships between climate elements and ecosystem components. Continuous climate monitoring and satellite-based data collection will play a comprehensive role. Without such relationships, credible high-resolution scenarios will merely be platforms, albeit elaborate ones, from which to continue to launch what are at best educated guesses about arctic ecosystems in a changing climate.

VII. Summary

Two major climate elements determining ecosystem location and character— temperature and precipitation—are expected to change substantially in the Arctic under future global warming. GCM projections of climate conditions under a doubling of the present-day concentration of atmospheric CO_2 indicate an average annual global warming of 1.9–5.2° C and a global annual precipitation increase of 15% before the middle of the next century. In the Arctic, winter warming would be 2–2.4 times the annual global value, and summer warming would be 0.5–0.7 times the global average; precipitation would be 5–10% greater than the annual global value in all seasons. Increased winter soil moisture should result, but summer soil moisture changes are less clear.

It seems likely that such arctic climate changes will result in the lengthening of the snow-free season by a month or more, a significantly deepened active soil layer combined with a northward movement of the permafrost boundary, disappearance of sea ice in summer from many areas where it currently persists (such as the Canadian arctic islands) and thinning of sea ice in winter leading to less multiyear ice; and melting of land ice combined with thermal expansion of ocean water leading to a global mean sea-level rise of up to one meter.

Ecosystems within the Arctic exist in a delicate balance with the region's climate and, on the basis of their reaction to different climates in the past, are likely to change significantly under any future climatic change. Work on regional-scale impacts involving more sophisticated models and better definition of relationships between climate elements and ecosystem components will be essential to help quantify the magnitude and direction of ecosystem change with climate warming.

References

Atmospheric Environment Service (AES). (1985). Past climatic change in the Canadian Arctic. Canadian Climate Centre Rep. 85–14. Downsview, Ontario.

Atmospheric Environment Service (AES). (1986). "Climatic Atlas—Canada." Map Series 2, Precipitation. En 56-63/2-1986. Supply and Services Canada, Ottawa.

Barry, R. G. (1982). Snow and ice indicators of possible climatic effects of increasing atmospheric carbon dioxide. *In* "Proceedings of the DOE Workshop on First Detection of CO_2 Effects" (N. B. Beatty, ed.), pp. 207–236. U.S. Dep. of Energy, Office of Energy Research, Washington, DC.

Boer, G. J., McFarlane, N. A., Blanchet, J.-P., and Lazare, M. (1990). Greenhouse gas–induced climatic change simulated with the CCC second-generation GCM. *In* "Application of the Canadian Climate Centre General Circulation Model Output for Regional Climate Impact Studies: Guidelines for Users" (S. J. Cohen, ed.), pp. 2–5. Atmospheric Environment Service, Canadian Climate Centre, Downsview, Ontario.

Boer, G. J., McFarlane, N. A., and Lazare, M. (1991). Greenhouse gas–induced climate change simulated with the Canadian Climate Centre second-generational general circulation model. Atmospheric Environment Service, Canadian Climate Centre, Downsview, Ontario (in press).

Central Intelligence Agency (CIA). (1978). "Polar Regions Atlas." GC78-10040. U.S. Government Printing Office, Washington, DC.

Cess, R. D., Potter, G. L., Blanchet, J. P., Boer, G. J., Ghan, S. J., Kiehl, J. T., Le Treut, H., Li, Z.-X., Liang, X.-Z., Mitchell, J. F. B., Morcrette, J.-J., Randall, D. A., Riches, M. R., Roeckner, E., Schlese, U., Slingo, A., Taylor, K. E., Washington, W. M., Wetherald, R. T., and Yagai, I. (1989). Interpretation of cloud–climate feedback as produced by 14 atmospheric general circulation models. *Science* **245**, 513–516.

Cohen, S. J. (1990). Bringing the global warming issue closer to home: The challenge of regional impact studies. *Bull. Am. Meteorol. Soc.* **71**, 520–526.

Dickinson, R. (1986). The climate system and modelling of future climate. *In* "The Greenhouse Effect, Climatic Change, and Ecosystems" (B. Bolin, B. Doos, J. Jager, and R. A. Warrick, eds.), pp. 207–270. Wiley, Chichester.

Dickinson, R. E., Errico, R. M., Giorgi, F., and Bates, G. T. (1989). A regional model for the western United States. *Clim. Change* **15**, 383–422.

Edlund, S. A. (1986). Modern arctic vegetation distribution and its congruence with summer climate patterns. *In* "Impact of Climatic Change on the Canadian Arctic" (H. A. French, ed.), pp. 84–99. Atmospheric Environment Service, Canadian Climate Centre, Downsview, Ontario.

Egginton, P. A., and Andrews, J. T. (1989). Sea levels are changing: Current and future relative sea levels in Canada. *Geos* **1989(2)**, 15–22.

Eley, F. J., and Findlay, B. F. (1977). Agroclimatic capability of southern portions of the Yukon Territory and Mackenzie District, NWT. Meteorol. Appl. Branch Proj. Rep. 33. Atmospheric Environment Service. Downsview, Ontario.

Etkin, D. A. (1990a). Greenhouse warming: Consequences for arctic climate. *J. Cold Reg. Eng.* **4**, 54–66.

Etkin, D. A. (1990b). Equivalency of greenhouse gases. CO_2/Climate Rep. 90-01, 5-6. Atmospheric Environment Service, Canadian Climate Centre, Downsview, Ontario.

Foster, J. L. (1989). The significance of the date of snow disappearance on the arctic tundra as a possible indicator of climate change. *Arct. Alp. Res.* **21**, 60–70.

Frei, A., MacCracken, M. C., and Hoffert, M. I. (1988). Eustatic sea level and CO_2. *Northeast. Environ. Sci.* **7**, 91–96.

French, H. A., ed. (1986). "Impact of Climatic Change on the Canadian Arctic." Atmospheric Environment Service, Canadian Climate Centre, Downsview, Ontario.

Gammon, R. H., Sundquist, E. T., and Fraser, P. J. (1985). History of carbon dioxide in the atmosphere. *In* "Atmospheric Carbon Dioxide and the Global Carbon Cycle" (J. R. Trabalka, ed.), pp. 25–62. DOE/ER-0239. Natl. Tech. Info. Serv., Springfield, Virginia.

Gates, W. L. (1985). Modelling as a means of studying the climate system. *In* "The Potential Climatic Effects of Increasing Carbon Dioxide" (M. C. MacCracken and F. M. Luther, eds.), pp. 57–80. DOE/ER-0237. Natl. Tech. Info. Serv., Springfield, Virginia.

Grotch, S. L. (1988). "Regional Intercomparisons of General Circulation Model Predictions and Historical Climate Data." DOE/NBB-0084. U.S. Dep. of Energy, Office of Energy Research, Washington, DC.

Hansen, J., Lacis, A., Rind, D., Russell, G., Stone, P., Fung, I., Ruedy, R., and Lerner, J. (1984). Climate sensitivity: Analysis of feedback mechanisms. *In* "Climate Processes and Climate Sensitivity" (J. E. Hansen and T. Takahashi, eds.), pp. 130–163. American Geophysical Union, Washington, DC.

Hansen, J., Lacis, A., and Prather, M. (1989). Greenhouse effect of chlorofluorocarbons and other trace gases. *J. Geophys. Res.* **94**, 16417-16421.

Harington, C. R. (1986). The impact of changing climate on some vertebrates in the Canadian Arctic. *In* "Impact of Climatic Change on the Canadian Arctic" (H. A. French, ed.), pp. 100–113. Atmospheric Environment Service, Canadian Climate Centre, Downsview, Ontario.

Harley, W. S. (1980). Northern Hemisphere monthly mean 50-kPa and 100-kPa height charts. Rep. CLI 1-80. Atmospheric Environment Service, Downsview, Ontario.

Houghton, J. T., Jenkins, G. J., and Ephraums, J. J., eds. (1990). "Climate Change: The IPCC Scientific Assessment." Cambridge Univ. Press, Cambridge.

Jaeger, J. (1988). Developing policies for responding to climatic change. WMO/TD 225. World Meteorological Organization, Geneva.

Lashof, D. A. (1989). The dynamic greenhouse: Feedback processes that may influence future concentrations of atmospheric trace gases and climatic change. *Clim. Change* **14**, 213–242.

Manabe, S., and Stouffer, R. J. (1980). Sensitivity of a global climate model to an increase of CO_2 concentration in the atmosphere. *J. Geophys. Res.* **85**, 5529–5554.

Manabe, S., and Stouffer, R. J. (1988). Two stable equilibria of a coupled ocean–atmosphere model. *J. Clim.* **1**, 841–866.

Manabe, S., and Wetherald, R. T. (1975). The effects of doubling the CO_2 concentration on a general circulation model. *J. Atmos. Sci.* **32**, 3–15.

Manabe, S., Wetherald, R. T., and Stouffer, R. J. (1981). Summer dryness due to an increase of atmospheric CO_2 concentration. *Clim. Change* **3**, 347–385.

Maxwell, J. B. (1980). "The Climate of the Canadian Arctic Islands and Adjacent Waters," Vol. 1. Climatological Studies 30. En 57-7/30-1. Supply and Services Canada, Ottawa.

Maxwell, J. B. (1982). The Climate of the Canadian Arctic Islands and Adjacent Waters," Vol. 2. Climatological Studies 30. En 57-7/30-2. Supply and Services Canada, Ottawa.

Maxwell, J. B., and Barrie, L. A. (1989). Atmospheric and climatic change in the Arctic and Antarctic. *Ambio* **18**, 42–49.

Meehl, G. A., and Washington, W. M. (1988). A comparison of soil-moisture sensitivity in two global climate models. *J. Atmos. Sci.* **45**, 1476–1492.

Miller, F. L., Russell, R. H., and Gunn, A. (1977). Distributions, movements, and numbers of Peary caribou and muskoxen on western Queen Elizabeth Islands, Northwest Territories, 1972–74. *Can. Wild. Serv. Rep. Ser.* **40**, 1–55.

Mitchell, J. F. B. (1989). The "greenhouse" effect and climate change. *Rev. Geophys.* **27**, 115–139.

National Research Council. (1985). "Glaciers, Ice Sheets, and Sea Level: Effects of a CO_2-induced Climate Change." DOE/ER/60235-1. U.S. Dep. of Energy, Office of Energy Research, Washington, DC.

Nelson, F., and Outcalt, S. I. (1983). A frost index number for spatial prediction of ground frost zones. *In* "Permafrost: Fourth International Conference Proceedings," pp. 907–911. National Academy Press, Washington, DC.

Palutikof, J. (1986). Scenario construction for regional climate change in the warmer world. *In* "Impact of Climatic Change on the Canadian Arctic" (H. A. French, ed.), pp. 2–14. Atmospheric Environment Service, Canadian Climate Centre, Downsview, Ontario.

Ramanathan, V., Callis, L., Cess, R., Hansen, J., Isaksen, I., Kuhn, W., Lacis, A., Luther, F., Mahlman, J., Reck, R., and Schlesinger, M. (1987). Climate–chemical interactions and effects of changing atmospheric trace gases. *Rev. Geophys.* **25**, 1441–1482.

Roots, E. F. (1989). Climate change: High-latitude regions. *Clim. Change* **15**, 223–253.

Schlesinger, M. E., and Mitchell, J. F. B. (1985). Model projections of equilibrium climatic response to increased CO_2 concentration. *In* "The Potential Climatic Effects of Increasing Carbon Dioxide" (M. C. MacCracken and F. M. Luther, eds.), pp. 81–147. DOE/ER-0237. Natl. Tech. Info. Serv., Springfield, Virginia.

Schlesinger, M. E., and Zhao, Z. (1987). Seasonal climate changes induced by doubling CO_2 as simulated by the OSU atmospheric GCM/mixed layer model. *Climate Institute Rep. 70.* Oregon State University, Corvallis.

Smith, J. B., and Tirpak, D., eds. (1990). "The Potential Effects of Global Climate Change on the United States." Hemisphere, New York.

Stuart, A. (1986). A spatial permafrost model for northern Canada and its application to scenarios of climate change. Canadian Climate Centre Rep. 86-15. Atmospheric Environment Service, Downsview, Ontario.

Tegart, W. J. McG., Sheldon, G. W., and Griffiths, D. C., eds. (1990). "Climate Change: The IPCC Impacts Assessment." Australian Government Printing Service, Canberra.

Wang, W.-C., Wuebbles, D. J., Washington, W. M., Isaacs, R. G., and Molnar, G. (1986). Trace gases and other potential perturbations to global climate. *Rev. Geophys.* **24**, 110–140.

Warrick, R. A., Shugart, H. H., Antonovsky, M. Ja. with Tarrant, J. R., and Tucker, C. J. (1986). The effects of increased CO_2 and climatic change on terrestrial ecosystems. *In* "The *Greenhouse Effect, Climatic Change, and Ecosystems" (B. Bolin, B. Doos, J. Jager, and R. A. Warrick, eds.), pp. 363–392. Chichester.

Washington, W. M., and Meehl, G. A. (1984). A seasonal-cycle experiment on the climate sensitivity due to a doubling of CO_2 with an atmospheric circulation model coupled to a simple mixed-layer ocean model. *J. Geophys. Res.* **89**, 9475–9503.

Washington, W. M., and Meehl, G. A. (1989). Climate sensitivity due to increased CO_2: Experiments with a coupled atmosphere and ocean general circulation model. *Clim. Dyn.* **4**, 1–38.

Wetherald, R. T., and Manabe, S. (1986). An investigation of cloud cover change in response to thermal forcing. *Clim. Change* **8**, 5–24.

Whalen, S. C., and W. S. Reeburgh. (1990). Consumption of atmospheric methane by tundra soils. *Nature* **346**, 160–162.

Wheaton, E. E., and Singh, T. (1989). Exploring the implications of climatic change for the boreal forest and forestry economics of western Canada. Climate Change Digest CCD 89-02. Atmospheric Environment Service, Canadian Climate Centre, Downsview, Ontario.

Wilson, C. A., and Mitchell, J. F. B. (1987). A doubled CO_2 climate sensitivity experiment with a GCM including a simple ocean. *J. Geophys. Res.* **92**, 13315–13343.

Zoltai, S. C. (1988). Ecoclimatic provinces of Canada and man-induced climatic change. Canada Committee on Ecological Land Classification Newsletter 17, pp. 12–15. Supply and Services Canada, Ottawa.

3

Arctic Hydrology and Climate Change

Douglas L. Kane, Larry D. Hinzman, Ming-ko Woo, and Kaye R. Everett

I. Introduction

The implications of climatic warming extend well beyond milder winters and warmer, longer summers. The oceans, the land, and the atmosphere will all be affected. Arctic hydrology will undergo two major changes in response to global warming. First, as in tropical and temperate watersheds, the mass and energy inputs into the system will change. For some regions these changes will mean an increase in precipitation, for others a decrease (Etkin, 1990). Generally, surface air temperatures will increase, but this increase will not be uniform. Second, in response to warmer air temperatures and warmer ground surface temperatures, the physical structure of watersheds will change, with the active layer increasing in thickness.

In terrestrial regions of the Arctic, permafrost, a zone of permanently frozen soil, is continuous under the surface. The presence of this permafrost (Fig. 1) is the primary factor distinguishing arctic from temperate watersheds. Each summer a shallow active layer melts in response to a positive flux of energy at the surface. Permafrost persists because this active layer completely refreezes during most winters. This refreezing removes geothermal heat from the terrestrial system and allows the permafrost to remain in relative equilibrium. Although the active layer is shallow, it plays a crucial role in arctic hydrology (Kane and Hinzman, 1988; Kane *et al.*, 1989; Woo *et al.*, 1981; Woo, 1986; Roulet and Woo, 1986, 1988; McCann and Cogley, 1972). All water traveling downslope moves either over or through this layer, which is the major source of moisture for plants. Soils of the active layer can be quite complex, particularly where organic soils overlie mineral soils. Where they do, thermal and hydraulic properties can vary by orders of magnitude over very short vertical distances (Hinzman *et al.*, 1991). In response to climatic warming, the active layer will thicken, increasing the soil's water storage capacity. If a large volume of ice in the soils melts, runoff would increase and additional surface storage might be created in depressions caused by the subsiding surface. Evaporation and transpiration will likely increase in those areas receiving more precipitation and decrease in those receiving less.

Quantifying the magnitude of hydrologic change due directly to climate change in the Arctic is going to be difficult because of the limited existing data base. From a quick examination of arctic hydrologic literature, one finds that most studies are of limited duration; many field studies start after snowmelt; most studies concentrate only on one or two hydrologic processes; and the quality of some of the data is compromised because of harsh environmental conditions. Many of the publications are in the gray literature, and because of the short duration of record, the stochastic variability of the hydrologic data is unknown. How are we going to separate changes in the hydrology induced by climatic change from natural variation?

Permafrost and snowmelt dominate the hydrology of the Arctic—permafrost because it dictates the structure of arctic watersheds, and snowmelt because it produces a high percentage of annual runoff and near-saturated conditions in the active layer. Wide fluctuations occur annually in surface energy balance, that is, the net energy at the ground or snow surface from solar and longwave radiation, atmospheric convection (sensible warming or cooling), latent energy (evaporation, transpiration, melting, and freezing), heat conduction into snow and soil, and mass transfer of heat by precipitation. The hydrologic role of surface energy balance is much more pronounced

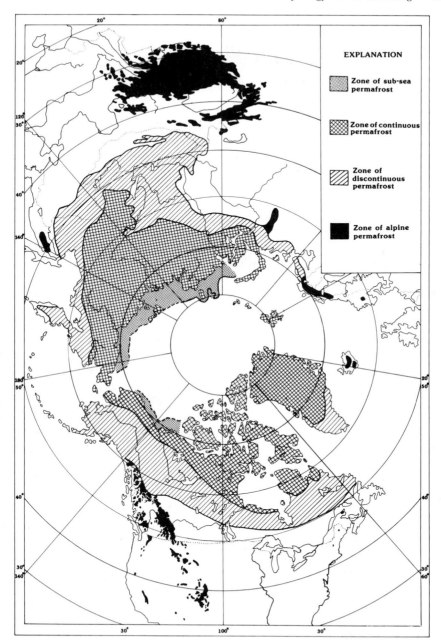

Figure 1   Distribution of permafrost in the Northern Hemisphere. Some isolated areas of alpine permafrost lie off the map; these, and similar areas within the map limits are not shown. [Reprinted from Péwé (1983) with permission from the Institute of Arctic and Alpine Research.]

in the Arctic because of the predominance of snow and ice and phase changes (snow ablation, soil freezing and thawing, ice freezing and thawing). During snowmelt the surface albedo (reflected solar radiation/incoming solar radiation) changes from a high of 0.8 to a low of 0.2 at a time when the incoming solar radiation is near an annual maximum.

II. Watershed Structure

The hydrologic importance of permafrost, which can be hundreds of meters thick, lies in its ice content and its potential to freeze infiltrating fluids. Generally, the presence of ice in the frozen soil makes permafrost relatively impermeable (Mackay, 1983) and leads to the development of a saturated zone in the active layer following snowmelt and major rainfall events. In some parts of the Arctic, the presence of permafrost guarantees that a saturated layer will prevail throughout the summer, even though mean annual precipitation is low. Without permafrost, the soil moisture regime would be much different (Slaughter and Kane, 1979), and consequently, the vegetation patterns would change drastically. The geographic extent of wetlands may also decrease because permafrost will no longer maintain the perched water table.

The hydrology of arctic watersheds is simplified because subsurface water dynamics are confined to the shallow active layer. During and following snowmelt, small quantities of meltwater move down through the active layer and refreeze (Mackay, 1983). Although the overall amount is small, it is significant over the long term because when it refreezes it produces relatively impermeable conditions at the base of the active layer. Except for surface water bodies, all precipitation falls on the active layer. During snowmelt, the amount of infiltration depends not only on the soil type, but also on soil conditions (Kane and Stein, 1983). Ice blocking the soil pores can severely limit infiltration. Desiccation of surface organic soils in response to the natural temperature gradient during the winter months provides storage for some early meltwater. As spring meltwater infiltrates open pores in the cold soil, it refreezes, and the soil quickly becomes isothermal and saturated. The position of the water table thus depends on the soil type and the conditions that exist during freezing. Once the soil is saturated, water will start to move downslope, where it may be intercepted by a water track that rapidly conveys the water to the valley bottom.

During snowmelt, the suprapermafrost groundwater table is close to the surface; in fact, the flow alternates between surface and subsurface flow through the organic layer. During the summer rainfall periods, a similar scenario occurs when the active layer is saturated. Water moves downslope, and, where the surface topography is irregular, surface flow again occurs in the shallow depressions.

Precipitation input into arctic watersheds may be either snowfall and subsequent melt or rainfall; the major watershed outflows are evaporation, evapotranspiration, and runoff. Runoff dominates during the short snow-melt period, but evapotranspiration is the major mechanism for water loss during the summer.

III. Watershed Processes

A. Precipitation

As one proceeds northward, summer becomes progressively shorter, and a larger proportion of annual precipitation falls as snow. Redistribution of snow by wind is a very important process, for it tends to concentrate the snow in depressions on the ground (Benson, 1982). In the Low Arctic, the water content of the snowpack has been measured to be as little as 24% of the annual precipitation (Kane *et al.*, 1991a); in the High Arctic, the snowpack has represented as much as 60 to 80% of annual precipitation (Woo *et al.*, 1983). This statistic varies considerably among years for various spatial scales.

In both the High and Low Arctic, precipitation increases throughout the summer, with maximums in July and August. In the Low Arctic, the wettest months are July, August, and September (Kane *et al.*, 1989), corresponding to the period of minimum sea ice cover on the Beaufort and Chukchi seas (LaBelle *et al.*, 1983). Rainfall intensities are generally light. Results from a Wyoming precipitation gage located in Imnavait Creek in Northern Alaska for a seven-year period (1982–1988) showed that the maximum daily precipitation was 33 mm; the next two highest daily rainfall levels were 25 mm and 23 mm. In the Canadian High Arctic, a single storm produced 26.6 mm of precipitation over a 48-hr period on Devon Island (McCann and Cogley, 1972). Maximum daily snowmelt rates often exceed these rainfall rates.

Precipitation data are generally good for rainfall but poor for snowfall. Snowfall measurements at Barrow and Barter Island in Alaska, for example, have at times been underestimated by a factor of three (Benson, 1982). Clagett (1988) reports that the catch of Wyoming shielded gages was 200% greater than the gages used by the U.S. National Weather Service and that the total catch efficiency of this gage was 80–90% of total snowfall. Basins in the High Arctic can have 130–300% more water in the snowpack than what is measured by weather stations (Woo *et al.*, 1982). Much of the variation is due to the poor performance of various gages being used in windy environments. Researchers need to be aware of the quality of existing data sets.

B. Snowmelt

In the Low Arctic, snowmelt takes place from early May to early June. In the High Arctic, snowmelt generally occurs in June, although large snowfields may persist into the summer (Young and Lewkowicz, 1988).

Snowmelt is the dominant arctic hydrologic event each year. Not only do large volumes of runoff leave a basin over a short period, but the annual peak

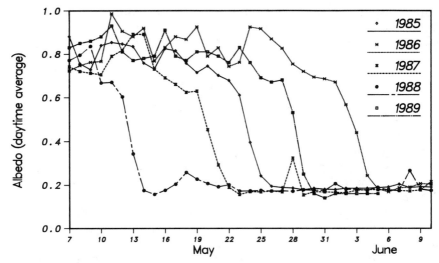

Figure 2 Change in surface albedo during snowmelt and variation in the timing of melt for a low-arctic watershed over five years. [Reprinted from Kane *et al.* (1991a) with permission from the American Geophysical Union.]

discharge usually occurs during snowmelt. At the same time, the energy balance changes significantly. Before snowmelt, surface albedo is around 0.8 (Fig. 2), but once the ground is bare, this value drops drastically (Woo and Dubreuil, 1985; Kane *et al.*, 1991a). This decline means that the amount of solar energy absorbed increases sharply by three to four times. The amount of daily snowmelt is controlled by the surface energy balance.

Considerable variation exists in the density of arctic snowpacks before melt. Wind is the primary factor responsible for increasing the density of the shallow snowpacks. Maximum densities of 540 kg m^{-3} have been measured, but typical values are usually 250–300 kg m^{-3} (Woo, 1982; Kane *et al.*, 1991a). The erosion of snow by wind from ridges and slopes and the deposition of this snow in valleys and along water tracks is an important process. This redistribution can occur during the time of deposition or during wind events without precipitation. This process may leave some areas with no snow and other areas with water contents much higher than average. Most of the redeposited snow accumulates in wind-protected areas along the leeward side of ridges and in depressions. This means that the areas closest to surface channels will have more snow; therefore, the watershed's snowmelt runoff index (the ratio of total runoff volume to snowpack water content of the basin) will be higher than if the snowpack is uniformly distributed throughout the watershed. This process ensures that overall snowmelt runoff volume will be greater than for a uniformly distributed snowpack and it can dictate vegetation patterns. During strong wind events, sublimation can also constitute a significant loss from the snowpack (Benson, 1982). In the High Arctic of Canada, for example, sublimation increased from 0.03 mm day^{-1} in late April to 0.6 mm day^{-1} during the last days of the dry snow period (Ohmura, 1982).

Snowmelt timing and intensity are directly related to surface energy balance. Heat conduction into the active layer is zero or slightly negative during snowmelt. Evaporation becomes an important mechanism of energy transfer when bare ground, covered by meltwater, emerges on the slopes. On exposed slopes and ridges, wind reduces snowpack thickness, exposing vegetation with a relatively low albedo, which accelerates snowmelt and also provides areas for enhanced evaporation. In the Low Arctic, snowmelt occurs about a month before the potential maximum solar radiation; in the High Arctic, snowmelt can coincide with potential maximum incoming solar radiation. Maximum ablation rates in the Low Arctic as high as 30 mm day^{-1} (Kane *et al.*, 1991a) are common. In the High Arctic, 60 mm day^{-1} has been reported as the maximum melt, although a rate from 20 to 30 mm day^{-1} is more common (Woo, 1983).

The snowpack generally disappears over a one- to two-week period, but the melt can be arrested by a cold spell. At Imnavait Creek, for example, ablation stopped when the air flow changed from southerly to northerly (Kane *et al.*, 1991a). In this case, most of the incoming solar radiation is used as sensible heat to warm the atmosphere. Similar observations have been made at Churchill on Hudson Bay (Rouse, 1984).

When snowmelt begins, temperatures at both the bottom of the snowpack and in the soils are well below 0° C. This means that the first meltwater freezes as it infiltrates the snow and the active layer. This process helps to warm both the snowpack and the near-surface soils to 0° C, but at the same time it delays snowmelt runoff (Marsh and Woo, 1984). The freezing meltwater can also form an impermeable layer of ice at the interface between the soil and snow (Woo *et al.*, 1982), reducing infiltration and increasing runoff.

C. Active Layer Processes

The role of the active layer in storing and transmitting soil water varies, depending on the soil type, its moisture content, and whether it is frozen. The depth of thaw in the active layer depends on surface temperature, thermal properties of active layer soils, and ice content. Ice content influences thermal conductivity, but it also requires large quantities of energy for the phase change from ice to water.

Hinzman *et al.* (1991) showed how thermal conductivity varies in the active layer where organic soils (15 cm thick) overlie mineral soils. Temperatures examined in their study ranged from −20° to +20° C, and the soil moisture ranged from dry to saturated. Organic soils with low bulk densities can be very good insulators, having a thermal conductivity almost an order of magnitude less than that of frozen, nearly saturated mineral soils.

In areas of continuous permafrost, the thickness of the active layer in natural undisturbed sites seldom exceeds one meter. The maximum thaw depth is usually found on hill slopes, where soils are well drained. Minimum thaw

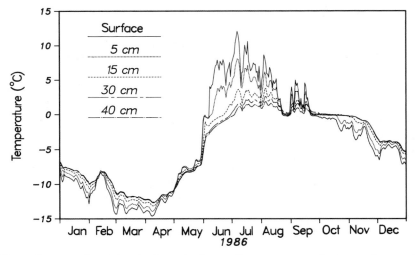

Figure 3   Annual variation in the active layer soil temperatures of a low-arctic watershed. [Reprinted from Hinzman *et al.* (1991) with permission from Elsevier Scientific Publishing Co.]

depths (approximately 25 cm) occur in poorly drained wetland sites, where the ice content is quite high. Typical active layer temperatures at a low-arctic site range from minimums between $-10°$ and $-15°$ C in late winter to maximums of 10° to 15° C in the summer (Fig. 3). Positive temperatures last only from early June until mid-September.

D. Snowmelt Runoff

Snow ablation and subsequent runoff in the Arctic is another major hydrologic event each year. McCann and Cogley (1972) reported that 75% of the runoff leaving a small high-arctic watershed on Devon Island was snowmelt. Woo *et al.* (1981) reported that more than 70% of the water entering a high-arctic lake came from snowmelt. In the Low Arctic, Roulet and Woo (1988) reported that 75% of the snowpack left as runoff in a small 1.36-km^2 drainage. In a small low-arctic watershed in Alaska, runoff generated from snowmelt represented 52–66% of the snowpack over five years, and the annual peak runoff event each year was generated by snowmelt (Kane *et al.*, 1991a).

In general, it appears that the High Arctic has a higher snowmelt runoff index than the Low Arctic, for several reasons: As one proceeds north, more of the annual precipitation occurs as snowfall; snowmelt takes place nearer to summer solstice; vegetation is less; and soils are generally coarser, with shallower active layers. The quality of runoff data collected during spring breakup varies from poor to good because it is difficult to make measurements of high flows entrained with ice, often in adverse weather conditions. Snow damming can also increase the snowmelt runoff peak (Woo, 1979).

E. Summer Runoff

Summer in the Low Arctic may last as long as four months or more, but in the High Arctic, snowmelt can be delayed into July, with snow accumulation beginning again in late August. Although snowfall is possible during any summer month, it seldom remains on the ground for more than several days.

Although rainfall intensities can physically exceed snowmelt rates, the peak runoff events are caused by snowmelt (Woo, 1986). Thawing of the active layer and ongoing evapotranspiration during the summer provide more available storage in the active layer than exists during snowmelt. In one low-arctic watershed, peak flows generated from summer rainfall approached one-half the value of the snowmelt peak flows (Kane *et al.*, 1989; Kane *et al.*, 1991a). The intensities of these rainfall events were low, with maximum daily values near 33 mm day^{-1} and typical values about 5 mm day^{-1}.

F. Evaporation and Transpiration

Water losses to the atmosphere by evaporation and transpiration are the least-known components of the arctic hydrologic cycle. In many cases, water balances are studied with evapotranspiration considered to be the residual unmeasured term (Kane and Carlson, 1973; Woo *et al.*, 1983; Kane *et al.*, 1990). This technique can nevertheless give meaningful results if all the other components of the hydrologic cycle are measured with minimal error. Evaporation pans have also been used to indicate evaporation; although they are good for defining an upper limit, a pan coefficient needs to be determined to estimate actual evapotranspiration. For the Low Arctic, Kane *et al.* (1990) determined the pan coefficient to be 0.49 over a four-year period. Two other methods have been used in the Arctic to estimate evapotranspiration: the energy balance equation and the equation of Priestley and Taylor (1972). The energy balance method requires costly instrumentation to measure all the energy fluxes and is therefore rarely used. The Priestley–Taylor method requires considerably less energy-related data but does need an estimate of an evaporability parameter. Both Rouse and Mills (1977) and Marsh *et al.* (1981) have estimated this parameter for a few varying conditions in the Arctic.

Different researchers have applied these methods, but seldom was more than one method used. Figure 4 compares various methods applied to one low-arctic watershed. For this watershed, evapotranspiration has varied over three years from 34 to 66% of the annual precipitation, so it is an important component of the hydrologic cycle. In the High Arctic, however, Woo *et al.* (1983) found that over a six-year period, evaporative loss represented only 20% of the annual precipitation.

G. Water Balance

Numerous field studies have been carried out in both the Low and High Arctic of North America where some calculations of the hydrologic water balance could be made (Tables I and II). In addition, Dingman (1973) provides an

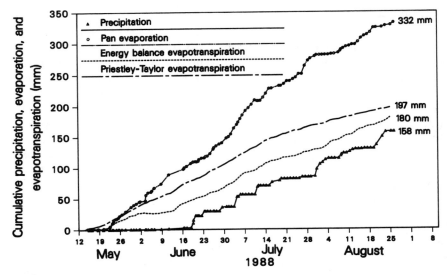

Figure 4   Measured cumulative summer precipitation and pan evaporation and calculated evapotranspiration for northern Alaska. [Reprinted from Kane *et al.* (1990) with permission from the Nordic Association for Hydrology.]

annotated bibliography of the water balance of arctic and subarctic regions throughout the world.

A higher percentage of total precipitation leaves high-arctic basins as runoff than in more temperate watersheds because of the considerable amount of spring freshet produced during the melt season, when the active layer is largely frozen. It is also apparent that total precipitation, particularly snowfall, is underestimated in many water balance studies. Runoff indices should not exceed 1.0 unless there is an additional source of water such as glaciers, perennial snowpacks, or the melting of ice in the active layer.

Tables I and II show that hydrologic processes vary considerably on small spatial scales. The results for a watershed integrate processes from many different terrain types, and because of this averaging, the extremes one may encounter for an individual terrain setting will not be evident. This principle is nicely illustrated by Lewkowicz and French (1982), Landals and Gill (1973) and Kane *et al.* (1991a).

H. Unique Hydrologic Features

In many watersheds in the Low Arctic, shallow water tracks efficiently convey water from the slopes to the valley bottoms. These surface drainage features, which result from the permafrost, are oriented parallel to the slopes. The channels have very small differences in topographic relief and are difficult to see when one is on the ground. The best time to observe them is during major runoff events. Flow ceases in these channels shortly after snowmelt or significant rainfall.

The presence of snow dams along the channels delays streamflow response to snowmelt. The wind-packed snow is hard and dense, and the flow must cut a path through this snow (Woo and Sauriol, 1980; Heginbottom, 1984; and Kane *et al.*, 1991a). Snow damming introduces timing difficulties when one attempts to model snowmelt runoff (Hinzman and Kane, 1991).

Another noteworthy process is the redistribution of soil water in response to temperature gradients. During winter, water will migrate in response to a temperature gradient. This migration is upward to the freezing front, where ice lenses may develop if the soil is frost-susceptible and contains enough soil moisture. Mackay (1983), observed that frost heave occurred during snowmelt. He concluded that meltwater was moving through the active layer and that ice growth was occurring at the permafrost table. This ice growth is hydrologically important because it favors the development of saturated conditions in the active layer during the summer.

IV. Impact of Climatic Warming on Watershed Structure

A. Heat Transfer

Climatic warming will alter surface energy balance, causing more heat to be transferred to the soil. In permafrost areas, this heat transfer may lead to a thickening of the active layer and a decrease in permafrost thickness as melting also occurs at the base because of geothermal heat. This heat transfer is transient, and the soil will not reach thermal equilibrium. Thickening of the active layer will take place over several decades, whereas degradation of the permafrost will take centuries.

B. Thermal Modeling

Numerous mathematical models for the prediction of soil temperatures have been developed out of engineering needs in cold regions. These models range from very simple to very complex depending upon their application (Lunardini, 1981). Models based solely on heat conduction adequately predict near-surface temperatures in arctic soils. Kane *et al.* (1991b) used a two-dimensional, non-steady-state heat conduction model with phase change to predict the thermal regime of the active layer. Their model results simulated active layer temperatures over a four-year period quite well, thereby supporting their assumption that conduction was the main mechanism of heat transfer within the active layer and permafrost. This conclusion corroborates those reached by Nixon (1975) and Lachenbruch and Marshall (1986).

Using surface temperatures measured over a four-year period, Kane *et al.* (1991b) fitted a truncated Fourier series to these time series. To simulate climatic warming, the Fourier series was changed gradually and continuously to reflect a slight warming, and these data were used to drive the thermal model.

Table I Water Balance Data from Low-Arctic Watersheds of North America

Location	Area (km²)	Notes	Study period	Precipitation Snow (mm)	Precipitation Rain (mm)	Precipitation Total (mm)	Runoff Snow (mm)	Runoff Rain (mm)	Runoff Total (mm)	Evapo- trans- piration (mm)	Change in soil storage (mm)	Snowmelt runoff index	Rainfall runoff index	Total runoff index	Reference
Imnavait watershed, Alaska	2.2	Watershed	1985	102	163	272	66	62	119	153		.65	.38	.44	Hinzman (1990)
	2.2	Watershed	1986	109	272	380	57	179	250	130		.52	.66	.66	
	2.2	Watershed	1987	108	252	330	71	72	110	219		.66	.29	.33	
	2.2	Watershed	1988	78	257	412	39	78	172	240		.50	.30	.42	
	2.2	Watershed	1989	155			94					.61			
	0.00009	Plot	1985	138			109					.79			
	0.00009	Plot	1985	117			44					.38			
	0.00009	Plot	1985	97			21					.22			
	0.00009	Plot	1985	106			33					.31			
	0.00009	Plot	1986	147			100					.68			
	0.00009	Plot	1986	124			62					.50			
	0.00009	Plot	1986	81			23					.28			
	0.00009	Plot	1986	102			51					.50			
	0.00009	Plot	1987	98			68					.69			
	0.00009	Plot	1987	107			35					.33			
	0.00009	Plot	1987	102											
	0.00009	Plot	1987	100			20					.20			
	0.00009	Plot	1988	72											
	0.00009	Plot	1988	73			12					.16			
	0.00009	Plot	1988	84			36					.43			
	0.00009	Plot	1988	69			25					.36			
	0.00009	Plot	1989	135											
	0.00009	Plot	1989	116			42					.36			
	0.00009	Plot	1989	117			82					.70			
	0.00009	Plot	1989	137											
Boot Crk., NWT.	31	Watershed	1973	140	145	285	119		210			.85		.74	Anderson (1974)
Putuligayuk River basin, Alaska	568	Watershed	1970	54	40	94			75					.80	Kane and Carlson (1973)

Site	Area	Type	Year							Reference
Barrow watershed, Point Barrow, Alaska	1.6	Watershed	1963	42			20		.46	Brown et al. (1968)
		Watershed	1964	8			0.08		.01	
		Watershed	1965	14			0.55		.04	
		Watershed	1966	42			2.2		.05	
Baker Lake, NWT.	0.15	Plot	1983	146	39	185	217	265		Roulet and Woo (1986)
Baker Lake, NWT.	1.1	Watershed	1982	135	108	243				Roulet and Woo (1988)
	1.1	Watershed[a]	1983	201	39	240				
		Plot	1983	106	42				.40	
Yellowknife, NWT.	0.00015	Plot[b,d]	1969	46	1				.01	Landals and Gill (1973)
	0.00024	Plot[c,d]	1969	80	65				.82	
	0.00260	Plot[c,e]	1969	80	54				.68	
	0.01068	Plot[b,e]	1969	74	53				.72	
	0.05110	Plot[f]	1969	80	30				.37	
	0.08850	Plot[f]	1969	61	34				.56	
	0.02670	Plot[g]	1969	66	50				.77	
	0.16000	Plot[g]	1969	94	47				.50	

[a] Basin has four lakes and a small pond.
[b] South-facing site.
[c] North-facing site.
[d] Bare rock.
[e] Vegetated ridges.
[f] Muskeg depressions.
[g] Compound sites.

Table II Water Balance Data from High-Arctic Watersheds of North America

Location	Area (km²)	Notes	Study period	Precipitation Snow (mm)	Precipitation Rain (mm)	Precipitation Total (mm)	Runoff Snow (mm)	Runoff Rain (mm)	Runoff Total (mm)	Evapo-transpiration (mm)	Change in soil storage (mm)	Snowmelt runoff index	Rainfall runoff index	Total runoff index	Reference
McMaster River basin, Cornwallis Island, NWT.	33	Watershed	1976	163	28	191			161	31	−1			0.84	Woo (1983)
	33	Watershed[a]	1977	97	23	120			155	31	−37			1.29	
	33	Watershed	1978	212	61	273			213	38	23			0.78	
	33	Watershed	1979	136	42	178			143	30	5			0.80	
	33	Watershed	1980	128	37	165			130	51	−16			0.79	
	33	Watershed	1981	148	67	215			148	47	21			0.69	
	33	Watershed	1982	162	27	189									
Resolute, Cornwallis Island, NWT.	0.5	Sub-basin	1976	121	31	152			137	46				0.90	Marsh and Woo (1979)
	0.5	Sub-basin	1977	87											
	0.5	Sub-basin	1978	151	54	205								1.08	
	10	Sub-basin	1976	115	31	146			157	40					
	10	Sub-basin	1977	90											
	10	Sub-basin	1978	194	54	248								1.03	
	21	Sub-basin	1976	124	31	155			160	39					
	21	Sub-basin	1977	98											
	21	Sub-basin	1978	186	54	240								1.06	
	33	Watershed	1976	121	31	152			161	41					
	33	Watershed	1977	95											
	33	Watershed	1978	185	54	239			213	40				0.89	
Resolute, Cornwallis Island, NWT.	1.5	Watershed	1978	166	57	223				61					Woo et al. (1981)
	1.5	Watershed	1979	121	27	148				52					
Baffin Island, NWT.	125		1965			375			430					1.15	Ostrem et al. (1967)
	12.8		1965			375			252					0.67	
	7.2		1965			461			450	10				0.98	
F2 Watershed, Baffin Is.	2.5	Watershed	1967			616			462	154				0.75	Church (1974)

Location	Area (km²)	Type	Year							Reference
Lewis River, Baffin Is., NWT.	208	Watershed	1963–66		550	500	50		.90	Anonymous (1967)
Thomsen River, Banks Island	0.00015	Plot	1977	71	6			.09		Lewkowicz and French (1982)
	0.00044	Plot	1979	85	19			.22		
	0.00052	Plot	1977	260	68			.26		
	0.00052	Plot	1978	234	23			.10		
	0.00052	Plot	1979	218	4			.02		
	0.00026	Plot	1978	710	511			.74		
	0.00007	Plot	1978	912	301			.33		
Weir River, Ellesmere Island, NWT.	29.4	3	1973		143	146	0		1.02	Walker et al. (1973)
					104	146			1.40	
Jason's Crk., Devon Island, NWT.	2.3	Watershed	1970	73	63	136				McCann and Cogley (1972)
Axel Heiberg Island, NWT.		Point[d]	1969, 1970, 1972	65			84			Ohmura (1982)
				57			90			
				59			97			
				75			90			
				106			102			
				110			85			

[a] Dry year; storage deficit probably balanced by the melting of semipermanent snowbanks.
[b] Includes net glacial ice ablation.
[c] Precipitation data partially estimated from Eureka data.
[d] Values are three-year averages.

49

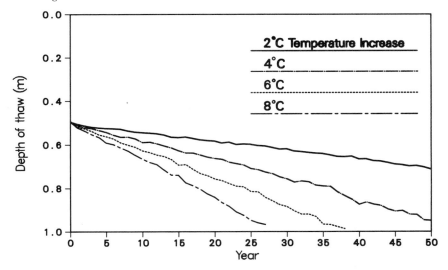

Figure 5 Predicted thaw depth of the active layer near Toolik Lake, Alaska, in response to various scenarios of warming. [Reprinted from Kane *et al.* (1991b) with permission from Elsevier Scientific Publishing Co.]

Although results from general circulation models (GCMs) indicate that warming at high latitudes will be greater in the winter than in the summer (Chap. 2), the model assumes that warming is uniform throughout the year. In a subsequent publication, scenarios other than uniform temperature increase were examined and found not to alter the results significantly (Kane and Hinzman, 1991).

The results of Kane *et al.* (1991b) predict that the active layer will increase in thickness by 20 to more than 50 cm in response to a 2–8° C warming of the ground surface over a 50-year period (Fig. 5). In the two warmest scenarios, an unfrozen zone (talik) develops above the permafrost after 38 years with a 6° C warming and after 27 years with a warming of 8° C. This result means that the winter frost does not extend down to the permafrost before summer thawing occurs. For increases in the mean annual surface temperature of 2° and 4° C, the active layer thickness increases by 20 and 40 cm, respectively, over an original thickness of 50 cm. Ecologically, an additional thaw depth of this magnitude could significantly change existing vegetation patterns. Also, increased vertical and horizontal drainage could increase nutrient availability from higher rates of weathering.

Figure 6 shows differences in soil temperatures over a 50-year period with a 4° C warming (Kane *et al.*, 1991b). Although maximum surface warming was limited to 4° C in the case shown, it can be seen that warming in excess of 4° C is possible; at 30 m, it was computed that the permafrost warmed more than 1.5° C. The surface temperatures in October show little change over the

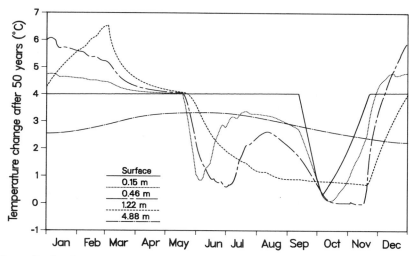

Figure 6   Predicted changes in near-surface soil temperatures after 50 years of 4° C surface climatic warming. Each line represents the difference in temperature between year 0, simulated immediately before warming, and year 50, after 50 years of a simulated 4° C rise in temperature.

50-year period. During freeze-up, the soils cool to the phase-change temperature and remain isothermal until phase change is complete (see Fig. 6). The phase-change temperature will remain the same under climatic warming, so the least difference will occur during the period of overlap during fall freeze-up and spring snowmelt. The greatest soil warming will occur during the winter. Following snowmelt, the soils will generally remain warmer throughout the summer, by about 3° C at 15 cm depth.

V. Hydrologic Response of an Arctic Watershed to Global Warming

A. Available Hydrologic Models for Cold Regions

Our ability to predict hydrologic responses to climatic change in the Arctic is controlled by the available models. Suitable models must have a snowmelt component. The World Meteorological Organization (1986) compared 11 snowmelt runoff models from eight different countries. Data sets from six countries were used in the comparison; in some cases model runs were not made with particular data sets because they were incompatible. The comparison indicates that no model was superior in every application; rather, each model performed best at the purpose for which it was designed (e.g., models designed to calculate snowmelt runoff performed best at that application). Shafer and Skaggs (1983) also evaluated a large number of hydrologic models developed in the United States to determine whether they would be suitable

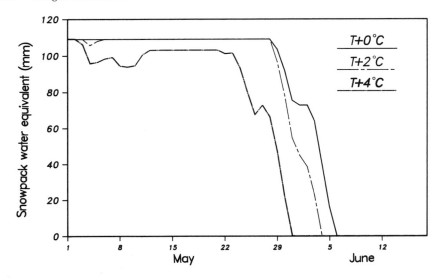

Figure 7 Predicted rates of snow ablation, predicted by the Swedish HBV hydrologic model, with present climate and with 2° and 4° C increases in surface temperatures. [From Hinzman (1990).]

for detecting climatic change on a watershed scale. All the models they examined had a snowmelt routine. They did not select a single best model.

Hinzman and Kane (1991) have used the Swedish HBV hydrologic model (Bergström, 1976) to simulate snowmelt runoff for five years, and Hinzman (1990) used the same model for continuous simulations throughout the summer for a small (2.2 km^2) arctic watershed at Imnavait Creek. Designed for continuous calculation of runoff from snowmelt or rainfall, this model has been used extensively in cold regions. The HBV model simulated the overall hydrologic processes of the watershed quite well; the timing and volume of runoff and snowpack ablation were modeled exceptionally well (Hinzman and Kane, 1991).

B. Hydrologic Modeling to Reflect Climatic Change

Using the HBV model, Hinzman (1990) studied the impact of increasing the annual precipitation by 0, 15, 20, and 30% at 2° and 4° C warming. The first obvious hydrologic response to climate warming would be earlier ablation of the snowpack (Fig. 7). An increase of 2° C in air temperature would initiate melting only a couple of days earlier, and 4° C of warming would advance the appearance of snow-free ground by seven days. This minor increase lies well within the natural existing variation of the completion of snowmelt ablation and is not really significant (see Fig. 2). If precipitation increases so that the water content of the snowpack increases, lengthening of the snow-free season by climatic warming will be partially negated. In autumn, the onset of freez-

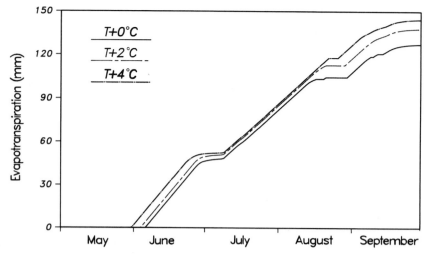

Figure 8   Increase predicted by the HBV model for cumulative evapotranspiration in response to climatic warming for a low-arctic watershed. [From Hinzman (1990).]

ing should be delayed substantially because of the slow cooling rate. In the spring, based on average conditions, plants would likely have a chance to become active just a few days earlier, but in the fall other variables, such as light, control plant activity. Summer precipitation, which may be snowfall under the current thermal regime, may change to rainfall or to a greater percentage of rain mixed with snow.

Cumulative evapotranspiration is likely to increase because of both warmer air temperatures and a longer summer (Fig. 8). In addition, total leaf area of the plants subject to evaporation may be greater. Without any additional precipitation, and assuming that there is no net gain or loss of soil water storage, the volume of runoff would decrease (Fig. 9).

We have not discussed the possibility that precipitation would decrease, but GCMs do predict reduced precipitation over some land areas. Less precipitation would produce less evapotranspiration, less soil moisture in the active layer, and less runoff, although in exactly what proportions is unknown. The proportions would also depend on changes in the plant cover. In fact, plant response to climatic warming and possible precipitation changes are major unknowns and very important to the hydrologic response.

C. Hydrologic Response to Warming

The response of arctic watersheds to global warming is likely to appear first as a change in the structure of watersheds: the thickening of the active layer. Second, precipitation input and possibly the pattern of precipitation is also likely to change, although the magnitude of this change remains uncertain.

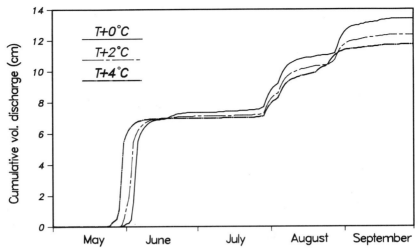

Figure 9   Decrease in the cumulative volume of discharge as predicted by the HBV model for a low-arctic watershed for two scenarios of climatic warming. [From Hinzman (1990).]

These changes, combined with the increased energy into the system, would induce changes in the hydrologic response of arctic watersheds. Hydrologic simulations show that the volume of evapotranspiration would increase, runoff would decrease, and the spring melt would be earlier and less intense, given no changes in precipitation. Additional precipitation in the form of snow could offset the effect of warming, thereby producing little change in the initial period of snow-free ground. Climatic warming should lengthen the snow-free period in the fall because the slope of the cooling rate at this time is less than the slope of the warming trend in the spring.

Evapotranspiration will increase with climatic warming; the greater the warming the greater the increase. Under present precipitation patterns, soil moisture is at times too low for evapotranspiration to proceed at the potential rates. Even with increased precipitation associated with climatic warming, evapotranspiration losses will still increase because soil water storage within the active layer would increase, and additional precipitation would reduce the soil water deficit. Soil moisture, which depends heavily on precipitation patterns, would increase or decrease in concert with evapotranspiration.

VI. Summary

Hydrology in the Arctic differs substantially from that of temperate regions because of the presence of permafrost. Permafrost limits the subsurface system to the active layer; it promotes saturated conditions that increase water availability for plants and evaporation, and increases the proportion of runoff and evapotranspiration by restricting the migration of water to deeper soils. Permafrost is also responsible for the presence of water tracks, which quickly and efficiently remove excess water from the hill slopes, resulting in a rapid response to precipitation, both rainfall and snowmelt.

The hydrology of the Arctic is dominated by snowmelt as the surface energy balance rapidly goes from a net loss of energy to a net gain during a period of near maximum incoming solar radiation. The hydrologic regime can be broken into three distinct periods: winter, snowmelt, and summer. Winter can last from 8 to 10 months; its dominant processes are snowfall and accumulation, redistribution of snow by wind, and seasonal freezing of the active layer with redistribution of soil water. The snowmelt period is dominated by runoff; snow ablation is completed within 10–14 days. After snowmelt, evapotranspiration is usually the major mechanism of water loss throughout the rest of the summer. Major rainfall events will produce significant runoff, but the greatest volume of runoff is produced by snowmelt.

Climatic warming will affect arctic hydrology through its affect on the active layer. With sufficient climatic warming, an unfrozen layer will develop between the seasonal frost and the permafrost. This unfrozen zone, or talik, will substantially alter surface hydrology by increasing the amount of soil moisture storage; improving subsurface drainage and, consequently, decreasing the amount of soil moisture within the soil profile; and decreasing the total areal extent of wetlands in the Arctic.

Acknowledgments

This research was sponsored in part by the U.S. Department of Energy, Office of Health and Environmental Research, Ecological Research Division, as part of the R_4D program in Arctic Tussock Tundra. We would like to acknowledge Elizabeth K. Lilly and Robert E. Gieck for their assistance in data collection, reduction, and analysis.

References

Anderson, J. C. (1974). Permafrost–hydrology studies at Boot Creek and Peter Lake Watershed, N.W.T. *In* "Permafrost Hydrology," pp. 39–44. Canadian National Committee, Ottawa.

Anonymous. (1967). Hydrology of the Lewis Glacier, north-central Baffin Island, N.W.T., and discussion of reliability of the measurements. *Geogr. Bull.* **9**, 232–261.

Benson, C. S. (1982). Reassessment of winter precipitation on Alaska's Arctic Slope and measurements on the flux of wind-blown snow. Rep. UAG R-288. Geophysical Institute, University of Alaska, Fairbanks.

Bergström, S. (1976). Development and application of a conceptual runoff model for Scandinavian catchments. Rep. RH07. Swedish Meteorological and Hydrological Institute, Norrköping, Sweden.

Brown, J., Dingman, S. L., and Lewellen, R. I. (1968). Hydrology of a drainage basin on the Alaska Coastal Plain. Res. Rep. 240. U.S. Army Cold Regions Research and Engineering Laboratory, Hanover, New Hampshire.

Church, M. (1974). Hydrology and permafrost with reference to northern North America. *In* "Permafrost Hydrology," pp. 7–20. Canadian National Committee, Ottawa.

Clagett, G. P. (1988). The Wyoming Windshield–An evaluation after 12 years of use in Alaska. *In* "Proceedings: Western Snow Conference," Vol. 56, pp. 113–123. Western Snow Conference, Beaverton, Oregon.

Dingman, S. L. (1973). The water balance in arctic and subarctic regions. Spec. Rep. 187. U.S. Army Cold Regions Research and Engineering Laboratory, Hanover, New Hampshire.

Etkin, D. (1990). Greenhouse warming: Consequences for arctic climate. *J. Cold Reg. Eng.* **4**, 54–66.

Heginbottom, J. A. (1984). The bursting of a snow dam, Tingmisut Lake, Melville Island, Northwest Territories. *Geol. Surv. Can. Pap.* **84-1B**, 187–192.

Hinzman, L. D. (1990). The interdependence of the thermal and hydrologic processes of an arctic watershed and their response to climatic change. Ph.D. diss., University of Alaska, Fairbanks.

Hinzman, L. D., and Kane, D. L. (1991). Snow hydrology of a headwater arctic basin. 2. Conceptual analysis and computer modeling. *Water Resour. Res.* **27**(6), 1111–1121.

Hinzman, L. D., Kane, D. L., Gieck, R. E., and Everett, K. R. (1991). Hydrologic and thermal properties of the active layer in the Alaskan Arctic. *Cold Reg. Sci. Technol.* **19**(2), 95–110.

Kane, D. L., and Carlson, R. F. (1973). Hydrology of the central arctic river basins of Alaska. Rep. 41. Institute of Water Resources, University of Alaska, Fairbanks.

Kane, D. L., and Hinzman, L. D. (1988). Permafrost hydrology of a small arctic watershed. *In* "Permafrost: Fifth International Conference" (K. Senneset, ed.), pp. 590–595. Tapir Press, Trondheim, Norway.

Kane, D. L., and Hinzman, L. D. (1991). Potential response of an arctic watershed during a period of global warming. *J. Geophys. Res.* (in press).

Kane, D. L., and Stein, J. (1983). Water movement into seasonally frozen soils. *Water Resour. Res.* **19**, 1547–1557.

Kane, D. L., Hinzman, L. D., Benson, C. S., and Everett, K. R. (1989). Hydrology of Imnavait Creek, an arctic watershed. *Holarct. Ecol.* **12**, 262–269.

Kane, D. L., Gieck, R. E., and Hinzman, L. D. (1990). Evapotranspiration from a small Alaskan arctic watershed. *Nord. Hydrol.* **21**, 253–272.

Kane, D. L., Hinzman, L. D., Benson, C. S., and Liston, G. E. (1991a). Snow hydrology of a headwater arctic basin. 1. Physical measurements and process studies. *Water Resour. Res.* **27**(6), 1099–1109.

Kane, D. L., Hinzman, L. D., and Zarling, J. P. (1991b). Thermal response of the active layer in a permafrost environment to climatic warming. *Cold Reg. Sci. Technol.* **19**(2), 111–122.

LaBelle, J. C., Wise, J. L., Voelker, R. P., Schulze, R. H., and Wohl, G. M. (1983). "Alaska Marine Ice Atlas." Arctic Environmental Information and Data Center, University of Alaska, Anchorage.

Lachenbruch, A. H., and Marshall, B. V. (1986). Changing climate: Geothermal evidence from permafrost in the Alaskan Arctic. *Science* **234**, 689–696.

Landals, A. L., and Gill, D. (1973). Differences in the volume of surface runoff during the snowmelt period: Yellowknife, Northwest Territories. *In* "The Role of Snow and Ice in Hydrology," pp. 927–942. IAHS Pub. No. 107. World Meteorological Organization, Geneva.

Lewkowicz, A. G., and French, H. M. (1982). The hydrology of small runoff plots in an area of discontinuous permafrost, Banks Island, N.W.T. *In* "Proceedings of the Fourth Canadian Permafrost Conference," (H. M. French, ed.) pp. 151–162. National Research Council of Canada, Ottawa.

Lunardini, V. J. (1981). "Heat Transfer in Cold Climates." Van Nostrand Reinhold, New York.

McCann, S. B., and Cogley, J. G. (1972). Hydrological observations on a small arctic catchment, Devon Island. *Can. J. Earth Sci.* **9**, 361–365.

Mackay, J. R. (1983). Downward water movement into frozen ground, Western Arctic Coast, Canada. *Can. J. Earth Sci.* **10**, 120–134.

Marsh, P., and Woo, M. K. (1979). Annual water balance of small high arctic basins. *In* "Canadian Hydrology Symposium: 79," pp. 536–546. National Research Council of Canada, Ottawa.

Marsh, P., and Woo, M. K. (1984). Wetting front advance and freezing of meltwater within a snow cover. 1. Observations in the Canadian Arctic. *Water Resour. Res.* **20**, 1853–1864.

Marsh, P., Rouse, W. R., and Woo, M. K. (1981). Evaporation at a high arctic site. *J. Appl. Meteorol.* **20**, 713–716.

Nixon, J. F. (1975). The role of convection heat transport in the thawing of frozen soils. *Can. Geotech. J.* **12**, 425–429.

Ohmura, A. (1982). Evaporation from the surface of the arctic tundra on Axel Heiberg Island. *Water Resour. Res.* **18**, 291–300.

Ostrem, G., Bridge, C. W., and Rannie, W. F. (1967). Glaciohydrology, discharge, and sediment transport in the Decade Glacier area, Baffin Island, N.W.T. *Geogr. Ann.* **49A**, 268–282.

Péwé, T. L. (1983). Alpine permafrost in the contiguous United States: A review. *Arct. Alp. Res.* **15**, 145–156.

Priestley, C. H. B., and Taylor, R. J. (1972). On the assessment of surface heat flux and evaporation using large-scale parameters. *Mon. Weather Rev.* **100**, 81–92.

Roulet, N. T., and Woo, M. K. (1986). Hydrology of a wetland in the continuous permafrost region. *J. Hydrol.* **89**, 73–91.

Roulet, N. T., and Woo, M. K. (1988). Runoff generation in a low arctic drainage basin. *J. Hydrol.* **101**, 213–226.

Rouse, W. R. (1984). Microclimate at arctic tree line. 3. The effects of regional advection on the surface energy balance of upland tundra. *Water Resour. Res.* **20**, 74–78.

Rouse, W. R., and Mills, P. F. (1977). Evaporation in high latitudes. *Water Resour. Res.* **13**, 909–914.

Shafer, J. M., and Skaggs, R. L. (1983). Identification and characterization of watershed models for evaluation of impacts of climatic change on hydrology. Rep. DE84-005049. U.S. Dep. of Energy, Pacific Northwest Laboratory, Richland, Washington.

Slaughter, C. W., and Kane, D. L. (1979). Hydrologic role of shallow organic soils. *In* "Canadian Hydrology Symposium: 79," pp. 380–389. National Research Council of Canada, Ottawa.

Walker, E. R., Lewis, E. L., and Lake, R. A. (1973). Runoff from a small high arctic basin (abstract). *Eos: Trans. Am. Geophys. Union* **54**, 1090.

Woo, M. K. (1979). Breakup of streams in the Canadian High Arctic. *In* "Proceedings: Eastern Snow Conference, Alexandria Bay, NY." (B. E. Goodison, ed.) pp. 95–107.

Woo, M. K. (1982). Snow hydrology of the High Arctic. *In* "Proceedings: Joint Eastern and Western Snow Conference, Reno, NV," pp. 63–74.

Woo, M. K. (1983). Hydrology of a drainage basin in the Canadian High Arctic. *Ann. Assoc. Am. Geogr.* **73**, 577–596.

Woo, M. K. (1986). Permafrost hydrology in North America. *Atmos. Ocean* **24**, 201–234.

Woo, M. K., and Dubreuil, M. A. (1985). Empirical relationship between dust content and arctic snow albedo. *Cold Reg. Sci. Technol.* **10**, 125–132.

Woo, M. K., and Sauriol, J. (1980). Channel development in a snow-filled valley, Resolute, N.W.T., Canada. *Geogr. Ann.* **62A**, 37–56.

Woo, M. K., Heron, R., and Steer, P. (1981). Catchment hydrology of a high arctic lake. *Cold Reg. Sci. Technol.* **5**, 29–41.

Woo, M. K., Heron, R., and Marsh, P. (1982). Basal ice in high arctic snowpacks. *Arct. Alp. Res.* **14**, 251–260.

Woo, M. K., Marsh, P., and Steer, P. (1983). Basin water balance in a continuous permafrost environment. *In* "Permafrost: Fourth International Conference Proceedings." pp. 1407–1411. National Academy Press, Washington, DC.

World Meteorological Organization (1986). "Intercomparison of Models of Snowmelt Runoff." Oper. Hydrol. Rep. 23. World Meteorological Organization, Geneva.

Young, K. L., and Lewkowicz, A. G. (1988). Measurement of outflow from a snowbank with basal ice. *J. Glaciol.* **34**, 358–362.

4

Circumpolar Arctic Vegetation

L. C. Bliss and N. V. Matveyeva

I. Introduction

Ecologists are fundamentally interested in the patterns and processes of organisms and how organisms respond to physical and biological phenomena. Present concerns about global climatic change add a new dimension to this interest, particularly in the far north. This book aims at a synthesis of the physiological adaptations of plants in the Arctic, but without background knowledge of the floristic and vegetation patterns within this vast region, detailed studies of plant processes become less meaningful. The paleobotanical record at high latitudes, information on subarctic and arctic floras, and the distribution of these floras into vegetation patterns increase our ability to interpret the significance of ecophysiological studies. This chapter, thus lays the groundwork for better understanding of pattern and process within the Arctic.

Biologists generally accept the definition of the Arctic as those lands beyond the climatic limit of trees dominating the mesic habitats of upland areas between river drainages (also called interfluves or placors). It is more difficult, however, to divide these huge landscapes—roughly 360,000 km^2 in Alaska, 2,480,000 km^2 in Canada, 2,167,000 km^2 in Greenland and Iceland, and 2,560,000 km^2 in the Soviet Union and Scandinavia—into biogeographic or vegetation units or zones. For many years arctic ecologists and biogeographers have viewed the circumpolar arctic as one floristic unit; they also believed the vegetation patterns were very similar on the North American and Eurasian continents. Traditionally, though, Soviet scientists have recognized a larger number of vegetation zones in their north than has been common in North America. With the opportunity for a greater exchange of people and ideas, it is now possible to bring together these diverse schemes of classification and to better understand the similarities and differences in the geomorphology, climate, and resulting patterns of vegetation in these northern lands (Fig. 1).

In the Soviet Arctic most of the land is continental, with groups of islands within the Arctic Ocean. The Taimyr Peninsula extends as a low-elevation (30–100 m) landmass north to Cape Chelyuskin (78° N), broken only by the Byrranga Mountains (200–1000 m). Consequently, there is a gradual change in climate (Table 1) and a corresponding gradual change in floristics and vegetation from tree line near Norilsk (69° N) in the southwest and Khatanga (72° N) in the northeast to the Laptev Sea. Eastward, tree line ranges from 70–71° N on the Lena River and the Indigirka River to 60° N on the Kamchatka Peninsula. Eastern Siberia has extensive mountain ranges, unlike the low relief of the western and central Siberian Arctic. The vegetation zones are based upon a strong northward gradient in temperature, floristic composition, and plant life-forms (Aleksandrova, 1971; 1980).

The geomorphology of the North American Arctic is very different because the Brooks Range in Alaska and the Richardson Mountains in the Yukon Territory limit the southward extent of the Arctic, and the Beaufort Sea lies to the north. Extensive glaciation in the Northwest Territories (NWT), Canada, has limited soil development, thereby limiting vegetation on the central and eastern mainland. The most influential factor limiting a gradual change in vegetation and wildlife northward is the presence of the Canadian Arctic Archipelago, which creates a climatic and biotic boundary between the mainland and the extensive islands to the north (Bliss, 1988). These land–sea relationships are very different from the mainland. An additional factor is the low relief of the western islands (mostly <200 m), compared with the mountainous eastern islands (200–2000 m), and the strong

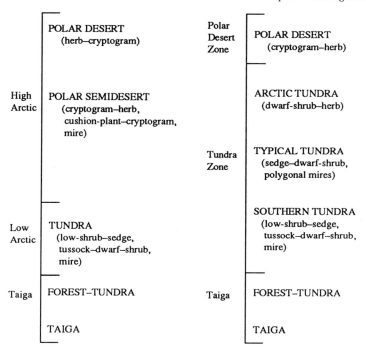

Figure 1 Classification schemes for arctic vegetation in North America (left) (Bliss, 1975, 1978) and Eurasia (right) (Gorodkov, 1935; Chernov and Matveyeva, 1979).

climatic influence of the warmer waters of the Davis Strait and the Greenland ice cap with its "continental" influence. Thus, in North America no unbroken landmass extends to 78° N as in the Soviet Arctic. Consequently, the vegetation patterns often lie in a coarse-grained mosaic rather than in belts or zones. Temperature and precipitation also show a mosaic pattern in the High Arctic as well (Table 1).

Polunin (1951) divided the North American Arctic into three zones based upon floristics. Young (1971) and Edlund and Alt (1989) have proposed dividing the Canadian Arctic Archipelago into zones, based mainly on floristics and mean monthly temperature, although they also consider vegetation in comparable habitats. Although temperature is the regional control (Bliss, 1975, 1988; Rannie, 1986), the mosaic pattern of mires, cushion plant, cryptogam–herb, and barren polar desert vegetation that is found on these arctic islands is a product of topography, temperature, soil water balance, and geologic substrate (Bliss *et al.*, 1984; Bliss and Svoboda, 1984; Bliss, 1988). In this book, the Arctic is divided into the Low and High Arctic (Fig. 2), with subdivisions within each major zone.

Table I Climatic Data for Selected Stations in the Circumpolar Low and High Arctic[a]

Station (latitude)	Temperature (°C) Mean monthly June	July	Aug	Mean annual	Degree-days >0°C	Precipitation (mm) Mean monthly June	July	Aug	Annual
Low Arctic									
Baker Lake, NWT (65° N)	3.2	10.8	9.8	-12.2	1251	16	36	34	213
Tuktoyaktuk, NWT (70°)	4.7	10.3	8.8	-10.4	903	13	22	29	130
Umiat, AK (69°)	9.1	11.7	7.3	-11.7	993	43	24	20	119
Godthaab, Greenland (64°)	5.7	7.6	6.9	-0.7	809	46	59	69	515
Umanak, Greenland (71°)	4.8	7.8	7.0	-4.0	682	12	12	12	201
Khatanga, USSR (72°)	3.9	12.2	8.9	-12.8	838	20	40	45	228
Dickson, USSR (73°)	-0.6	4.4	5.0	-10.6	385	18	20	40	165
Ustye, USSR (76°)	0.2	4.4	2.8	-13.3	244	20	38	50	162
High Arctic									
Barrow, AK (71°)	1.3	3.8	2.2	-12.1	288	8	22	20	100
Sachs Harbour, NWT (72°)	2.2	5.5	4.4	-13.6	458	8	18	22	103
Resolute, NWT (75°)	-0.3	4.3	2.8	-16.4	222	13	26	30	136
Eureka, NWT (80°)	2.2	5.5	3.6	-19.3	318	4	13	9	58
Scoresbysund, Greenland (70°)	2.4	4.7	3.7	-6.7	333	26	38	33	428
Nord, Greenland (81°)	-0.4	4.2	1.6	-11.1	202	5	12	19	204
Cape Zhelaniya, USSR (76°)	-1.1	1.7	2.2	-8.3	134	13	25	38	128
Cape Chelyuskin, USSR (78°)	-1.1	1.7	1.1	-13.9	117	18	28	28	113

[a] Data are from various sources; values are generally 10-year averages.

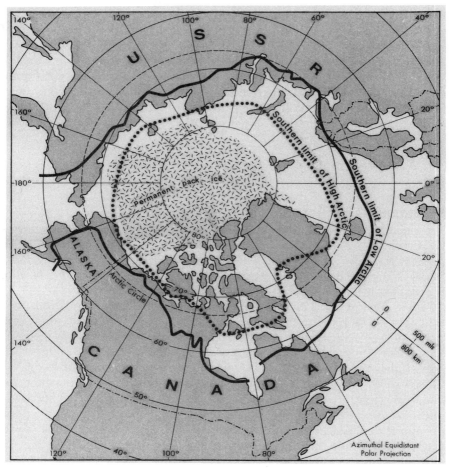

Figure 2 Circumpolar Arctic showing the limits of the Low and High Arctic. [Reprinted from L. C. Bliss (1979) with permission of the National Research Council of Canada.]

II. Forest–Tundra

As the climate becomes more severe poleward, permafrost is continuous, and the shallower and colder active soil layer provides a rooting zone that is less favorable for trees. Consequently, forests become more open and trees grow shorter, finally occurring as narrow ribbons along rivers (Color Plate 1) and as scattered individuals on protected slopes. In many upland areas, the transition from forest–tundra to tundra results in only the loss of *Picea, Larix,* or *Betula* (Drew and Shanks, 1965; Larsen, 1965; Black and Bliss, 1978, 1980). Fire plays a major role in the lichen woodlands and forest-tundras, effectively eliminating tree establishment during years of cooler summers (Nichols, 1975;

Black and Bliss, 1978). Because relatively high surface soil temperatures are required for *Picea mariana* seed germination, massive fires during periods of cooler and drier summers can effectively force tree line 50–100 km south (Black and Bliss, 1980). It is assumed that fire, climate, and tree seedling establishment show similar relationships across Siberia. Because shrub tundra contains many heath and other shrub species common to the taiga, some Soviet ecologists have considered shrub tundra (but not open forest) a part of the Subarctic rather than the Arctic (Aleksandrova, 1970, 1971, 1980; Andreyev and Aleksandrova, 1981).

III. Low Arctic

A. Introduction

The mainland Arctic (tundra) in Eurasia and North America is characterized by an abundance of low shrubs (*Alnus, Betula, Salix*), dwarf shrubs of heath species (*Arctostaphylos, Empetrum, Ledum, Vaccinium, Cassiope*), and various forbs. Many of these species are also typical of the northern taiga and have thus been called hypoarctic species by Tolmachev (1932) and Yurtsev (1966, 1972). Another general feature is the limited amount of bare ground from frost action. In North America the major vegetation types are low-shrub tundra just north of tree line, tussock tundra on imperfectly drained lands, and wet sedge–moss tundra, or mires, on poorly drained lands, especially on the coastal plain in Alaska and the Tuktoyaktuk Peninsula in the Northwest Territories (NWT). In Eurasia, low-shrub tundras cover the placors north of tree line, with large areas of tussock tundra in central and eastern Siberia and mires (wet sedge–moss tundra) in limited areas in western Siberia, except on the Yamal Peninsula, where mires are more extensive. North of these southern tundras lie the "typical tundras" of the Taimyr Peninsula (see Fig. 1), defined as those lands where the southern shrub species and forbs (hypoarctic species) are minor on placors and where *Carex ensifolia* ssp. *arctisibirica* and *Dryas* species predominate. East of the Indigirka River, however, *C. ensifolia* ssp. *arctisibirica* is minor, and it is absent from Chukotka (Yurtsev *et al.*, 1975).

B. Shrub Tundras

Beyond the forest–tundra in Alaska, the Yukon Territory, and the Mackenzie River Delta region, shrub communities are dominated by *Betula nana* ssp. *exilis* (=*B. exilis*), *Salix glauca* ssp. *acutifolia, S. pulchra,* and *S. lanata* ssp. *richardsonii* in various combinations (Hanson, 1953; Corns, 1974). The ground cover includes *Carex consimilis, Eriophorum vaginatum,* numerous dwarf heath shrubs, grasses, and forbs; and an abundance of lichens and mosses. The heath shrubs (= dwarf shrubs) are generally dominated by *Ledum palustre* ssp. *decumbens, Vaccinium vitis-idaea* ssp. *minus, Empetrum hermaphroditum* with lesser amounts of *Rubus chamaemorus, Arctostaphylos alpina (Arctous alpina), A. rubra,* and *Cassiope tetragona.* The most common mosses include *Hylocomium spendens*

Figure 3   Low-shrub–dwarf-shrub heath tundra northeast of Inuvik, NWT. The dominant shrubs are *Betula nana* ssp. *exilis, Salix glauca,* ssp. *acutifolia, S. pulchra;* the dwarf shrubs include *Vaccinium uliginosum, V. vitis-idaea* ssp. *minus, Empetrum hermaphroditum, Ledum palustre* ssp. *decumbens,* and scattered clumps of *Carex bigelowii* and *Eriophorum vaginatum.*

ssp. *alaskanum, Aulacomnium turgidum, Polytrichum juniperinum,* and several species of *Sphagnum.* Important fruticose lichens are *Cladina mitis, C. rangiferina, Cladonia gracilis, Cetraria cucullata, C. nivalis, C. rangiferina,* and *Thamnolia vermicularis.* The open shrub canopy is 40–60 cm high, and the heath shrubs and forbs are 10–20 cm in height; the cryptogams provide a complete ground cover (Fig. 3).

It is this type of vegetation that predominates at Toolik Lake, Alaska (Fig. 4) and in the uplands north of Inuvik, NWT. (Color Plate 2). Much of the mainland in central and eastern Canada contains relatively little low-shrub tundra, probably because of limited winter snow and abrasive winter winds (Larsen, 1965, 1972). In northern Quebec, southern Baffin Island, and the Fort Churchill area, dwarf-shrub heaths and low-shrub tundra with the cryptogams listed above predominate (Polunin, 1948; Payette and Filion, 1984; Payette *et al.,* 1989). Similar limited low-shrub tundra is found from west to central Chukotka and in much of arctic Yakutia east of the Olenek River.

In Greenland, low-shrub tundras occur in the inner fiord regions that are warmer in summer. Shrub communities dominated by *Betula glandulosa, B. nana,* or *Salix glauca* ssp. *calicarpaea* extend to 74° N in west Greenland but only to 62° N in east Greenland because of the colder climate away from the Labrador Current (Böcher, 1959). *Alnus crispa* and *Sorbus groenlandica* are part

Figure 4   A close-up of low-shrub tundra at Toolik Lake, Alaska. Dominant species include *Betula nana* ssp. *exilis, Salix planifolia* ssp. *pulchra, Ledum palustre* ssp. *decumbens, Carex lugens,* and *Sphagnum* mosses.

of the western coast shrub communities. Many forbs and various combinations of dwarf-shrub heaths are commonly found as an understory community along with mosses and lichens.

Southern low-arctic tundra is distinguished by the wide distribution and dominance of hypoarctic species, together with a rich variety of boreal species. Shrub tundras extend across much of the U.S.S.R., on rolling uplands and placors north of the forest–tundra zone (Figs. 5, 6; Color Plate 3). Shrub communities are dominated by *Betula nana* in the west, from the European North to the central part of Taimyr, and by *B. exilis* in the east, with a mixture of *Salix lanata, S. glauca, S. phylicifolia,* and *S. pulchra* (Aleksandrova, 1980). The ground cover includes *Carex ensifolia* ssp. *arctisibirica,* the heath shrubs *Ledum palustre* ssp. *decumbens, Vaccinium uliginosum* ssp. *microphyllum, V. vitis-idaea* ssp. *minus, Empetrum hermaphroditum,* and *Arctous alpina,* as well as *Dryas punctata, D. octopetala,* and *Cassiope tetragona.* There is a continuous moss cover comprising *Hylocomium spendens* ssp. *alaskanum, Tomenthypnum nitens, Aulacomnium turgidum, Dicranum congestum, D. spadiceum,* and the liverwort *Ptilidium ciliare* (Aleksandrova, 1980; Matveyeva and Zanokha, 1986). In the European North, *Pleurozium schreberi* and *Polytrichum commune,* as well as *Carex bigelowii* and *C. globularis,* occur in birch and willow thickets (Rebristaya, 1977). Many fruticose lichens, such as *Cladina arbuscula, C. rangiferina, Cladonia amaurocraea,* and *Cetraria cucullata,* occur within the mosses. The shrub

Figure 5   Low-shrub tundra with dwarf heath shrubs and *Carex ensifolia* ssp. *arctisibirica* on the placors, with *Salix lanata, Alnaster fruticosa* and a few scattered *Larix sibirica* in the drainages, Taz Peninsula, USSR.

canopy is 50–80 cm high. The tallest shrubs (up to 2 m), form thickets on the interfluves in the Siberian low-arctic tundra, are *Alnus fruticosa,* which is very important in many places from the Ural Mountains to Chukotka. Thickets of *Salix lanata,* 1–2 m tall, occur everywhere in drainages (Pospelova, 1972; Chernov and Matveyeva, 1979; Matveyeva and Zanokha, 1986). In eastern Siberia, the amphiatlantic shrub species disappear and are replaced by amphiberingian species such as *Salix alaxensis, S. fuscescens,* and others (Aleksandrova, 1980). In the Anadyr–Penzhina area, shrub communities occur on gentle mountain slopes, dominated by *Pinus pumila, Betula middendorffii,* and *Alnus fruticosa,* with an admixture of *Spiraea beauverdiana, Rhododendron aureum,* and *Potentilla fruticosa* (Yurtsev, 1974).

C. Sedge–Dwarf-Shrub Tundras

Northern low-arctic "typical tundra" (Gorodkov, 1935; Chernov and Matveyeva, 1979) is characterized by the loss of many boreal species from the flora and the decreasing role of those hypoarctic species that were so important in the south. Most of the shrubs are still present, but they do not occur on the interfluves. *Betula* species grow mainly in depressions and on raised polygons in mires. *Alnus fruticosa* is absent except on sheltered slopes in Chukotka; thickets of *Salix lanata,* no more than 50–80 cm tall, occur exclusively in valleys. Only prostrate *Betula nana, B. exilis,* and *Salix pulchra,* as well

Figure 6   Low-shrub tundra with *Betula nana, Salix lanata, Vaccinium uliginosum, V. vitis idaea, Carex ensifolia* ssp. *arctisibirica*, and *Eriophorum vaginatum* on the Yamal Peninsula, USSR.

as the arctic willow *S. reptans*, are common on the mesic interfluves, but they do not form a shrub canopy. On the Taimyr Peninsula the majority of heath shrubs, such as *Ledum palustre* ssp. *decumbens, Empetrum hermaphroditum, Arctous alpina*, and *Vaccinium uliginosum* ssp. *microphyllum*, become very rare, growing mainly on valley slopes; they are present in Chukotka and Yakutia.

In the vast Siberian Arctic, the main species are the mosses *Hylocomium spendens* ssp. *alaskanum, Tomenthypnum nitens*, and *Aulacomnium turgidum*, and the liverwort *Ptilidium ciliare*, with *Carex ensifolia* ssp. *arctisibirica* and *Dryas punctata* in the ground layer. *Vaccinium vitis-idaea* ssp. *minus* is still common, as is *Cassiope tetragona*. Among the more abundant lichens are *Cladina arbuscula, C. rangiferina, C. uncialis, Cladonia amaurocraea, C. gracilis*, and some foliose species such as *Peltigera aphthosa, P. rufescens*, and *Nephroma expallidum* (Chernov and Matveyeva, 1979). In the European North and Yamal, *Dryas octopetala* dominates in many communities. Plant cover may be continuous on the placors in typical tundras, but frost-boil tundras with regular patches of bare ground (~20%) are more typical for this subzone ("spotty tundras" in the Russian literature). In Taimyr there are from 30 to 60 patches of bare ground per 100 m² in low-arctic frost-boil tundras. A specific community exists on these bare patches, with small vascular plants such as *Juncus biglumis, Sagina intermedia, Minuartia rubella*, and *Epilobium davuricum*; crustose lichens such as *Toninia lobulata, Solorina saccata, Baeomyces carneus, Lecanora verrucosa*, and *Rinodina turfacea*; and mosses such as *Psilopilum laevigatum, Ceratodon pur-*

pureus, Ditrichum flexicaule, Myurella julacea, Orthothecium chryseum, and *Hypnum bambergeri* (Matveyeva *et al.*, 1975).

In deep, narrow valleys with late snowmelt grow communities dominated by *Cassiope tetragona, Salix polaris,* and *Drepanocladus uncinatus* with a mixture of herbs, *Oxyria digyna, Ranunculus sulphureus, R. nivalis, R. pygmaeus,* and *Saxifraga hyperborea. Dryas* mats are still common on snowless fell-fields with some herb species, such as *Oxytropis middendorffii, O. nigrescens, Pedicularis dasyantha,* and *P. oederi.* An interesting plant community inhabits the south-facing slopes of river banks in the central Siberian Arctic, particularly in Taimyr. These are true mesic meadows, with a dense cover of numerous grasses such as *Poa alpigena, Festuca cryophila,* and *Alopecurus alpinus* and abundant forbs such as *Astragalus umbellatus, A. subpolaris, Hedysarum arcticum, Pedicularis verticillata, Silene paucifolia, Sanguisorba officinalis, Cerastium maximum, Myosotis asiatica, Polemonium boreale, Valeriana capitata,* and many others. Mosses and lichens, so typical for tundras, are absent or rare under the dense herb canopy (Matveyeva *et al.*, 1975; Matveyeva and Zanokha, 1986).

D. Steppe Tundras

Tundra steppes appear in Chukotka, where they have been thoroughly described by Yurtsev (1974); they seem to be relicts from the cold, dry Pleistocene. *Carex duriuscula,* otherwise distributed over the steppes of southern Siberia and northern Mongolia as well as on the Canadian prairies, occurs here, as do *Helictotrichon krylovii* and other steppe and mountain-steppe species.

E. Tussock–Dwarf-Shrub Tundras

Tundras dominated by *Eriophorum vaginatum* tussocks with *Carex lugens* in the west and *C. consimilis* to the east, along with the common dwarf-shrub heath species and an abundance of mosses and lichens, occupy large areas north of the shrub tundras in Alaska and the Yukon Territory (Hanson, 1953; Churchill, 1955; Johnson *et al.*, 1966; Wein and Bliss, 1974). *Betula nana* ssp. *exilis, Salix glauca,* and *S. pulchra* are common along with numerous forbs, grasses, and an abundance of mosses, including *Hylocomium splendens* ssp. *alaskanum, Dicranum elongatum, Aulacomnium turgidum, A. palustre,* and *Tomenthypnum nitens.* Common lichens include *Cetraria cucullata, C. nivalis, Cladonia amaurocraea, Cladina rangiferina, C. mitis, Dactylina arctica,* and *Thamnolia vermicularis. Sphagnum* mosses are abundant where drainage is poor and the soils developed from nonsedimentary rocks (Fig. 7). Tussock tundra is more limited east of the Anderson River in the heavily glaciated Northwest Territories (Larsen, 1965), probably because of the lack of soil.

Although tussocks of *Eriophorum vaginatum* form a small component of the subarctic tundras (as defined by Aleksandrova) in western Siberia, large tracks of land dominated by this species, along with the common heath shrubs and an abundance of lichens and mosses, are found in eastern Siberia and Chukotka (Aleksandrova, 1980). This predominance probably relates to the region's finely textured soils with their impeded drainage, unlike the sandy

Figure 7   Tussock–dwarf-shrub tundra in northern Alaska. *Eriophorum vaginatum* predominates with scattered clumps of *Betula nana, Salix pulchra, S. glauca* and the major heath species. Taller shrubs of *Salix alaxensis, Alnus crispa*, and other willows occur on steep slopes and along the river.

soils of the Yamal and Gydan peninsulas and the low-elevation landscapes of much of the Taimyr Peninsula. From the Khatanga River eastward, however, tussock tundras occupy less well-drained habitats with admixtures of *Carex ensifolia* ssp. *arctisibirica* or, east of the Indigirka River, *Carex lugens*; low-growing shrubs of *Betula exilis, Salix pulchra,* and *S. glauca*; and the common heath species. Tussock tundra is well developed in the regions of the Indigirka and Kolyma rivers, again on imperfectly drained soils as in Chukotka (Yurtsev, 1972). *Salix pulchra* and *Betula exilis* occur as low, scattered shrubs.

F. Mires

Wetland plant communities are limited in the mountains of northern Alaska and the Yukon Territory but gain importance in Alaska's Foothill Province and dominate on the Coastal Plain Province and the flat coastal areas in the Yukon (Britton, 1957; Webber, 1978; Komarkova and Webber, 1980). Comparable mires (muskeg, bog) occur on islands in the Mackenzie River Delta and eastward on the Tuktoyaktuk Peninsula (Corns, 1974). On this peninsula and on the Alaskan coastal plain, thaw lakes may merge or drain, providing new surfaces for secondary succession (see Chap. 6). Peats are often 1–5 m thick, with only a shallow (15–25 cm) active layer developing in these cold, wet, nutrient-poor soils.

Figure 8   *Arctophila fulva* in shallow water with *Carex aquatilis, Eriophrum scheuchzeri, E. angustifolium, Dupontia fisheri,* and *Carex bigelowii* beyond on drier sites at Barrow, Alaska.

The dominant sedges are *Carex aquatilis, C. membranacea, C. chordorrhiza, C. rariflora, C. rotundata, Eriophorum angustifolium,* and *E. scheuchzeri.* The grasses *Arctagrostis latifolia, Dupontia fisheri,* and *Arctophila fulva* occur in a gradient from drier to wetter habitats (Fig. 8). Growing in the shallow (50–60 cm) waters of ponds and lake margins are *Menyanthes trifoliata, Equisetum limnosum,* and *Arctophila fulva; Potentilla palustris* and *Hippuris vulgaris* are found in waters only 20–30 cm deep (Britton, 1957; Hettinger *et al.,* 1973; Corns, 1974). Lichens are a minor component of these wetlands, but mosses are abundant, including species of *Aulocomnium, Calliergon, Ditrichum, Drepanocladus,* and *Sphagnum* as well as *Hylocomium spendens* ssp. *alaskanum, Meesia triquetra,* and *Tomenthypnum nitens.*

Sedge-dominated mires occur in lowlands across the Canadian Shield but in limited extent (Polunin, 1948; Larsen, 1965, 1972). Common species include *Carex rariflora, C. membranacea, C. stans, Eriophorum angustifolium,* and *E. scheuchzeri.* Mires are also present in Greenland, but as in large areas of central and eastern Canada, they are uncommon. The most important species are *Carex rariflora, C. vaginata, C. holostoma, Eriophorum angustifolium,* and *E. triste* (Böcher, 1954, 1959).

Across the Soviet Arctic, mires are a common feature of the landscape. They are especially well developed in the central part of the Yamal Peninsula and in the lowlands of the Yana, Indigirka, and Kolyma river basins. Three main types can be distinguished: homogeneous mires; flat-top hillock mires with wet depressions and peat hillocks; and polygonal mires with low-center polygons about 10 m in diameter. Wet depressions in all types of mires are dominated by *Carex stans, C. chordorrhiza, C. rariflora, Eriophorum angustifolium, E. medium, Caltha arctica,* and *Comarum palustre* as well as the mosses *Drepanocladus revolvens, D. vernicosus, Meesia triquetra, Calliergon giganteum, C. sarmentosum, Polytrichum jensenii,* and *Cinclidium latifolium.* Dense thickets of *Betula nana* or *B. exilis* are typical for large peat hillocks up to 30 m in diameter, with *Dicranum elongatum, D. angustum,* and *Polytrichum alpestre* in the ground cover. *Carex ensifolia* ssp. *arctisibirica* or *C. lugens, Dryas punctata, Vaccinium vitis-idaea* ssp. *minus, Salix pulchra,* and in the southern part of the Low Arctic *Betula nana,* also grow on the rims of polygonal mires. The mesic tundra mosses *Tomenthypnum nitens, Aulacomnium turgidum, Hylocomium spendens* ssp. *alaskanum,* as well as the liverwort *Ptilidium ciliare,* form a thick cover. Some *Sphagnum* species, such as *S. squarrosum, S. fimbriatum, S. warnstorffii,* occur on rims. Typical small peat hummocks accompany these *Sphagnum* species, as does *Rubus chamaemorus. Carex stans, Arctophila fulva,* and *Hippuris vulgaris* are very common in numerous shallow lakes as well as on wet, silty river banks where *Eriophorum scheuchzeri* is also abundant.

G. Coastal Salt Marshes

Coastal salt marshes are a minor feature in the North American Arctic, yet where they occur, there are interesting combinations of species. Such marshes are limited in extent because of a lack of sands and silts, the annual reworking of shorelines by ice, a limited tidal amplitude (0.5–2.0 m), a short growing season, and low salinity. The most common species include *Puccinellia phryganodes, Carex ursina, C. subspathacea, C. ramenskii, C. glareosa, Cochlearia officinalis,* and *Stellaria humifusa* (Polunin, 1948; Jefferies, 1977; Vestergaard, 1978). Similar coastal marshes with the same species are well developed along the whole coast of the Arctic Ocean, including in the Siberian Arctic. The much more extensive *Puccinellia* marshes along Hudson Bay are heavily grazed by lesser snow geese, which play a significant role in nutrient release into these meadows and thereby help maintain their productivity (Jefferies *et al.,* 1979; Jefferies, 1988); Tikhomirov, 1957; Chap. 19).

IV. High Arctic

A. Introduction

In North America, the major change in the structure of arctic vegetation as well as floristics occurs in the shift from mainland Canada to the Canadian Arctic Archipelago. The exceptions are the area around Barrow, Alaska; the northern part of the District of Keewatin, including the Boothia Peninsula; and northern Quebec. These changes include a shift from the predominance of low-shrub (*Betula, Salix*), dwarf-shrub (heath species), and cottongrass–tussock–dwarf-shrub tundras to landscapes dominated by cushion plants (*Dryas integrifolia, Saxifraga oppositifolia*), prostrate shrubs of *Salix arctica,* and rosette species of *Saxifraga, Draba,* and *Minuartia.* Many other biological and environmental characteristics separate the Arctic into Low and High Arctic (Bliss 1981; 1988). Mires (muskegs) are found in both regions, although they are much more common in the Low Arctic because of the extensive coastal plain in Alaska and the Mackenzie River Delta.

Because the Soviet Arctic is nearly all continental and spans so many latitudes, one observes more subtle shifts in species northward with a gradual change in climate. The low shrub (hypoarctic) species drop out, and dwarf shrub heath species decrease significantly in the transition from the subarctic tundra to the arctic tundra (Aleksandrova, 1980; Andreyev and Aleksandrova, 1981). *Carex ensifolia* ssp. *arctisibirica, Dryas punctata, D. octopetala,* and *Salix reptans* are still common on placors in the northern part of typical tundras, but they are insignificant in the true arctic tundra, where *Salix polaris, S. arctica, Alopecurus alpinus, Deschampsia borealis,* and *Luzula confusa* dominate (Chernov and Mateyeva, 1979; Aleksandrova, 1980). Comparable landscapes in North America have been called polar semideserts (Bliss, 1975, 1981, 1988; Bliss and Svoboda, 1984).

The High Arctic is as diverse in its vegetation as the Low Arctic. It includes small areas of tall shrubs (*Salix alaxensis, S. pulchra*) in protected river valleys and near lakes on Banks and Victoria islands (Kuc, 1974), and limited areas of low-shrub and dwarf-shrub–heath tundra, cottongrass-tussock tundra, and extensive mires on these same islands. Low-shrub and dwarf-shrub tundras and mires occur on southern Baffin Island as well. Cushion plant–cryptogam vegetation predominates on better-drained lands in the southern islands, areas of Melville Island, and limited lowlands of the central and eastern islands (typically, former coastal beach ridges and valleys). On the northwestern islands, cryptogam–herb and graminoid–steppe vegetation predominates on fine sand to clay loam soils (Bliss and Svoboda, 1984). Finally, where coarse materials occur at sea level as well as at 300 to 1000+ m on plateaus, there is very little plant cover. These landscapes have been called polar deserts (Bliss *et al.,* 1984).

B. Mires

High-arctic communities are structurally and floristically related to mires in the Low Arctic. These sedge–moss and grass–moss tundras cover 5–40% of the southern islands, except Baffin Island, in the Canadian Archipelago. In the Queen Elizabeth Islands, less than 2% of the area contains mires, yet these oases are the major habitat for muskox and the breeding grounds for waterfowl and shorebirds. *Carex stans* is the dominant graminoid, with lesser amounts of *C. membranacea, Eriophorum scheuchzeri, E. triste, Dupontia fisheri,* and *Alopecurus alpinus* (Fig. 9) in the eastern and southern islands (Muc, 1977; Sheard and Geale, 1983; Bliss and Svoboda, 1984; Muc *et al.,* 1989). *Dupontia fisheri* dominates in wetlands of the northwestern islands and northern Melville Island; associated species include *Juncus biglumis* and *Eriophorum triste.* Shallow ponds in the northern islands contain *Pleuropogon sabinei,* the ecological equivalent of *Arctophila fulva* in the southern islands and on the mainland

Figure 9   Mire dominated by *Carex stans*, with lesser amounts of *Eriophorum membranacea* and *Salix arctica*, on the Truelove Lowland, Devon Island, NWT.

of the Low Arctic (Bliss and Svoboda, 1984). Moss species include *Orthothecium chryseum, Campylium arcticum, Tomenthypnum nitens, Drepanocladus* spp, *Ditrichum flexicaule*, and *Cinclidium arcticum*. Species of cyanobacteria, especially *Nostoc commune*, play an important role in nitrogen fixation in these wetlands (Chap. 14) and in mire formation (Chap. 6). High-arctic mires occur only where water remains on the landscape much of the summer because drainage is blocked at snowmelt by raised beach ridges (Sabine Lowland, Melville Island; Truelove Lowland, Devon Island) or where the land is kept moist by glacial meltwaters, as on Ellesmere Island (Alexandra Fiord, Sverdrup Pass) (Henry *et al.*, 1986; Muc *et al.*, 1989).

Comparable mire communities occur in the Soviet Arctic. The polygonal mires with low-center polygons are similar to those in the Low Arctic but without *Betula nana, Salix pulchra*, and *Vaccinium vitis-idaea* ssp. *minus*. There are the same vascular plants (*Carex stans, Dupontia fisheri, Eriophorum agustifolium*) and mosses (*Drepanocladus revolvens, D. vernicosus, Meesia triquetra, Cinclidium latifolium, Polytrichum jensenii*) on the wet polygons; *Carex ensifolia* ssp. *arctisibirica* and *Dryas punctata* occur on rims with *Tomenthypnum nitens, Aulacomnium turgidum* and *Hylocomium splendens* ssp. *alaskanum* in troughs. But these polygonal mires are rare in this area. More typical are homogeneous mires in wide valleys where the above-mentioned graminoids predominate, with the addition of *Cardamine pratensis, Cerastium regelii, Eriophorum scheuchz-*

Figure 10   Cushion-plant community dominated by *Dryas integrifolia*, with lesser amounts of *Salix arctica, Carex rupestris, Saxifraga oppositifolia,* and *Pedicularis lanata* with *Cassiope tetragona* in the depressions, northern Banks Island, NWT.

eri, Poa arctica, and *Chrysoplenium alternifolium.* The main moss species are *Drepanocladus vernicosus, Calliergon giganteum, C. sarmentosum, Campylium zemliae, Cinclidium arcticum,* and *Mnium rugicum.* Small moss–peat hummocks sometimes develop along the edge of wet depressions, representing a successional sere from *Sphagnum* hummocks with *S. squarrosum, S. warnstorffii, S. rubellum, S. fimbriatum,* and *S. lenense* to moss–peat hummocks with *Polytrichum alpestre* and *Dicranum elongatum.* The common vascular plants for such hummocks are *Poa arctica, Calamagrostis holmii, Hierochloe pauciflora, Saxifraga cernua, Stellaria ciliatosepala, Arctagrostis latifolia,* and *Cardamine pratensis.* On the hummock tops the lichens *Cladonia macroceras, C. amaurocraea, C. gracilis, Cetraria islandica, Dactylina arctica, Ochrolechia frigida,* and *Thamnolia vermicularis* are common.

C. Polar Semideserts and Polar Deserts

1. Cushion Plant (or Dwarf-Shrub)–Cryptogam

In the Low Arctic, mats of *Dryas* with associated species of *Draba* and *Saxifraga, Salix arctica,* and *Carex rupestris* are limited to exposed uplands and mountain slopes. These communities are common in the Brooks Range and the northern mountains of the Yukon Territory, with extensions south into the central Rocky Mountains. This type of vegetation is the most common in the southern

Canadian arctic islands (Fig. 10), and it predominates in the coastal lowlands and mountain valleys in the northeastern islands and in dry, continental valleys in Greenland. Communities dominated by *Dryas integrifolia* with associated *Salix arctica, Saxifraga oppositifolia, S. caespitosa, S. cernua, Draba corymbosa, D. oblongata, Cerastium alpinum, Papaver radicatum,* and several species of *Minuartia* and *Stellaria* cover vast areas of well-drained neutral to slightly alkaline soils. Common graminoids include *Carex rupestris, C. nardina,* and *Alopecurus alpinus.* Where ice-wedge depressions occur, *Cassiope tetragona* is often present. Plant communities of this general type have been described from Ellesmere Island (Color Plate 4; Brassard and Longton, 1970; Muc *et al.,* 1989), Devon Island (Svoboda, 1977), Bathurst (Sheard and Geale, 1983; Bliss *et al.,* 1984), Cornwallis and Somerset islands (Bliss *et al.,* 1984), and South Hampton Island (Reznicek and Svoboda, 1982). These communities have limited snow cover (2–10 cm) in winter and a relatively deep active layer (50–100+ cm) in summer.

Plant cover is seldom continuous on the interfluves within the High Arctic in the Soviet Union. In high-arctic frost-boil tundras bare ground equals or even exceeds plant cover; in Taimyr, for example, there are about 150 patches of bare ground per 100 m^2 (Matveyeva, 1979). The active soil layer in summer is no more than 50 cm. Graminoids (*Alopecurus alpinus, Deschampsia borealis, Luzula confusa,* and sometimes *Carex ensifolia* ssp. *arctisibirica*) account for 5% cover; willows (*Salix polaris* and *S. arctica*), 15%; and cryptogams, mainly mosses (*Hylocomium splendens* ssp. *alaskanum, Aulacomnium turgidum,* and *Tomenthypnum nitens*), 30–40%. Many *Saxifraga* and *Draba* species become very abundant here, especially *S. oppositifolia, S. hirculus, S. caespitosa, Draba oblongata, D. alpina,* and *D. fladnizensis,* as well as *Papaver polare.* The fruticose lichens are less abundant than farther south. *Cladina arbuscula, C. rangiferina,* and *C. uncialis,* so typical farther south, are rare, but *Cladonia amaurocraea* and *C. gracilis,* as well as the foliose species *Peltigera aphthosa* and *P. rufescens,* are still common. The liverwort *Ptilidium ciliare* is also present but less abundant. Such communities are common in the northern part of the Taimyr and Gydan peninsulas, along a narrow strip of mainland bordering the Arctic Ocean in eastern Siberia, and on the large islands of Novaya Zemlya, Novosibirskie Ostrova, and Wrangel. Their structure and floristic composition are sometimes similar.

Mats of *Dryas punctata* in central Siberia and *D. punctata* and *D. integrifolia* in Chukotka are limited to fell-fields, with plant cover no more than 40%. Numerous lichens, in particular crustose species of *Pertusaria, Ochrolechia* and *Rinodina,* as well as mosses such as *Polytrichum piliferum, Tortula ruralis, Rhacomitrium canescens, Rhytidium rugosum,* and *Ditrichum flexicaule,* are typical, but they are not abundant. Mats of *Dryas* are widely distributed on Wrangel Island (Yurtsev, 1987).

On the Taimyr Peninsula there are numerous snow beds on the lower parts of northern slopes with the dwarf shrub *Salix polaris;* the moss *Drepanocladus uncinatus;* the lichens *Cetraria delisei, Stereocaulon rivulorum,* and *S. alpinum;* and the liverwort *Anthelia juratzkana* predominating. In deep, narrow valleys where the snow melts only in the second half of August, sparse vegetation is

Figure 11   Cryptogram–herb community at Rea Point, Melville Island, NWT. The dominant species include *Alopecurus alpina, Luzula confusa, Papaver radicatum, Salix arctica, Oxyria digyna, Saxifraga oppositifolia,* and abundant lichens and mosses.

formed with *Phippsia concinna, Cerastium regelii, Campylium polygamum,* and *Gymnomitrion corallioides.*

Meadows with grasses and herbs can still be found on southern slopes, but species composition differs from that in the Low Arctic. The main species in Taimyr are *Poa alpigena, Alopercurus alpinus, Deschampsia borealis, Draba glacialis, D. subcapitata, Saxifraga hirculus, Lloydia serotina, Myosotis asiatica, Polygonum viviparum,* and *Pedicularis sudetica.*

2. Cryptogam–Herb

Communities in which *Alopecurus alpinus, Luzula confusa,* and *L. nivalis* dominate along with several species of *Draba, Saxifraga, Minuartia, Papaver radicatum, Cerastium alpinum,* and *Juncus albescens* predominate on sandy to clay loam soils (Fig. 11) in the central and western Queen Elizabeth Islands (Bliss and Svoboda, 1984). Bryophytes are abundant, including *Aulacomnium turgidum, Tomenthypnum nitens, Dicranoweisia crispula, Polytrichum juniperinum, Pogonatum alpinum, Rhacomitrium lanuginosum, R. sudeticum, Schistidium holmenianum,* and in wetter sites, *Gymnomitrion corallioides.* Lichens are also common, including *Cladonia gracilis, Cetraria cucullata, C. delisei, C. nivalis, Parmelia omphalodes,* and *Dactylina ramulosa.* Seedling establishment is much greater in moss mats and desiccation cracks than on bare soil or crustose lichen mats because of soil drying in late summer and spring and needle-ice formation in the fall (Sohlberg and Bliss, 1984). Vascular plants contribute 5–20% cover and mosses and lichens 50–80%. In some areas there is little bare ground, whereas in others 20–50% of the ground can be bare, often as soil polygons and stripes. Summer temperatures and precipitation are lower and cloud cover is higher here than in the eastern Queen Elizabeth Islands (Bliss and Svoboda, 1984; Edlund and Alt, 1989).

At Cape Chelyuskin, there are diverse plant communities with total cover varying from 5 to 40%, sometimes even to 60%. Mosses and lichens account for most of the cover, with a few vascular plants within the cryptogam mats. *Rhacomitrium lanuginosum*, *Aulacomnium turgidum*, *Orthothecium chryseum*, *Ditrichum flexicaule*, *Bryum tortifolium*, and *Dicranoweisia crispula* are the most common mosses, forming a net-like cover along the cracks in the ground. The most abundant lichens are *Cetraria islandica* var. *polaris*, *C. delisei*, *C. cucullata*, *Thamnolia subuliformis*, *Stereocaulon rivulorum*, *Dactylina ramulosa*, and *Parmelia omphalodes*. The first four of these can dominate and also form a net-like pattern of vegetation. There are 57 vascular plant species but only ten of them occur with any frequency. These are *Alopecurus alpinus*, *Deshampisia glauca*, *Phippsia algida*, *Stellaria edwardsii*, *Cerastium regelii*, *Papaver polare*, *Draba oblongata*, *D. subcapitata*, *Saxifraga cernua*, *S. foliolosa*, *S. hyperborea*, *S. oppositifolia*, and *Eritrichium villosum*. There are no mires, meadows, or snow-bed communities near Cape Chelyuskin (Matveyeva, 1979). Plant communities with comparable floristics and structure, albeit generally more plant cover, have been called semidesert in the central and western High Arctic of Canada (Bliss and Svoboda, 1984).

The Franz Josef Land Archipelago consists mostly of glaciers with only small ice-free areas. *Phippsia algida* is the only common grass, along with scattered plants of *Papaver polare*, *Cerastium arcticum*, *Saxifraga hyperborea*, *Draba oblongata*, *D. subcapitata*, and *Stellaria edwardsii*. Lichen crust species of *Pertusaria* and *Ochrolechia* predominate, but *Cetraria islandica* var. *polaris*, *C. delisei*, *C. cucullata* and *C. nivalis* are also abundant. Mosses and lichens provide most of the cover and most of the flora. There are 57 species of vascular plants and 272 species of cryptogams. Total plant cover for 15 relevés (stand samples) averaged 21% (3-46%) (Aleksandrova, 1988).

The vegetation of Severnaya Zemlya Archipelago resembles that of Cape Chelyuskin, although its vascular flora is richer (90 vs. 57 species). The flora and vegetation differ between northern and southern islands in this archipelago. Plant life is poorer on the northern islands of Komsomolets, Pioneer, Sedova, and Diabasovye; only 17 vascular plants have been found. An insignificant surface area is occupied by netlike vegetation along the fissures separating polygons. A moss–lichen sward consisting predominantly of *Cetraria delisei*, *C. cucullata*, and crustaceous lichens is most common. Among the mosses, *Distichum capillaceum* and *Ditrichum flexicaule* predominate, *Saxifraga oppositifolia* is relatively abundant; other vascular plants such as *Cerastium regelii*, *Draba pauciflora*, *Papaver polare*, *Phippsia algida*, *Saxifraga caespitosa*, *S. cernua*, and *Stellaria edwardsii*, play an insignificant role (Korothkevich, 1958).

Color Plate 1 Black spruce extending down a small river with abundant tall shrubs of *Salix alaxensis*, *S. lanata*, and *Alnus crispa*. Low-shrub tundra in the uplands dominated by *Betula nana*, *Salix glauca*, *S. pulchra*, dwarf heath species, and *Eriophorum vaginatum* tussocks. Northeast of Inuvik, NT.

Color Plate 2 A large pingo developed in a drained lake. Wetlands dominated by sedges, low shrubs of *Betula* and *Salix* in the uplands, along with cottongrass tussocks (brown) and scattered green alder (*Alnus crispa*). Northeast of Inuvik, NT, with east channel of the Mackenzie River in the background.

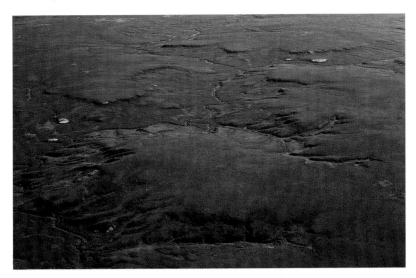

Color Plate 3 Low-shrub tundra with dwarf heath shrubs and *Carex ensifolia* ssp. *arctisibirica* on the interfluves with *Salix lanata*, *Alnaster fruticosa*, and a few scattered *Larix sibirica* in the drainages. Taz Peninsula, USSR.

Color Plate 4 Polar semidesert dominated by *Dryas integrifolia* in flower. Small clumps of *Oxyria digyma* and *Papaver radicatum* in foreground and large glaciers beyond. Alexander Fiord, east coast of Ellesmere Island, NT.

The southern islands, Ostrov Oktyabrskoi Revolyutsii and Ostrov Bolshevik, are considerably less glaciated and about 90 vascular species are now known. In addition to vegetation similar to Cape Chelyuskin, *Salix polaris* plays a significant role in plant cover at several places, and here the vegetation is more like tundra or semidesert than polar desert (Safronova, 1976). Habitats with *S. polaris* are restricted to warm, south-facing slopes. Ostrov Oktyabrskoi Revolyutsii harbors a climatic oasis at Zhiloi Peninsula with grass–lichen–moss communities; plant cover ranges from 50 to 85%, with mosses representing 35–60%, lichens 10–40%, and herbs 12–50%. The main vascular plants in these communities are *Poa abbreviata*, *P. alpigena*, *Alopecurus alpinus*, *Papaver polare*, *Saxifraga oppositifolia*, *S. caespitosa*, *Cerastium bialynickii*, *Draba oblongata*, *D. subcapitata*, and *D. macrocarpa*. Associations with considerable quantities of *Salix polaris* can also be found. Khodachek (1986) reported that summer temperatures were higher here, possibly the result of föehn winds.

In small group of islands Ostrova De Longa, species more typical of lands farther south, such as *Dryas octopetala*, *Nardosmia frigida* and *Saussurea tilesii*, occur in river valleys, but the whole pattern and composition of vegetation otherwise resembles that of other arctic islands. There are some extrazonal types of vegetation with a continuous moss cover in protected habitats. Here *Alopecurus alpinus*, *Saxifraga caespitosa*, *S. cernua*, *S. foliolosa*, *Poa arctica*, *Salix polaris*, *Ranunculus sulphureus*, and some lichens predominate. A polygonal herb–dwarf-shrub–moss tundra can be found in which *Dryas octopetala* provides 50% of the cover (Kartushin, 1963, in Aleksandrova, 1988).

With such diversity of species and the predominance of mosses and lichens, most of these Soviet landscapes contain much more vegetation than the barren polar deserts of the Canadian Arctic; large portions of the islands in these archipelagos thus fit the concept of polar semideserts (Bliss and Svoboda, 1984).

3. Polar Deserts

Thousands of square kilometers of the Queen Elizabeth Islands are almost devoid of plants. These barren landscapes occur from sea level to plateaus above 200–300 m. Although lower summer temperatures and a shorter growing season are important controlling factors, the additional control is a geologic substrate with surface soils that dry in midseason. At most sites, vascular plants are more important in cover and species richness than are cryptogams, a significant difference from adjacent lands with semidesert vegetation. In a study of 23 barren sites on six islands, altogether 17 vascular species and 14 species of lichens and mosses were found. Vascular plant cover averaged 1.8%, visible cryptogams 0.7%, and bare soil and rocks 98%. Soils are medium-grained sands to clay loams, often with a thin veneer of patena, a black crust of lichens, cyanobacteria, and mosses (Bliss *et al..* 1984). The most common vascular plants are *Draba corymbosa*, *D. subcapitata*, *Papaver radicatum*, *Saxifraga oppositifolia*, *Puccinellia angustata*, and *Minuartia rubella*. The most important cryptogams include *Lecanora epibryon*, *Dermatocarpon hepaticum*, *Thamnolia*

subulifomis, Hypnum bambergeri, and *Tortula ruralis.* Much of the land is orga-
nized into sorted and nonsorted polygons and stripes. Many areas are covered
with a veneer of small rocks; other vast areas have few surface stones.

Within these barren landscapes small areas of more lush vegetation often
exist. These are either snowflush communities fed by melting snowbanks or
areas where the soils remain moist for much of the summer because of water
seepage. In these habitats, most vascular plants are found on moss carpets and
crustose lichens, often with considerable amounts of *Nostoc commune* (Fig. 12).
Species richness for vascular plants, mosses, and lichens is much higher in
these habitats, a clear indication that available water and nutrients, rather
than temperature alone, are the limiting factors. *Luzula confusa, Salix arctica,*
and *Ranunculus sulphureus* can be found above 500 m in seepage and moist
sites, yet they are seldom found above 200 m elsewhere.

According to Aleksandrova (1988), polar deserts (polar barrens) in the
Soviet Arctic are restricted to the large arctic islands of Novaya Zemlya (only
the northern island), Severnaya Zemlya (North Land Archipelago), and
Zemlya Franz-Josef; some smaller islands such as Ostrova De Longa; and on
the mainland only Cape Chelyuskin in northern Taimyr. The mean temperature of the warmest month (July) does not exceed 2° C, the active layer is no
more than 40 cm, plant cover is very limited, and there is no soil profile development. By her definition and map, only the northern tip of Greenland, Axel

Figure 12   Polar desert on the plateau (320 m), Devon Island, NWT. Edge of a snowflush site
with mosses and crustose lichens plus scattered plants of *Saxifraga oppositifolia, Alopecurus alpinus,
Papaver radicatum,* and *Cerastium alpinum.*

Heiberg, Ellef Ringnes, and Prince Patrick islands fall within the polar desert region, a much more limited land area than that recognized by most North American ecologists. Precipitation appears to be higher (100–250 mm) in the polar deserts of the Soviet Arctic compared with North America (75–150 mm) with its more continental climate.

In all Eurasian polar deserts, both in the arctic islands and at Cape Chelyuskin, mosses and lichens predominate and fill small desiccation cracks. Thus polygonal patterns of vegetation are the most typical; cushions of mosses and lichens are found within the cracks. In the Soviet polar deserts, the moss and lichen flora is richer than the vascular plant floras. The ratio of vascular plants to mosses to lichens is 1:1.3:2.8.

V. Arctic Carbon Reserves

Studies of global carbon balance include estimates of the area covered by major vegetation types plus estimates of standing crop, carbon storage in the soil, and net annual production. Little attention was given to arctic lands in the past, for they have low standing crops and net annual production, yet they do contain large reserves of soil carbon, and they cover large areas. Thus the Arctic, especially the Low Arctic, stores considerable carbon in the vegetation and in soil organic matter (Chap. 7). This carbon reserve may play an important role with anticipated climate warming and increased release of CO_2, assuming that soil respiration will exceed net plant production, which it may not.

The estimates presented below and in Tables II and III are based on data compiled at the 1980 workshop "Carbon Balance in Northern Ecosystems and the Potential Effect of Future Increased Global Temperatures by Year 2000." We have modified these data considerably to include more recent information and information related to this chapter. Arctic lands devoid of ice total 5.60×10^6 km^2, with Eurasia contributing 45%, Canada 42%, Greenland and Iceland 6.5%, and Alaska 6.5%. Area covered by the major vegetation types ranges from 0.174×10^6 km^2 for tall-shrub tundra to 1.363×10^6 km^2 for semideserts (Table II). Net annual plant production of vascular plants (above and below ground), bryophytes, and lichens ranges from 1 g m^{-2} in the polar deserts to 1000 g m^{-2} in the tall shrub communities of the Low Arctic (Table III).

Our estimates of soil organic matter (Table III) are much higher than previous estimates. For low-arctic mires (38.75 kg m^{-2} soil organic matter), we have used the data from Barrow, Alaska (summarized by Bliss, 1986), for a peat depth of 20 cm. Because peat profiles of Alaskan mires are often much deeper than 20 cm, this estimate should be conservative. Our estimates for high-arctic mires should also be conservative. They are based on figures from Muc (1977): 28.6 kg m^{-2} of soil organic matter in hummock-sedge–moss communities and 13.4 kg m^{-2} in frost-boil sedge communities, both measured to a depth of 23 cm. We assumed that these two *Carex* communities are equal in area, to give an average of 21.0 kg m^{-2} of soil organic matter in high-arctic mires. Similarly, our

Table II   Areal Extent ($\times$ 10^6 km^2) of Major Arctic Vegetation Types

Vegetation type	Alaska	Canada	Greenland, Iceland	Eurasia	Total area
Low Arctic					
Tall shrub	0.018	0.026	0.018	0.112	0.174
Low shrub	0.090	0.264	0.032	0.896	1.282
Tussock, sedge–dwarf shrub	0.126	0.088	0.036	0.672	0.922
Mire (wet sedge)	0.104	0.176	0.040	0.560	0.880
Semidesert	0.018	0.326	0.014	—	0.358
Ice caps	—	—	0.776	—	0.776
High Arctic					
Mire (wet sedge)	0.004	0.096	—	0.032	0.132
Semidesert	—	0.720	0.093	0.192	1.005
Polar desert	—	0.640	0.127	0.080	0.847
Ice caps	—	0.144	1.031	0.016	1.191
Total land	0.360	2.336	0.368	2.544	5.600
Total land plus ice caps	0.360	2.480	2.167	2.560	7.567

estimates for high-arctic polar semideserts include equal contributions by *Dryas* cushion-plant communities (1.7 kg m^{-2} soil organic matter) and cryptogam–herb communities (0.4 kg m^{-2}) (Bliss and Svoboda, 1984).

Overall, then, we estimate arctic carbon reserves at 26.33 $\times$ 10^{15} g (Table IV), significantly less than the 68 $\times$ 10^{15} g reported by Oechel (1989; cf. Chap. 7). Of this total, mires contribute 70%; tussock–heath and sedge–dwarf-shrub tundra, 13%; and low-shrub tundra, so extensive in the Soviet Arctic, 9%. The Low Arctic comprises 65% of the Arctic's ice-free land area but 91% of the carbon stored in the upper 20–25 cm of soil and above ground components. These estimates differ substantially from those of Miller *et al.* (1983) and Oechel (1989), whose estimates were higher for tussock tundra and semidesert but much lower for mires (wet sedge tundra).

VI. Arctic Climate Change and Vegetation Patterns

At present it is not possible to accurately predict what arctic climate changes will result from global change (Chap. 2). Global models predict that winter temperature will rise more than summer temperature and that precipitation will be reduced (Gammon *et al.*, 1985), but we believe precipitation will rise because of more open water. If so, cloud cover will increase and thus somewhat reduce the projected rise in temperature. The model presented in Figure 13 indicates that the growing season will lengthen, plant growth will be delayed in spring because of increased spring snow, and effective degree-days

Table III Above- and Belowground Net Annual Production, Standing Crop, and Soil Organic Matter (O.M.) for Major Arctic Vegetation Types[a]

Vegetation type	Net annual production (g m^{-2})			Standing crop (kg m^{-2})	Soil O.M. (kg m^{-2})	Soil depth (cm)	Total (kg m^{-2})
	Above	Below	Total				
Low Arctic							
Tall Shrub[b]	400*	600*	1,000*	5.800*	0.40*	20*	6.200
Low Shrub[c]	125	250*	375*	3.100*	1.00*	25*	4.100
Tussock, Sedge–dwarf shrub[d]	125	100*	225*	7.400	1.00*	40	8.400
Mire[e]	70	150	220	4.560	38.75	20	43.310
Semidesert[f]	28	17	45	1.470	2.54	20	4.010
High Arctic							
Mire[g]	60	80	140	2.360	21.00	23	23.360
Semidesert[h]	25	10	35	1.155	1.03	20–30	2.190
Desert[i]	0.7	0.3	1	0.024	0.022	15	0.046

[a]*denotes estimates.

[b]Komarkova and Webber (1980). Values for streamside deciduous shrubs: aboveground production doubled for more productive sites elsewhere; belowground estimated (1.5 × aboveground).

[c]Komarkova and Webber (1980). Deciduous shrub savanna on polygons; belowground estimated (2 × aboveground).

[d]Bliss and Richards (1982), Bliss (1986).

[e]Miller et al. (1980), Chapin et al. (1980).

[f]Svoboda (1977: Table 12, Banks and Victoria Islands). Measurements from high-arctic sites used since low-arctic semideserts have not been adequately measured.

[g]Muc (1977), Bliss (1977: Table 3).

[h]Bliss and Svoboda (1984: Table 7, all stands) for cryptogam–herb; Svoboda (1977: Table 12); Bliss (1977: Tables 2, 3).

[i]Bliss et al. (1984: Table 4, all stands).

Table IV   Area, Organic Matter (O.M.), and Total Carbon Reserves in Major Arctic Vegetation Types

Vegetation type	Area ($\times 10^6$ km^2)	Standing crop per land area		
		O. M. (kg m^{-2})	O. M. (10^{15} g)	Carbon (10^{15} g)
Low Arctic				
Tall shrub	0.174	6.20	1.0788	0.475
Low shrub	1.282	4.10	5.2562	2.316
Tussock, sedge–dwarf shrub	0.922	8.40	7.7448	3.412
Mire	0.880	43.31	38.1128	17.129
Semidesert	0.358	4.01	1.4356	0.645
High Arctic				
Mire	0.132	23.36	3.0835	1.358
Semidesert	1.005	2.19	2.2010	0.978
Polar desert	0.847	0.046	0.0390	0.017
Total				26.33

will rise only 5–15%.

With these assumptions, tree line will advance northward, shrub tundra will spread at the expense of tussock tundra, and mires will increase in extent where topography permits. In the High Arctic, mires will increase where water remains on the landscape; polar semideserts (arctic tundra and polar deserts in the USSR) will increase in area, for mats of mosses and lichens will expand, permitting vascular plants to spread into more barren lands. Polar deserts will change little except where snowbanks become more extensive, with the resultant increase in snowflush communities. Where winter snow is deeper and melts later in summer, the limited plant cover may show little change, perhaps even be reduced.

VII. Summary

Many of the species dominating the various plant communities throughout the circumpolar Arctic are the same or ecological equivalents. Their organization into major vegetation types, however, is not always the same in North America and Eurasia. Shrub–sedge tundras, tussock–dwarf-shrub tundras, and wet graminoid–moss tundras, or mires, are very similar in structure and floristic composition throughout the Low Arctic. The midarctic region (typical tundra on the Taimyr Peninsula and the cushion plant–cryptogam and cryptogam–gerb semideserts of the Canadian Arctic Archipelago) differs in structure and floristics between the two continents. The arctic tundra zone of Siberia, with dwarf willows, herbs, and cryptogams, is similar to portions of the semidesert landscapes of Canada. The Soviet "polar deserts" of the northern

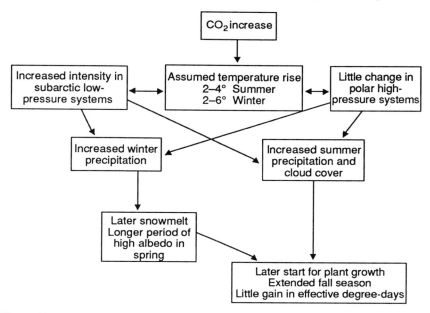

Figure 13   Hypothesized implications of increased atmospheric CO_2 for arctic vegetation.

tip of the Taimyr Peninsula and the several Eurasian arctic archipelagos are similar to the more barren polar semidesert landscapes of Canada, for mosses and lichens dominate. In sharp contrast, the polar deserts in the Canadian Arctic have few cryptogams, and vascular plant cover is highly limited. The predominance of sedimentary rocks, mostly calcareous; excessively drained soils in summer; and summers with limited precipitation appear to be more important than temperature alone in the landscape patterning of vegetation in the Canadian High Arctic.

The Taimyr Peninsula, with its relatively uniform landscape, continuous landmass from 69° to 78° N, and gradual reduction in summer temperature, is in sharp contrast to the abrupt change from continent to islands in North America. The more sudden change in physical and biological parameters across this latitudinal gradient in the New World results in the lack of a zone or belt of vegetation that can be clearly identified as typical tundra or midarctic tundra as in the Yamal, Gydan, and Taimyr peninsulas. If such a zone does occur, it would be the *Dryas*-dominated uplands on Banks, Victoria, and Baffin islands.

The northern islands of Canada at 75° to 82° N harbor oases where the vegetation and wildlife are lush relative to the surrounding landscapes. These are lands with higher summer temperature and greater summer water supply. Further study and the exchange of arctic ecologists should provide a better understanding of the similarities and differences in the organization of plants on the two continents. Physiological studies of dominant species will continue to provide a fuller understanding of the pattern and process of plants on these environmentally stressful landscapes.

Acknowledgments

Major support for arctic research has come from the Canadian Research Council, the Polar Continental Shelf Project, the Arctic Petroleum Operators Association, the National Science Foundation (U.S.A.), and the Soviet Academy of Sciences. We thank Boris Yurtsev for his helpful additions to the text on eastern Siberia.

References

Aleksandrova, V. D. (1970). The vegetation of the tundra zones in the U.S.S.R. and data about its productivity. *In* "Productivity and Conservation in Northern Circumpolar Lands" (W. A. Fuller and P. G. Kevan, eds.), pp. 93–114. N.S. 16. International Union for the Conservation of Nature, Morges, Switzerland.

Aleksandrova, V. D. (1971). Principles for the zonal division of the vegetation of the arctic (in Russian). *Bot. Zh.* **56**, 1–21.

Aleksandrova, V. D. (1980). "The Arctic and Antarctic: Their Division into Geobotanical Areas." Cambridge Univ. Press, Cambridge.

Aleksandrova, V. D. (1988). "Vegetation of the Soviet Polar Deserts." Cambridge Univ. Press, Cambridge.

Andreyev, N. N., and Aleksandrova, V. D. (1981). Geobotanical division of the Soviet Arctic. *In* "Tundra Ecosystems: A Comparative Analysis" (L. C. Bliss, D. W. Heal, and J. J. Moore, eds.), pp. 25–37. Cambridge Univ Press, Cambridge.

Black, R. A., and Bliss, L. C. (1978). Recovery sequence of *Picea mariana–Vaccinium uliginosum* forests after fire near Inuvik, Northwest Territories, Canada. *Can. J. Bot.* **56**, 2020–2030.

Black, R. A., and Bliss, L. C. (1980). Reproductive ecology of *Picea mariana* (Mill.) BSP., at treeline near Inuvik, Northwest Territories, Canada. *Ecol. Monogr.* **50**, 331–354.

Bliss, L. C. (1975). Tundra, grasslands, herblands, and shrublands and the role of herbivores. *Geosci. Man* **10**, 51–79.

Bliss, L. C. (1977). General summary, Truelove Lowland ecosystem. *In* "Truelove Lowland, Devon Island, Canada: A High Arctic Ecosystem" (L. C. Bliss, ed.), pp. 657–675. Univ. of Alberta Press, Edmonton.

Bliss, L. C. (1979). Vascular plant vegetation of the southern circumpolar region in relation to the antarctic, alpine, and arctic vegetation. *Can. J. Bot.* **57**, 2167–2178.

Bliss, L. C. (1981). North American and Scandinavian tundras and polar deserts. *In* "Tundra Ecosystems: A Comparative Analysis" (L. C. Bliss, O. W. Heal, and J. J. Moore, eds.), pp. 8–24. Cambridge Univ. Press, Cambridge.

Bliss, L. C. (1986). Arctic ecosystems: Their structure, function, and herbivore carrying capacity. *In* "Grazing Research at Northern Latitudes" (O. Gundmundeson, ed.), pp. 5–25. Plenum, New York.

Bliss, L. C. (1988). Arctic tundra and polar desert biome. *In* "North American Terrestrial Vegetation" (M. G. Barbour and W. D. Billings, eds.), pp. 1–32. Cambridge Univ. Press, New York.

Bliss, L. C., and Richards, J. H. (1982). present-day arctic vegetation and ecosystems as a predictive tool for the arctic–steppe mammoth biome. *In* "Paleoecology of Beringia" (D. M. Hopkins, J. V. Matthews, C. E. Scheweger, and S. B. Young, eds.), pp. 241–257. Academic Press, New York.

Bliss, L. C., and Svoboda, J. (1984). Plant communities and plant production in the western Queen Elizabeth Islands. *Holarc. Ecol.* **7**, 324–344.

Bliss, L. C., Svoboda, J., and Bliss, D. I. (1984). Polar deserts, their plant cover, and plant production in the Canadian High Arctic. *Holarc. Ecol.* **7**, 304–324.

Böcher, T. W. (1954). Oceanic and continental vegetational complexes in southwest Greenland. *Medd. Grønl.* **148(1)**, 1–336.

Böcher, T. W. (1959). Floristics and ecological studies in middle west Greenland. *Medd. Grønl.* **156(5)**, 1–68.

Brassard, G. R., and Longton, R. E. (1970). The flora and vegetation of Van Hauen Pass, northwestern Ellesmere Island. *Can. Field Nat.* **84**, 357–364.

Britton, M. E. (1957). Vegetation of the arctic tundra. *In* "Arctic Biology" (H. P. Hansen, ed.), pp. 26–61. Oregon State Univ. Press, Corvallis.

Chapin, F. S., III, Miller, P. C., Billings, W. D., and Coyne, P. I. (1980). Carbon and nutrient budgets and their control in coastal tundra. *In* "An Arctic Ecosystem: The Coastal Tundra at Barrow, Alaska" (J. Brown, P. C. Miller, L. L. Tieszen, and F. L. Bunnell, eds.), pp. 458–482. Dowden, Hutchinson & Ross, Stroudsburg, Pennsylvania.

Chernov, Y. I., and Matveyeva, N. V. (1979). The zonal distribution of communities on Taimyr (in Russian). *In* "Arctic Tundras and Polar Deserts of Taimyr" (V. D. Aleksandrova and N. V. Matveyeva, eds.) pp. 166–200. Nauka, Leningrad.

Churchill, E. D. (1955). Phytosociological and environmental characteristics of some plant communities in the Umiat region of Alaska. *Ecology* **36**, 606–627.

Corns, I. G. W. (1974). Arctic plant communities east of the Mackenzie Delta. *Can. J. Bot.* **52**, 1730–1745.

Drew, J. V., and Shanks, R. E. (1965). Landscape relationships of soils and vegetation in the forest–tundra, Upper Firth River Valley, Alaska–Canada. *Ecol. Monogr.* **35**, 285–306.

Edlund, S. A., and Alt, B. T. (1989). Regional congruence of vegetation and summer climate patterns in the Queen Elizabeth Islands, Northwest Territories, Canada. *Arctic* **42**, 3–23.

Gammon, R. H., Sundquist, E. T., and Fraser, P. J. (1985). History of carbon dioxide in the atmosphere. *In* "Atmospheric Carbon Dioxide and the Global Carbon Cycle" (J. R. Trabalka, ed.), pp. 25–62. DOE/ER-0239. Natl. Tech. Info. Serv., Springfield, Virginia.

Gorodkov, B. N. (1935). "Vegetation of the Tundra Zone of the U.S.S.R" (in Russian). Academy of Sciences of the U.S.S.R., Leningrad.

Hanson, H. C. (1953). Vegetation types in northwestern Alaska and comparisons with communities in other arctic regions. *Ecology* **34**, 111–140.

Henry, G., Freedman, B., and Svoboda, J. (1986). Survey of vegetated areas and muskox populations in east-central Ellesmere Island. *Arctic* **39**, 78–61.

Hettinger, L., Janz, A., and Wein, R. W. (1973). Vegetation of the Northern Yukon Territory. *Arct. Gas Biol. Rep. Ser.* **1**, Canadian Arctic Gas Study, Calgary.

Jefferies, R. L. (1977). The vegetation of salt marshes at some coastal sites in arctic North America. *J. Ecol.* **65**, 661–672.

Jefferies, R. L. (1988). Pattern and process of arctic coastal vegetation in response to foraging by lesser snow geese. *In* "Plant Form and Vegetation Structure" (M. J. A. Werger, P. J. M. van der Aart, H. J. During, and J. T. A. Verhoeven, eds.), pp. 281–300. SPB Academic, The Hague.

Jefferies, R. L., Jensen, A, and Abraham, K. F. (1979). Vegetational development and the effect of geese on vegetation at La Perouse Bay, Manitoba. *Can. J. Bot.* **57**, 1439–1450.

Johnson, A. W., Viereck, L. A., Johnson, R. E., and Melchior, H. (1966). Vegetation and flora. *In* "Environment of the Cape Thompson Region Alaska" (N. J. Wilimovsky and J. N. Wolfe, eds.), pp. 277–354. U.S. Atomic Energy Commission, Division of Technical Information, Oak Ridge, Tennessee.

Kartushin, V. M. (1963). On the vegetation of Ostrov Bennetta (in Russian). *Trans. Ark. Antark. Inst.* **224**, 177–179.

Khodachek, E. A. (1986). The main plant communities in the western part of October Revolution Island (Severnaya Zemlya) (in Russian). *Bot. Zh.* **71**, 1628–1638.

Komarkova, V., and Webber, P. J. (1980). Two low arctic vegetation maps near Atkasook, Alaska.

Arct. Alp. Res. **12**, 447–472.

Korotkevich, E. S. (1958). The vegetation of Severnaya Zemlya (in Russian). *Bot. Zh.* **43**, 644–663.

Kuc, M. (1974). Noteworthy vascular plants collected in southwestern Banks Island, N.W.T. *Arctic* **26**, 146–150.

Larsen, J. A. (1965). The vegetation of the Ennadai Lake area, N.W.T.: Studies in subarctic and arctic bioclimatology. *Ecol. Monog.* **35**, 37–59.

Larsen, J. A. (1972). The vegetation of northern Keewatin. *Can. Field Nat.* **86**, 45–72.

Matveyeva, N. V. (1979). The structure of the plant cover in the polar deserts of the Taimyr Peninsula (Mys Chelyuskin) (in Russian). *In* "The Arctic Tundras and Polar Deserts of Taimyr" (V. D. Aleksandrova and N. V. Matveyeva, eds.), pp. 5–27. Nauka, Leningrad.

Matveyeva, N. V., Polozova, T. G., Blagodatskykh, L. S., and Dorogostaiskaya, E. V. (1975). A brief essay on the vegetation in the vicinity of the Taimyr biogeocoenological station. International Tundra Biome Transl. 13. Fairbanks, Alaska.

Matveyeva, N. V., and Zanokha, L. L. (1986). Vegetation of southern tundras on Western Taimyr (in Russian). *In* "Southern Tundras of Taimyr" (Y. I. Chernov and N. V. Matveyeva, eds.), pp 5–67. Nauka, Leningrad.

Miller, P. C., Kendall, R., and Oechel, W. C. (1983). Simulating carbon accumulation in northern ecosystems. *Simulation* **40**, 119–131.

Miller, P. C., Webber, P. J., Oechel, W. C., and Tieszen, L. L. (1980). Biophysical processes and primary production. *In* "An Arctic Ecosystem: The Coastal Tundra at Barrow, Alaska" (J. Brown, P. C. Miller, L. L. Tieszen, and F. L. Bunnell, eds.), pp. 66–101. Dowden, Hutchinson & Ross, Stroudsburg, Pennsylvania.

Muc, M. (1977). Ecology and primary production of the Truelove Lowland sedge–moss meadow communities. *In* "Truelove Lowland, Devon Island, Canada: A High Arctic Ecosystem" (L. C. Bliss, ed), pp. 157–184. Univ. of Alberta Press, Edmonton.

Muc, M., Freedman, B., and Svoboda, J. (1989). Vascular plant communities of a polar oasis at Alexandra Fiord (79° N), Ellesmere Island, Canada. *Can. J. Bot.* **67**, 1126–1136.

Nichols, H. (1975). Palynological and paleoclimatic study of the Late-Quaternary displacement of the boreal forest–tundra ecotone in Keewatin an Mackenzie, N.W.T., Canada. *Arct. Alp. Res. Occas. Pap.* 15.

Oechel, W. C. (1989). Nutrient and water flux in a small arctic watershed: An overview. *Holarct. Ecol.* **12**, 229–237.

Payette, G., and Filion, L. (1984). White spruce expansion at the tree line and recent climatic change. *Can. J. For. Res.* **15**, 241–251.

Payette, S., Morneau, C. Sirosis, L., and Desponts, M. (1989). Recent fire history of the northern Quebec biomes. *Ecology* **70**, 656–673.

Polunin, N. (1948). Botany of the Canadian eastern Arctic. Part III. Vegetation and ecology. *Nat. Mus. Can. Bull.* **104**.

Polunin, N. (1951). The real arctic: Suggestions for its delimitation, subdivision, and characterization. *J. Ecol.* **39**, 308–315.

Pospelova, E. B. (1972). Vegetation of the Agapa station and productivity of the main plant communities. *In* "Tundra Biome" (F. E. Wielgolaski and T. Roswall, eds.), pp. 204–208. Tundra Biome Steering Committee, Stockholm.

Rannie, W. F. (1986). Summer air temperature and number of vascular species in arctic Canada. *Arctic* **39**, 133–137.

Rebristaya, O. V. (1977). "Flora of Eastern Bol'shezemelskaya Tundra" (in Russian). Nauka, Leningrad.

Reznicek, S. A., and Svoboda, J. (1982). Tundra communities along a microenvironment gradient at Coral Harbour, Southampton Island, N.W.T. *Nat. Can.* **109**, 583–595.

Safranova, I. N. (1976). On the flora and vegetation of Ostrov Oktyabrskoy Revolyutsii (in Russian). *In* "Biological Problems of the North: VII Symposium (Botany)", pp. 191–193. Petrozavodsk.

Sheard, J. W., and Geale, D. W. (1983). Vegetation studies of Polar Bear Pass, Bathurst Island, N.W.T. I. Classification of plant communities. *Can. J. Bot.* **61**, 1618–1636.

Sohlberg, E., and Bliss, L. C. (1984). Microscale pattern of vascular plant distribution in two high arctic communities. *Can. J. Bot.* **62**, 2033–2042.

Svoboda, J. (1977). Ecology and primary production of raised beach communities, Truelove Lowland. *In* "Truelove Lowland, Devon Island, Canada: A High Arctic Ecosystem" (L. C. Bliss, ed.), pp. 185–216. Univ. of Alberta Press, Edmonton.

Tikhomirov, B. A. (1957). Dynamic phenomena in the vegetation of the spotty tundras of the Arctic. *Bot. Zh.* **42**, 1691–1717.

Tolmachev, A. I. (1932). The flora of the central part of eastern Taimyr. *Tran. Polyarn Komiksii* **8** and **13**.

Vestergaard, P. (1978). Studies on vegetation and soil of coastal salt marshes in the Disko area, West Greenland. *Medd. Gronl.* **204(2)**, 1–51.

Webber, P. J. (1978). Spatial and temporal variation in the vegetation and its production, Barrow, Alaska. *In* "Vegetation and Production Ecology of an Alaskan Arctic Tundra" (L. L. Tieszen, ed.), pp. 37–112. Springer-Verlag, New York.

Wein, R. W., and Bliss, L. C. (1974). Primary production in arctic cottongrass tussock tundra communities. *Arct. Alp. Res.* **6**, 261–274.

Young, S. B. (1971). The vascular flora of Saint Lawrence Island with special reference to floristic zonation in the arctic regions. *Contrib. Gray Herb. Harv. Univ.* **201**, 11–115.

Yurtsev, B. A. (1966). "The Hypoarctic Botanical–geographical Belt and the origin of its Flora" (in Russian). Nauka, Leningrad.

Yurtsev, B. A. (1972). Phytogeography of northeastern Asia and the problems of Transberingian floristic interrelations. *In* "Floristics and Paleofloristics of Asia and Eastern North America" (A. Graham, ed.), pp. 19–54. Elsevier, New York.

Yurtsev, B. A. (1974). "Problems of Botanical Geography in Northeastern Asia" (in Russian). Nauka, Leningrad.

Yurtsev, B. A. (1987). The effect of historical factors on the adaptation of plants to extreme environmental conditions of the arctic tundra subzone (Wrangel Island flora taken as an example) (in Russian). *Bot. Zh.* **72**, 1436–1947.

Yurtsev, B. A., Tolmatchev, A. I., and Rebristaya, O. V. (1975). The floristic delineation and subdivision of the Arctic *In* "The Arctic Floristic Region" (B. A. Yurtsev, ed.), pp. 9–106. Nauka, Leningrad.

5

Phytogeographic and Evolutionary Potential of the Arctic Flora and Vegetation in a Changing Climate

W. D. Billings

I. Introduction

Among the earth's terrestrial environments, none has less biologically usable heat or fewer species of plants than the tundras and cold deserts of the Arctic. What are the environmental stresses, the genetic signatures, the metabolic processes and rates, and the evolutionary pathways that have allowed plants to exist and reproduce in this seemingly hostile part of the world? As the Arctic warms through the next century (Chap. 2), how might the present flora and vegetation adjust? Should we expect a new wave of species migrations causing a northward shift in the boundaries of current ecosystems and biomes?

Plant metabolism, growth, and reproduction in the Arctic are controlled by the interactions of the physical environment and the genetic structure of an old, small, cold-adapted flora derived in large part from the high mountains of the Northern Hemisphere. In North America, the long Cordillera of the Rocky Mountains, from New Mexico to Alaska and Yukon Territory, has been the main avenue of migration and evolution for alpine–arctic taxa between the middle latitudes and the Arctic. Other pathways include the even more ancient Appalachians and the younger mountains of Europe and northeastern Asia, which connect southward to the high alpine systems of the Alps, Caucasus, Hindu Kush, and Himalayas. The Sierra Nevada of California is even younger (Pliocene and Pleistocene; Axelrod, 1962), and a very rich alpine flora is evolving there from the deserts below (Chabot and Billings, 1972; Billings, 1988). These taxa have not yet moved northward into the Arctic because of the Feather River lowland gaps, weak migratory connections through the Cascades from the Sierra (Billings, 1978); insufficient time is also a factor.

Although there are some relatively old alpine and arctic taxa, the accelerated uplift of mountain ranges during the Tertiary—especially during the late Miocene, Pliocene, and Pleistocene—has allowed a more recent enrichment of middle-latitude alpine floras by upward migration and evolution from lower elevations. Once these populations become genetically cold-adapted on the summit ridges, numerous opportunities become available for northward migration and adaptive radiation into and around the Arctic. The fluctuations in Pleistocene continental glaciations have alternately opened the way for such migrations or blocked them. Interglacial refugia on the summits may thus have been more important than glacial refugia in allowing the consolidation of evolutionary changes in alpine taxa in response to decreasing summer temperatures toward the Arctic. Adaptation northward to longer summer day lengths, which control growth and flowering, is probably more easily accomplished than adaptation to low temperatures. Billings *et al.* (1965) have shown such correlation with day length for latitudinal ecotypes of *Oxyria digyna* for both dormancy and flowering. Once arrived in ice-free parts of the Arctic and Subarctic during favorable times, some plant taxa have migrated around the Arctic and become prominent in the mosaic of tundra vegetation.

Arctic tundra ecosystems are dominated by plants with one trait in common that differentiates them from all others: the ability to grow, metabolize, and reproduce at temperatures only slightly above freezing (Billings and Mooney, 1968). Perennial herbaceous taxa or dwarf shrubby vascular plants predominate, along with lichens and bryophytes. All photosynthesize by the C_3 pathway (or CAM in *Sedum*), with peak rates between 15° and 20° C or lower, and all can acclimate to temperature changes (Billings *et al.*, 1971). Their growth and metabolism are closely coupled to the flux of solar radiation during the continuous daylight of the arctic summer.

It is the physical environment that controls most plant growth and establishment in the Arctic. The rigors of low air and soil temperatures, winds, and snow in both winter and summer outweigh most biological factors in the environment. The vegetation lives and dies in only a shallow layer of air and soil about one meter deep between the leaf tips of the grasses and the top of the permafrost. As a microenvironment for plant growth, reproduction, and establishment, this layer is ephemeral, available only from thaw in mid-June to freeze-up in late August. The soil is cold and nutrient-poor, and yet most of the plant biomass extends below ground in the form of roots and rhizomes (Billings *et al.*, 1978). In wet sediments, the active soil layer is underlain with polygonal wedges or lenses of white ice. These are quite susceptible to melting and thermokarst, which can follow mechanical disturbance, climatic change, or overrunning by the warmer waters of streams and lakes (Billings and Peterson, 1991).

II. Status and History of the Arctic Flora

The Arctic can be treated as a single floristic region as Good (1964) has done. As such, for convenience, its entire vascular flora can be considered as a unit even though, as Hultén (1937, 1962), Böcher (1951), and Polunin (1959) have shown, the distributions of many vascular plant species in the Arctic are uneven, and there are many subregions with their own characteristic floras. Polunin (1959) lists 892 species in 230 genera and 66 families in his *Circumpolar Arctic Flora*. Many of these, of course, occur only in certain arctic regions and are not circumpolar, and many range far south of the Arctic into alpine regions. In an attempt to determine how many of these 892 species are arctic–alpine in distribution, I have reached the preliminary figure of 354 species. The number probably exceeds 400 species, or about 45% of the total arctic flora. One can say with some assurance that the arctic flora is made up of about 900 vascular species—certainly less than 1000—and that this relatively small flora is distributed over a land area of about 13 million km^2, or almost 15% of the earth's land surface.

Floras have been compiled for a number of smaller regions within the Arctic that vary considerably in size. Among them are those of Holmen (1957) for Peary Land in northern Greenland (96 species in 63,000 km^2), Rønning (1979) for Svalbard (171 species in 62,000 km^2), Porsild (1957) for the Canadian Arctic Archipelago (340 species in 1,400,000 km^2), Wiggins and Thomas (1962) for the North Slope of Alaska (435 species in 212,000 km^2), and Johnson *et al.* (1966) for Ogoturuk Creek Valley and the Cape Thompson region in northwestern Alaska (295 species in about 500 km^2). Northern and western Alaska are relatively rich floristically, partly because these regions were not glaciated. Tikhomirov *et al.* (1969) list 281 species in the vicinity of Tiksi-harbour in arctic Siberia at 72°N within an area of perhaps 500 km^2, thus making this flora comparable in diversity to that of the Cape Thompson region in unglaciated Alaska. Nevertheless, Young's (1971) vascular flora of Saint Lawrence Island (5200 km^2), a remnant of the Bering Land Bridge in the Bering Sea, totals only 235 species, considerably less rich than the Siberian and Alaskan mainland arctic floras to the west and east. Why? Could it be isolation or the present environment? Isolation seems an unlikely cause since the Chukchi Peninsula in Siberia lies only about 65 km to the west.

There are total terrestrial flora data for a number of smaller areas in the Arctic that can be useful as benchmarks from which to evaluate the effects of global warming. For example, Barrow, Atkasook, and Prudhoe Bay form a right triangle of research sites, each of about the same size (ca. 200 km^2), on the Alaskan North Slope. The floras are quite well known at these sites from International Biological Programme research at Barrow and Prudhoe and the RATE (Research on Arctic Tundra Environments) project at Atkasook during the 1970s. The sides of the triangle are climatic gradients. One is north–south (coastal to inland) from Barrow to Atkasook. A second is coastal from Barrow southeast to Prudhoe. The third is west–east (inland to coastal) from Atkasook to Prudhoe.

At Barrow (71°23′N), the vascular taxa number approximately 125 (Murray and Murray, 1978), and at Prudhoe Bay (70°15′N), some 335 km southeast of Barrow on the Beaufort Sea coast, the number of such taxa is about 172 (Murray and Murray, 1978). The vascular flora at Atkasook (70°29′N), approximately 100 km south of Barrow and 350 km west of Prudhoe, is around 246 species, considerably richer than at the other two sites (Komárková and Webber, 1977). This relative richness reflects its inland location in the tussock tundra on the Meade River.

In the bryophytes, as of 1978 there were approximately 57 taxa of hepatics known from Barrow and only 15 from Prudhoe Bay (Murray and Murray, 1978; Steere and Inoue, 1978). The Prudhoe Bay figure for hepatics may reflect the much shorter time than at Barrow that bryophytes have been collected there. As of 1978, 181 taxa of mosses were known from Barrow and 109 from Prudhoe Bay (Steere, 1978; Murray and Murray, 1978). Lichens numbered about 106 taxa at Barrow and 71 at Prudhoe as reported by Murray and Murray in 1978. Thus, the flora at Barrow consists overwhelmingly of terrestrial nonvascular taxa: 73.3% to 26.7% vascular. This ratio is very much the same as that at Truelove Lowland on the north coast of Devon Island (75°33′N) in the High Arctic, where mosses and lichens constitute 77% of the total flora of 414 taxa, compared with the 23% represented by vascular taxa (Bliss, 1977). At Prudhoe Bay, the ratio of nonvascular taxa to vascular is roughly 53.1 to 46.9%, or nearer one-half rather than three-quarters of the total flora as at Barrow or Truelove Lowland. Whether this reflects relative lack of collecting at Prudhoe Bay or more severe summer climates at Barrow and Truelove Lowland cannot be said; it is possible that both these factors are involved. Also, the limestone substrate at Prudhoe Bay differentiates that region from Barrow.

These are the basic facts, then, on floristic diversity as they appear toward the end of the twentieth century in the Alaska tundra. This is near the end of six to seven millennia on the never-glaciated Alaskan North Slope. During this time, atmospheric CO_2 concentrations have remained relatively steady at well below 300 ppm until near the end of the nineteenth century; now, however, they are rising, as the measurements of CO_2 concentration over the last 20 years at Barrow attest (see Chap. 2: Fig. 5).

III. History of the Tundra Vegetation on the Alaskan North Slope

Because of soil frost activity and the thaw-lake cycle (Chap. 6), it is difficult in the tundra to get dated pollen cores for vegetational history. Colinvaux (1964), however, has obtained two partial cores from Barrow. These indicate that a sedge–grass–willow tundra occupied the site from 14,000 to 9500 B.P., when the climate was similar to the present but with slightly colder summers. The Barrow climate warmed from 9500 through 5000 B.P., judging from the increased abundance of *Betula* pollen.

During the last 5000 years, the climate and vegetation in the Barrow region have remained relatively steady as a cold, wet tundra ecosystem (Murray, 1978). Similarly, the palynological record at Teshekpuk Lake (140 km southeast of Barrow) indicates that the Holocene tundra vegetation there was not too different from the wet coastal tundra as it now exists around the lake (Anderson, 1982).

Inland and southwest of Ocean Point, about 240 km southeast of Barrow, Nelson (1982) has done a thorough pollen and macrofossil analysis of a section exposed along the Titaluk River at 69°42′N, 155°12′W, in the northern part of the present tussock tundra zone. His pollen Section II is through Holocene thaw-lake sediments in which the radiocarbon date at the bottom is 2500 ± 50 years B.P. The pollen sequence is essentially modern tundra, with grasses, sedges, *Salix, Alnus, Betula,* and *Artemisia,* similar to the present surrounding vegetation. The plant macrofossils near the base of the thaw-lake sediments include *Rubus chamaemorus, Potentilla* spp., *Dryas* spp., *Ledum, Vaccinium, Betula nana,* and *Cassiope tetragona*; these are typical of the modern surrounding tundra.

There is a well-dated pollen profile from the tussock tundra in a valley fill near Umiat (Livingstone, 1957). This section is about eight meters deep and dated at the bottom at 8125 ± 250 years B.P. with rather rapid filling until 5900 ± 200 B.P. at about one meter. The top meter has filled only slowly. The floristic composition through the profile shows a rather typical tussock tundra with low shrubs throughout, except for an increase in *Alnus* in the last 5000 years. Apparently, the tussock tundra vegetation and climate along the middle Colville River have changed little during the last 6000 years.

There are few fossil pollen profiles from anywhere in the arctic or subarctic regions where the present tundra vegetation is as well known on a local topographic scale as that at Prudhoe Bay. This area is wet coastal tundra near its eastern boundary in Alaska. Walker *et al.* (1981) have taken three pollen profiles in peat within 10 km of the sea along the Kuparuk, Putuligayuk, and Sagavanirktok rivers near latitude 70°18–20′N. These data are presented along with some radiocarbon dates indicating that the peat and pollen were deposited during the last 3700–1200 years. The authors conclude that in the Prudhoe Bay region, the pollen record is not a faithful indication of local vegetational conditions through time because important local floristic elements, such as the insect-pollinated willows, are underrepresented. Also, certain pollens have come from afar, *Picea* and *Pinus*, for example. Still, the pollen profiles were often dominated by members of the Cyperaceae, and grass pollen was also abundant. So even if there are difficulties in making inferences about past vegetations and climates from the pollen profiles, the vegetation and the climate of the Prudhoe Bay region during the last 4000 years would still be classified as wet coastal tundra.

It seems safe to conclude that during the last 5000–6000 years, the wet coastal tundra on the North Slope of Alaska has changed little floristically and vegetationally. Albeit with less evidence, this conclusion also seems to hold for the tussock tundra.

IV. History of the Subarctic Vegetation in Central Alaska

The late Pleistocene and Holocene vegetational history of the Tanana Valley in central Alaska has been studied by Matthews (1968, 1970, 1074), Ager (1975, 1983), Hamilton *et al.* (1983), Billings (1987a), and others. This region was never glaciated. The present vegetation consists of four principal forest site types: (1) upland mesic white spruce, (2) floodplain white spruce, (3) upland black spruce, and (4) lowland black spruce (Viereck *et al.*, 1986). In addition, there are large, relatively open peatlands underlain with permafrost, particularly in the Fairbanks region, which are dominated by several species of *Sphagnum* (Luken, 1984; Luken *et al.*, 1985; Billings, 1987a). Sedge meadows dominated by *Carex aquatilis* also occur in these peatlands. These meadows begin as floating mats following stream thermokarst in bogs and are eventually succeeded by black spruce (see Chap. 6).

Table I (adapted from Ager, 1983) summarizes the late Quaternary vegetational history of the Tanana Valley in the general vicinity of Fairbanks. Before about 11,000 years B.P., the vegetation of these central Alaskan lowlands was tundra of one kind or another. The herbaceous tundra before 14,000 years B.P. was most likely tussock tundra with *Eriophorum vaginatum* still present in the valley. Spruce probably arrived after 10,000 years B.P. and existed as a gallery forest or open woodland. Ager's pollen records (1975, 1983) for the lowland taiga sites in central Alaska show little vegetational change during the last 6000 years and indicate an open *Picea–Alnus–Betula* woodland occupying the Tanana Valley peatlands.

A 6000-year time span, however, may not show the fluctuations caused by climatic changes, fire in the taiga, or both. For example, a frozen core in dwarf black spruce woodland in the middle of the College Road peatland, Fairbanks, showed that the vegetation of this bog, and certainly other peatlands in the Tanana region, alternated between sedge meadows and spruce woodlands within a few centuries (Table II; Billings, 1987a). *Sphagnum* itself seems to be a relative latecomer to the complex system. The sedge meadows were probably tussock vegetation of large *Eriophorum vaginatum*.

Table I Late Quaternary Vegetational History of the Tanana Valley,
Central Alaska[a]

^{14}C years (B.P.)	Vegetation
Holocene	
1,000	
2,000	
3,000	*Picea–Alnus–Betula*
4,000	Boreal forest
5,000	
6,000	
7,000	*Picea* decline
8,000	*Picea–Alnus–Betula* forest
9,000	*Picea–Betula* gallery forest and/or parkland
Late Wisconsinan	
10,000	*Populus–Salix* scrub forest and shrub tundra
11,000	
12,000	
13,000	*Betula–Ericaceae* shrub tundra
14,000	
15,000	Herbaceous tundra ("steppe tundra")

[a] Adapted from Ager (1983).

In summary, for the taiga forests and bogs of central Alaska, the regional floristic composition has changed very little in the last 6000 years. At any given site, however, it has changed a great deal. For example, it has gone from *Sphagnum* bog to spruce–birch forest to sedge meadow to thermokarst ponds and back again in oscillations ranging from several centuries to well over a thousand years. Such oscillations may well be expected in the future as climates change. Past situations provide useful models for possible futures in both tundra and taiga.

One must not overlook barriers to migration other than climate and micro-topographic site differences. The east–west Brooks Range has stood as a physical barrier for millions of years, and it will continue to do so no matter how the climate changes. It is simply too high for most taiga tree species, except *Populus balsamifera* (Wiggins and Thomas, 1962: Viereck and Little, 1972) and possibly *Picea glauca* (Cooper, 1986), to cross without aid from people or vehicles carrying the seeds to the North Slope. The Dalton Highway crosses Atigun Pass at 1450 m elevation from the timberline in the Dietrich Valley at about 800 m (Densmore, 1980). The low, windswept vegetation at the pass and on the surrounding heights of the Brooks Range is arctic-alpine tundra. The spruces and larch probably cannot cross this expanse to the foothills on the

Table II General Peat Stratigraphy, Postulated Vegetation, Radiocarbon Age, and Carbon Contents of a Frozen Core in Dwarf Black Spruce Woodland in the College Road Bog, Fairbanks, Alaska[a]

Depth (cm)	Radiocarbon age (B.P.)	Carbon (g C dm³)	Peat characteristics	Vegetation
0–7		24.01	Loose, live moss	Moss–shrubs–dwarf spruce
7–12			Fibrous mat of roots in peat	
12–17		49.20	Woody, fibrous peat with many roots	
17–22			Black, tight, graminoid peat	Sedges and some low shrubs; essentially a meadow
22–27		62.75	Tight, dark peat with roots	
27–32			Tight, dark peat; *Eriophorum*, *Carex*; roots	
32–37		75.59	Tight, dark peat	
37–42			Tight, dark peat; *Eriophorum*; roots	
42–47	1760 ± 80			
47–52		74.51	Dark, amorphous, firm peat; many graminoids	
52–57			Compact, firm peat; dark	
57–62		68.97	Compact, firm peat; dark	Spruce–birch forest or woodland
62–67		60.63	Dark, firm peat; big piece of wood; spruce needles; birch	
67–72		56.06	Firm, dark peat; big pieces of wood; spruce needles	
72–77			Fragmented woody peat; much conifer wood; birch bark	
77–82	3810 ± 100			
82–87		75.18	Tight, dark woody peat; conifer wood; birch bark	
87–92			Woody peat; mostly spruce wood (a forest)	
92–97			Big piece of wood (black spruce) ca. 7–8 cm diameter	
97–102			Tight, dark, woody peat; spruce needles; birch	
102–107		71.19	Firm, dark, woody peat; all wood and twigs; charcoal	
107–112			Dark, fibrous, loose peat; spruce, birch, graminoid	
112–117		62.06	Firm, fibrous, dark peat; wood; spruce needles	
117–122			Dense, but friable, dark peat; much spruce wood	
122–127		67.96	Firm, dark peat; fragmented wood	
127–132			Friable woody peat; spruce bark; much spruce wood	
132–137	4430 ± 110			

137–142	45.21	Very loose woody peat	
142–147		Fragmented woody peat; many spruce needles; birch bark	
147–152	40.37	Fragmented woody peat; much wood; spruce needles	
152–157		Dark, twiggy peat; lots of spruce needles; birch bark	
157–162	42.48	Very friable peat; lots of spruce needles	
162–167		Woody, twiggy peat; chunks of wood; much birch bark	
167–172	40.33	Fibrous dark peat; lots of graminoid roots; spruce needles	Sedge meadow with low shrubs
172–177		Fibrous, dark, smelly peat; many Carex fruits, roots	
177–182	48.61	Fibrous, but consolidated, dark peat; twigs; Carex leaves	
182–187	4790 ± 100		
187–192			

[a] Frozen core extracted in 1983 with a SIPRE ice auger to a depth of 192 cm and kept frozen until after sawing into 5-cm sections for radiocarbon dating and macrofossil analysis. [Reprinted from Billings (1987a) with permission from Pergamon Press.]

North Slope step by step; the migration must be in one great jump, probably with human aid. Cooper (1986), however, reported a small white spruce about 27 cm high at an elevation of 1465 m, well above tree line, on the Arrigetch Peaks and near the continental divide of the Brooks Range. He also found 11 other small white spruces beyond tree line in the same region at elevations between 770 and 1067 m; none of these trees were cone-bearing. Although this area is within striking distance of the North Slope tundra, the seeds would still have to come over the range from tree line far below. Similar physical barriers exist elsewhere. Yet because of insufficient collecting, we may be unaware of the full range of these tree species. Odasz (1986), for example, has collected 40 vascular plant species in the Alatna River drainage beyond their known distributions on the south slopes of the Brooks Range.

V. Floristic Richness along Latitudinal Gradients

Of the approximately 250,000 known species of vascular plants on Earth, only about 1,000, or 0.4% occur in the Arctic. In contrast, about 165,000, or 66% exist in the lowland tropics. This steep descent in floristic richness along the northward latitudinal gradient is at least as steep as the drop in temperature and number of degree-days above 0° C per year. This latter parameter drops even more steeply than do temperatures along an alpine–arctic gradient from the tropical high Andes to the Arctic (see Billings, 1973).

Because of the Gulf Stream Drift, the temperature and floristic gradients in western Europe are not as steep in the middle latitudes and subarctic as they are in North America. In northern Norway, forests of birch and pine occur north of 70°N with associated herbaceous plants and shrubs characteristic of more southerly latitudes in North America. The growing season in Norway north of 70°N, during which the daily mean temperature is above 6° C, is between 110 and 120 days; such temperatures are quite warm by North American arctic standards. Beyond the North Cape, in the jump to Svalbard, however, the gradients become very steep and are truly arctic. During full-glacial times in the past, all climatic gradients in Scandinavia were shortened and steepened because of the ice and because the Gulf Stream Drift was pushed far to the south.

Even so, populations of certain vascular plant taxa probably remained on Svalbard during the Wachselian glaciation of 100,000 years B.P. or even earlier. *Pedicularis dasyantha* (Traut.) Hadac, an endemic on Svalbard and on Novaya Zemlya, has been studied both in the field and in the laboratory by Odasz 1990, 1991). Its populations on Svalbard are restricted to inner-fjord substrata around Kongsfjorden that have been ice free for up to 290,000 years. The species has frost tolerance to −70° C, the ability to take up nutrients through a number of pathways, resistance to desiccation, and pubescence that traps heat. It has also evolved self-fertilization, although it belongs to a small group of species that are otherwise obligately pollinated by bumblebees

(Odasz, 1991). Moreover, on Svalbard geographically isolated populations of *P. dasyantha* show pronounced phenotypic variation (clear polymorphism at the 6-PGO enzyme locus, Odasz, 1990) that suggests migration from sites of glacial ice-free refugia. Variations in allele frequencies between these isolated populations suggest limestone ridges as sites of survival during glaciations. The recent range expansion of *P. dasyantha* appears in the present distribution of overlapping "ancient" gene pools.

Evolution of arctic and alpine plant taxa north of 30°N requires adaptation to chilling and freezing temperatures during the growing season and to bitterly cold temperatures during the winter dormant season. The very low number of vascular plant species and the floristic zonation in the Arctic attest to low growing-season temperature as a stressor of first rank. Young (1971) agrees. In an essay on floristic zonation in the Arctic, he concludes that "floristic zones are correlated with the amount of summer warmth to the point that other ecological factors are insignificant in comparison. . . ." And yet most of these taxa are so well-adapted to functioning at low temperatures that they are really limited in their local distributions and activity by other factors, such as soil nitrogen, phosphorus, and depth of the water table (Ulrich and Gersper, 1978; Chapin, 1978; Billings, 1987b).

Nevertheless, floristic richness (expressed as the number of species of vascular plants per 10,000 km^2) is correlated with mean annual temperature in the humid regions of the earth (Cailleux, 1961). At mean annual temperatures above 20° C, floristic richness ranges from about 2500 species to approximately 4500 species of vascular plants per 10,000 km^2 (Fig. 1). For geographical regions with mean annual temperatures below $-10°$ C, Figure 1 predicts only 60 to 160 species of vascular plants per 10,000 km^2. These estimates correspond well with the situation in the tundra at Barrow (71°23'N), where the mean annual temperature is $-12.9°$ C, the July mean is 3.9° C, and the vascular plant flora consists of about 125 species within a 25-km radius of the weather station. At Alert (82°30'N) on Ellesmere Island and at Nord (81°36'N) near the northern tip of Greenland, the mean annual temperatures at the weather stations are -17.8 and $-16.5°$ C, respectively. From Figure 1, one would predict about 63 species around Alert and 71 species near Nord. These two stations are immediately west and east of Pearyland (63,000 km^2), where the vascular plant flora consists of about 96 species (Holmen, 1957), so these figures for floristic richness are about right.

Data on number of species are usually imprecise. The numbers tend to increase through time as new species are described and intensive field collecting expands the knowledge of species ranges. Moreover, "richness" depends on the size of the region: Cailleux's "richesse aréale" is not independent of area. A small, homogeneous region will be richer in species per 10,000 km^2 than a large, environmentally diverse region at the same latitude. An index of floristic richness that takes size of area into consideration is more precise for determining floristic richness of geographic regions, large or small (Billings, 1987b):

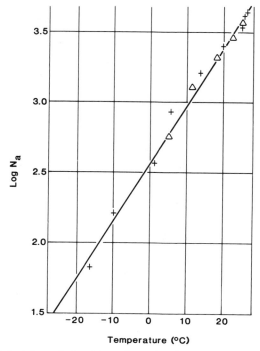

Figure 1   Floristic richness in humid continental regions of the Northern (+) and Southern (Δ) hemispheres. The logarithm of the number of vascular plant species 10,000 km² (N_a) is plotted as a function of mean annual temperature. [Redrawn from Cailleux (1961).]

Floristic richness index (FRI) = (number of species/area in km²) × 10,000

This equation allows a more accurate comparison of floristic richness among areas unequal in size (Billings, 1987b). Examples of how this method works in relatively small geographic areas along a latitudinal gradient as compared with relatively large areas along the same gradient appear in Table III. In the Arctic, by this index, Peary Land, with only 63,000 km² and 96 species, is 7.5 times richer in species per 10,000 km² than the Canadian Arctic Archipelago (1,422,000 km²) with its 340 species. The mean number of species per genus also drops off with increasing latitude, probably for three reasons. First, the time during which the Arctic has been warm enough may have been insufficient for it to have been colonized by more than one or two taxa of the common genera. Second, in many genera speciation is slow in cold climates. Third, relatively few of the species in even large genera have the genetic potential that could fit them for life in cold summer climates.

A quantitative attempt to test the temperature effect on the local numbers of vascular plant species in the American Arctic has been done by Rannie

Table III Relative Richness of Vascular Floras at Different Latitudes in America[a]

	Latitude (°N)	Area (km²)	Number of species	Number of genera	Floristic richness index[b]	Mean number of species per genus
Relatively small regions						
Costa Rica (various sources)	8–11	60,660	ca. 9000	1525	1484	5.9
Indiana (Deam, 1940)	38–41.5	93,994	2140	690	228	3.1
West Virginia (Strasbaugh and Core, 1978)	37–40.5	62,629	2155	693	344	3.1
Peary Land, North Greenland (Holmen, 1957)	80.5–83.6	63,095	96	53	15	1.8
Relatively large regions						
Venezuela (Pittier et al., 1945–47)	1–12	912,068	ca. 15,000	—	164	—
Southeast United States (Small, 1933)	25–37	898,139	5557	—	62	—
Canadian Arctic Archipelago (Porsild, 1957)	61–83	1,421,910	340	—	2	—

[a] Reprinted from Billings (1987b) with permission from the Institute of Arctic and Alpine Research.

[b] $FRI = \dfrac{\text{No. of species}}{\text{Area in km}^2} \times 10,000$

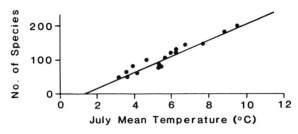

Figure 2   Relationship between number of arctic vascular plant species and mean daily temperature in July. Floristic data are from 35 local floras in the Canadian Arctic Archipelago and 3 in northwest Greenland. Temperature data are from published weather-station records, long-term field records, or interpolation from a map of July mean temperature isotherms (N, 24.2T–29.1) (Maxwell, 1980). [Redrawn from Rannie (1986).]

(1986) for 35 local floras in the Canadian Arctic Archipelago and 3 such floras in northwest Greenland (Fig. 2). Correlation coefficients were calculated between total number of species at each site and various indices of summer temperatures; these were very high, ranging between $r = 0.90$ and 0.97. Highest correlations ($r = 0.97$) were with mean July temperature and number of degree-days above 0° C in July. Regression equations between mean July temperature and number of species show a diversity gradient of about 25 species per degree C.

VI. Warmer Climates and Future Migrations

Even though the effects of temperature on plant growth and distribution in the Arctic are obvious on a broad geographical scale, temperature is only one in a suite of interacting factors affecting the growth of different species in a community at the local level (Scott and Billings, 1964; Chapin and Shaver, 1985). Scott and Billings concluded (p. 268) that plant production is not controlled by "one or even a few factors, but is the aggregate response of many species reacting more or less independently to many environmental factors. Factors important to one species need not be important to another." The same conclusion has been reached by Chapin and Shaver (1985) in the tussock tundra at Toolik Lake on the Alaskan North Slope.

Certainly, populations of each species in a community will respond individualistically to climatic change or to increases in ultraviolet-B radiation. Their behavior will depend on inherent physiological pathways, reproductive structures, and leaf morphology. Acclimation is likely to play a role, as are mutation rates and possible hybridization. Mechanisms and vectors involved in migration rates will also be important.

In a globally warming climate, it will be necessary to know the potential migration rates of the principal herbaceous taxa and species of dwarf shrubs within the Arctic and Subarctic. We will also need to know about the vectors (wind, birds, mammals) that carry the propagules (seeds, pieces of rhizomes). Quantitative data are needed on the interactions of physiological processes, flowering, fruiting, and the availability of vectors moving northward or around the pole at the right time of year. We have little information on these parameters now. There are good data on some of these interactions from palynological research in more southerly latitudes, but these come primarily from the present pollen rain and fossil pollen of tree taxa in relation to deglaciation. Only some of this information will be applicable to movements of tundra vegetation.

If the predicted rates of climatic warming during the twenty-first century hold in the Arctic, the change may be far more rapid than any climatic changes during deglaciation or even during the Little Ice Age. Will this warming be much faster than the ability of plant migration and soil formation to keep up? Perhaps. Davis (1989) suggests that for tree taxa in the Great Lakes states, future range extensions will have to occur at least 10 times as fast as the average Holocene rate, or 200 km per century, to track the expected temperature rise. We must also expect that the northward movement of bird and mammal vectors may lag behind climatic warming if some of their favorite food plants are left behind. Some of these vectors, such as blue jays and Clark's nutcrackers, can carry beechnuts, acorns, and pine seeds several kilometers beyond the present limits of the respective tree taxa (Webb, 1986), but such bird and mammal vectors cannot wait around for these seed caches to produce fruiting trees. The complexity of migrations in a rapidly changing climate challenges ecological imagination and techniques.

The principal plant communities in both the American and Eurasian Arctic provide appropriate subjects for studies of migrations and relative rates. These communities and their plant species are fairly well known, as are the physical gradients of temperatures, soils, nutrients, the thaw-lake cycle, and depths of thaw. Will species of the tussock tundra react differently from those in the wet coastal tundra? How will the components of the coastal tundra react to invasions from the south? Some of the plant populations in this coastal ecosystem may be replaced by populations of taxa not present there now. Such adventive species have not yet appeared at Barrow, for example, and diversity there remains low. But competition from alien taxa is a real possibility there for the near future. Will taxa in the tundras react differently to invasions than will those in the polar deserts? To answer such questions, we will need more precise information on possible rates and kinds of climatic changes expected in different parts of the Arctic. Also needed will be controlled experiments on competition and on the physiological ecology of the important plant species both alone and in natural microcosms along present climatic and soil gradients.

VII. Summary

Derived in large part from the high mountains of the Northern Hemisphere, the flora of the Arctic is old, small, and cold-adapted. It comprises about 900 vascular species distributed over about 13 million km^2. Floristic diversity varies considerably throughout this vast area and is relatively higher in regions that were not glaciated in the Pleistocene. The regional floras of the wet coastal tundra on the North Slope of Alaska and of the taiga forests and bogs of central Alaska have changed little over the past 5000–6000 years; at any given site, however, the flora can change a great deal in response to environmental oscillations. Floristic richness per 10,000 km^2 as well as numbers of species present per genus, decreases markedly as one moves from low to high latitudes, partly because of low temperatures, but also because of various barriers to northward migration, including mountain ranges. At the predicted rates of accelerated global warming, we do not know if plant migration and soil formation can keep up with a changing climate. It is nevertheless critical that we understand the potential migration rates of individual plant taxa under possible warmer climates in the future. Doing so will involve understanding physiological and reproductive processes and traits, as well as the vectors responsible for carrying the propagules. The readiness and availability of sites is also of prime importance.

References

Ager, T. A. (1975). Late Quaternary environmental history of the Tanana Valley, Alaska. Rep. 54. Institute of Polar Studies, Ohio State University, Columbus.

Ager, T. A. (1983). Holocene vegetational history of Alaska. In "Late Quaternary Environments of the United States" (H. E. Wright, Jr., ed.), Vol. 2, The Holocene, pp. 128–141. Univ. of Minnesota Press, Minneapolis.

Anderson, P. M. (1982). Reconstructing the past: The synthesis of archeological and palynological data, northern Alaska and northwestern Canada. Ph.D. diss., Brown University, Providence, Rhode Island.

Axelrod, D. I. (1962). Post-Pliocene uplift of the Sierra Nevada, California. Bull. Geol. Soc. Am.**73**, 183–198.

Billings, W. D. (1973). Arctic and alpine vegetations: Similarities, differences, and susceptibility to disturbance. Bioscience **23**, 697–704.

Billings, W. D. (1978). Alpine phytogeography across the Great Basin. Great Basin Nat. Mem. **2**, 105–117.

Billings, W. D. (1987a). Carbon balance of Alaskan tundra and taiga ecosystems: Past, present, and future. Quat. Sci. Rev. **6**, 165–177.

Billings, W. D. (1987b). Constraints to plant growth, reproduction, and establishment in arctic environments. Arct. Alp. Res. **19**, 357–365.

Billings, W. D. (1988). Alpine plant communities and vegetational gradients. In "North American Terrestrial Vegetation" (M. G. Barbour and W. D. Billings, eds.), pp. 391–420. Cambridge Univ. Press, New York.

Billings, W. D., and Mooney, H. A. (1968). The ecology of arctic and alpine plants. Biol. Rev. Camb. Philos. Soc. **43**, 481–529.

Billings, W. D., and Peterson, K. M. (1991). Some possible effects of climatic warming on arctic tundra ecosystems of the Alaskan North Slope. *In* "Consequences of the Greenhouse Effect for Biological Diversity" (R. L. Peters, ed.). Yale Univ. Press, New Haven, Connecticut (in press).

Billings, W. D., Godfrey, P. J., and Hillier, R. D. (1965). Photoperiodic and temperature effects on growth, flowering, and dormancy of widely distributed populations of *Oxyria* (abstract). *Bull. Ecol. Soc. Am.* **46**, 189.

Billings, W. D., Godfrey, P. J., Chabot, B. F., and Bourque, D. P. (1971). Metabolic acclimation to temperature in arctic and alpine ecotypes of *Oxyria digyna. Arct. Alp. Res.* **3**, 277–289.

Billings, W. D., Peterson, K. M., and Shaver, G. R. (1978). Growth, turnover, and respiration rates of roots and tillers in tundra graminoids. *In* "Vegetation and Production Ecology of an Alaskan Arctic Tundra" (L. L. Tieszen, ed.), pp. 415–434. Springer-Verlag, New York.

Bliss, L. C. (1977). Vascular plants of Truelove Lowland and adjacent areas, including their relative importance. *In* "Truelove Lowland, Devon Island, Canada: A High Arctic Ecosystem" (L. C. Bliss, ed.), pp. 697–698. Univ. of Alberta Press, Edmonton.

Böcher, T. W. (1951). Distributions of plants in the circumpolar area in relation to ecological and historical factors. *J. Ecol.* **39**, 376–395.

Cailleux, A. (1961). "Biogéographie Mondiale," 4th ed. Presses Universitaires de France, Paris.

Chabot, B. F., and Billings, W. D. (1972). Origins and ecology of the Sierran alpine flora and vegetation. *Ecol. Monogr.* **42**, 163–199.

Chapin, F. S., III. (1978). Phosphate uptake and nutrient utilization by Barrow tundra vegetation. *In* "Vegetation and Production Ecology of an Alaskan Arctic Tundra" (L. L. Tieszen, ed.) pp. 483–507. Springer-Verlag, New York.

Chapin, F. S. III, and Shaver, G. R. (1985). Individualistic growth response of tundra plant species to environmental manipulations in the field. *Ecology* **66**, 564–574.

Colinveaux, P. A. (1964). Origin of the Ice Ages: Pollen evidence from arctic Alaska. *Science* **145**, 707–708.

Cooper, D. J. (1986). White spruce above and beyond treeline in the Arrigetch Peaks region, Brooks Range, Alaska. *Arctic* **39**, 247–252.

Davis, M. B. (1989). Lags in vegetation response to greenhouse warming. *Clim. Change* **15**, 75–81.

Deam, C. C. (1940). "Flora of Indiana." Indiana Dep. of Conservation, Div. of Forestry, Indianapolis.

Densmore, D. (1980). Vegetation and forest dynamics of the upper Dietrich River valley, Alaska. M. S. thesis, North Carolina State University, Raleigh.

Hamilton, T. D., Ager, T. A., and Robinson, S. W. (1983). Late Holocene ice wedges near Fairbanks, Alaska, U.S.A.: Environmental setting and history of growth. *Arct. Alp. Res.* **15**, 157–168.

Holmen, K. (1957). The vascular plants of Peary Land, North Greenland. *Medd. Grønl.* **124(9)**, 1–149.

Hultén, E. (1937). "Outline of the History of Arctic and Boreal Biota during the Quaternary Period." Bokförlags Aktiebolaget Thule, Stockholm.

Hultén, E. (1962). The circumpolar plants. I. Vascular cryptogams, conifers, monocotyledons. *K. Sven. Vetenskapsakad. Handl.* **8**, 1–275.

Johnson, A. W., Viereck, L. A., Johnson, R. E., and Melchior, H. (1966). Vegetation and flora. *In* "Environment of the Cape Thompson Region, Alaska" (N. J. Wilimovsky and J. N. Wolfe, eds.), pp. 277–354. U.S. Atomic Energy Commission, Division of Technical Information, Oak Ridge, Tennessee.

Komarkova, V., and Webber, P. J. (1977). "Annotated List of Vascular Plants from the Meade River Area near Atkasook, Alaska." National Science Foundation, Washington, D.C.

Livingstone, D. A. (1957). Pollen analysis of a valley fill near Umiat, Alaska. *Am. J. Sci.* **255**, 254–260.

Luken, J. O. (1984). Net ecosystem production in a subarctic peatland. Ph.D. diss., Duke University, Durham, North Carolina.

Luken, J. O., Billings, W. D., and Peterson, K. M. (1985). Succession and biomass allocation as controlled by *Sphagnum* in an Alaskan peatland. *Can. J. Bot.* **63**, 1500–1507.

Matthews, J. V., Jr. (1968). A paleoenvironmental analysis of three late Pleistocene coleopterous assemblages from Fairbanks, Alaska. *Quaest. Entomol.* **4**, 202–224.

Matthews, J. V., Jr. (1970). Quaternary environmental history of interior Alaska: Pollen samples from organic colluvium and peats. *Arct. Alp. Res.* **2**, 241–251.

Matthews, J. V., Jr. (1974). Wisconsin environment of interior Alaska: Pollen and macrofossil analysis of a 27-meter core from the Isabella Basin (Fairbanks, Alaska). *Can. J. Earth Sci.* **11**, 828–841.

Maxwell, J. B. (1980). "The Climate of the Canadian Arctic Islands and Adjacent Waters," Vol. 1. Climatological Studies 30. En 57-7/30-1. Supply and Services Canada, Ottawa.

Murray, D. F. (1978). Vegetation, floristics, and phytogeography of northern Alaska. *In* "Vegetation and Production Ecology of an Alaskan Arctic Tundra" (L. L. Tieszen, ed.), pp. 19–36. Springer-Verlag, New York.

Murray, B. M., and Murray, D. F. (1978). Checklists of vascular plants, bryophytes, and lichens for the Alaskan U.S. IBP Tundra Biome study areas—Barrow, Prudhoe Bay, Eagle Summit. *In* "Vegetation and Production Ecology of an Alaskan Arctic Tundra" (L. L. Tieszen, ed.), pp. 647–67. Springer-Verlag, New York.

Nelson, R. E. (1982). Late Quaternary environments of the western Arctic Slope, Alaska. Ph.D. diss., University of Washington, Seattle.

Odasz, A. M. (1986). Distributions of 40 rare vascular plants in the Alatna River drainage of the central Brooks Range, Alaska. *Northwest Sci.* **60**, 104–107.

Odasz, A. M. (1990). Vascular plants on Svalbard since the Eemian: Isozyme evidence (abstract). International Conference on Climate of the Northern Latitudes: Past, Present, and Future. University of Tromsø, Norway.

Odasz, A. M. (1991). Distribution and ecology of herbaceous *Pedicularis dasyantha* (Scrophulariaceae) in Spitsbergen, Svalbard Archipelago: Relation to present and past environments. *Striae* (in press).

Pittier, H., *et al.* (1945–47). "Catalogo de la Flora Venezolana," 2 vols. Vargas, Caracas.

Polunin, N. (1959). "Circumpolar Arctic Flora." Clarendon Press, Oxford.

Porsild, A. E. (1957). Illustrated flora of the Canadian Arctic Archipelago. *Nat. Mus. Can. Bull.* **146**.

Rannie, W. F. (1986). Summer air temperature and number of vascular species in arctic Canada. *Arctic* **39**, 133–137.

Rønning, O. I. (1979). Svalbards flora. Polarhåndbok 1, Norsk Polarinstitutt, Oslo, Norway.

Scott, D., and Billings, W. D. (1964). Effects of environmental factors on standing crop and productivity of an alpine tundra. *Ecol. Monogr.* **34**, 243–270.

Small, J. K. (1933). "Manual of the Southeastern Flora." Univ. of North Carolina Press, Chapel Hill.

Steere, W. C. (1978). "The Mosses of Arctic Alaska." J. Cramer, Lehre, Germany.

Steere, W. C., and Inoue, H. (1978). The Hepaticae of arctic Alaska. *J. Hattori Bot. Lab.* **44**, 207–345.

Strasbaugh, P. D., and Core, E. L. (1978). "Flora of West Virginia," 2nd ed. Seneca Books, Grantsville, West Virginia.

Tikhomirov, B. A., Petrovskii, V. V., and Yurtsev, B. A. (1969). Flora of the vicinity of Tiksi-harbour, arctic east Siberia. *In* "Vascular Plants of the Siberian North and Northern Far East," pp. 7–40. Jerusalem, Israel, Prog. for Scientific Translations, Nat. Tech. Info. Serv., Springfield, Virginia.

Ulrich, A., and Gersper, P. L. (1978). Plant nutrient limitations of tundra plant growth. *In* "Vegetation and Production Ecology of an Alaskan Arctic Tundra" (L. L. Tieszen, ed.), pp. 457–481. Springer-Verlag, New York.

Viereck, L. A., and Little, E. L., Jr. (1972). "Alaska Trees and Shrubs." Agriculture Handbook No. 410. U. S. Government Printing Office, Washington, DC.

Viereck, L. A., Van Cleve, K., and Dyrness, C. T. (1986). Forest ecosystem distribution in the taiga environment. *In* "Forest Ecosystems in the Alaskan Taiga" (K. Van Cleve, F. S. Chapin III, P. W. Flanagan, L. A. Viereck, and C. T. Dyrness, eds.), pp. 22–43. Springer-Verlag, New York.

Walker, D. A., Short, S. K., Andrews, J. T., and Webber, P. J. (1981). Late Holocene pollen and present-day vegetation, Prudhoe Bay and Atigun River, Alaskan North Slope. *Arct. Alp. Res.* **13**, 153–172.

Webb, Thompson, III. (1986). Is vegetation in equilibrium with climate? How to interpret late-Quaternary pollen data. *Vegetatio* **67**, 75–91.

Wiggins, I. L., and Thomas, J. H. (1962). A flora of the Alaskan arctic slope. *Arct. Inst. N. Am. Spec. Publ.* **4**.

Young, S. B. (1971). The vascular flora of Saint Lawrence Island with special reference to floristic zonation in the arctic regions. *Contrib. Gray Herb. Harv. Univ.* **201**, 11–115.

6

Plant Succession, Competition, and the Physiological Constraints of Species in the Arctic

L. C. Bliss and K. M. Peterson

I. Introduction

Although the principle of plant succession leading to some level of community and ecosystem stability is a cornerstone in ecology, the magnitude of successional interactions varies from biome to biome. In tropical and temperate forests, patterns of directional and cyclic change are often very evident, largely the result of major changes in plant growth forms and associated shifts in the physiological characteristics of species during succession (Bazzaz, 1979). In highly stressed biomes, such as deserts and tundras, these patterns are less evident and may be absent in many habitats.

Succession and the processes controlling successional change have been less studied in the Arctic than in other biomes for several reasons. As one flies over the Arctic, one sees little evidence of natural disturbances (fire, massive erosion, permafrost melt, recent retreat of glaciers) or of human-induced disturbances, except locally (villages, mining and petroleum exploration and their development, temporary or permanent roads). Agriculture is not practiced, so there are no abandoned cultivated fields and pastures. Thus there is little disturbed land that plants can invade. Moreover, the flora is relatively small (Chap. 5), and few species serve the role of ruderal, or "*r*-selected," plants. The growing season is short, resulting in slower rates of plant establishment and slower potential change in community composition. Finally, only in the past 50 years has significant human disturbance of soils opened up sites for plant invasion.

Intense frost action processes (congeliturbation) have a profound influence on soil and plant community development. For this reason, Hopkins and Sigafoos (1950), Sigafoos (1951), Raup (1951), Britton (1957), and others questioned the validity of using concepts of succession and climax in the Arctic as they are used in temperate regions. Churchill and Hanson (1958) reviewed the earlier arctic literature and concluded that environmental change along macro-, meso-, and microgradients occurs that results in repetitious patterns of communities and that the vegetation is in dynamic equilibrium with the environmental complex, including soil-frost phenomena. They considered phasic cycles to be within the concept of regional climax communities, as long as the cycles are not directional as they are in succession. They recognized both progressive and retrogressive succession, leading to steady-state, or climax, communities.

Relatively little has been written about arctic plant succession in recent years, until publication of the symposium, "Restoration and Vegetation Succession in Circumpolar Lands" (Salzberg *et al.*, 1987). Several of these papers provide important information for this chapter. Our objectives here are (1) to present representative examples of plant establishment and maintenance; (2) to use physiological and plant life-history data from species that typify the diverse arctic environments; and (3) to discuss the limited data on plant competition.

II. Arctic Landscapes

The arctic biome is as diverse ecologically as temperate biomes. In the Low Arctic, riparian zones along rivers and streams are occupied by herbaceous and tall shrub vegetation. Rolling uplands north of the tree line generally contain low shrubs of *Salix* and *Betula* along with the dwarf shrubs of various heath species that fit within a matrix of upland sedges, including cottongrass tussocks and forbs. Beyond these landscapes are large areas dominated by sedges, dwarf shrubs of heath species, and scattered low shrubs of *Salix* and *Betula*. Lowlands that are imperfectly drained and the more expansive coastal plain lowlands are dominated by wetland sedges, grasses, and mosses forming extensive mires. Elevated habitats generally contain low or dwarf shrubs, cushion plants, lichens, and graminoids; species adapted to well-drained or intermediately drained soils. These are the major vegetation types within the Low Arctic of Alaska, mainland Canada, and much of mainland Siberia (Chap. 4).

In the High Arctic of the Canadian Archipelago, the northern portion of the Taimyr Peninsula, and the Soviet archipelagoes, the vegetation is shorter and sparser. There are a few areas of mire, but much of the landscape is covered with cushion plants, lichens, and mosses or with scattered vascular plants in a substrate of mosses and lichens. These are the polar semideserts. Other vast areas—the polar deserts—have almost no plant cover. Here vascular plants form the major biomass, with lichens and mosses very limited. (For details on these patterns of vegetation, see Chap. 4.)

III. Models of Succession

Succession generally involves changes in species composition or in the relative numbers of individuals through time. The changes are directional, leading to a more stable community with increased structure. Arctic environments—so limited in heat, nutrients, and often water—are not characterized by classical succession, where species replacement and directional change are paramount. With increasing latitude and environmental severity, succession shifts from species replacement to species establishment and survival (Svoboda and Henry, 1987). Under these conditions the life history and physiology of species become central to understanding the limited successional processes that do occur.

The succession model used here is adapted from Svoboda and Henry (1987)—a schematic model that illustrates the dynamic patterns found in limited areas within tundras, polar semideserts, and polar deserts (Fig. 1). Directional change with species replacement characterizes riparian vegetation in the Low Arctic and limited coastal areas within the High Arctic. Directional change and nonreplacement of species is common in some low-arctic tundras and in the polar semideserts. The model for nondirectional succession with nonreplacement of species fits the polar deserts, where biotic interactions are limited.

A. Index of Potential Succession

Svoboda and Henry (1987) developed an index of successional potential to demonstrate that the actual time available for revegetation since deglaciation varies latitudinally in terms of length of the growing season and mean seasonal

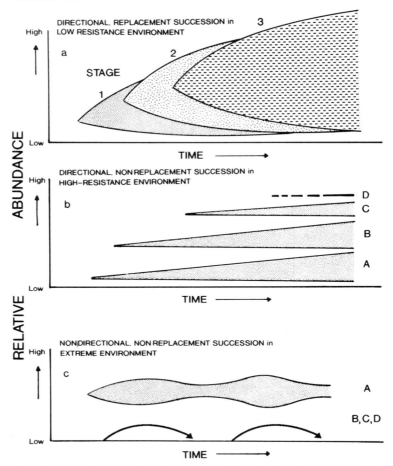

Figure 1   Schematic diagrams of arctic succession. Directional–replacement model of succession (a) within low-resistance environments characteristics of some low-arctic communities. Directional–nonreplacement model of succession (b) within the polar semideserts of the High Arctic. Nondirectional–nonreplacement model of succession (c) typical of some polar semidesert and most polar desert communities. Invading species (A, B, C, D) establish successfully in type b, but species, B, C, and D fail in type c, the polar deserts. [Modified from Svoboda and Henry (1987).]

degree-days. For mesic habitats, suitable for zonal soil development, an index of relative successional potential (*RSP*) was developed,

$$RSP = Yn \times iY \times jDG \qquad (1)$$

where *Yn* represents the number of years since deglaciation; *iY* is the fraction of the year represented by the growing season; and *jDG* is the ratio of the actual mean degree-days for the growing season and the optimal degree-day value for that vegetation.

Here we modify the equation to express simply the favorableness of the present climate. The index of potential succession (*IPS*) becomes

$$IPS = iY \times ADD \qquad (2)$$

where iY is the fraction of the year represented by the growing season in days, and *ADD* is the accumulated degree-days above the 0° C threshold for the growing season. Where possible, we have used the temperature records for the 1970s in these calculations. We recognize that the climate has warmed in the past 10,000 years (Chap. 2) and that retrogression probably occurred during the Little Ice Age (Svoboda, 1982) and no doubt earlier. Such climatic changes and their impact on succession are difficult to calculate with our current knowledge. Changes in vegetation may best be assessed along ecotones at the limit of trees and shrub tundra, along major rivers that run north–south, and in sites of surface disturbance where "*r*-selected" species are more likely to establish.

IV. Succession in the Low Arctic

A. Primary Succession

1. Directional Change with Species Replacement

a. Riparian Habitats. Major rivers that drain the Low Arctic are characterized by meandering channels with massive gravel and sand bars. Alluvial plant zonation and presumed succession are typical. At Umiat the broad valley of the Colville River, with its abundant gravel bars and river terraces, provides diverse habitats into which herbs and shrubs establish (Bliss and Cantlon, 1957). On the bare gravels where pockets of sand and silt accumulate, a variety of pioneer herbs establish as seedlings and stranded plants. Clumps of *Salix alaxensis* wash downstream and become stranded. These clumps, and shrub growth from them, accelerate the deposition of sand and the establishment of additional forbs and grasses. At least 25 herbaceous species are commonly present; all flower and fruit abundantly. Mosses cover 20–40% of the surface where extensive herb mats occur (Fig. 2).

On higher ground (1–2 m), extensive stands of *Salix alaxensis* predominate: shrubs 3–4 m tall with an understory of greenleaf willow species 1–1.5 m high. The ground layer includes 10–15 of the herbaceous species found in the pioneer herb communities (averaging 10–50% cover) and a 20–40% cover of mosses. The active layer is >1.0 m in late July.

On older gravel bars, decadent stands of feltleaf willow predominate, with an understory of willows. *Alnus crispa* is present in small numbers, but the herbaceous layer is species poor: 2–5 species in mesic habitats and 6–8 species in drier microsites. Moss cover increases to 70–100%, but herb cover drops to 10–30%. An organic layer develops, providing insulation that results in an active layer 50–70 cm in late July.

Where there is a gradual change in elevation from the floodplain to the first terrace, there is a gradual shift from decadent *Salix alaxensis*, often <2 m

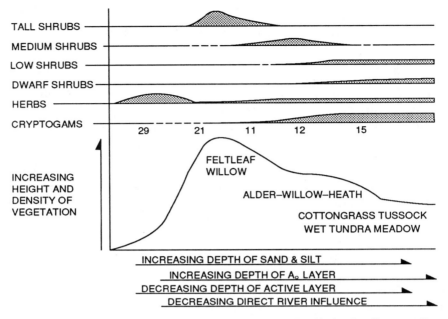

Figure 2   Plant succession along the Coville River near Umiat, Alaska, that illustrates directional species replacement. With time, the average number of vascular species decreases along with an increase in organic matter and a decrease in the active layer. [Modified from Bliss and Cantlon (1957).]

high, to the greenleaf willows, often only 0.5–1.0 m high. Associated with these changes in deciduous shrubs and herbs are an increase in heath species, an increase in organic matter, and a decrease in the active layer (20–30 cm in late July). These transition areas generally grade into open stands of *Alnus crispa* with the dwarf heath shrubs and graminoids; on higher ground a cotton-grass–heath vegetation predominates with scattered *Alnus* and *Salix* shrubs; and in depressions, mires (wet sedge–moss communities) predominate.

This pattern of species replacement—from tall shrubs to intermediate shrubs and finally low or dwarf shrubs—and the associated increase in cryptogams but great reduction in herbs results from both physical and biotic changes (Bliss and Cantlon, 1957) (see Fig. 1a). The young and decadent stands of feltleaf willow have the highest biomass, and probably the highest net annual production, but are low in species richness. Incorporation of fine-grained alluvium and the development of an organic layer associated with a reduction in the active layer, colder soils, and wetter soils illustrate the initial role of facilitation and allogenic processes. Over time, with the development of peat, and presumed lower pH and less available nutrients, heath shrubs, sedges, and a few low deciduous shrubs dominate. Here autogenic processes and competition are more important than allogenic processes. This pattern is more typical of primary succession in forest rather than in tundra landscapes. But there is one major difference from temperate regions, where the nutrient regime typically increases with time and increased vegetation structure (Odum, 1979; Tilman, 1987; and others). In riparian and mire habitats, nutrients appear to decrease with time, and competition increases for the limited nitrogen and phosphorus.

The index of potential succession for the Umiat, Alaska, riparian landscape shows that, among the sites examined, this is one of the most favorable for plant development (Table I). Accumulated degree days above 0° C are three times that of most other arctic sites. It is in these landscapes that plant succession most resembles that of cold temperate forest regions.

b. Sand Dune Habitats. In the vicinity of Atkasuk, Alaska, along the Meade River and west across a broad expanse of tundra to the lower reaches of the Colville River, sand dunes are common features of the landscape (Tedrow and Brown, 1967). Active dunes are largely restricted to river margins, but they are occasionally associated with leeward shores of the larger thaw lakes (Walker, 1973). Successional sequences associated with active dunes along rivers intergrade with riparian sequences on sand bars similar to those discussed above, with *Salix alaxensis* giving way to the willows of smaller stature (*Salix glauca* and *Salix lanata*) as age and distance from the river increase (Peterson, 1978). *Elymus mollis* (*=Elymus arenarius* ssp. *mollis, Leymus mollis*) is a common pioneer in open sands of both riparian and upland dunes. In upland sites with moderate sand deposition, *Carex bigelowii* and *Arctostaphylos rubra* dominate along with *Dryas* in seral communities. In slightly less-exposed sites where some snow cover and small depressions afford winter protection, *Cassiope tetragona* becomes codominant with *Dryas*, and eventually a lichen and dwarf heath community replaces the *Dryas* (Peterson and Billings, 1980). The long-stabilized dunes of the region support *Eriophorum vaginatum* tussock tundra with its associated dwarf shrubs, forbs, and numerous cryptogams (Komarkova and Webber, 1980). Although tussock tundra may be expected to develop eventually following most successional sequences in this region, soil moisture during summer and snow depth during winter strongly influence the ultimate community composition, and extremes of these factors prevent the establishment of tussocks in some sites (Peterson, 1978; Komarkova and Webber, 1980). Differences in the colonizing abilities of plants and the degree of riparian influence are important factors during early successional community development. The mechanisms of species replacements during succession undoubtedly vary among species, but on dunes, facilitation (through binding of sand by initial pioneers and accumulation of organic matter) and competition (for nutrients and soil water) are probably the most important processes. Such autogenic succession takes hundreds of years to develop stable communities, in part because wind plays an important role (Peterson and Billings, 1978).

These landscapes at Atkasuk have been available for plant colonization for a long time, but the summer climate is less equitable than at Umiat (Table I). The flora is smaller and shrubs are shorter than farther south.

Table I   Index of Potential Succession (IPS) for Sites in the North American
Low and High Arctic

Site	° N	Growing season (iY)	Accumulated degree-days $>0°$ C (ADD)	IPS
Umiat, AK	69	0.25	993	248
Atkasuk, AK	71	0.29	618	179
Barrow, AK	71	0.21	301	63
Truelove Lowland, NWT	76			
Central lowland		0.21	278	58
Coastal site		0.21	210	44
Plateau site		0.15	220	33
King Christian Island, NWT	78	0.17	203	35

c. Mire Habitats. Lowlands tend to be poorly drained. In these habitats vegetation develops many bog-like characteristics, and soils generally accumulate peat. Like the previous examples where geomorphic processes (fluvial
and aeolian erosion) initiate open habitats for colonization and succession, a
geomorphic process known as the thaw-lake cycle is responsible for initiating
successional sequences throughout many arctic coastal lowlands. The classic
thaw-lake cycle (Britton, 1957; Billings and Peterson, 1980), which takes hundreds of years, results from cycles of lake drainage → plant establishment →
development of low-center polygons with massive ice wedges → thaw and coa-
lescence of polygons → thaw-lake formation, and eventual drainage. Initial
plant invaders and successional sequences vary among regions over the range
of this geomorphic process, and even within regions, some floristic variations
are associated with soil moisture. The variation due to moisture is often pronounced early in succession since the degree of drainage varies considerably
between individual basins and even within a single basin. Ovenden (1986)
noted that more than 50 species of plants had invaded a lake basin in the
vicinity of the Mackenzie River Delta, Northwest Territories (NWT) Canada,
seven years after drainage.

Near Barrow, Alaska, the succession of vascular plant and bryophyte species
in drained lake basins (Fig. 3) can be considered largely as a species replacement pattern (at the scale of entire drained lake basins). Mosses are generally the first to establish in all but the driest sites, with initial invasion by
Psilopilum cavifolium yielding in wet sites to *Calliergon sarmentosum* and *Drepanocladus aduncus.* With time these species gave way to various *Sphagnum* species.
Vascular plant dominance in these basins is initially by *Dupontia fisheri* with
either *Eriophorum scheuchzeri* or *Arctophila fulva* or both as codominants
(Billings and Peterson, 1991). These species produce abundant seeds that are
dispersed into newly drained basins by wind or water during spring thaw
(unpubl. data). Once established, these plants grow rapidly, covering lake
basins through vegetative propagation. These rapidly growing species estab-

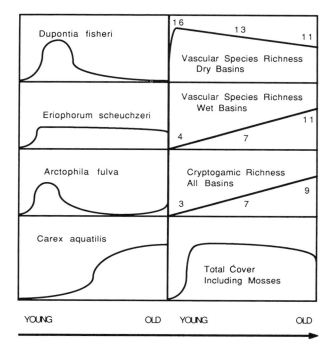

Figure 3   In drained lake basins near Barrow, Alaska, well-dispersed plants dominate early in succession. Tolerance of low fertility and increased competition characterize late successional species. Numbers associated with species richness curves represent average numbers of species with cover exceeding approximately 1%.

lish a closed vegetation in which the clones of less commonly established and more slowly growing species such as *Eriophorum angustifolium* and *Carex aquatilis* produce expanding circles of tillers within the vegetation (Billings and Peterson, 1980). Vegetative growth of *Carex aquatilis* clones eventually leads to nearly continuous cover throughout the basins. On a local scale, areas are either dominated by *Dupontia* or *Carex*, but through time the cover of *Carex* increases and *Dupontia* decreases. Changes in the physical environment accompanying this succession include decreased availability of phosphorus, decreased pH, decreased depth of thaw, and the accumulation of soil organic matter (Billings and Peterson, 1991; unpubl. data).

The more severe climate at Barrow (Table I) and the much wetter landscape than Atkasuk and Umiat explain the considerable reduction in species richness and plant growth forms. Although the general successional model is the same (Fig. 1a), the predominance of graminoids throughout succession contrasts sharply with the previous example.

2. Morphological and Physiological Adaptations

a. Riparian and Upland Habitats. The available data indicate some parallels in the suite of characteristics for pioneer versus stable species in these arctic habitats compared with temperate forests species (Bazzaz, 1979; Chapin, 1980). Pioneer herbs and deciduous *Salix* shrubs grow rapidly; produce large amounts of seed, often with high viability (Bliss, 1956, 1958; Densmore and Zasada, 1983); and have relatively high rates of photosynthesis (12–14 µmol $m^{-2}s^{-1}$) (Tieszen, 1973; Oberbauer and Oechel, 1989) and stomatal conductance (0.008–0.017 mol $m^{-2}s^{-1}$) (Stoner and Miller, 1975). Those species more typical of river terrace and upland habitats that appear to form stable or climax communities have a different suite of characteristics. They grow more slowly, produce less seed with somewhat lower rates of germination (Bliss, 1956, 1958; Chapin and Shaver, 1985), and their rates of photosynthesis and stomatal conductance are also lower, especially in the evergreen heath species, than in the deciduous shrubs and herbs of more pioneer communities (Chapin, 1980; Chapin and Shaver, 1985).

There are considerable shifts in the life-form and general morphology of plants characteristic of primary succession along rivers and sand deposits. The pioneer perennial herbs along rivers may live 10–20(?) years, whereas the deciduous shrub *Salix alaxensis* has individual stems for 30–50 years (Bliss and Cantlon, 1957). In contrast, *Eriophorum vaginatum* tussocks and dwarf heath shrubs are longer-lived (120–190 years for tussocks; Mark *et al.*, 1985). The evergreen heath shrubs and the deciduous *Betula* and *Salix* low shrubs are probably 50–100+ years old.

The pioneer *Elymus mollis* has the ability to germinate and establish on shifting sand, and its rapid rhizome growth helps to stabilize sand. Although tillers are widely spaced and provide limited cover (generally <10%), the dead leaves are retained several years, which further stabilizes shifting sand (Peterson, 1978). Where litter accumulates, *Dryas integrifolia* becomes established and can form extensive mats over time. Its slow growth rates in contrast with *Elymus*, however, does not enable it to tolerate burial. Photosynthetic rates are also in sharp contrast; maximum rates are 19–20 µmol $m^{-2}s^{-1}$ in *Elymus*, the pioneer (Tieszen, 1973) but only 2–4 µmol $m^{-2}s^{-1}$ in *Dryas* (Mayo *et al.*, 1973, 1977). Associated heath species and *Eriophorum vaginatum* have maximum rates of 6–10 and 13–17 µmol $m^{-2}s^{-1}$, respectively (Tissue and Oechel, 1987; Oberbauer and Oechel, 1989), again illustrating differences in morphological and physiological traits of early versus later-establishing species.

b. Mire Habitats. Graminoids and mosses predominate in these wetlands, and their relatively rapid growth rates permit considerable accumulation of organic matter. Aboveground:belowground ratios (shoots:rhizomes and roots) of live plants are typically 1:5 to 1:8 (Dennis, 1977). *Dupontia fisheri, Arctophila fulva,* and *Eriophorum scheuchzeri* produce abundant and readily dispersed seeds, typical of pioneer species. In contrast, *Carex aquatilis* is a poor seeder, and seedlings are uncommon. The pioneer species grow more rapidly, and their tillering habit promotes rapid vegetative spread in open habitats. In contrast, *Carex aquatilis* grows more slowly and is better adapted to sites with extremely low nutrient supply. Although these graminoids are capable of using the cold, low-phosphate and -nitrogen environment in which they live, it appears that *Carex,* the more stable, climax species, can tolerate more stressful habitats (Ulrich and Gersper, 1978; Chapin, 1978; Billings and Peterson, 1991; unpubl. data). Here, as in riparian habitats, nutrients decrease with time—the opposite of succession dynamics in most temperate regions (Odum, 1979; Tilman, 1987). Rates of photosynthesis (10–12 μmol m^{-2}s^{-1}) (Oberbauer and Oechel, 1989) and rates of stomatal conductance (0.012–0.20 mol m^{-2}s^{-1}) (Miller *et al.*, 1978) do not vary significantly between the pioneer and more climax species.

B. Secondary Succession

Examples of secondary succession are relatively few, the result of only limited areas of disturbance resulting from fires or industrial development in the last 20 years. Nevertheless, studies conducted do show diverse patterns of recovery on different landscapes.

1. Fires

Tundra fires are much less common in arctic tundra than in boreal forest or forest–tundra, and they are usually small (Wein, 1976; Johnson and Viereck, 1983). Fires are more common in western Alaska, which averages 12 fires per 5000 km^2 per 23 years on the Seward Peninsula and 7 fires per 5000 km^2 per 23 years in the Noatak River Valley, but only 0.4 fires per 5000 km^2 per 23 years along the Kokolik River on the North Slope (Racine *et al.*, 1987). The unusually dry summer of 1977 resulted in 3800 km^2 of burned tundra in western and northern Alaska. Tussock tundra with scattered shrubs of *Salix* and *Betula* coupled with the common dwarf shrub heath species appears to be the major vegetation type that burns. This susceptibility results from the flammability of litter and the high lipid content of heath shrubs (Racine *et al.*, 1987; Auclair, 1983).

Studies of the 1977 burns on the Seward Peninsula and the Noatak and Kokolik rivers found that vascular plant cover returned to prefire levels in 6–10 years. At all sites, most of the regrowth was from the fire-adapted *Eriophorum vaginatum,* with lesser amounts contributed by resprouting of low shrubs (Racine *et al.*, 1987). Flowering increased significantly in cottongrass after burning, and seedlings were abundant at the above sites, at a tundra site on the Elliott Highway in central Alaska and at three sites near Inuvik, NWT (Wein and Bliss, 1973). At the Eliott site, *Eriophorum vaginatum, Carex bigelowii, Betula nana, Calamagrostis canadensis,* and the common heath species were most prominent 13 years after the burn (Fetcher *et al.*, 1984). Net plant production had recovered but not the relative abundance of the original species. Although many seedlings established the first year, few survived more than two summers. Laboratory experiments with heat treatment showed that few seeds would survive within the burned profile and that seedling emergence was minimal for seeds buried one centimeter below the surface in four soil substrates. Thus, even though many seeds are contained within the seedbank (McGraw, 1980), seedling establishment results mainly from plants stimulated to flower and fruit after the fire (Wein and Bliss, 1973). These studies indicate that few new pioneer species enter these tussock communities after fire (*Epilobium angustifolium, Senecio congestus*). The grasses *Calamagrostis canadensis, Poa arctica*, and *Arctagrostis latifolia*, however, are minor in undisturbed vegetation but seed abundantly in some burned sites.

A nondirectional pattern of succession thus predominates after fires, with the residual species maintaining their populations, although the density of species may shift for a number of years. The exceptions are the bryophytes and lichens, with lichens entering later in succession. The rapid recovery of tussock tundra after fire results from an increased depth of the active layer, warmer soils, and increased nutrient availability and uptake (Wein and Bliss, 1973; Racine *et al.*, 1987).

2. Human-Disturbed Soils

Off-road vehicle tracks, well pads, and borrow pits related to petroleum exploration and development have provided disturbed surfaces for plant establishment. The vehicle tracks and well pads occur in upland tussock tundra, low shrub–dwarf shrub heath tundra, and in lowlands with sedge mires and both raised-center and depressed-center polygons. In all of these habitats, secondary succession with directional change and nonreplacement of species predominates.

Bladed surfaces and vehicle tracks reduce the layer of surface peat and often expose mineral soil. In all of these microsites, the active layer deepens, soils are warmer, and nutrient availability and uptake by surviving plants often increase (Bliss and Wein, 1972; Hernandez, 1973; Gersper and Challinor, 1975; Chapin and Shaver, 1981). Plant recovery, measured as increased plant cover and net primary production, is generally rapid. Where ice wedges are exposed, melt-ponds form with aquatic plants predominating. The most obvious change is a lowering of the soil surface where ground ice has melted. Such scars today occur infrequently because off-road vehicles are limited to winter when the ground is frozen. Surface disturbance is minimal, but grasses often increase in cover where small amounts of mineral or peat surfaces are exposed, and flowering of sedges may increase dramatically where mineralization rates increase from soil warming.

V. Succession in the High Arctic

Studies of plant succession are even less common in the High Arctic because areas of surface disturbance are limited. The extreme severity of the environment, where abiotic rather than biotic factors are dominant and the flora is small with very few ruderal species, further limits succession.

A. Primary Succession:

1. Directional Change with Species Replacement

a. Coastal Lowlands. Tundra vegetation, similar to that at Barrow and Prudhoe Bay, is restricted to 1–2% of the Queen Elizabeth Islands (Babb and Bliss, 1974; Muc *et al.*, 1989). These scattered oases in a sea of polar desert or semidesert landscape result from higher solar radiation, and therefore higher summer temperatures, and a more constant water supply from drainage of melting snow that has been blocked by raised beaches caused by isostatic rebound or from glacial meltwater and runoff from surrounding highlands. These coastal lowlands are dominated by cushion plants on the better-drained sites and mires in the wetter habitats. In the mires *Carex stans* and *Eriophorum triste* predominate, with lesser amounts of other graminoids.

It is in these ecosystems that cyanobacteria play a central role in nitrogen fixation and carbon accumulation (see Chap. 14). They are important in maintaining these systems, as well as initiating them. Current research on the Truelove Lowland, Devon Island, indicates that directional change, species replacement, and facilitation play a role in succession in coastal habitats (Bliss *et al.*, 1990). These habitats are considerably younger than those in the central part of the lowland. In addition, the index of potential succession for the coastal site is low, much lower than at Barrow where the degree-days are more numerous (see Table I).

Marine algae wash into shallow lagoons or onto coarse gravel beaches and are invaded by cyanobacteria and green algae. At the shoreline of brackish water lagoons, the developing mats of organic matter are invaded by *Puccinellia phryganodes* and small numbers of other salt-tolerant species. Upslope from these small brackish marshes are meadows of *Dupontia fisheri*, with some *Salix arctica* where meltwaters keep the soils moist all summer. Mosses are abundant in these habitats.

Farther upslope where meltwaters are more abundant, and presumably later in succession, meadows of *Carex stans* dominate. *Salix arctica* is always present, as are limited numbers of forbs. Mosses are also abundant in these sites.

On drier rocky sites near the coast and along the drier margins of all the meadows, the soils often form hummocks on which *Salix arctica*, *Saxifraga oppositifolia*, *Juncus biglumis*, and species of *Draba* predominate. Where winter snows are deeper and summer temperatures higher, mats of *Dryas integrifolia* dominate.

Several patterns of plant succession are present. Where long, gentle slopes rise from brackish lagoons, and where meltwater is sufficient to keep the soils wet much of the summer, the succession sequence is marine algae → *Puccinellia*–algal marsh → *Dupontia*–moss meadow → *Carex*–moss meadow. The controlling factors are maintenance of wet soils, accumulation of peat, and the gradual reduction of salinity, pH, and possibly phosphorus over time. *Dupontia fisheri* is only abundant in these most coastal areas, indicating that with time, it is outcompeted by *Carex stans*. Indirect evidence for this hypothesis is the establishment of *Carex stans* rather than *Dupontia* directly on algal mats at the margins of shallow ponds and lakes in older portions of the lowland.

In many coastal areas marine algae accumulate little, or if present, algal mats have washed up on coarse gravels. Here and in the margins of the graminoid meadows, rosette and cushion plants predominate, along with mosses and lichens. In these habitats the succession sequence is marine or terrestrial algae → rosette plant–moss or cushion plant–moss. Succession is arrested by the lack of water and further soil development.

Maximum rates of nitrogen fixation, measured as acetylene reduction, occur in early summer, shortly after the snow and ice melt. The highest rates of fixation occur in brackish meadows, intermediate rates in the *Dupontia* and *Carex* meadows, and the lowest rates in the cushion plant– and rosette plant–moss communities. The marine algal substrate, higher levels of phosphorus, and abundant water appear to be the controlling factors for the high nitrogen fixation rates (Chap. 14; Chapin *et al.*, 1991). The low fixation rates in the shallow soils of the beach ridges result from limited water and the limited time in the summer that the soils are saturated.

It is only in these coastal or valley lowlands that a directional change with species replacement occurs in the High Arctic. The patterns of succession depend on topographic position, the relative water and nutrient regime of the soils, the abundance of cyanobacteria, and the percentage of the growing season (40–60 days) when plant growth and nitrogen fixation can take place (Bliss *et al.*, 1990). The initial incorporation of organic matter and phosphorus from marine algae followed by nitrogen fixation and further addition of organic matter by cyanobacteria contrasts with succession patterns in the Low Arctic.

2. Morphological and Physiological Adaptations

a. Coastal Lowlands. The photosynthetic data (Mayo *et al.*, 1977) and water relations data (Addison, 1977) for *Carex stans* indicate similar response patterns to its closely related species, or ecotype, *C. aquatilis* of the Low Arctic. The limited data on *Dupontia fisheri* also show parallels with its low-arctic counterpart; rapid rhizome development, occurrence in more nutrient-rich soils, and an adaptation to tolerate slightly saline soils (Bliss *et al.*, 1990). These physiological characteristics, and those described for *Salix, Dryas*, and *Saxifraga* in the next section, help explain the roles these species play in the species replacement patterns (primary succession) of the graminoids in these high-arctic coastal lowlands.

3. Directional Change with Nonreplacement of Species

a. Western Islands. In the western Queen Elizabeth Islands, the soils are often fine sands to silt loams. Bryophytes and lichens (crustose and foliose) predominate, with scattered individuals or small clumps of vascular plants.

These landscapes have been termed polar semideserts (Bliss and Svoboda, 1984). Where open soil polygons and stripes occur, there are fewer species of vascular plants (Sohlberg and Bliss, 1984). The two most common species on bare soil or lichen crusts are *Phippsia algida* and *Juncus biglumis*. *P. algida* produces large amounts of seed, and the plant has roots that spiral, an important adaptation where there is considerable needle-ice formation and soil movement (Bell and Bliss, 1978). Microclimate analyses showed that cryptogam–herb habitats have higher temperatures, lower wind speeds, higher levels of soil moisture, and higher soil nitrate levels, all factors more favorable to seed germination and seedling establishment than open soil–barren habitats (Sohlberg and Bliss, 1984). Although *Phippsia* functions as a ruderal species in disturbed soils, it is often an important species within the cryptogam mats as well. Water stress and low soil and air temperatures reduce growth in this species (Grulke and Bliss, 1988), but it functions over a wide range of habitats.

Although plant succession was not studied explicitly, evidence from several sources indicates that cryptogam crusts of lichens, cyanobacteria, and mosses establish, and, with time and the buildup of bryophytes, vascular plants follow. The cryptogams are not replaced, and there is no evidence that any replacement sequence exists among the vascular plant species, only that vascular plant cover increases in more mesic habitats. This pattern fits the second model of succession in Figure 1b. The climate on King Christian Island and other northwestern islands is very severe, significantly reducing species richness, plant cover, and the index of potential succession (see Table I).

b. Eastern Islands. In the eastern and southern islands, plant succession has not been studied in detail except on the Truelove Lowland, Devon Island. As outlined in Section V, A, 2, a, cyanobacteria establish in moist coastal areas. As these lands are elevated with isostatic rebound, mosses and lichens establish in the drier sites and, with time, are invaded by cushion and rosette species (*Salix*, *Dryas*, and species of *Saxifraga*, *Draba*, and *Minuartia*).

The raised beach ridges in the older portions of the lowland are dominated by these communities (Svoboda, 1977). The active layer is generally 50–120 cm deep, and the rooting-zone soils are relatively warm (5–8° C) compared with the soils in the western islands (2–5° C). Plant cover, standing crop, and the number of species increases with increased age of beach ridges on the Truelove Lowland. Plant cover averaged 10% on the crest and 44% in the transition zone of beach ridges 3000–6000 years old, and beach ridges 6000–9000 years old averaged 18 and 72% cover, respectively. The corresponding data for standing crop were 80 and 145g m^{-2} for the crest and transition zones of younger beach ridges, increasing to 275 and 425 g m^{-2}, respectively, on the older beach ridges (Svoboda, 1977). These data illustrate slow but significant plant changes with time, an example of directional but nonreplacement succession (Svoboda and Henry, 1987; (Fig. 1b). The mesoclimate within this lowland is warmer, and one finds greater species richness, plant cover, and net plant production than in the colder coastal area and thus a somewhat greater index of potential succession (see Table I).

4. Morphological and Physiological Adaptations

a. Western Islands. The limited flora in the northwestern Queen Elizabeth Islands is characterized by small plants—often 1–3 cm and seldom 5–8 cm high in anthesis. Most species maintain some wintergreen leaves and are therefore able to initiate photosynthesis with snowmelt and higher temperatures (Bell and Bliss, 1977). Many species produce viable seed, and seedlings are often present in the cryptogam mats, although seldom in the open soil where needle-ice formation and summer surface drying in some summers limit establishment (Bell and Bliss, 1980; Sohlberg and Bliss, 1984). Woody species are restricted to the warmest mesohabitats. Plant growth is slow, and most species are characterized as stress tolerators. Many herbaceous plants live 25–100 years, including clumps of *Luzula confusa.* Root growth is generally restricted to the upper 5–10 cm, and shoot:root ratios are low (1:0.2–1:0.7) (Bell and Bliss, 1978).

Luzula confusa is a dominant graminoid in many cryptogam–herb communities but is uncommon in dry habitats. Its photosynthetic rates are similar to those other arctic graminoids (5–12 μmol m^{-2}s^{-1}), but these rates are very sensitive to changes in air temperature and to xylem water potential (Addison and Bliss, 1984). With reduced soil water potential in late summer, some osmotic adjustment takes place. *Phippsia algida* and *Puccinellia vaginata* grow even more slowly but produce considerable viable seed, unlike *Luzula. Phippsia* and *Puccinellia* plants are relatively long-lived (15–34 and 20–74 years, respectively), develop very low xylem water potentials (−2.5 to −4.5 MPa), and can maintain relatively high rates of stomatal conductance (50–450 μmol m^{-2}s^{-1}) (Grulke and Bliss, 1988). *Phippsia algida* is one of the few species that can successfully invade bare soil. In these nutrient-poor soils, however, *Phippsia* and *Puccinellia* are not sexually mature until plants are at least 10 years old. These data help explain why so few species are able to pioneer in bare soil compared with establishment once moss–lichen mats have developed.

b. Eastern and Southern Islands. Species representative of the cushion, or mat, form predominate in the southern and eastern arctic islands in well-drained habitats. These species are generally wintergreen or evergreen, although the very successful deciduous shrub, *Salix arctica,* is almost always present. *Salix, Dryas integrifolia,* and *Saxifraga oppositifolia* all produce large amounts of viable seed, yet they are slow-growing (except *Salix*) perennials that live a long time (50–100+ years).

Salix arctica is a dioecious shrub, with female plants more abundant in wet, higher-nutrient habitats and males in xeric, nutrient-poor habitats (Dawson and Bliss, 1989). Female plants have a more positive carbon isotope ratio than males, indicting a possibly higher water-use efficiency. Photosynthetic rates resemble those of graminoids from mesic to wet habitats, but respiration rates are much higher. These observations may help explain the more rapid growth of *Salix* than any other woody species at high latitudes. The ability to osmoregulate; maintain a high tissue elasticity; and, for female plants, to occupy cold, wet soils and for male plants to occupy warm, dry soils enables this species to live in a wide range of habitats. As we learn more about the eco-physiology of other high-arctic species that also occur in diverse habitats, we may find similar suites of characteristics. If so, their presence would help explain why species replacement and a directional pattern of succession is so limited, especially in the High Arctic.

Dryas is an exceedingly important species of mesic to xeric habitats in the southern and eastern arctic islands. It is the most abundant plant in terms of cover and biomass in large areas (Svoboda, 1977; Reznicek and Svoboda, 1982; Bliss and Svoboda, 1984). Unlike *Salix, Dryas* is a very slow growing, winter-green, cushion plant. Both photosynthetic ($2–4\ \mu mol\ m^{-2}\ s^{-1}$) and respiratory ($1–2\ \mu mol\ m^{-2}\ s^{-1}$) rates are very low compared with other species (Mayo *et al.*, 1973, 1977). These plants also develop very low water potentials (-1.0 to -4.0 MPa); stomatal conductance is low, and the plants have high cell wall elasticity. All these traits characterize stress-tolerant species, and we believe that numerous species inhabiting these northern lands have them. As clumps of *Dryas, Saxifraga oppositifolia,* and *S. caespitosa* age, rosette species of *Draba, Saxifraga, Minuartia,* and *Cerastium* become established in the more nutrient-rich stable soils. Nevertheless, there is little or no indication of species replacement.

5. Nondirectional Change with Species Nonreplacement

Vast areas of the Canadian Arctic Archipelago are essentially devoid of plants, vascular and/or cryptogam. These polar deserts result from a combination of low temperature; limited available soil moisture in surface soils for part of the summer; often coarse-textured substrates; and very low soil nutrients, especially nitrogen and phosphorus. Large sections of the upland landscapes were covered with thin ice or permanent snow during the Little Ice Age only a few decades ago (Svoboda, 1982), which has allowed only a short time for succession to take place. These factors, coupled with a 40–50-day growing season, leave large areas from sea level to 300–400 m with less than 2–3% plant cover and essentially no biomass (Bliss *et al.*, 1984). Under these extreme conditions, it is apparently impossible for a significant plant cover to develop. In a sampling of 23 stands on six islands, only 17 species of vascular plants were found, with species richness averaging 6.3 ± 2.4 species per stand (60 m^2; Bliss *et al.*, 1984). Most vascular plants were found in desiccation cracks. Bryophytes and lichens were seldom closely associated with the vascular species, indicating that chance, rather than "safe-site," patterning of the cryptogams was involved. Here, the third successional model in Figure 1 applies (Fig. 1c).

Only in sites with more soil moisture (snowflush or soil-water seepage habitats) were vascular plants and cryptogams more abundant (Bliss *et al.* 1984, 1991). In these habitats *Salix arctica* and even *Ranunculus sulfureus* can be found—species more typical of "warmer," more mesic habitats. In all cases where vascular plants are more numerous, there are also mats of cyanobacteria (*Nostoc commune* and other species), indicating that nitrogen and soil moisture are more abundant. In most of the polar desert landscapes observed on 12 northern islands, cyanobacteria appear to be very minor except in snowflush and streamside habitats.

The only places where plant succession now occurs are in the limited snowflush sites. Here mosses and cyanobacteria appear to be pioneers, followed by a limited number (20) of vascular species (Bliss *et al.*, 1991). It is only in these habitats that directional but nonreplacement succession occurs in the polar deserts. With climate change and an increase in temperature and precipitation, snowflush communities could assume a larger percentage of the landscape, provided that snowbanks melted to expose plants in early summer. Otherwise, summer climate is severe in these sites, greatly limiting succession (Table I).

B. Secondary Succession and Human Disturbance

Human disturbance in the High Arctic, with the potential for secondary succession, results largely from petroleum and mining activities that began in the 1970s. Observations at numerous well sites, limited seedling experiments, and detailed research on several graminoid species have demonstrated that there are few ruderal species capable of establishing in large numbers following natural or human-induced disturbance (Bliss and Grulke, 1988). The high-arctic flora is small, the growing season is short and severe, and the limited extent of past disturbances has resulted in little selection for species able to play this successional role.

Saxifraga oppositifolia, Dryas integrifolia, Papaver radicatum, Minuartia rubella, and several species of *Draba* produce large amounts of seed, yet viable seed is not produced each year, and seedlings on undisturbed surfaces are generally uncommon (Barrett and Schulten, 1975; Bell and Bliss, 1980). Some of these species are commonly found around animal dens; they are sometimes found in larger areas of disturbed soils.

Rhizomes of *Carex stans* extend into vehicle tracks that occur within sedge–moss meadows, but slow vegetative growth limits the role of this species in succession (Barrett, 1975). Flowering and seed production in this species are limited, and seedlings are seldom found (Muc, 1977; Bliss and Grulke, 1988). *Luzula confusa* is another important graminoid, yet it seldom, if ever, sets viable seed, and tiller development is limited (Bell and Bliss, 1980; Addison and Bliss, 1984). *Alopecurus alpinus* occurs in many habitats, and this rhizomatous species does spread into disturbed soils, especially nutrient-rich sites (Nosko, 1984; Bliss and Grulke, 1988). We have seldom found seedlings of *Alopecurus*, and viable seed is rarely produced. *Phippsia algida* is the only truly ruderal species that readily invades disturbed soils; it produces abundant viable seed, seedlings are common in disturbed soils, and the species is quite tolerant of soils with low water potential. *Phippsia* is also long-lived (Grulke and Bliss, 1988). Species able to play a significant role in secondary succession in these northern lands are thus very limited: *Carex stans* in wetlands, *Alopecurus alpinus* in uplands via rhizome development, and *Phippsia algida* in uplands via seedlings.

Model *b* in Figure 1 best describes the successional pattern after disturbance in lowlands with mires and in some uplands where cryptogam–herb or cushion plant–cryptogam communities predominate. Human disturbances in the polar desert have little biological impact because plants are so minor in these landscapes.

VI. Plant Competition

A. Low Arctic

Although poorly documented, interactions among plants, including competition, are undoubtedly important in structuring the communities of the Low Arctic. In shrub-dominated communities, and to a lesser extent in graminoid dominated communities, competition for light and nutrients contribute to successional patterns. Although photoperiods are long, solar angles are always low, resulting in reduced intensities and contributing to mutual shading by plants.

Cloudy or foggy conditions are frequent during the summer, especially near the arctic coast, further reducing light intensity. Field photosynthetic rates of *Dupontia fisheri* at Barrow are seldom light-saturated (Tieszen, 1973). Light reduction by shrub canopies may actually benefit hydrophilic mosses like *Sphagnum* ssp. in competition with other mosses (Murray *et al.*, 1989). Bryophytes and vascular species apparently compete for light in some tundra habitats, with mosses actually smothering vascular species in some instances. Lush bryophytic mats may often have a few leaf tips of species such as *Rubus chamaemorus* or *Vaccinium vitis-idaea* emerging from otherwise buried plants. In other cases, stems or rhizomes of vascular species provide an historic record of growth that has essentially kept pace with the accumulating moss medium in which the plants are rooted.

Vascular plants all draw upon the limited pool of mineral nutrients available each year. Although some evidence of spatial partitioning of rooting exists (Shaver and Billings, 1975), essentially all species have the potential to interfere with one another's nutrient uptake. *Dupontia fisheri*, *Arctophila fulva*, and *Carex aquatilis* have distinct physiological and life-history adaptations that lead to patterns of coexistence corresponding to competitive sorting along spatial patterns of habitat. *Dupontia* and *Arctophila* grow rapidly, and if growth is used to evaluate competitive advantage, these species grow very well in almost all conditions except extreme nutrient deprivation compared with *Carex aquatilis*. *Carex* can tolerate low nutrient supply rates and, under such conditions, appears to eliminate other graminoids through slow growth and nutrient accumulation, reducing nutrient availability below the threshold required for maintenance of more rapidly growing species (Billings and Peterson, 1991, and unpubl. data). Mixed stands of *Arctophila, Dupontia,* and other graminoids, including *Eriophorum scheuchzeri,* are common near Barrow, but within clones of *Carex aquatilis,* other graminoids are generally absent. This pattern probably reflects competition for nutrients, although one may argue that stress tolerance is also involved (Grime, 1977). Certainly, as one moves to the High Arctic, most species are stress tolerators for nutrients and cold soils, and often for soil drought.

Eriophorum vaginatum and *E. scheuchzeri* exhibit competitive relationships that are roughly analogous to those of *Carex aquatilis* and *Dupontia fisheri.* Assemblages with *E. vaginatum* often represent mature communities with limited fertility, whereas *Eriophorum scheuchzeri* and *Dupontia fisheri* are found in early successional sites with relatively favorable nutrient status. Competition between the two *Eriophorum* species is largely influenced by the nutritional characteristic of the site (McGraw and Chapin, 1989). Adaptations such as high nitrogen-use efficiency in *E. vaginatum* may be contrasted with adaptations such as high nitrogen-uptake efficiency in *E. scheuchzeri* to account for their competitive relations and field distributions. In these upland habitats, the limited species replacement that occurs in succession appears to be nutrient-driven.

Competition among individual plants of a single species may be important to population regulation and population genetics in many tundra habitats. Intraspecific competition for nutrients is implicated in the regular spacing of *Alnus crispa* in regions of shrub tundra on Alaska's North Slope. Shrub growth is apparently limited by the availability of mineral nutrients, and both mineral accumulation and growth increase in areas of experimentally reduced shrub density (Chapin *et al.,* 1989). In these experiments, removal of alders did not enhance the growth or nutritional levels of other species in the community, indicating alders may compete for resource pools unavailable to other species. Intraspecific competition is likely to be important in maintaining spatial segregation of ecotypes along short environmental gradients. Locally, steep gradients of nutrition, moisture, and snow cover are common within the Arctic. Competition may be important in maintaining fell-field and snowband ecotypes of *Dryas octopetala* (McGraw, 1985). Ecotypic differentiation along environmental gradients has been suggested for *Carex aquatilis* (Shaver *et al.,* 1979), but given the importance of clonal expansion in this species, these patterns may represent differential survival of clones following intraspecific competition.

B. High Arctic

Within the polar barrens, there is no evidence that competition occurs because in most areas plant cover averages <1%. Where plant cover is greater in snowflush habitats, vascular plant establishment appears to be facilitated by cryptogamic mats, as it is in the polar semideserts.

In the polar semideserts, where cryptogams play such an important role in cover and biomass, plant interactions are more evident. In these habitats, four to seven species of *Saxifraga* commonly grow sympatrically and thus compete directly for nutrients, water, and light. The seven species appear differentiated by reproductive strategies rather than by specific growth forms (Grulke and Bliss, 1985). Most vascular species are more common on moss turf and fruticose lichens than on bare soil or crustose lichen mats (Sohlberg and Bliss, 1984). In another study, the effect of moss removal was examined over a two-year period for *Ranunculus sabinei* and *Papaver radicatum* (Sohlberg and Bliss, 1987). Shoots of *Papaver* were significantly larger in plots from which moss had been removed or removed and replaced than in control plots. *Ranunculus* shoots were significantly larger in plots from which moss had been removed or clipped. Removal and clipping of mosses enhanced soil temperature and probably increased nutrient availability. From previous research we know that dry moss turfs are poor heat conductors because they dissipate much of the net radiation received, thus reducing soil heat flux (Addison and Bliss, 1980). Nitrate nitrogen and potassium contents were higher within moss mats than in bare soil at adjacent microsites (Sohlberg and Bliss, 1984). Although seeds of *Ranunculus* germinate more frequently in bare soil and crustose lichen substrates than in moss turf, adult plants do better in moss turf, suggesting that this species is less tolerant of water stress (soil water potentials of -4.0 to -5.0 MPa) within the rooting zone. *Papaver* plants respond somewhat differently, for growth was greater in moss-removal and removal-replacement plots. Under these conditions, surface soils are both warmer and drier. *Papaver radicatum* is a ubiquitous species but infrequent in snowflush sites in the cryptogam–herb communities where soils are waterlogged much of the summer (Bliss and Svoboda, 1984).

Thus the bryophyte and bryophtye–fruticose lichen mats of the High Arctic show a commensal rather than a competitive role with vascular species. The direct benefits of these mats are increased soil water potential during drought and prevention of needle-ice formation. The detrimental effects are reduced soil temperature and reduced availability of soil nutrients, especially nitrate nitrogen. We also know that these moss mats enable cyanobacteria to reach a larger biomass than they do on bare soil, thus enhancing nitrogen fixation. On beach ridges and other habitats where crustose lichens predominate, these crusts may prevent vascular plant establishment (Svoboda, 1977; Sohlberg and Bliss, 1987), although these "sealed surfaces" do retain more soil moisture for the vascular plants that are present (Addison, 1977; Addison and Bliss, 1980).

The limited areas of sedge or grass–moss meadows are believed to present a different pattern of development, at least in coastal landscapes. The development of *Puccinellia phryganodes* marshes generally occurs on decayed organic mats of marine algae and terrestrial cyanobacteria. With isostatic rebound of the land, and a change from brackish to fresh water from snowmelt, *Puccinellia* is replaced by *Dupontia fisheri* or *Carex stans*. Bryophytes appear to play a minimal role until the graminoid, nonbrackish meadow develops. As organic soil thickens and the habitat becomes wetter from water from upslope ponds and snowbanks, it appears that *Carex* replaces *Dupontia*, for older meadows in the lowland contain no *Dupontia*. Where contact occurs, *Dupontia* invades *Puccinellia* but not established mats of *Carex*.

Facilitation is involved in the establishment of *Puccinellia* on the marine algal mats in brackish lagoons. Facilitation is also central in the establishment of *Carex stans* on mats of cyanobacteria along the edges of ponds and lakes. The replacement of *Dupontia* by *Carex* however, may result more from changes in depth of organic matter, reduction in salinity, and changes in soil nutrient regime and pH rather than a direct competitive interaction.

VII. Summary

Within the High Arctic, there are examples of directional but nonreplacement succession in snowflush habitats of the polar deserts and in the cryptogam–herb and the cushion plant–cryptogam communities of the semideserts. Directional change with species replacement occurs in lowland coastal meadows, but we cannot yet say whether the driving forces are restricted to facilitation or whether plant competition also plays a major role in these northern most extensions of tundra. Only by understanding the life history and physiology of dominant species can we determine the role of plant establishment and succession in these stressful environments.

Within the Low Arctic, there are examples of directional replacement succession, but successional processes similar to those of the High Arctic also exist. Such successional sequences are simple, involving fewer species than better-known examples in temperate regions. The Low Arctic represents a transition zone between the vegetation of the High Arctic, where the stresses of the physical environment control vegetational development, and the closed vegetation of boreal forests, where biological factors, including competition, influence vegetational structure. Physiological and morphological adaptations allow tundra plants of various species to segregate along steep local gradients of environmental factors, contributing to the remarkable patterning associated with arctic tundra. Natural disturbances, often associated with geomorphic activity, create openings in the vegetation, leading to succession. Certain morphological and physiological traits are associated with the rapid invasion and success of some tundra plants in newly opened habitats, whereas other plant adaptations appear to have evolved in those species found in the closed communities, often with infertile soils, that are associated with late successional habitats.

Acknowledgments

Much of the information reported here results from the research of numerous students at the universities of Alberta and Washington over the past two decades and students and associates at

Clemson University, Duke University, and the Duke Phytotron. We gratefully acknowledge grant support from the National Research Council of Canada, the member companies of the Arctic Petroleum Operators Association, and the National Science Foundation. Most of the logistic support in the High Arctic was provided by the Polar Continental Shelf Project.

References

Addison, P. A. (1977). Studies on evapotranspiration and energy budgets on the Truelove Lowland. *In* "Truelove Lowland, Devon Island, Canada: A High Arctic Ecosystem" (L. C. Bliss, ed.), pp. 281–300. Univ. of Alberta Press, Edmonton.

Addison, P. A. and Bliss, L. C. (1980). Summer climate, microclimate, and energy budget of a polar semidesert on King Christian Island, Northwest Territory, Canada. *Arct. Alp. Res.* **12**, 161–170.

Addison, P. A., and Bliss, L. C. (1984). Adaptations of *Luzula confusa* to the polar semidesert environment. *Arctic* **37**, 121–132.

Auclair, A. N. D. (1983). The role of fire in lichen-dominated tundra and forest–tundra. *In* "The Role of Fire in Northern Circumpolar Ecosystems" (R. W. Wein, ed), pp. 235–256. Wiley, New York.

Babb, T. A., and Bliss, L. C. (1974). Susceptibility to environmental impact in the Queen Elizabeth Islands. *Arctic* **27**, 234–237.

Barrett, P. (1975). Preliminary observations of off-road vehicle disturbance to sedge meadow tundra at a coastal lowland location, Devon Island, N. W. T. ALUR 1973-74-71. Ministry of Indian and Northern Affairs, Ottawa.

Barrett, P., and Schulten, R. (1975). Disturbance and the successional response of arctic plants on polar desert habitats. *Arctic* **28**, 70–73.

Bazzaz, F. A. (1979). The physiological ecology of plant succession. *Annu. Rev. Ecol. Syst.* **10**, 351–371.

Bell, K. L., and Bliss, L. C. (1977). Overwintering phenology of plants in a polar semidesert. *Arctic* **30**, 118–121.

Bell, K. L., and Bliss, L. C. (1978). Root growth in a polar semidesert environment. *Can. J. Bot.* **56**, 2470–2490.

Bell, K. L., and Bliss, L. C. (1980). Plant reproduction in a high arctic environment. *Arct. Alp. Res.* **12**, 1–10.

Billings, W. D., and Peterson, K. M. (1980). Vegetational change and ice-wedge polygons through the thaw-lake cycle in arctic Alaska. *Arct. Alp. Res.* **12**, 413–432.

Billings, W. D., and Peterson, K. M. (1991). Some possible effects of climatic warming on arctic tundra ecosystems of the Alaskan North Slope. *In* "Consequences of the Greenhouse Effect for Biological Diversity" (R. L. Peters, ed.). Yale Univ. Press, New Haven, Connecticut (in press).

Bliss, L. C. (1956). A comparison of plant development in microenvironments of arctic and alpine tundras. *Ecol. Monogr.* **26**, 303–337.

Bliss, L. C. (1958). Seed germination in arctic and alpine species. *Arctic* **11**, 180–188.

Bliss, L. C., and Cantlon, J. E. (1957). Succession on river alluvium in northern Alaska. *Am. Midl. Nat.* **58**, 452–469.

Bliss, L. C., Chapin, D. M., Leggett, A. S., Lennihan, R., Dickson, L. G., Bledsoe, C., and Bledsoe, L. S. (1990). Ecosystem development in a coastal lowland of the Canadian High Arctic. *In* "Arctic Research Advances and Prospects" (V. M. Kotlyakov and V. E. Sokolov, eds.), Vol. 2, pp. 165–173. Nauka, Moscow.

Bliss, L. C., and Grulke, N. E. (1988). Revegetation in the High Arctic: Its role in reclamation of surface disturbance. *In* "Northern Environmental Disturbances" (P. Kershaw, ed). Boreal Institute for Northern Studies, University of Alberta, Edmonton., Occ. Pub. Vol. 24, pp. 43–55.

Bliss, L. C., and Svoboda, J. (1984). Plant communities and plant production in the western Queen Elizabeth Islands. *Holarc. Ecol.* **7**, 324–344.

Bliss, L. C., Svoboda, J., and Bliss, D. I. (1984). Polar deserts and their plant cover and plant production in the Canadian High Arctic. *Holarct. Ecol.* **7**, 304–324.

Bliss, L. C., Henry, G. H. R., Svoboda, J., and Bliss, D. I. (1991). Substrate and plant diversity within two polar desert landscapes. (unpubl. manuscript).

Bliss, L. C., and Wein, R. W. (1972). Plant community response to disturbances in the western Canadian arctic. *Can. J. Bot.* **50**, 1097–1109.

Britton, M. E. (1957). Vegetation of the arctic tundra. *In* "Arctic Biology" (H. P. Hansen, ed.), pp. 26–61. Oregon State Univ. Press, Corvallis.

Chapin, D. M., Bliss, L. C., and Bledsoe, L. J. (1991). Environmental regulation of nitrogen fixation in a high arctic lowland ecosystem. *Can. J. Bot.* (in press).

Chapin, F. S., III. (1978). Phosphate uptake and nutrient utilization by Barrow tundra vegetation. *In* "Vegetation and Production Ecology of an Alaskan Arctic Tundra" (L. L. Tieszen, ed., pp. 483–507. Springer-Verlag, New York.

Chapin, F. S., III. (1980). The mineral nutrition of wild plants. *Annu. Rev. Ecol. Syst.* **11**, 233–260.

Chapin, F. S., III, and Shaver, G. R. (1981). Changes in soil properties and vegetation following disturbance of Alaskan arctic tundra. *J. Appl. Ecol.* **18**, 605–617.

Chapin, F. S., III, and Shaver, G. R. (1985). Individualistic growth response of tundra plant species to environmental manipulations in the field. *Ecology* **66**, 564–576.

Chapin, F. S., III, McGraw, J. B., and Shaver, G. R. (1989). Competition causes regular spacing of alder in Alaskan shrub tundra. *Oecologia* **79**, 412–416.

Churchill, E. D., and Hanson, H. C. (1958). The concept of climax in arctic and alpine vegetation. *Bot. Rev.* **24**, 127–191.

Dawson, T. E., and Bliss, L. C. (1989). Pattern of water use and the tissue water relations in the dioecious shrub, *Salix arctica*: The physiological basic for habitat partitioning between the sexes. *Oecologia* **79**, 332–343.

Dennis, J. G. (1977). Distribution patterns of belowground standing crop in arctic tundra at Barrow, Alaska. *Arct. Alp. Res.* **9**, 111–125.

Densmore, R., and Zasada, J. (1983). Seed dispersal and dormancy patterns in northern willows: Ecological and evolutionary significance. *Can. J. Bot.* **61**, 3207–3216.

Fetcher, N., Beatly, T. F., Mullinax, B., and Winkler, D. S. (1984). Changes in arctic tussock tundra thirteen years after fire. *Ecology* **65**, 1332–1333.

Gersper, P. L., and Challinor, J. L. (1975). Vehicle perturbation effects upon a tundra soil–plant system: 1. Effects on morphological and physical environmental properties of the soils. *Soil Sci. Soc. Am. Proc.* **39**, 737–744.

Grime, J. P. (1977). Evidence for the existence of three primary strategies in plants and its relevance to ecological and evolutionary theory. *Am. Nat.* **111**, 1169–1194.

Grulke, N. E., and Bliss, L. C. (1985). Growth forms, carbon allocation, and reproductive patterns of high arctic saxifrages. *Arct. Alp. Res.* **17**, 241–250.

Grulke, N. E., and Bliss, L. C. (1988). Comparative life-history characteristics of two high arctic grasses, Northwest Territories. *Ecology* **69**, 484–496.

Hernandez, H. (1973). Natural plant recolonization of surface disturbances, Tuktoyaktuk Peninsula Region, Northwest Territories. *Can. J. Bot.* **51**, 2177–2196.

Hopkins, D. M., and Sigafoos, R. S. (1950). Frost action and vegetation patterns on Seward Peninsula, Alaska. *Geol. Surv. Bull.* **974-C**, 51–101.

Johnson, L., and Viereck, L. (1983). Recovery of active layer changes following a tundra fire in northwestern Alaska. *In* "Permafrost: Fourth International Conference Proceedings," pp. 543–547. National Academy Press, Washington, DC.

Komarkova, V., and Webber, P. J. (1980). Two low arctic vegetation maps near Atkasook, Alaska. *Arct. Alp. Res.* **12**, 447–472.

McGraw, J. B. (1980). Seed bank size and distribution of seed in cottongrass tussock tundra. *Can. J. Bot.* **58**, 1607–1611.

McGraw, J. B. (1985). Experimental ecology of *Dryas octopetala* ecotypes: Relative response to competitors. *New Phytol.* **100**, 233–241.

McGraw, J. B., and Chapin, F. S., III. (1989). Competitive ability and adaptation to fertile and infertile soils in two *Eriophorum* species. *Ecology* **70**, 736–749.

Mark, A. F., Fetcher, N., Shaver, G. R., and Chapin, F. S., III. (1985). Estimated ages of mature tussocks of *Eriophorum vaginatum* along a latitudinal gradient in central Alaska. *Arct. Alp. Res.* **17**, 1–5.

Mayo, J. M., Despain, D. G., and van Zinderen Bakker, E. M., Jr. (1973). CO_2 assimilation by *Dryas integrifolia* on Devon Island, Northwest Territories. *Can. J. Bot.* **51**, 581–588.

Mayo, J. M., Hartgerink, A. P., Despain, D. G., Thompson, R. G., van Zinderen Bakker, E. M., Jr., and Nelson, S. D. (1977). Gas exchange studies of *Carex* and *Dryas*, Truelove Lowland. *In* "Truelove Lowland, Devon Island, Canada: A High Arctic Ecosystem" (L. C. Bliss, ed.), pp. 265–280. Univ. of Alberta Press, Edmonton.

Miller, P. C., Stoner, W. A., and Ehleringer, J. R. (1978). Some aspects of water relations of arctic and alpine regions. *In* "Vegetation and Production Ecology of an Alaskan Arctic Tundra" (L. L. Tieszen, ed.), pp. 341–357. Springer-Verlag, New York.

Muc, M. (1977). Ecology and primary production of sedge–moss meadow communities, Truelove Lowland. *In* "Truelove Lowland, Devon Island, Canada: A High Arctic Ecosystem" (L. C. Bliss, ed.), pp. 157–184. Univ. of Alberta Press, Edmonton.

Muc, M., Freedman, B., and Svoboda, J. (1989). Vascular plant communities of a polar oasis at Alexandra Fiord (79° N), Ellesmere Island, Canada. *Can. J. Bot.* **67**, 1126–1136.

Murray, K. J., Tenhunen, J. D., and Kummerow, J. (1989). Limitations on *Sphagnum* growth and net primary production in the foothills of the Philip Smith Mountains, Alaska. *Oecologia* **80**, 256–262.

Nosko, P. (1984). A comparative study of plant adaptations of *Alopecurus alpinus* in the Canadian High Arctic. Ph.D. diss., University of Alberta, Edmonton.

Oberbauer, S. F., and Oechel, W. C. (1989). Maximum CO_2-assimilation rates of vascular plants on an Alaskan arctic tundra slope. *Holarct. Ecol.* **12**, 312–316.

Odum, E. P. (1979). The strategy of ecosystem development. *Science* **164**, 262–270.

Overden, L. (1986). Vegetation colonizing the bed of a recently drained thermokarst lake (Illisarvils), Northwest Territories. *Can. J. Bot.* **64**, 2688–2692.

Peterson, K. M. (1978). Vegetational successions and other ecosystemsic changes in two arctic tundras. Ph.D. diss., Duke University, Durham, North Carolina.

Peterson, K. M., and Billings, W. D. (1978). Geomorphic processes and vegetational change along the Meade River sand bluffs in northern Alaska. *Arctic* **31**, 7–23.

Peterson, K. M., and Billings, W. D. (1980). Tundra vegetational patterns and succession in relation to microtopography near Atkasuk, Alaska. *Arct. Alp. Res.* **12**, 473–482.

Racine, C. H., Johnson, L. A., and Viereck, L. A. (1987). Patterns of vegetation recovery after tundra fires in northwestern Alaska, U.S.A. *Arct. Alp. Res.* **19**, 461–469.

Raup, H. M. (1951). Vegetation and cryoplanation. *Ohio J. Sci.* **51**, 105–116.

Reznicek, S. A., and Svoboda, J. (1982). Tundra communities along an environmental gradient at Coral Harbour, Southampton Island, N.W.T. *Nat. Can.* **109**, 583–595.

Salzberg, K. A., Fridriksson, S., and Webber, P. J., eds. (1987). Restoration and vegetation succession in circumpolar lands. *Arct. Alp. Res.* **19**, 337–577.

Shaver, G. R., and Billings, W. D. (1975). Root production and root turnover in a wet tundra ecosystem, Barrow, Alaska. *Ecology* **56**, 401–409.

Shaver, G. R., Chapin, F. S., III, and Billings, W. D. (1979). Ecotypic differentiation in *Carex aquatilis* on ice-wedge polygons in the Alaskan coastal tundra. *J. Ecol.* **67**, 1025–1046.

Sigafoos, R. S. (1951). Soil stability in tundra vegetation. *Ohio J. Sci.* **51**, 281–298.

Sohlberg, E., and Bliss, L. C. (1984). Microscale pattern of vascular plant distribution in two high-arctic communities. *Can. J. Bot.* **62**, 2033–2042.

Sohlberg, E. H., and Bliss, L. C. (1987). Responses of *Ranunculus sabinei* and *Papaver radicatum* to removal of the moss layer in a high-arctic meadow. *Can. J. Bot.* **65**, 1224–1228.

Stoner, W. A., and Miller, P. C. (1975). Water relations of plant species in the wet coastal tundra at Barrow, Alaska. *Arct. Alp. Res.* **7**, 109–124.

Svoboda, J. (1977). Ecology and primary production of raised beach communities, Truelove Lowland. *In* "Truelove Lowland, Devon Island, Canada: A High Arctic Ecosystem" (L. C. Bliss, ed.), pp. 185–216. Univ. of Alberta Press, Edmonton.

Svoboda, J. (1982). Due to the Little Ice Age climatic impact, most of the vegetative cover in the Canadian High Arctic is of recent origin: A hypothesis (abstract). *Proc. Alsk. Sci. Conf.* **33**, 206.

Svoboda, J., and Henry, G. H. R. (1987). Succession in marginal arctic environments. *Arct. Alp. Res.* **19**, 373–384.

Tedrow, J. C. F., and Brown, J. (1967). Soils of arctic Alaska. *In* "Arctic and Alpine Environments" (H. E. Wright Jr. and W. H. Osburn, eds.), pp. 238–294. Indiana Univ. Press, Bloomington.

Tieszen, L. L. (1973). Photosynthesis and respiration in arctic grasses: Field light intensity and temperature responses. *Arct. Alp. Res.* **5**, 239–251.

Tilman, D. (1987). Secondary succession and the pattern of plant dominance along experimental nitrogen gradients. *Ecol. Monogr.* **57**, 189–214.

Tissue, D. T., and Oechel, W. C. (1987). Response of *Eriophorum vaginatum* to elevated CO_2 and temperature in Alaskan tussock tundra. *Ecology* **68**, 401–410.

Ulrich, A., and Gersper, P. L. (1978). Plant nutrient limitations of tundra plant growth. *In* "Vegetation and Production Ecology of an Alaskan Arctic Tundra" (L. L. Tieszen, ed.), pp. 457–481. Springer-Verlag, New York.

Walker, H. J. (1973). Morphology of the North Slope. *Arct. Inst. N. Am. Tech. Pap.* **25**, 49–92.

Wein, R. W. (1976). Frequency and characteristics of arctic tundra fires. *Arctic* **29**, 213–222.

Wein, R. W., and Bliss, L. C. (1973). Changes in arctic *Eriophorum* tussock communities following fire. *Ecology* **54**, 845–852.

Part II

Carbon Balance

7

Effects of Global Change on the Carbon Balance of Arctic Plants and Ecosystems

Walter C. Oechel and W. D. Billings

I. Introduction

Because of the large carbon stocks present in northern soils and the presumed sensitivity of soil carbon accumulation or loss to climate change, northern ecosystems may be particularly important to global carbon balance in the future (Miller *et al.*, 1983; Billings, 1987). Between 250 and 455 petagrams (1 Pg = 10^{15} g) of carbon are present in the permafrost and seasonally thawed soil layers (Miller *et al.*, 1983; Post *et al.*, 1985; Gorham, 1991). This amount is about one-third the total world pool of soil carbon (Post *et al.*, 1985). Warmer soils could deepen the active layer and lead to thermokarst erosion and the eventual loss of permafrost over much of the Arctic and the boreal forest.

These changes could in turn alter arctic hydrology, drying the upper soil layers and increasing decomposition rates (Chap. 3, 15). As a result, much of the carbon now stored in the active soil layer and permafrost could be released to the atmosphere, thereby increasing CO_2 emissions and exacerbating CO_2-induced warming (Billings *et al.*, 1982). Alternatively, elevated atmospheric CO_2 and changed nutrient availabilities could change plant communities and vegetation. New communities might be taller and have higher rates of primary productivity than does extant vegetation. The net result could be higher primary productivity, increased carbon storage in plant biomass, and a negative feedback on global atmospheric CO_2.

Arctic and boreal forest ecosystems are unique in their potentially positive and negative response to elevated CO_2 and associated climate change (Manabe and Stouffer, 1979). Few other ecosystems have the capacity for massive, continuing, long-term carbon accumulation that permafrost-dominated northern ecosystems do, and few systems are as sensitive to global warming. The uncertain effects of global change on arctic and boreal forest carbon balance make the study of northern ecosystem response a key to understanding and predicting future global atmospheric CO_2 patterns. Focusing on processes likely to change with a doubling of CO_2 over the next 50–60 years, this chapter describes the major controls on carbon cycling in arctic ecosystems and the likely effects of elevated atmospheric CO_2 and concomitant climate change on carbon storage.

II. Current Net Ecosystem Carbon Storage and Flux

Gorham (1991) estimates that, overall, boreal and subarctic peatlands (areas with a soil peat layer of 40 cm or more) contain about 455 Pg of carbon in the soil and have accumulated carbon since the last glaciation at the rate of 0.096 Pg per year. He estimates current accumulation to be slower, about 0.076 Pg yr^{-1}, based on overall peat accumulation of 0.5 mm yr^{-1}, peatlands of 3.3×10^{12} m^2, and a peat carbon content of 58×10^3 g m^{-3}. Carbon pools and rates of carbon accumulation vary, of course, depending on vegetation type and environmental conditions.

More than 90% of the carbon in arctic ecosystems is located in soils, with even higher percentages (98%) in soils of northern peatlands. In upland boreal forest, in contrast, only about 55% of the ecosystem carbon is found in the soil (Miller, 1981; Table I). Not only is the proportion of soil carbon substantial, but so are the absolute amounts. Arctic tundra has 55 Pg of carbon stored as soil organic matter in the A horizon, compared with 87.5 Pg in nonpeatland boreal forest and 122 Pg in forest peatlands (see Table I) and Gorham's (1991) estimates of 455 Pg C are considerably higher. Tussock and wet sedge tundra soils account for the bulk of circumpolar tundra carbon stores because they have large amounts of carbon per square meter and cover large areas (Miller *et al.*, 1983; Table I). Although per unit area, carbon storage in

Table I Areal Distribution, Productivity, and Carbon Accumulation of Tundra and Boreal Forest Ecosystems[a]

	Tundra							Boreal forest				
	Polar desert	Semi-desert	Wet sedge	Tussock	Low shrub	Tall shrub	Total	Northern	Middle	Southern	Peat-lands	Total
Area (× 10^6 km^2)	0.80	1.50	1.00	0.90	1.28	0.23	5.71	3.34	3.57	3.18	1.1	11.19
Standing crop (kg C m^{-2})												
Biomass (above- & belowground)[b]	0.002	0.29	0.95	3.33	0.77	2.61		3.6	6.8	10.9	2.7	
Dead organic matter[b]	0.091	7.2	13.4	29.0	3.8	0.4		10.8	7.5	7.7	110.9	
Total	0.093	7.49	14.35	32.33	4.57	3.01		14.4	14.3	18.6	113.6	
World total (× 10^{15} g)												
Biomass	0.00	0.43	0.95	3.00	0.98	0.60	5.84	11.9	23.3	34.7	3.0	74.10
Dead organic matter[b]	0.07	10.8	13.4	26.1	4.86	0.09	55.13	36.2	26.8	24.5	122	209.50
Total	0.07	11.23	14.35	29.10	5.85	0.69	61.06	48.1	51.1	59.2	125	283.40
Net photosynthesis (g C m^{-2} yr^{-1})	2	139	741	2163	431	965		794	1603	2391		
Growth (g C m^{-2} yr^{-1})	1	19	144	114	176	355		107	450	317		
Net production												
Region (g C m^{-2} yr^{-1})	1	14	108	90	135	315		90	340	250		
World (× 10^{15} Pg C yr^{-1})	<0.001	0.021	0.108	0.081	0.173	0.072	0.46	0.29	1.22	0.80		2.31
Current accumulation rate (g C m^{-2} yr^{-1})	0	4	27	23	10	29		22	40	10		
Accumulation times (years)[c]		1800	496	1261	380	138		491	188	770		

[a] Adapted from Miller et al. (1983) with permission from the Society for Computer Simulation.

[b] These values include only A horizons and organic horizons. Enmixed organic material is probably significant in wet sedge and tussock tundra to about 20% of the stated values. Thus, 7.9 × 10^{15} Pg should be added to the world tundra total, bringing it to 68.96 × 10^{15} Pg. Organic carbon frozen in permafrost is not included.

[c] Standing crop of dead organic matter/current accumulation rate.

polar semideserts is only about one-half that in the wet sedge tundra, the greater extent of these semideserts results in carbon stores approaching those of wet sedge tundra. The remaining arctic ecosystems—polar desert and shrub tundra—account for only about 10% of arctic soil carbon.

For these large stores of soil organic matter to have accumulated in northern ecosystems, production must have exceeded decomposition at some time in the past. Recent estimates (Post, 1990; Gorham, 1991) indicate that northern ecosystems still constitute a small net sink for atmospheric carbon (about $0.1-0.3$ Pg yr^{-1}); current accumulation rates, however, are difficult to assess. Because the rates vary with conditions and ecosystem type, soil carbon accumulation is positive in some areas and negative in others. The overall balance is still uncertain.

A. Wet Coastal Tundra

Many studies have indicated that wet and moist tundras and northern bogs may be net carbon sinks for atmospheric CO_2 under present conditions (Miller, 1981; Post, 1990). At the current accumulation rates for wet coastal tundra (about 27 g C m^{-2} yr^{-1}), this soil carbon pool of 14 kg m^{-2} would have required about 500 years to develop (see Table I). Rates of soil decomposition and accumulation change as the soil organic layer develops, however, and decomposition slows considerably as organic matter is buried more deeply. These factors, combined with the high degree of spatial heterogeneity in the content of soil organic matter and the small database, make the available estimates only rough approximations at best.

Chapin *et al.* (1980) report higher carbon accumulation rates (109 g C m^{-2} yr^{-1}) for the upper 20 cm of soil and, therefore, less time required to accumulate current soil stocks. Gersper *et al.* (1980) estimate soil carbon at 34 kg m^{-2} for the upper 35 cm of wet sedge tundra at Barrow, Alaska. This estimate may be conservative, for the bottom sample, taken at the permafrost table, nevertheless contained 44 kg C m^{-2}. It is therefore possible that additional carbon was entombed in permafrost that had formed since the surface began accumulating carbon. At a net accumulation rate of 109 g m^{-2} yr^{-1} (a reasonable assumption since decomposition in the deep, cold, anaerobic soil layers is likely to be very slow), the belowground carbon pool could have taken approximately 310 years to accumulate; this estimate is comparable to those of Miller *et al.* (1983; see Table I). Given the time spans of the thaw-lake cycle (Billings and Peterson, 1980), 300–500 years for carbon accumulation seems realistic.

These rates of carbon accumulation are much higher than those obtained from ^{14}C dating of peat profiles in arctic Alaska, which indicate accumulation rates of 1.4 to 7.9 g m^{-2} g^{-1} (Marion and Oechel, unpubl. data) to 7.4 to 18.3 g C m^{-2} yr^{-1} (Schell, unpubl. data). The peat profile at Prudhoe Bay contains 25.6 kg C m^{-2} to 6000 yrs B.P. which yields a mean accumulation rate of 4.3 g C m^{-2} yr^{-1}, and in the absence of thaw lake activity, peat is much older and the period of peat accumulation much shorter than those assumed in the above calculations (6000 versus <500 yrs).

The discrepancy between estimates of rates of carbon accumulation made by analysis of peat cores and from direct measurement of CO_2 flux may be due to differences in time scales inherent in the measurements. ^{14}C peat profile information integrates the rates of carbon accumulation over centuries, whereas carbon flux measurements are based on measurements made over one or a few growing seasons.

Wet coastal tundra ecosystems thus have the potential for long-term carbon sequestering in the soil. As the permafrost table moves upward because of the insulating effects of the graminoid–moss vegetation and its accumulating peat, decomposer activity decreases dramatically below depths of 25–30 cm, conserving soil carbon indefinitely at depth and in the permafrost. As Brown (1969) has illustrated, buried peat horizons are commonly found embedded deep in the permafrost in the coastal tundra at Barrow. Such burial is due either to a combination of thaw-lake cycle activity, wind deposition, solifluction, and frost heaving or to the independent influence of any one of these factors. Organic horizons do not remain near or on the tundra soil surface more than 1500 or 2000 years before these processes bury them. Radiocarbon dates as old as 10,600 B.P. at a depth of 3 m suggest a large reservoir of old soil carbon, with a cycling rate to the atmosphere that is very slow despite superficial melting at certain stages in the thaw-lake cycle.

Although terrestrial surfaces seem to be accumulating carbon, carbon is nevertheless being exported to aquatic ecosystems. Tundra ponds and streams release CO_2 to the atmosphere, and peat is exported to the oceans (Hobbie, 1980). The net carbon balance of the land alone—including export of organic matter to streams, ponds, and marine ecosystems—is thus uncertain. If one accounts for the material exported to aquatic systems, which may be substantial, then the overall carbon budget of the wet coastal tundra may be roughly in balance (Kling *et al.*, 1991). Paleontological evidence indicates that the vegetation in the wet coastal tundra from Barrow to Prudhoe Bay has changed little during the last 5,000–10,000 years (Chap. 5). This evidence suggests that past tundra ecosystems were accumulating carbon and serving as a carbon sink during the Holocene. Coastal areas appear to be continuing this process. However, little is known about past carbon export to aquatic systems; the overall carbon balance for coastal areas may actually have been negative, despite evidence of carbon accumulation on land. Indeed, aquatic systems today may well be a source of carbon to the atmosphere, for some of the terrestrial carbon coming into lakes and streams (Billings and Peterson, in press) is consumed and released to the atmosphere as CO_2 and other carbon-containing gases (Schell, 1983; Kling *et al.* 1991)

B. Tussock Tundra

The current carbon balance of tussock tundra is more problematic; recent changes in climate may have significantly altered carbon storage in this ecosystem type. Alaska's tussock tundra is better drained than wet coastal tundra and thus has improved soil aeration, higher overall decomposition rates, and less carbon storage. According to the calculations of Miller *et al.* (1983) for net photosynthesis, net plant production, soil respiration (excluding root respiration), and net carbon accumulation rate, it would take 1261 years to acquire the 29 kg C m^{-2} contained in the A horizon (Table I). Miller *et al.* (1984) estimated net ecosystem accumulation rates in a number of tussock tundra communities to be higher, ranging from 111 g dry matter m^{-2} yr^{-1} at Toolik Lake, Alaska, to 156 g dry matter m^{-2} yr^{-1} at Cape Thompson.

Recent information, however, indicates that historic and long-term carbon accumulation rates may have decreased in the last few decades at Toolik Lake and possibly elsewhere (Oechel and Riechers, 1987). Direct field measurements of net ecosystem carbon flux over four growing seasons suggest that rather than accumulating carbon at 23 g m^{-2} yr^{-1} or faster, as previously calculated (Miller, 1981; Miller *et al.*, 1983, 1984), the tussock tundra in this region is actually losing carbon at 180–360 g m^{-2} yr^{-1} (Oechel and Reichers, 1987; Grulke *et al.*, 1990; Oechel *et al.*, unpubl. data). These results are consistent with recent measurements of carbon flux along a latitudinal transect from Toolik Lake to Prudhoe Bay (Oechel *et al.*, unpubl. data) and with conclusions drawn from analyses of atmospheric CO_2 concentrations and ocean fluxes that point to a high-latitude terrestrial (arctic or taiga) source of CO_2 to the atmosphere (Tans *et al.*, 1990).

Temperature profiles of bore holes in the permafrost on Alaska's North Slope indicate that air temperatures have increased 2–4° C in the last century, possibly within the last few decades (Lachenbruch and Marshall, 1986; Lachenbruch *et al.*, 1988). Canadian weather records show similar increases in surface air temperatures for certain regions, including northwest Canada adjacent to Alaska (see Chap. 2). Thus warming of surface layers over the last few decades may be related to the apparently recent loss of carbon from tussock tundra ecosystems. At this time, neither the regional extent of carbon loss, nor the reduction in carbon accumulation rates is the existence or magnitude of recent corresponding changes at the coast.

If past peat accumulation in tussock tundra was on the order of 30 g C m^{-2} yr^{-1}, 1000 years' accumulation would equal 30 kg C m^{-2}, or about the current soil stores. If carbon is now being lost at 200 g m^{-2} yr^{-1}, this amount could be lost over 15 years. If this average rate applied over all tussock tundra, 0.2 Pg C yr^{-1} would be lost to the atmosphere worldwide. An additional 0.04 Pg C yr^{-1} might be lost from the terrestrial tundra to the atmosphere through arctic lakes and streams (Kling *et al.*, 1991). At present, the regional pattern of carbon fluxes is still poorly understood and requires further investigation.

C. Taiga

Little information exists on whole-system carbon budgets of taiga vegetation. Some data are available on soil respiration (Moore, 1987), rates of peat accumulation (Armentano and Menges, 1986; Ovenden, 1980), and primary productivity (Van Cleve *et al.*, 1983b). Flux rates of CO_2 from boreal forest soils average about 165 g C m^{-2} yr^{-1} (Moore, 1987). The seasonal pattern of atmo-

Table II Net Primary Production of Two Principal Vegetation Types in the Taiga Bog at Fairbanks, Alaska[a]

Vegetation	Dry weight $(g\ m^{-2}\ yr^{-1})$	Carbon content $(g\ m^{-2}\ yr^{-1})$
Bog forest		
Black spruce leaves and stems	77.3	44.8
Coarse tree roots	10.6	6.2
Fine tree roots	21.0	12.2
Shrubs and herbs	74.2	43.0
Sphagnum	20.3	9.1
Other mosses	4.0	1.8
Total net production		+ 117.1
Carbon loss from soil		− 110.0
Ecosystem carbon balance		+ 7.1
Carex mat		
Herbs	111.7	64.8
Sphagnum	194.9	87.7
Total net production	+ 306.6	+ 152.5
Carbon loss from soil		− 60.4
Ecosystem carbon balance		+ 92.1

[a] Assembled from data of Luken (1984) and Mortensen (1983) by Billings (1987). Reprinted from Billings (1987) with permission from Pergamon Press.

spheric CO_2 concentrations in the Northern Hemisphere may partially reflect seasonal patterns of taiga photosynthesis and respiration. Increased productivity in the taiga, resulting from atmospheric increases in CO_2, may help explain the higher seasonal amplitude of atmospheric CO_2 recently recorded at Barrow and elsewhere (D'Arrigo *et al.*, 1987).

Using a mass balance approach, Luken (1984) measured net primary production in the 18-km^2 College Road taiga bog at Fairbanks, Alaska, for two of the bog's principal vegetation types—bog forest and successional sedge (*Carex*) mat vegetation (Table II). Soil CO_2 loss from microbial activity and root respiration was also measured through an entire season (Luken and Billings, 1985). The results show a positive carbon balance for the *Carex* mat that was 13 times that of the old black spruce bog forest. This disparity reflects the wet, nutrient-rich habitat at the edge of a thaw pond, in contrast to the cold, relatively nutrient-poor and drier conditions in the forest, which permits high productivity and carbon capture by *Sphagnum* and the graminoids.

These limited data for the taiga nevertheless agree well with those from the wet coastal tundra, particularly in the positive carbon balances of the graminoid communities at the two sites, which share a dominant species, *Carex aquatilis*. Perhaps it is not surprising that carbon flux in the "climax" bog forest is near a steady state, whereas the successional sedge community has a relatively high rate of net carbon capture.

In order to help evaluate whether the taiga either has been or is now a carbon source or sink, we have plotted past rates of carbon accumulation against depth, age, and postulated vegetations (as interpreted from the fossil data) (Fig. 1). The fastest rates of carbon accumulation in the peat (61 g C m^{-2} yr^{-1}) were associated with spruce–birch forest or woodland. These forests and woodlands, dating to more than 3800 B.P., differed from today's spruce forest because *Sphagnum* was scarce. Billings (1987) hypothesizes that as *Sphagnum* begins to take control of the ground cover in a bog forest, carbon storage in that forest drops to 25–50% of peak. Such a drop may have occurred in the last 1000 years, since *Sphagnum* is now dominant on the bog forest floor.

The slowest carbon accumulation in the bog's past (11–14 g m^{-2} yr^{-1}) appears to have occurred during the dominance of *Eriophorum vaginatum* in patches of tussock semitundra. Not all sedge meadows are as slow in storing carbon, as noted above for *Carex aquatilis*. But such *Carex* vegetation lasts only a short time near thermokarst ponds and streams; it is soon replaced by other vegetation such as *Andromeda* meadow and open groves of *Larix* (Billings, 1987).

During the last 4800 years, spruce–birch forests appear to have occupied the College Road site less than half the time and in two periods (see Fig. 1). The first was from about 4600 B.P. to 3500 B.P., when the forest was fast-growing; the second lasted from about 800 B.P. to the present, when the site

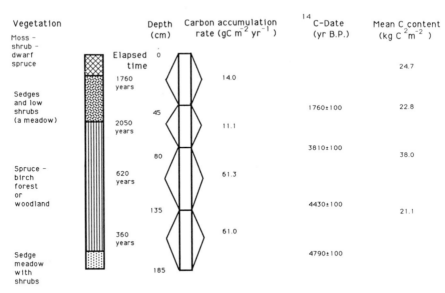

Figure 1 Vegetational changes, carbon accumulation rates, and mean carbon contents between radiation dates for the last 4800 years in the bog at Fairbanks, Alaska, as interpreted from the deep core at the College Road site (cf. Chap. 5). Total C, 106.6K. [Adapted from Billings (1987) with permission from Pergamon Press.]

has been occupied by a slow-growing spruce woodland. The vegetation between these episodes was primarily sedge–low shrub meadow, which lasted for about 2500 years and produced little peat. This meadow was probably *Eriophorum vaginatum* tussock, and although the peat produced was not very thick, it was dense and carbon-rich.

After reviewing the paleontological, macrofossil, and carbon balance evidence for central Alaska, in particular for the Fairbanks bog, we conclude that this region has been a carbon sink throughout the last 5000 years. It remains a carbon sink but with different carbon balances for the various vegetational communities resulting from the thermokarst and fire cycles inherent in the subarctic environment. Rates of carbon that accumulate for the Fairbanks bog are similar to the $10–35$ $gC\,m^{-1}\,yr^{-1}$ reported for Canadian subarctic boreal forest and temperate peatlands (Ovenden, 1990).

III. Effects of Global Change on Photosynthesis and Net Primary Productivity

Expected increases in atmospheric CO_2 and temperature could directly increase photosynthetic rates (Fig. 2). In general, short-term photosynthetic rates are limited by CO_2 supply, and higher CO_2 concentrations could substantially increase photosynthesis (Oechel and Strain, 1985). Photosynthesis is also constrained by temperature, and in the Arctic the photosynthetic temperature optimum is usually above measured air temperatures (Miller *et al.*, 1976; Oechel, 1976; Stoner *et al.*, 1978). Rising air temperatures could thus translate to greater photosynthesis. Growth is limited by temperature in arctic and boreal forest ecosystems as well (Van Cleve *et al.* 1983a; Kummerow and Ellis, 1984). Warmer temperatures can increase the growth of roots and shoots, even without an increase in photosynthetic rates.

Since total canopy photosynthesis is the product of canopy area and photosynthetic rate, changes in leaf area can alter or totally offset the effect on primary production of changes in photosynthetic rate. Whole-plant carbon balance includes carbon losses to various sinks, including respiration. Changes in allocation patterns to leaves, shoots, and roots and alterations in amounts and rates of respiration can offset or even completely counterbalance gains made through increased photosynthetic rates.

Furthermore, any increases in photosynthetic rates because of higher atmospheric CO_2 concentration or other changes in climate may not necessarily be realized. In the Arctic, where plant carbohydrate supply may outstrip intrinsic sink activity, changes in environmental factors may have little long-term effect on photosynthetic rates unless they also alter sink activity or plant growth potential. Because of photosynthetic feedback, including end-product inhibition of photosynthesis, factors that increase the growth potential and sink activity of plants should result in higher photosynthetic rates. Alteration in photosynthetic

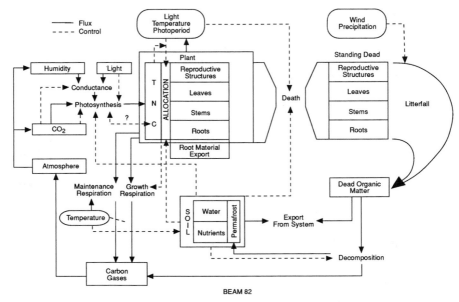

Figure 2   Major fluxes and controls important in predicting plant growth response and ecosystem response to elevated CO_2. TNC denotes total nonstructural carbohydrates. [Modified from Strain and Bazzaz (1983) with permission from Westview Press, Inc.]

potential is unlikely to have a long-lasting impact on productivity unless growth potential is also affected, for example, through greater nutrient supply or higher temperatures (Shaver and Chapin, 1980; Chapin and Shaver, 1985; Oechel and Strain, 1985; Sage and Sharkey, 1987; Tissue and Oechel, 1987).

A. Photosynthesis

1. Elevated Temperature

Daily temperatures in the Arctic are suboptimal for photosynthesis in vascular plants (Miller *et al.*, 1976; Limbach *et al.*, 1982; Chap. 8) and mosses (Oechel, 1976; Oechel and Sveinbjörnsson, 1978; Stoner *et al.*, 1978; Sveinbjörnsson and Oechel, 1981, 1983). However, because of the broad temperature range for photosynthesis suboptimal temperatures in nature reduce photosynthesis below maximal rates by less than 15% and therefore cause a fairly minor reduction in photosynthesis both in mosses (Oechel, 1976; Stoner *et al.*, 1978) and vascular plants (Miller *et al.*, 1976). This broad temperature range in most species, photosynthetic acclimation (Mooney and Billings, 1961; Tieszen, 1978; Limbach *et al.*, 1982; Chapin and Oechel, 1983; Sveinbjörnsson and Oechel, 1983), the actual frequency distribution of leaf temperatures, and the environmental conditions occurring at various temperatures minimize the effective limitation of temperature on photosynthesis (Miller *et al.*, 1976; Stoner *et al.*, 1978).

Photosynthetic acclimation to temperature can occur over days to weeks (Oechel, 1976), thereby minimizing the effect of elevated temperature on canopy photosynthesis. Ecotypic differentiation further tunes the photosynthetic temperature response to the local environment (Mooney and Billings, 1961; Hicklenton and Oechel, 1976; Chapin and Oechel, 1983; Sveinbjörnsson and Oechel, 1983), potentially limiting the longer-term response to otherwise suboptimal temperatures. Moreover, there is growing evidence that photosynthesis and growth in arctic plants are limited more by nutrient availability than by photosynthate production (Chapin and Shaver, 1985; Oberbauer *et al.*, 1986a; Tissue and Oechel, 1987; Chapin, 1991) and that photosynthetic capacity may exceed need (Tissue and Oechel, 1987).

For these reasons, warmer temperatures may have little direct effect on photosynthetic rates. Rather, the main effects may be indirect, through increased mineralization rates and nutrient availability (Bigger and Oechel, 1982; Post, 1990; Chap. 13) and greater sink activity and sink strength (Lawrence and Oechel, 1983a,b; Kummerow and Ellis, 1984), which can relieve end-product inhibition or other negative feedbacks on photosynthesis (Lawrence and Oechel, 1983b; Tissue and Oechel, 1987; Chapin, 1991; see Fig. 2), thus allowing higher photosynthetic rates and, through the development of greater leaf area, greater total canopy photosynthesis. If leaf areas of the community do increase, or if current species are replaced by more photosynthetically active species, whole-ecosystem carbon uptake could rise.

2. Elevated Atmospheric CO_2

Currently available data indicate that elevated atmospheric CO_2 by itself will provide little long-term direct stimulation of net photosynthesis in arctic ecosystems (Billings *et al.*, 1982, 1983, 1984; Oberbauer *et al.*, 1986a; Tissue and Oechel, 1987; Grolke *et al.*, 1990). The response of photosynthesis to elevated CO_2 appears to depend largely on water and nutrient availability. In general, plants respond most strongly to elevated CO_2 when either nutrients are abundant (Oechel and Strain, 1985), temperatures are warm (Kimball *et al.*, 1986; Idso *et al.*, 1987), or water is limited (Oechel and Strain, 1985). These conditions are not common in most regions of the tundra.

For these reasons, initial increases in photosynthesis of *Eriophorum vaginatum*, the dominant species in tussock tundra, are not sustained under elevated CO_2. Following a brief but dramatic increase in photosynthetic rates, individuals exposed to elevated CO_2 adjust so that within three weeks the rate of photosynthesis at twice ambient CO_2 levels does not differ from that at ambient CO_2 (Fig. 3). There is complete homeostatic adjustment of photosynthesis to the new, elevated CO_2 levels (Tissue and Oechel, 1987). This pattern presumably reflects the carbohydrate sufficiency in these and many other arctic plants (Shaver and Chapin, 1980; Chapin *et al.*, 1986; Tissue and Oechel,

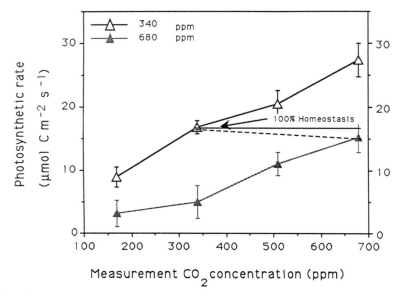

Figure 3   Homeostatic adjustment of photosynthesis in *Eriophorum vaginatum* after three weeks' exposure in the field to ambient (340 ppm) and elevated (680 ppm) atmospheric CO_2. The dashed line connecting actual photosynthetic rates at ambient and elevated CO_2 shows nearly perfect homeostatic adjustment (indicated by solid horizontal line) to the higher CO_2 levels. [Adapted from Tissue and Oechel (1987) with permission from the Ecological Society of America.]

1987). Given the adequacy of carbohydrate supply relative to other resources required for growth, these sedges appear to reduce photosynthetic capacity to maintain a more or less constant photosynthetic rate in the face of increased atmospheric CO_2 supply.

Similar results have been found in controlled-environment studies of arctic plants (Oberbauer *et al.*, 1986a) and on intact cores of wet tundra vegetation (Billings, 1987), where elevated atmospheric CO_2 stimulates photosynthesis relatively little. Even with adequate nutrients, arctic plant species respond to elevated CO_2 to only a limited degree (Oberbauer *et al.*, 1986a; Fig. 4), suggesting that they have inherent physiological limitations to their ability to use additional CO_2.

Among species, considerable variation exists in the pattern of interactions between nutrient availability and photosynthesis. In most arctic species, however, the magnitude of the postacclimation photosynthetic response to elevated CO_2 is negligible over the range of nutrient availabilities expected in the field. In only one species (*Carex bigelowii*) of the six studied by Oberbauer *et al.* (1986a) is there a sustained CO_2-induced enhancement of photosynthesis of more than 20%. *C. bigelowii* showed a >60% CO_2-induced enhancement of photosynthesis at the higher nutrient treatment level (see Fig. 4); the reason

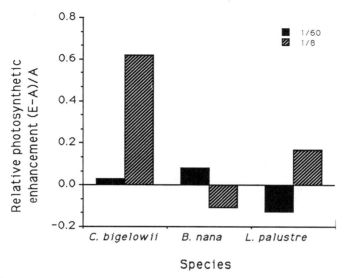

Figure 4 Relative photosynthetic enhancement, (E-A)/A, caused by a doubling of atmospheric CO_2 content, in three arctic plant species (*Carex bigelowii, Betula nana,* and *Ledum palustre*), where E=photosynthetic rate at elevated CO_2 (675 ppm) and A=photosynthetic rate ambient CO_2 (350 ppm). Plants were grown at low (1/60-strength) and moderate (1/8-strength) nutrient solutions. Three-way analysis of variance indicated a significant effect of CO_2 on photosynthesis for *Betula* and *Ledum.* [Data from Oberbauer *et al.* (1986a) with permission from the National Research Council of Canada.]

for this response is unknown. In simulations, however, *C. bigelowii* is one of the few arctic tussock tundra species whose growth is more limited by carbohydrate than by nutrient supply (Miller *et al.,* 1984; Oechel and Strain, 1985; Oechel and Blake-Jacobson, unpubl. data), and in field manipulations of elevated CO_2, it is one of the only vascular plant species to show increased growth with elevated CO_2 (Robberecht and Oechel, unpubl. data). In other cases, postacclimation photosynthetic rate decreases at elevated atmospheric CO_2. The reasons for this decrease are also unknown, but it may be due to nutrient limitation induced, at least in part, by dilution of essential nutrients after an initial simulation of growth and photosynthesis (Bigger and Oechel, 1982), end-product inhibition of photosynthesis (Oechel and Strain, 1985; Strain, 1987; Evans, 1989), or changes in photosynthetic capacity and enzyme systems at elevated CO_2 (Sage and Sharkey, 1987).

Nonetheless, if conditions change so as to increase temperature, resource availability, or sink activity, plants may be better able to respond to greater CO_2 availability. Synergistic effects, where other factors change as CO_2 concentration increases, may allow plants to respond to the combination of elevated atmospheric CO_2 and otherwise new environmental conditions. Therefore, even if plants do not respond to elevated CO_2 under current field conditions, they may do so under future warmer temperatures and concomitant environmental change.

No *in situ* manipulations or experiments have tested the ability of taiga trees to respond photosynthetically to long-term elevation of atmospheric CO_2, although the warmer temperatures and drier soils found in the upland taiga will be conducive to such a response. Also recent increases in atmospheric CO_2 might have already stimulated photosynthesis in the taiga (D'Arrigo *et al.*, 1987), but there are no direct measurements available for confirmation.

B. Net Primary Productivity

The factors controlling growth may differ from and act independently of the factors controlling photosynthesis. Environmental factors can stimulate growth directly rather than acting first through an increase in photosynthetic rate. Increased photosynthetic rate may follow as a result of greater sink activity. If growth is stimulated, whole-plant photosynthesis may increase because of greater leaf area development even if photosynthetic rates per unit of leaf area do not. Higher temperatures or increased nutrient availability are thus expected to increase growth rate and net primary productivity, even if the rate of photosynthesis remains unaffected.

1. Temperature

Temperature limitations on productivity in the Arctic are complex. Temperatures above normal ambient levels tend to increase the growth of arctic plants, partly because temperature directly limits the growth of at least some species (Kummerow and Ellis, 1984; Chapin and Shaver, 1985). Warmer temperatures act primarily by directly increasing growth (sink strength and sink activity). At moderately higher temperatures, growth, maintenance respiration, and growth respiration may all increase. Photosynthetic rate or total canopy photosynthesis generally also rises to compensate so that a net increase occurs in primary productivity. In addition elevated temperatures also increase nutrient availability through higher mineralization rates, a deeper active soil layer with more rooting volume, increased nutrient mobility, and greater root activity for nutrient absorption. Because of changes in the permafrost table, thermokarst erosion, and subsequent changes in drainage, higher temperatures can increase water movement and subsequent mass transport of nutrients to plant roots (Marion and Everett, 1989), thereby stimulating growth (Chapin *et al.*, 1988; Hastings *et al.*, 1989; Oberbauer *et al.*, 1989).

Through such indirect effects, the net result of increased temperature in the Arctic may be more canopy photosynthesis and higher net primary productivity. However, this increased productivity will likely be realized only in the short term. As temperatures rise, arctic species will probably be replaced by temperate and boreal species that should be more competitive under the new conditions. Productivity and ecosystem dynamics will then reflect the interaction of environment with these new communities.

2. Carbon Dioxide

Carbon dioxide alone causes little long-term stimulation of growth and primary productivity in the field (Oechel, 1987; Tissue and Oechel, 1987; Grulke *et al.*, 1990). Phytotron and growth-chamber studies, indicate a major increase in biomass and primary productivity from increases in nutrient availability (Billings *et al.*, 1984; Oberbauer *et al.*, 1986a) and from temperature (Kummerow and Ellis, 1984) but not from elevated atmospheric CO_2 (Billings *et al.*, 1984; Oberbauer *et al.*, 1986a; Fig. 5). Moreover, these studies, which found little statistically significant interaction between nutrient treatment and CO_2 concentration, imply that photosynthesis and carbohydrate supply are sufficient to match resource availability and growth potential over the resource ranges measured and that over this range productivity is more limited by factors, other than photosynthetic rate (for example, nutrient supply).

Nevertheless, elevated CO_2 may effect productivity differently in the future. Higher air and soil temperatures, drier soils, and faster decomposition may increase nutrient availability and growth potential. Consequently, future increases in CO_2 availability may stimulate primary productivity more than has been observed under current environmental conditions and community composition.

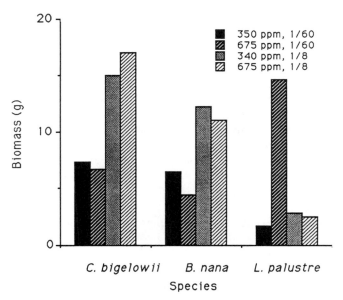

Figure 5   Biomass produced after three months' growth at two CO_2 levels (350 and 675 ppm CO_2) and two nutrient treatments (1/60- or 1/8-strength nutrient solution). [Adapted from Oberbauer *et al.* (1986a) with permission from the National Research Council of Canada.]

3. Interactions

Environmental factors will not change in isolation. Higher temperatures, drier soils, greater nutrient availability, and increased atmospheric CO_2 should, acting in concert, increase net primary productivity in much of the wet and moist Arctic. Differential effects on species will affect reproduction and competition, thereby altering species composition. Changes in species composition will have other effects on growth rates and net productivity, effects reflecting the new communities and their interactions with the new environment. In addition, vegetation boundaries are likely to shift gradually, and these shifts will affect patterns of biomass accumulation and distribution. As temperatures warm and soil active layers deepen, we can expect taiga tree species to move northward. Where they do, aboveground biomass storage and net primary productivity will increase. In other areas, increasing aridity may give rise to communities with lower primary productivity.

IV. Expected Effects of Global Change on Net Ecosystem Carbon Flux

The net ecosystem carbon budget includes net primary productivity, soil respiration, and carbon export of litter and other dissolved and particulate organic matter. Climate–vegetation relationships derived from paleontological records and observations along environmental gradients may provide important insight into the effect of climate change on vegetation composition and on soil and plant carbon content with climate change. But since atmospheric CO_2 concentration and air temperature affect the response of plants and ecosystems in concert with other environmental factors, predictions based on paleontological evidence are limited by the lack of information over the range of CO_2 concentrations and temperatures expected during the next century. Such predictions are further limited by the fact that past climatological changes have been relatively slow compared with the anticipated rate of climate change. Despite provocative projections and simulations (e.g. Emmanuel *et al.*, 1985a,b), one cannot predict ecosystem response to global change without manipulative experiments that include realistic rates of change in climate and CO_2.

A. Temperature

A strong correlation exists between temperature and rates of peat accumulation in the past as estimated from peat profiles. These profiles suggest that more carbon should accumulate with warmer temperatures, not less, as some researchers have predicted (Silvola and Hanski, 1979; Post, 1990). For example, because of warmer temperatures, peat formation was common in arctic Canada 8500 to 9000 years B.P. in areas where peat does not accumulate at present (Zoltai and Pollett, 1983). Based on these records, one would predict that bands of peat accumulation would shift northward with global warming

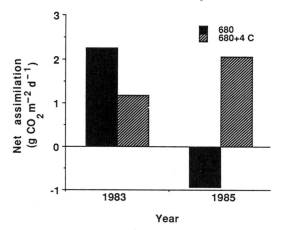

Figure 6   Daily net ecosystem carbon flux in tussock tundra at Toolik Lake, Alaska, under doubled atmospheric CO_2 (680 ppm) at ambient temperature (solid bars) and at elevated (ambient + 4° C) temperature (hatched bars). Positive values indicate net influx to the ecosystem; negative values indicate net flux to the atmosphere.

and that southern peatlands would dry and lose carbon to the atmosphere (Gorham, 1991).

1. Tundra

Experimental evidence is somewhat conflicting. In field experiments where atmospheric CO_2 was elevated and soil moisture was maintained at ambient levels, net ecosystem carbon influx to tussock tundra initially decreased at warmer temperatures (Oechel *et al.*, unpubl. data). After three years of adjustment to rising temperature, however, net influx was greater at elevated temperature than at ambient temperature (Fig. 6). This contradicts many published predictions for boreal forest and tundra (e.g., Moore, 1987; Post, 1990; Gorham, 1991); it corroborates predictions from simulation models (e.g., Miller *et al.*, 1983) and arctic soil and carbon accumulation rates (Miller, 1981; Zoltai and Pollett, 1983; Marion and Oechel, in prep.). With water level held constant, a 4° C rise in temperature increased net primary production more than it did soil-decomposition processes (Oberbauer *et al.*, 1986b; Oechel, 1987) and significantly raised net ecosystem carbon sequestering. The effect of elevated CO_2 and temperature was to shift the ecosystem from net carbon loss to the atmosphere to net carbon accumulation over the three years of measurement and manipulation.

In general, the stimulation of soil decomposition by elevated temperature may be more than offset by increased net primary productivity, partly because the wet and moist arctic tundra is nutrient-limited and because the carbon:nitrogen ratio is higher in plant material than it is in soil organic matter (Post, 1990). Any nutrients released through increased soil mineralization and taken up by plants might thus produce plant biomass more than equal to the soil organic matter mineralized. The net result would be increased carbon sequestering. In addition, elevated soil and air temperatures may stimulate plant sink activity and increase growth potential of the vegetation as described above.

In contrast to these experimental results, for microcosms from wet coastal tundra consisting of soil cores and intact vegetation in the phytotron, Billings *et al.* (1982, 1983, 1984) found that warming from 4° C (the mean July temperature at Barrow) to 8° C produced a 15–83% decrease in the rate of carbon accumulation at elevated temperature and a decrease in net ecosystem carbon storage over one simulated growing season. Significantly, this effect depended heavily on the position of the water table: The reduction in net carbon incorporation with increased temperature was relatively small (15%) when the water table was at the surface. When the microcosms were better drained and the water table was at −5 cm, an identical 4° C rise in temperature reduced carbon uptake by 73–83%. Respiration rate is apparently limited more by oxygen availability than by temperature in a saturated soil, and under these conditions, the potential impact of increasing temperature on soil respiration is also limited. Elevated temperature should thus have relatively little impact on soil respiration in wet tundra unless the temperature increase also lowers the water table.

The differences between the studies by Oechel *et al.* and Billings *et al.* may be due to differences in the ecosystems studied, in the experimental approach, or in the moisture content and water tables when the studies were done. One might also expect differences in response between studies on intact ecosystems and phytotron studies of transplanted cores. The net effect of any treatment on carbon balance results from an interaction between the soil and plant systems. The capacity of the plants in the microcosm study to respond to elevated temperature may have been limited by rooting volume, edge effects, or other factors.

2. Taiga

The effect of increasing temperature on soil respiration and net ecosystem carbon balance in tundra and taiga ecosystems may be similar in numerous ways. Studies of soil heating in the taiga of central Alaska and the wet tundra at Barrow revealed major changes in soil characteristics and processes with warming. In the taiga at elevated summer soil temperatures of 8–10° C, depth of thaw increased more than 100% to 115 cm and seasonal heat sum was increased by 100% to 1133 degree-days. This increase caused rapid oxidation of the soil organic matter in the forest floor, 20% of which was lost during the experiment (Hom, 1986; Van Cleve *et al.*, 1990). Nutrient availability also increased (Van Cleve *et al.*, 1983b), and the soil surface receded and slumped. The depth of soil available for root exploitation increased, as did plant production, photosynthesis, and nutrient content. Tissue nitrogen and phosphorus levels increased by 40 and 75%, respectively (Hom, 1986). Although the effect on whole-ecosystem carbon balance is unknown, soil heating had a positive effect on net primary productivity that tended to offset carbon losses through increased soil decomposition.

It should be borne in mind that although only temperature was manipulated in these studies, soil heating had indirect effects on depth of the active layer and water table (Van Cleve *et al.*, 1990). It is not yet known to what extent the increased decomposition rate was due to the direct effects of temperature as opposed to indirect effects on water table and thawed soil volume (see Sections III,B,1 and IV,B). Moreover, these results were in response to a step increase in soil temperature, with no corresponding increase in air temperature. Either chronic changes in temperature or new equilibrium conditions may produce different steady-state conditions. Instead of decomposition, elevated air temperature may favor increased primary productivity. Nevertheless, knowing the effect of temperature on net ecosystem carbon flux is critical to predicting future carbon stocks and flux rates.

B. Moisture

Rising temperatures in the Arctic can dry soils in several ways. Most climate models predict increasing soil aridity because of increased evapotranspiration and minimal change in precipitation (Chap. 2, 3). In addition, warming will likely lead either to greater thaw depth, the loss of permafrost, or to both. Deeper thaw will increase drainage and drying of the upper soil. For the reasons listed in Sections III,B,1, warmer, drier soils should thus lead not only to greater loss of carbon from the soil but also to increased nutrient availability and uptake and, therefore, to higher net primary productivity (Van Cleve *et al.*, 1983a, 1990; Hom, 1986).

Using intact cores of a wet tundra ecosystem, Billings *et al.* (1983) found that at either ambient or elevated atmospheric CO_2, a lowering of the water table increased soil decomposition more than it increased primary productivity. Net ecosystem carbon sequestering was greatly reduced under both CO_2 treatments, and the tundra core became a net source of CO_2 to the atmosphere rather than a net sink as it was at higher water tables (Fig. 7; Section IV,A,1).

Draining northern peatlands has been shown to convert them from a CO_2 sink to a CO_2 source (Bramryd, 1979, 1980). The flux of CO_2 from soils of Finnish peatlands may be tenfold higher after drainage than before (Silvola, 1986). Undisturbed peatlands can accumulate organic carbon at a rate of about 25 g m^{-2} yr^{-1}, but after drainage they lose carbon at a rate of about 150 g m^{-2} yr^{-1} (Silvola, 1986), and organic soils may subside at rates of 1–3 cm yr^{-1} (Armentano and Menges, 1986). About half of this subsidence appears to be due to oxidation of the organic matter to CO_2, which produces a CO_2 efflux from the soil of 200–700 g C m^{-2} yr^{-1} (Moore, 1987).

Over longer time scales, however, nutrient availability from increased mineralization rates may result in greater primary productivity and higher plant biomass (see Section IV,A,1). Since the carbon:nitrogen ratio of plant material

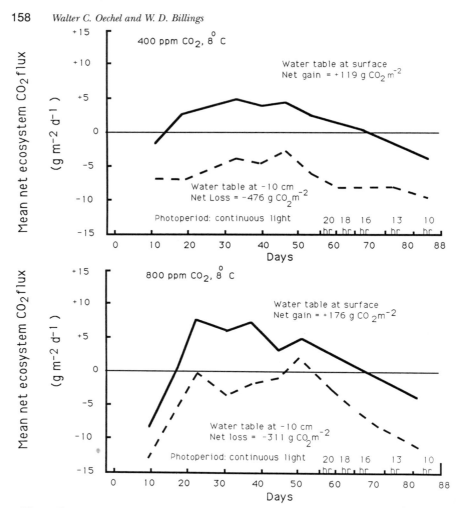

Figure 7   Daily mean net ecosystem CO_2 flux through a simulated wet tundra growing season at 8° C and two atmospheric CO_2 concentrations (400 ppm, top; 800 ppm, bottom) and two water table levels (surface, solid line; −10 cm, dashed line). Measurements are plotted for progressively shorter photoperiods starting with continuous light (left). Present-day July mean temperature at Barrow, Alaska, is 4° C; this July mean has been predicted to rise to 8° C when atmospheric CO_2 reaches approximately 800 ppm. [Adapted from Billings *et al.* (1983) with permission from Springer-Verlag.]

exceeds that of soil organic matter, an increase in soil mineralization and decomposition might further increase plant biomass, and in the medium term the system could become a sink rather than a source. Long-term net carbon sequestering would probably require conversion of tundra vegetation to shrub or forest species that have greater aboveground carbon storage potentials (see Section III,B,3).

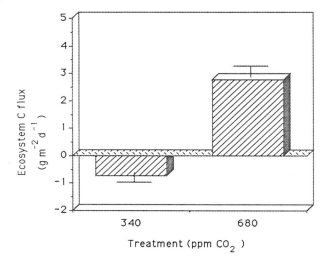

Figure 8 Net *in situ* ecosystem CO_2 flux of tussock tundra at Toolik Lake, Alaska, under ambient and elevated atmospheric CO_2 levels during the first season of treatment (16 June to 31 August 1987). [Adapted from Grulke *et al.* (1990) with permission from Springer-Verlag.]

C. Elevated CO_2

Because plants respond as integrated systems to multiple environmental controls, in the absence of change in other factors, one might expect minimal medium-term effects of elevated CO_2 on ecosystem CO_2 flux. Although in whole-ecosystem manipulations of tussock tundra at Toolik Lake, Alaska, for example, higher atmospheric CO_2 initially increased net ecosystem CO_2 uptake from the atmosphere (Fig. 8) resource availability did not permit this increase to sustain itself over the long term. In the first year of elevated CO_2 treatment, photosynthetic rates and net ecosystem carbon balance increased significantly. Early in the first growing season, elevated CO_2 changed the diurnal pattern from one of little positive CO_2 flux into the ecosystem to one in which the net carbon balance was positive throughout most of the 24-hour sunlit period (Grulke *et al.*, 1990; Fig. 9). Within the first season, however, the system acclimated rapidly to elevated CO_2. By late July of the first year, at 680 ppm CO_2, no stimulation of net ecosystem carbon flux occurred; homeostatic adjustment appeared complete. Later in the season, net ecosystem balance was again more positive in tundra exposed to higher CO_2 levels. Elevated CO_2 has been shown to delay senescence in plants (Oechel and Strain, 1985), and the higher late-season rates of carbon uptake in tussock tundra are thought to be due to a delay in the senescence of photosynthetic capacity in plants maintained at elevated CO_2 (Grulke *et al.*, 1990).

The within-season homeostatic adjustment of ecosystem CO_2 flux to elevated atmospheric CO_2 was confirmed using reciprocal treatments. When

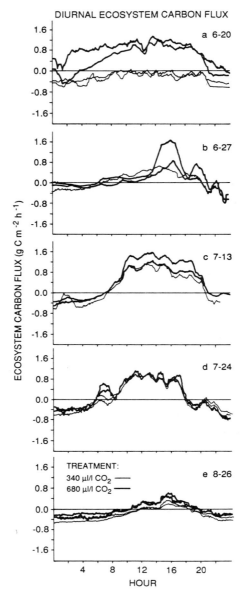

Figure 9   Seasonal ecosystem carbon flux at ambient and twice ambient atmospheric CO_2 concentration. The greatest stimulation in net ecosystem CO_2 flux was observed early in the season at the start of treatment. By midseason, significant acclimation of ecosystem CO_2 flux to elevated CO_2 had occurred. [Adapted from Grulke *et al.* (1990) with permission from Springer-Verlag.]

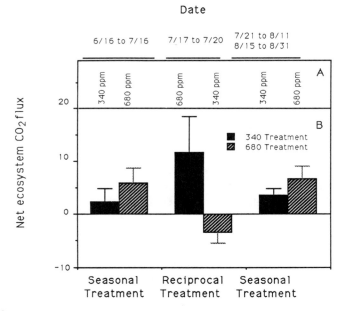

Figure 10   Homeostatic adjustment of ecosystem CO_2 flux to atmospheric CO_2 levels. Intact tussock tundra was exposed to ambient and elevated CO_2 treatments as indicated in (A). After four weeks (midseason, 7/17 to 7/20), the treatments were reversed to assess adjustment of each plot to the new CO_2 levels (B). Bars represent the means and standard errors for four days before, during, or immediately after the reciprocal treatments for each of two treatments. The differences in CO_2 flux between plots treated reciprocally and those treated and measured at the same CO_2 level represent homeostatic adjustment that occurred over the course of the treatment. Significantly different ecosystem flux rates occurred only immediately after reciprocal treatment. The differences between plots exposed to 340 ppm and those exposed to 680 ppm were significant in all measurements [Adapted from Grulke *et al.* (1990) with permission from Springer-Verlag.]

tundra that had been subjected to elevated CO_2 for four weeks was then subjected to ambient CO_2, carbon uptake rates fell below those of tundra maintained at ambient CO_2. Similarly, when tundra that had been exposed to ambient CO_2 was briefly subjected to elevated CO_2, carbon uptake rates were much greater than those of tundra exposed to prolonged elevated CO_2 (Fig. 10). These results demonstrate that ecosystem adjustment to elevated CO_2 occurs very quickly, within the first season of exposure.

Over several years, however, this stimulation of net ecosystem CO_2 uptake disappeared, so that by the third year, doubled atmospheric CO_2 had no effect on ecosystem carbon accumulation. This decrease resulted from a number of ecosystem-level adjustments, including decreased photosynthetic rates (see Section III,A,2), decreased leaf area (through either decreasing leaf production, decreasing leaf retention, or both), and altered shoot-to-root ratios.

Thus, in tussock tundra over the time scale observed, and in the absence of changes in other environmental variables, it appears that increasing atmospheric CO_2 will have little, if any, effect on ecosystem CO_2 flux. Nevertheless, several factors could change this outcome.

1. Time Scales

Carbon flux can vary considerably depending on the time scale considered (Fig. 11). After a week, increased atmospheric CO_2 increases net ecosystem carbon influx. Then, however, homeostatic adjustment of photosynthesis to the new CO_2 level rapidly decreases photosynthetic rates. For some species, such as *Eriophorum vaginatum*, complete homeostatic adjustment of photosynthesis occurs within three weeks (Tissue and Oechel, 1987). Other species respond more slowly or in other ways. Nevertheless, within three growing seasons, ecosystem flux adjusts to the new CO_2 level, and elevated CO_2 treatment and control fluxes are similar (Oechel, unpubl. data).

Over longer time scales, population changes may alter species composition, a change that would affect, and perhaps increase, whole-ecosystem CO_2 uptake. Plant species respond differently to elevated atmospheric CO_2 and climatic changes, resulting in differential growth, survival, and competition (Oberbauer *et al.*, 1986a; Bazzaz, 1990). In general, sexual reproduction

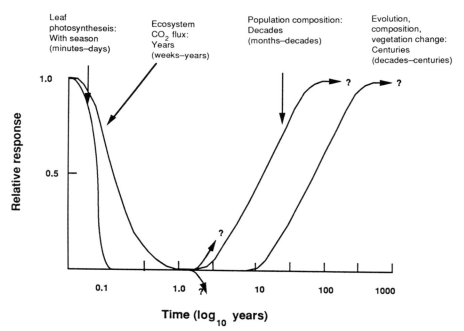

Figure 11   Hypothesized time scales for the effects of increased atmospheric CO_2 on key processes affecting net ecosystem carbon accumulation.

decreases and asexual reproduction increases in response to elevated CO_2. For example, in *Eriophorum vaginatum* grown at elevated CO_2, tillering can increase several hundred percent (Tissue and Oechel, 1987), as it does in certain temperate grasses (Sionit *et al.*, 1981). These reproductive and population characteristics are initiated within a few years, and over several decades they may affect community composition and, therefore, net ecosystem primary productivity and CO_2 flux. At still longer time scales, ecotypic differentiation and selection may produce populations that respond differently to the elevated CO_2, thereby altering net ecosystem carbon balance. Changing vegetation boundaries and distribution, such as replacement of tundra by boreal forest, will have major effects on net ecosystem primary productivity, carbon storage, and the distribution of carbon between above- and below-ground pools.

2. Vegetation Types

Response of net ecosystem flux to elevated CO_2 depends in part on the resource base of the ecosystem. Other factors being equal, a nutrient-poor environment is less likely to respond to elevated CO_2 than is an ecosystem limited either by carbohydrate availability or by water (Oechel and Strain, 1985; see Fig. 2). Warmer ecosystems also appear more likely to respond to elevated CO_2 than do cooler ecosystems (Idso *et al.*, 1987). One would expect areas of limited moisture—including certain polar desert, fell-field, dune, or heath communities—to respond with higher photosynthetic rates to the effects of increased water-use efficiency at elevated CO_2. Yet the low carbon storage in these areas indicates that such increases may have relatively little effect on northern carbon budgets, although relative effects of productivity may be appreciable. Relatively warm nutrient-rich arctic communities, such as riparian and water-track communities, that are now quite productive may also respond positively to elevated CO_2.

V. Summary and Conclusions

Photosynthesis and primary productivity in the Arctic are limited largely by low air temperatures and nutrient availability. Decomposition is constrained by low soil temperatures and poor drainage, factors associated with the presence of permafrost. Yet despite low rates of primary productivity, even lower rates of decomposition have resulted in massive accumulation of carbon as soil organic matter in arctic ecosystems. Whether the arctic tundra will retain its capacity for long-term sequestering of carbon will depend on the integrated responses of terrestrial and aquatic ecosystems to rising temperatures and levels of carbon dioxide.

The potential response of arctic carbon budgets to global change will depend strongly on changes in water and nutrient availability and on time scales. At present resource levels, rising CO_2 alone should have little effect on photosynthesis, net primary productivity, or ecosystem function, although carbon storage could increase because of effects on community composition. If other factors remain constant (and they probably would not), the combination of rising temperatures and rising CO_2 should increase photosynthesis, growth, and net primary productivity, thereby increasing the carbon sequestered in tundra ecosystems. Nevertheless, higher temperature could deepen the soil active layer, lower the water table, and dry the upper soil layers, thereby accelerating decomposition rates and increasing carbon loss to the atmosphere. Soil mineralization rates could also increase, however, thus improving conditions for plant growth and enhancing net primary productivity.

It is impossible to say at present whether increases in net ecosystem primary productivity will offset increased decomposition rates under future global conditions and, thus, whether carbon will continue to accumulate in arctic ecosystems. Furthermore, over the next several decades, other potentially important interactive effects may appear, including effects on phenology, allocation, reproduction, species composition, competition, and herbivory. The importance of these interactions will be impossible to determine solely from laboratory studies. *In situ* manipulation and experiments must therefore be done to help determine the most important plant and ecosystem responses to global change.

References

Armentano, T. V., and Menges, E. S. (1986). Patterns of change in the carbon balance of organic-soil wetlands of the temperate zone. *Ecology* **74**, 755–774.

Bazzaz, F. A. (1990). The response of natural ecosystems to the rising global CO_2 levels. *Annu. Rev. Ecol. Syst.* **21**, 164–196.

Bigger, C. M., and Oechel, W. C. (1982). Nutrient effect on maximum photosynthesis in arctic plants. *Holarct. Ecol.* **5**, 158–163.

Billings, W. D. (1987). Carbon balance of Alaskan tundra and taiga ecosystems: Past, present, and future. *Quat. Sci. Rev.* **6**, 165–177.

Billings, W. D., and Peterson, K. M. (1980). Vegetational change and ice-wedge polygons through the thaw-lake cycle in arctic Alaska. *Arct. Alp. Res.* **12**, 413–432.

Billings, W. D., and Peterson, K. M. (1991). *In* "Consequences of the Greenhouse Effect for Biological Diversity" (R. L. Peters, ed.) Yale Univ. Press, New Haven, Connecticut (in press).

Billings, W. D., Luken, J. O., Mortensen, D. A., and Peterson, K. M. (1982). Arctic tundra: A source or sink for atmospheric carbon dioxide in a changing environment? *Oecologia* **53**, 7–11.

Billings, W. D., Luken, J. O., Mortensen, D. A., and Peterson, K. M. (1983). Increasing atmospheric carbon dioxide: Possible effects on arctic tundra. *Oecologia* **58**, 286–289.

Billings, W. D., Peterson, K. M., Luken, J. D., and Mortensen, D. A. (1984). Interaction of increasing atmospheric carbon dioxide and soil nitrogen on the carbon balance of tundra microcosms. *Oecologia* **65**, 26–29.

Bramyrd, T. (1979). The effects of man on the biogeochemical cycle of carbon in terrestrial ecosystems. *In* "The Global Carbon Cycle" (B. Bolin, ed.), pp. 183–218. Wiley, New York.

Bramyrd, T. (1980). Effects on the carbon cycle due to human impact on forest ecosystems. *In* "Biogeochemistry of Ancient and Modern Environments" (P. A. Trudinger, M. R. Walter, B. J. Ralph, eds.), pp. 405–412. Springer-Verlag, New York.

Brown, J. (1969). Buried soils associated with permafrost. *In* "Symposium on Pedobiology and Quaternary Research," pp. 115–125. University of Alberta, Edmonton.

Chapin, F. S., III. (1980). The mineral nutrition of wild plants. *Annu. Rev. Ecol. Syst.* **11**, 233–260.

Chapin, F. S., III. (1991). Integrated responses of plants to stress. *Bioscience* **41**, 29–36.

Chapin, F. S., III, and Oechel, W. C. (1983). Photosynthesis, respiration, and phosphate absorption by *Carex aquatilis* ecotypes along latitudinal and local environmental gradients. *Ecology* **64**, 734–751.

Chapin, F. S., III. and Shaver, G. R. (1985). Individualistic growth response of tundra plant species to environmental manipulations in the field. *Ecology* **66**, 564–576.

Chapin, F. S., III, Miller, P. C., Billings, W. D., and Coyne, P. I. (1980). Carbon and nutrient budgets and their control in coastal tundra. *In* "An Arctic Ecosystem: The Coastal Tundra at Barrow, Alaska" (J. Brown, P. C. Miller, L. L. Tieszen, and F. L. Bunnell, eds.), pp. 458–482. Dowden, Hutchinson & Ross, Stroudsburg, Pennsylvania.

Chapin, F. S., III, Shaver, G. R., and Kedrowski, R. A. (1986). Environmental controls over carbon, nitrogen, and phosphorus fractions in *Eriophorum vaginatum* in Alaskan tussock tundra. *Ecology* **74**, 167–195.

Chapin, F. S., III, Fetcher, N., Kielland, N., Everett, K. R., and Linkins, A. E. (1988). Enhancement of productivity and nutrient cycling by flowing water in Alaskan tussock tundra. *Ecology* **69**, 693–702.

D'Arrigo, R., Jacoby, G., and Fung, I. (1987). The role of boreal forests in atmosphere–biosphere exchange of carbon dioxide. *Nature* **329**, 321–323.

Emanuel, W. R., Shugart, H. H., and Stevenson, M. P. (1985a). Climatic change and the broad-scale distribution of terrestrial ecosystem complexes. *Clim. Change* **7**, 29–43.

Emanuel, W. R., Shugart, H. H., and Stevenson, M. P. (1985b). Response to comment: Climate change and the broad-scale distribution of terrestrial ecosystem complexes. *Clim. Change* **7**, 457–460.

Evans, J. R. (1989). Nitrogen and photosynthesis in the flag leaf of wheat (*Triticum aestivum L.*). *Plant Physiol.* **72**, 297–302.

Gersper, P. L., Alexander, V., Barkley, S. A., Barsdate, R. J., and Flint, P. S. (1980). The soils and their nutrients. *In* "An Arctic Ecosystem: The Coastal Tundra at Barrow, Alaska" (J. Brown, P. C. Miller, L. L. Tieszen, and F. L. Bunnell, eds.) pp. 219–254. Dowden, Hutchinson & Ross, Stroudsburg, Pennsylvania.

Gorham, E. (1991). Northern peatlands: Role in the carbon cycle and probable responses to climatic warming. *Ecol. Applic.* **1**, 182–195.

Grulke, N. E., Riechers, G. H., Oechel, W. C., Hjelm, U., and Jaeger, C. (1990). Carbon balance in tussock tundra under ambient and elevated atmospheric CO_2. *Oecologia* **83**, 485–494.

Hastings, S. J., Luchessa, S. A., Oechel, W. C., and Tenhunen, J. D. (1989). Standing biomass and production in water drainages of the foothills of the Philip Smith Mountains, Alaska. *Holarct. Ecol.* **12**, 304–311.

Hicklenton, P. R., and W. C. Oechel. (1976). Physiological aspects of the ecology of *Dicranum fuscenscens* in the subarctic. I. Acclimation and acclimation potential of CO_2 exchange in relation to habitat, light, and temperature. *Can. J. Bot.* **54**, 1104–1119.

Hobbie, J. (1980). "Limnology of Tundra Ponds: Barrow Alaska." Dowden, Hutchinson & Ross, Stroudsburg, Pennsylvania.

Hom, J. (1986). Investigations into some of the major controls on the productivity of a black spruce (*Picea mariana* [Mill] B.S.P.) forest ecosystem in the interior of Alaska. Ph.D. diss., University of Alaska, Fairbanks.

Idso, S. B., Kimball, B. A., Anderson, M. G., and Mauney, J. R. (1987). Effects of atmospheric CO_2 enrichment on plant growth: The role of air temperature. *Agric. Ecosyst. Environ.* **20**, 1–10.

Kimball, B. A., Mauney, J. R., Radin, J. W., Nakayama, F. S., Idso, S. B., Hendrix, D. L., Akey, D. H., Allen, S. G., Anderson, M. G., and Hartung, W. (1986). Effects of increasing atmospheric CO_2 on the growth, water relations, and physiology of plants grown under optimal and limiting levels of water and nitrogen. *In* "Response of Vegetation to Carbon Dioxide 039." U.S. Dep. of Energy, Office of Energy Research, Washington, DC.

Kling, G. W., Kipphut, G. W., and Miller, M. C. (1991). Arctic lakes and streams as gas conduits to the atmosphere: Implications for tundra carbon budgets. *Science* **251**, 298–301.

Kummerow, J., and Ellis, B. A. (1984). Temperature effect on biomass production and root/shoot biomass ratios in two arctic sedges under controlled environmental conditions. *Can. J. Bot.* **62**, 2150–2153.

Kummerow, J., Mills, J. N., Ellis, B. A., Hastings, S. J., and Kummerow, A. (1987). Downslope fertilizer movement in arctic tussock tundra. *Holarct. Ecol.* **10**, 312–319.

Lachenbruch, A. H., and Marshall, B. V. (1986). Changing climate: Geothermal evidence from permafrost in the Alaskan Arctic. *Science* **234**, 689–696.

Lachenbruch, A. H., Cladouhos, T. T., and Saltus, R. W. (1988). Permafrost temperature and the changing climate. *In* "Permafrost: Fifth International Conference" (K. Senneset, ed.). Tapir Press, Trondheim, Norway.

Lawrence, W. T., and Oechel, W. C. (1983a). Effects of soil temperature on carbon exchange of taiga seedlings. I. Root respiration. *Can. J. For. Res.* **13**, 840–849.

Lawrence, W. T., and Oechel, W. C. (1983b). Effects of soil temperature on carbon exchange of taiga seedlings. II. Photosynthesis. *Can. J. For. Res.* **13**, 850–859.

Limbach, W. E., Oechel, W. C., and Lowell, W. (1982). Photosynthetic and respiratory responses to temperature and light of three Alaskan tundra growth forms. *Holarct. Ecol.* **5**, 150–157.

Luken, J. O. (1984). Net ecosystem production in a subarctic peatland. Ph.D. diss., Duke University, Durham, North Carolina.

Luken, J. O., and Billings, W. D. (1985). The influence of micro-topographic heterogeneity on carbon dioxide efflux from a subarctic bog. *Holarct. Ecol.* **8**, 306–312.

Manabe, S. and Stouffer, R. J. (1979). A CO_2 climate sensitivity study with a mathematical model of the global climate. *Nature* **28**, 491–493.

Marion, G. M., and Everett, K. R. (1989). The effect of nutrient and water additions on elemental mobility through small tundra watersheds. *Holarct. Ecol.* **12**, 317–323.

Miller, P. C., ed. (1981). "Carbon Balance in Northern Ecosystems and the Potential Effect of Carbon Dioxide–Induced Climate Change. CONF-8000318 Natl. Tech. Info. Serv., Springfield, Virginia.

Miller, P. C., Stoner, W. A., and Tieszen, L. L. (1976). A model of stand photosynthesis for the wet meadow tundra at Barrow, Alaska. *Ecology* **57**, 411–413.

Miller, P. C., Kendall, R., and Oechel, W. C. (1983). Simulating carbon accumulation in northern ecosystems. *Simulation* **40**, 119–131.

Miller, P. C., Miller, P. M., Blake-Jacobson, M., Chapin, F. S., III, Everett, K. R., Hilbert, D. H., Kummerow, J., Linkins, A. E., Marion, G. M., Oechel, W. C., Roberts, S. W., and Stuart, L. (1984). Plant–soil processes in *Eriophorum vaginatum* tussock tundra in Alaska: A systems modeling approach. *Ecological Monogr.* **54**, 361–405.

Mitchell, J. F. B., Manabe, S., Tokioka, T., and Meleshko, V. (1990). Equilibrium climate change. *In* "Climate Change: The IPCC Scientific Assessment." Cambridge Univ. Press, Cambridge.

Mooney, H. A., and Billings, W. D. (1961). Comparative physiological ecology of arctic and alpine populations of *Oxyria digyna*. *Ecol. Monogr.* **31**, 1–29.

Moore, T. R. (1987). A review of carbon dioxide and methane evolution from soils with particular reference to Canadian ecosystems. #09SE.KM171-6-1488. Atmospheric Environment Service, Ottawa, 36 pp.

Mortensen, D. A. (1983). Growth, biomass, and production of roots in a taiga bog. M. S. thesis, Duke University, Durham, North Carolina.

Oberbauer, S. F., Sionit, N., Hastings, S. J., and Oechel, W. C. (1986a). Effects of CO_2 enrichment and nutrition on growth, photosynthesis, and nutrient concentration of Alaskan tundra plant species. *Can. J. Bot.* **64**, 2993–2998.

Oberbauer, S. F., Oechel, W. C., and Riechers, G. H. (1986b). Soil respiration of Alaskan tundra at elevated atmospheric CO_2 concentrations. *Plant Soil* **46**, 145–158.

Oberbauer, S. F., Hastings, S. J., Beyers, J. L., and Oechel, W. C. (1989). Comparative effects of downslope water and nutrient movement on plant nutrition, photosynthesis, and growth in Alaskan tundra. *Holarct. Ecol.* **12**, 324–334.

Oechel, W. C. (1976). Seasonal patterns of temperature response of CO_2 flux and acclimation in arctic mosses growing *in situ*. *Photosynthetica* **10**, 447–456.

Oechel, W. C. (1987). Response of tundra ecosystems to elevated atmospheric carbon dioxide. Final Report. U.S. Dep. of Energy, Carbon Dioxide Research Division, Washington, DC.

Oechel, W. C., and Riechers, G. H. (1987). Response of a tundra ecosystem to elevated atmospheric carbon dioxide. *In* "Response of Vegetation to Carbon Dioxide 037. U.S. Dep. of Energy, Carbon Dioxide Research Division," Washington, DC.

Oechel, W. C., and Strain, B. R. (1985). Native species response to increased atmospheric carbon dioxide concentration. *In* "Direct Effects of Increasing Carbon Dioxide on Vegetation" (B. R. Strain and J. D. Cure, eds.), pp. 118–154. DOE/ER-0238. Natl. Tech. Info. Serv., Springfield, Virginia.

Oechel, W. C., and B. Sveinbjörnsson. (1978). Primary production processes in arctic bryophytes at Barrow, Alaska. *In* "Vegetation and Production Ecology of the Alaskan Arctic Tundra" (L. L. Tieszen, ed.), pp. 269–298. Springer-Verlag, New York.

Ovenden, L. (1990). Peat accumulation in northern wetlands. *Quat. Res.* **33**, 377–386.

Post, W. M., ed. (1990). Report of a workshop on climate feedbacks and the role of peatlands, tundra, and boreal ecosystems in the global carbon cycle. ORNL/TM-11457. Natl. Tech. Info. Serv., Springfield, Virginia.

Post, W. M., Pastor, J., Zinke, P. J., and Strangenberger, A. G. (1985). Global patterns of soil nitrogen. *Nature* **317**, 613–616.

Sage, R. F., and Sharkey, T. D. (1987). The effect of temperature on the occurrence of O_2- and CO_2-insensitive photosynthesis in field-grown plants. *Plant Physiol.* **84**, 658–664.

Shaver, G. R., and Chapin, F. S., III. (1980). Response to fertilization by various plant growth forms in an Alaskan tundra: Nutrient accumulation and growth. *Ecology* **61**, 662–675.

Schell, D. M. (1983). Carbon-13 and Carbon-14 abundances in Alaskan aquatic organisms: Delayed production from peat in arctic food webs. *Science* **219**, 1068.

Silvola, J. (1986). Carbon dioxide dynamics in mires reclaimed for forestry in eastern Finland. *Ann. Bot. Fenn.* **23**, 59–67.

Silvola, J., and Hanski, I. (1979). Carbon accumulation in a raised bog. *Oecologia* **37**, 285–295.

Sionit, N., Mortensen, D. A., Strain, B. R., and Hellmers, H. (1981). Growth response of wheat to CO_2 enrichment and different levels of mineral nutrition. *Agron. J.* **73**, 1023–1026.

Stoner, W. A., Miller, P. C., and Oechel, W. C. (1978). Simulation of the effect of the tundra vascular plant canopy on the productivity of four moss species. *In* "Vegetation and Production Ecology of the Alaskan Arctic Tundra" (L. L. Tieszen, ed.), pp. 371–487. Springer-Verlag, New York.

Strain, B. R. (1987). Direct effects of increasing atmospheric CO_2 on plants and ecosystems. *Trends Ecol. Evol.* **2**, 18–21.

Strain, B. R., and Bazzaz, F. A. (1983). Terrestrial plant communities. *In* "CO_2 and Plants: The Response of Plants to Rising Levels of Atmospheric Carbon Dioxide." (E. R. Leman, ed.) p. 177–222. Westview Press, Inc. Boulder.

Sveinbjörnsson, B., and Oechel, W. C. (1981). Controls on CO_2 exchange in two *Polytricum* moss species. I. Field studies on the tundra near Barrow, Alaska. *Oikos* **36**, 114–128.

Sveinbjörnsson, B., and Oechel, W. C. (1983). The effect of temperature preconditioning on the temperature sensitivity of net CO_2 flux in geographically diverse populations of the moss, *Polytrichum commune*. *Ecology* **64**, 1100–1108.

Tans, P., Fung, I., and Takahashi, T. (1990). Observational constraints on the global atmospheric CO_2 budget. *Science* **247**, 1431–1439.

Tieszen, L. L. (1978). Photosynthesis in the principal Barrow, Alaska, species: A summary of field and laboratory responses. *In* "Vegetation and Production Ecology of an Alaskan Arctic Tundra," (L. L. Tieszen, ed.), pp. 241–268. Springer-Verlag, New York.

Tissue, D. T., and Oechel, W. C. (1987). Response of *Eriophorum vaginatum* to elevated CO_2 and temperature in the Alaskan arctic tundra. *Ecology* **68**, 401–410.

Van Cleve, J., Dyrness, C. T., Viereck, L. A., Fox, J., Chapin, F. S., III, and Oechel, W. C. (1983a). Taiga ecosystems in interior Alaska. *Bioscience* **33**, 39–44.

Van Cleve, K., Oliver, L., Schlentner, R., Viereck, L. A., and Dyrness, C. T. (1983b). Productivity and nutrient cycling in taiga forest ecosystems. *Can. J. For. Res.* **13**, 747–766.

Van Cleve, K., Hom, J., and Oechel, W. C. (1990). Response of black spruce (Picea mariana) ecosystems to soil temperature modification in interior Alaska. *Can. J. For. Res.* **20**, 153–1535.

Zoltai, S. C., and Pollett, F. C. (1983). Wetlands in Canada. *In* "Mires: Swamp, Bog, Fen, and Moor," (A. J. P. Gore, ed.), pp. 245–268. Elsevier, Amsterdam.

8

Photosynthesis, Respiration, and Growth of Plants in the Soviet Arctic

O. A. Semikhatova, T. V. Gerasimenko, and T. I. Ivanova

I. Introduction

Considerable research has been done on the photosynthesis of arctic plants, but respiration and its relationship to plant carbon balance and growth have been studied much less. In the Laboratory of Photosynthesis at the Komarov Botanical Institute, Academy of Sciences of the USSR, both these processes have been studied using a consistent methodology developed in our laboratory. In this chapter we present some of this information for a variety of plants: dominants, widespread species, and minor components of plant communities. We emphasize diurnal and seasonal dynamics of these processes as well as their responses to varied experimental conditions.

We present data on both net photosynthesis (net CO_2 uptake at natural CO_2 concentrations) and potential photosynthesis (the rate in near-natural conditions of light and temperature but at saturating CO_2 concentrations). Our approach was dictated partly by the method we used (a stream through a leaf chamber of 1% CO_2 enriched with $^{14}CO_2$), and partly by the interest of our chief, Oleg Zalensky (1977), in potential photosynthesis, which he considered to be hereditary and constant for a species in any climatic zone. Zalensky also thought that without CO_2 limitation, the effect of other factors on photosynthesis would be expressed more clearly. Net photosynthesis was measured using an infrared gas analyzer (Infralyt-4) in a closed system. Respiration was mainly measured by manometric methods. Samples of leaves or whole plants (when indicated) were taken directly to the laboratory from natural habitats where the plants grew.

To provide a context for measurements of plant metabolism, we studied growth and separated total respiration into two components: construction (or growth) respiration and maintenance respiration. Measurements of photosynthesis were accompanied by measurements of pigment content and CO_2 output in the light. For each process investigated, we also examined the effect of major environmental factors at our study sites: Wrangel Island, Kola Peninsula, Western Taimyr, the polar Urals (photosynthesis only), and Heith Island (several measurements of respiration).

II. Photosynthesis

Most previous studies of photosynthesis have investigated only a few species. Our aim was to measure photosynthesis in a large number of species at different tundra sites (Gerasimenko and Zalensky, 1982; Gerasimenko *et al.* 1988, 1989). Maximum values of potential and net photosynthesis differed 4- to 10-fold among the species we measured (Fig. 1,2). The large variation among species, seen at all study sites, was more pronounced when expressed per unit of leaf dry weight than per unit area. This observation supports Zalensky's (1977) contention that every biome has a large interspecific variation in maximum photosynthetic rate. Nonetheless, the range of photosynthetic rates among arctic plants is narrower than in other biomes (Fig. 3).

There is some controversy as to whether arctic species have higher photosynthetic rates when growing in the Arctic than the same species growing in a temperate region. The conclusion may depend on the methods and indices for measuring photosynthesis. If rates are measured under ambient conditions, arctic plants have relatively low rates of photosynthesis (see Fig. 3). If, however, the comparison is made at the same temperature, arctic plants generally exhibit higher rates, as shown for plants from Wrangel Island (Piankov, 1980; Kislyuk and Sheinina, 1985).

Working at several tundra sites, Gerasimenko and Shvetzova (1989) compared potential photosynthesis in the same species growing in habitats of differing climatic severity. Potential photosynthesis was similar in the Taimyr and at Wrangel Island when expressed per unit of leaf area, but it was lower in the more northern site when expressed on the basis of leaf weight (Fig. 1). Kislyuk and Sheinina (1985) obtained similar results comparing Wrangel Island plants with temperate-zone plants. This observation indicates the significance of morphological changes in leaves as photosynthesis acclimates to the environment and supports Zalensky's (1977) idea that photosynthesis is a "sheltered process," in other words, that many compensations at different levels of a plant's organization contribute to a given photosynthetic rate.

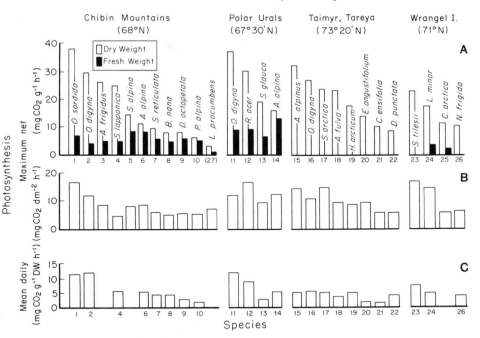

Figure 1   Maximum values (A,B) and maximum mean daily values (C) of net photosynthesis rates in tundra species at arctic and subarctic sites. Data were chosen (A,B) or calculated (C; mean for 24 hrs) from diurnal courses of photosynthesis (Gerasimenko and Shvetzova, 1989). Species are (1) *Oxytropis sordida*, (2) *Oxyria digyna*, (3) *Astragalus frigidus*, (4) *Soligado lapponica*, (5) *Saussurea alpina*, (6) *Arctous alpina*, (7) *Salix reticulata*, (8) *Betula nana*, (9) *Dryas octopetala*, (10) *Poa alpina*, (11) *Oxyria digyna*, (12) *Ranunculus acer*, (13) *Salix glauca*, (14) *Arctous alpina*, (15) *Alopecurus alpinus*, (16) *Oxyria digyna*, (17) *Salix arctica*, (18) *Arctophila fulva*, (19) *Hedysarum arcticum*, (20) *Eriophorum angustifolium*, (21) *Carex ensifolia*, (22) *Dryas punctata*, (23) *Saussurea tilesii*, (24) *Lagotis minor*, (25) *Caltha arctica*, (26) *Nardosmia frigida*, and (27) *Louseleuria procumbens*.

Despite the large variation in photosynthetic rates among species and the absence of exceptionally high rates in arctic plants, total daily photosynthesis of most arctic plants attains values as high as 100–200 mg CO_2 dm^{-2} d^{-1} (Fig. 4). Different species may have relatively similar mean daily rates of photosynthesis (see Figs. 1,2), but these similar values are achieved by different means. Some species exploit short periods of high light intensity; others photosynthesize at low light and therefore prolong the period of active assimilation into the night hours (e.g., *Alopecurus alpinus, Salix glauca, S. arctica*). The ability of northern plants to assimilate CO_2 actively during the polar night has long been known from the Kola Peninsula (Kostytschew *et al.*, 1930).

The proportion of daily photosynthesis that occurs at night can be as high as 25% or as low as -10% (and sometimes less; Gerasimenko and Zalensky, 1982; Fig. 4). This variation reflects both the weather conditions at night and

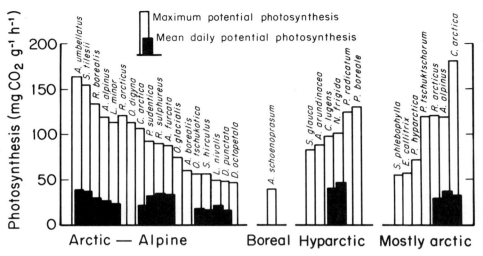

Figure 2   Potential photosynthesis (seasonal maximum and maximum of the mean daily values) on Wrangel Island. Data were chosen or calculated (mean for 24 hr) from diurnal courses (Gerasimenko and Zalensky, 1982; Gerasimenko *et al.*, 1989). Potential photosynthesis is defined as the rate measured at 1% CO_2. Species are arctic–alpine (*Astragalus umbellatus, Saussurea tilesii, Rhodiola borealis, Astragalus alpinus, Lagotis minor, Rumex arcticus, Oxyria digyna, Claytonia arctica, Pedicularis sudetica, Ranunculus sulphureus, Artemisia furcata, Oxygraphis glacialis, Artemisia borealis, Oxytropis tschukotica, Saxifraga hirculus, Luzula nivalis, Dryas punctata, Dryas octopetala*), boreal (*Allium shoenoprasum*), hyparctic (*Salix glauca, Arctagrostis arundinacea, Carex lugens, Nardosmia frigida, Papaver radicatum, Polemonium boreale*), and arctic (*Salix phlebophylla, Eriophorum callitrix, Potentilla hyparctica, Primula tschuktschorum, Rumex arcticus, Alopecurus alpinus, Caltha arctica*). *Rumex arcticus* and *Alopecurus alpinus* have been included in both arctic and arctic–alpine groups because their appropriate classification is debated.

during the day and the species' photosynthesis–light relations (Fig. 5). The photosynthesis–light curves of Wrangel Island plants differ among species but do not depend on the phenological stage of the plant. Among Kola Peninsula plants, some species had high light-saturation values (Fig. 5; Gerasimenko and Zalensky, 1982; Lukyanova *et al.*, 1986), but these species are also very responsive to light intensity below saturation (note the steepness of the light–photosynthesis curves of *Astragalus frigidus* and *Polemonium boreale* in Fig. 5). In most arctic species, photosynthesis reaches light saturation at moderate or even low light intensities relative to plants from the Temperate Zone. This enables them to assimilate at the low light intensities characteristic of the cloudy weather prevalent in the Arctic. Thus the assimilation rate of 20 out of 24 species studied was at least 40% of the light-saturated value when measured at 30 kilolux (klx), illustrating arctic plants' characteristic ability to exploit low light.

Availability of data on many species enabled Zalensky (1977) and Gerasimenko and Zalensky (1982) to look for correlations of photosynthetic rates and light relations with other plant features. They concluded that photosyn-

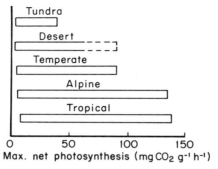

Figure 3 The range among species in maximum net photosynthesis in field-grown plants from different ecosystems (Zalensky, 1977; Vosnesensky, 1989). The dashed area for desert plants represents rates that are debated in the literature.

thetic characteristics were unrelated to the taxonomic affinities of a species, although several representatives of the Fabaceae, Asteraceae, Polygonaceae, and Poaceae had high rates of photosynthesis. Rather, the life-form of the plant and leaf longevity were the major factors correlated with photosynthetic rate (Table I). High rates of potential photosynthesis were frequently found in arctic–alpine and arctic species (see Fig. 2). Among the hypoarctic and boreal species, intermediate rates were more common. The light response of photosynthesis was even less correlated with other species characteristics. It is noteworthy that photosynthesis of hypoarctic and boreal species (e.g., *Vaccinium vitis-idaea, Comarum palustre,* and several *Salix* species) was saturated at the lowest light intensity.

Leaf ontogeny appears to play a more important role than environmental conditions in determining the seasonal pattern of photosynthesis in Wrangel Island plants. In spring, when the leaves have just emerged, photosynthetic rate is high despite low temperature. In some species these spring rates can be nearly as high as those observed in the most favorable period of the season, mid- and late July, when photosynthesis attains its maximum values, and the majority of species in the community are flowering (Gerasimenko *et al.,* 1988).

Species are more similar in their assimilation–temperature curves than in their light responses. In full agreement with the literature are the low optimum temperatures of photosynthesis in tundra plants from Wrangel Island and the Kola Peninsula. For Wrangel Island plants, the optimum is 10–20° C, for Kola Peninsula plants 15–20° C (Table II; Gerasimenko and Zalensky, 1973, 1982). For all species from these and other tundra regions, photosynthesis proceeds at near-maximum rates over a broad range of temperatures (Shvetzova, 1970; Gerasimenko, 1973; Gerasimenko and Zalensky, 1973, 1982). On Wrangel Island, within 7° C on either side of the optimum temperature, 7 out of 34 species maintained 90% of maximum assimilation rates, 17 species maintained 80%, and 26 species maintained 70%. These observations demonstrate the

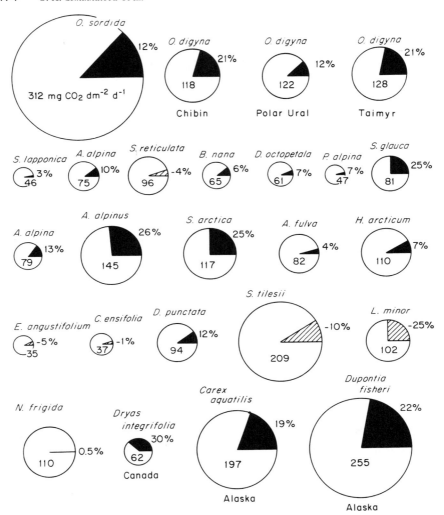

Figure 4   The percentage of maximum values of total daily photosynthesis occurring during the day (open part of each circle) or at night (from 2200 to 0400 hours: filled part of each circle). Hatched areas indicate net carbon loss at night. The size of each circle is proportional to total daily (24-hr) carbon gain, the value of which (mg CO_2 dm^{-2} h^{-1}) is given inside each circle. The species are the same as in Fig. 1 except for the three North American species: *Dryas integrifolia* (Mayo *et al.*, 1973), *Carex aquatilis*, and *Dupontia fisheri* (Tieszen, 1975).

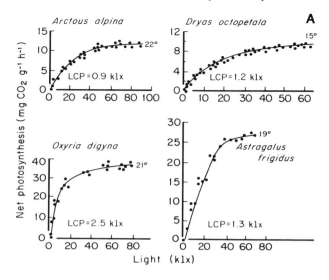

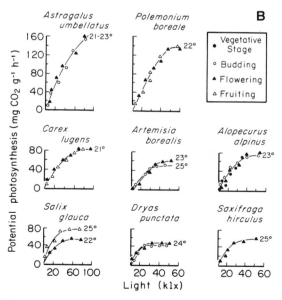

Figure 5   The light response of (A) net photosynthesis on Kola Peninsula and (B) potential photosynthesis on Wrangel Island. Irradiance (in klx) was measured at 15–25° C. LCP, light compensation point.

Table I Net Photosynthesis and Maximum Potential Photosynthesis in Different Life-Forms of Arctic, Subarctic, Antarctic, and Mountain Tundra[a]

	Net photosynthesis		Potential photosynthesis[b]	
	$(mg\ CO_2\ g^{-1}\ h^{-1})$	$(mg\ CO_2\ dm^{-2}\ h^{-1})$	$(mg\ CO_2\ g^{-1}\ h^{-1})$	$(mg\ CO_2\ dm^{-2}\ h^{-1})$
Monocotyledons[c]				
Grasses	6–37	5–31 (19)	32–88	49–57 (17)
Sedges	4–24	6–21 (13)	46–86	43–56 (18)
Dicotyledons				
Herbaceous perennials	2–35	5–27 (46)	27–193	37–154 (39)
Deciduous dwarf shrubs	6–30	5–26 (99)	33–80	38–47 (15)
Evergreen dwarf shrubs	2–7	4–8 (9)	6–24	38–45 (12)
Mosses (arctic, antarctic)	0.5–4.4	– (7)	1–2	–
Lichens	0.1–0.6	1.5–4.4 (10)	–	–

[a] Summarized by Gerasimenko and Shvetzova (1989).
[b] Measured at 1% CO_2.
[c] Numbers of species studied given in parentheses for each life-form.

Table II The Temperature Response of Net Photosynthesis in Some Tundra Species[a]

Species	Net photosynthesis at $+5°$ C (% of maximum)	Optimum temperature (° C)	Maximum temperature (° C)
Wrangel Island			
Caltha arctica	80	12–15	35
Oxytropis tschukotica	80	10	—
Artemisia furcata	30	20	—
Primula tschuktschorum	80	10–12	27
Kola Peninsula			
Dryas octopetala	35–50	10–20	37–47
Louseleuria procumbens	45–50	10–30	43
Arctous alpina	50	15–20	37
Oxyria digyna	75	10–20	38
Saussurea alpina	50	15–25	43
Oxytropis sordida	25	17–25	38
Astragalus frigidus	50	15–20	38
Oxygraphis glacialis	30	15–25	36–42
Salix glauca	45	15–20	43
Betula nana	—	12–30	44

[a] Measured at 0.2–0.25 cal cm^{-2} min^{-1} (photosynthetically active radiation; Gerasimenko and Zalensky, 1982).

ability of the plants studied to assimilate carbon at substantial rates over a broad temperature range.

The maximum temperature at which an arctic plant can assimilate CO_2 is typically 7–10° higher than the highest ambient temperature in the field. In other words, plants could continue to photosynthesize at higher temperatures than they ever experience (Zalensky, 1977).

Measuring daily air temperatures during the season, along with the temperature relations of photosynthesis, enabled us to calculate the degree of low-temperature limitation to photosynthesis for a given species. During half the season, air temperatures are significantly lower than the optimum (Fig. 6). In other words, even among low-temperature-adapted arctic plants, photosynthesis is still temperature-limited most of the time.

Comparing the temperature–photosynthesis relations of the same species at different tundra sites demonstrated that photosynthesis acclimates to temperature differently in different arctic species (Fig. 7). In *Dryas octopetala*, for example, the entire curve shifts to a higher temperature in the warmer site, as is commonly observed (Berry and Björkman, 1980). *Salix glauca* shows no shift in the optimum temperature, but at temperatures lower than optimum, photosynthesis is higher in plants growing in colder sites. *Oxyria digyna*, which is very conservative in its gas-exchange characteristics, shows no shift in the optimum, and only at temperatures above optimum is the curve displaced toward lower temperatures in northern sites.

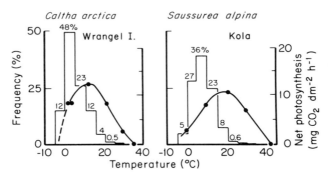

Figure 6   The temperature response of net photosynthesis (measured at saturating light intensity) in *Caltha arctica* and *Saussurea alpina* and the frequency distribution of average hourly air temperature as a percentage of the total number of hours in the season (Wrangel Island, 20 June to 25 August 1978; Kola Peninsula, 3 June to 3 October 1989). Air temperature was measured 10 cm above the ground and was generally similar to leaf temperature.

Thus, our data on net and potential photosynthesis confirm the results of other authors working on plants of several other tundra sites, but they also clearly demonstrate that tundra species differ greatly in their maximum rates and light responses of photosynthesis.

The patterns of net photosynthesis described above are the end result of several processes occurring simultaneously in the leaf: gross photosynthesis, photorespiration, dark respiration, and CO_2 reassimilation. These processes

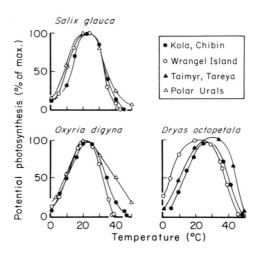

Figure 7   Differences in the temperature response of photosynthesis measured at saturating light in three species at four arctic sites, from warmest to coolest: polar Urals, Kola Peninsula, Taimyr, and Wrangel Island. Measurement accuracy is ±8%.

can be measured simultaneously using methods developed by Bykov (1962), Vosnesensky *et al.* (1984), and Vosnesensky (1989). The methods are based on a mathematical description of the CO_2-depletion curve with a leaf in a hermetically sealed chamber under 21% or 2% oxygen.

In Kola Peninsula plants, total CO_2 output in the light, including the portion generated by photorespiration, is proportional to the rate of net photosynthesis, as is typical of C_3 plants. If, however, we consider not only the rates of these processes but their percentage of gross photosynthesis, some interesting features of northern plants can be distinguished. In Kola Peninsula plants, both the CO_2 output in the light and the CO_2 compensation point are lower than in desert species. Photorespired CO_2 makes up about 20% of gross photosynthesis, which is less than observed in plants growing at high temperatures (Ledyaikina and Gerasimenko, 1983; Vosnesensky *et al.*, 1984).

If we assume that the respiratory CO_2 production in light and dark is the same, we can then calculate the portion of total CO_2 reassimilated by the leaf. Reassimilated CO_2 is less in tundra plants (7 species studied) than in warm-desert plants—some 50–60% instead of 75% of total CO_2 output in the light. Nevertheless, the CO_2 balance in the leaves of arctic plants can be considered very economical, since only a small part of the absorbed CO_2 is lost in the light. Certainly this is an important cause of the comparatively high daily total photosynthesis characteristic of northern plants.

The chlorophyll content of arctic plants is typically low (Fig. 8). On Wrangel Island, for example, the chlorophyll content of 24 species varied from 0.5 to 2.9 mg g^{-1} of fresh weight (Gerasimenko *et al.*, 1988). *Carex lugens* contained the most, *Saxifraga platysepala* and *Allium schoenoprasum* the least. In the less-severe conditions of the Kola Peninsula, chlorophyll content was higher (Lukyanova *et al.*, 1986). Low chlorophyll content is a feature usually considered to characterize sun plants. Yet even those Wrangel Island plants

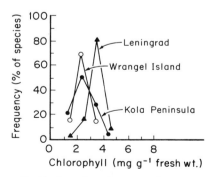

Figure 8   The frequency distribution of species with different chlorophyll contents in two arctic sites (Wrangel Island, n = 24 species; Kola Peninsula, n = 45) and one temperate site (Leningrad n = 220). Data are from Lubimenko (1916), Lukyanova *et al.* (1986), and Gerasimenko *et al.* (1988).

Table III   Percentage of Species with Different Chlorophyll: Carotenoid Ratios in Contrasting Botanical–Geographical Regions[a]

	Number of species	Chlorophyll: carotenoid ratio (percentage of species)					
		2:1–3:1	3:1–4:1	4:1–5:1	5:1–6:1	6:1–7:1	7:1–8:1
Gobi (Mongolian desert)	41	0	0	11	46	32	11
Karakum desert	37	0	0	14	50	23	13
Wrangel Island (Arctic)	20	0	5	75	5	5	10
Pamir (high mountains)	15	38	62	0	0	0	0

[a]Summarized by Popova *et al.* (1989).

that have a low light-saturation point have a low chlorophyll content. In contrast, the carotenoid content of northern plants is not unusually low (Gerasimenko *et al.*, 1989; Popova *et al.*, 1989). The ratio of chlorophyll to carotenoid is typically between 4:1 and 5:1. Lower values characterize only the extremely light-tolerant plants in the high mountains of East Pamir (Table III).

Another feature that arctic plants share with light-tolerant plants of sunnier habitats is the microstructure of their chloroplasts. Miroslavov and Jakovleva (1983) found that the thylakoid system of several species of Wrangel Island plants was typical of sun plants, with few, poorly developed grana.

III. Respiration and Growth

Respiratory capacity (i.e., respiration rate measured at a single temperature) of tundra plants is higher in the Arctic than in the Subarctic, and higher in the Subarctic than in the Temperate Zone (Fig. 9). This result, observed in a large number of species (Ivanova and Vaskovsky, 1976; Ivanova *et al.*, 1983),

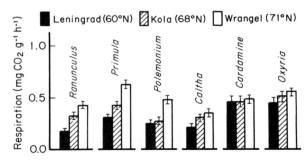

Figure 9   Leaf respiration rates of plants from different sites compared at the same temperature (15° C). Species names are given in Table IV.

Table IV   Leaf Respiration Rates of the Same or Related Species
Growing at Different Sites[a]

Species	Leaf respiration rate (mg CO_2 g^{-1} h^{-1})		
	Wrangel Island (10° C)	Kola Peninsula (15° C)	Leningrad (20° C)
Primula tschuktschorum	0.41		
P. elatior		0.43	0.42
Oxygraphis glacialis	0.35		
O. vulgaris		0.48	
Dryas punctata	0.42	0.54	
Astragalus umbellatus	0.62		
A. frigidus		0.48	
Cardamine pratensis	0.36	0.40	0.48
C. bellidifolia	0.32	0.34	
Polemonium boreale	0.36		
P. coeruleum		0.26	0.40
Claytonia arctica	0.18		
C. asarifolia		0.15	
Polygonum bistorta	0.58		
P. viviparum		0.52	
Papaver polare	0.29		
P. lapponicum		0.39	
Saussurea tilesii	0.38		
S. alpina		0.32	
Oxytropis tschukotica	0.60		
O. sordida		0.58	
Ranunculus sulphureus	0.28		
R. acris		0.32	0.30

[a] Respiration was measured at the mean ambient temperature (in parentheses) at each site.

confirms the data of other authors obtained on one or a few species. Laboratory data on plants grown at low temperature show that this increase in respiration rate is conditioned by cold. This response is compensatory; that is, at the mean temperature of a given habitat, plants from any habitat will have similar respiration rates. Stocker (1935) suggested that there is a general intensity of gas exchange that is the same for plants in different climatic zones at the temperature of each zone. Our results are consistent with this idea (Table IV).

A few species (e.g., *Oxyria digyna, Cardamine pratensis*), however, have the same respiratory capacity in habitats of different latitudes (Fig. 9). Such species inhabit special microsites, which may explain why some studies fail to find latitudinal differences in respiration (Scholander and Kanwisher, 1959). Plants of polar deserts are another exception. Our data on several species

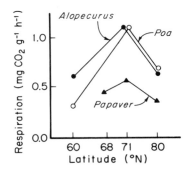

Figure 10   Latitudinal changes in leaf respiration measured at the same temperature (18–20° C). Species and sites are *Alopecurus pratensis* (Leningrad, 60°N), *A. alpinus* (Wrangel Island, 71°, and Heith Island, 81°), *Poa pratensis* (60°), *P. alpina* (71°, 81°), *Papaver lapponicum* (Kola, 68°), *P. polare* (71°, 81°).

from Heith Island (Semikhatova and Shukhtina, 1973) and results obtained by Wielgolaski (pers. comm.) show that the respiration rates of polar desert plants are typically lower than the respiration rates of plants of the same species from tundra regions. A plant's respiration–latitude curve thus has a descending as well as ascending part (Fig. 10).

It has been suggested that the physiological adaptation of plants to the arctic environment is reflected in the responses of respiration to temperature (Billings and Mooney, 1968; Tieszen, 1973). Studying these responses, we determined the optimum temperature, the temperature dependence of respiration (Q_{10}), and the critical temperature because these are the best indices of high-temperature tolerance. We define the critical temperature as that at which damage to respiration first occurs, as indicated by a drop in respiratory rate. Both the optimum and critical temperatures for respiration are lower in arctic plants than in temperate ones (Fig. 11). These values were lowest for Wrangel Island plants (the northernmost site) and highest for plants of warm climates (Fig. 12). Nevertheless, small but ecologically important interspecific differences do exist in the temperature tolerances of arctic plants (Table V).

The shapes of the respiration–temperature curves of Wrangel Island and Kola Peninsula plants are similar to those of temperate plants (Ivanova *et al.*, 1989; Fig. 11). These curves coincide when shifted to account for the differences in mean temperature of the corresponding habitats (Fig. 13). Thus, contrary to the idea of Tieszen (1973) and other authors that northern plants have a higher Q_{10} at low temperatures, we found Q_{10} values of arctic and temperate species to be equal.

To our knowledge, the seasonal respiration pattern in arctic plants has been measured only by Ivanova *et al.* (1985) and Lukyanova *et al.* (1986). Maximum respiration rates in Wrangel Island plants (always measured at 14° C) usually occur at the beginning of the growing season, the rate decreasing thereafter

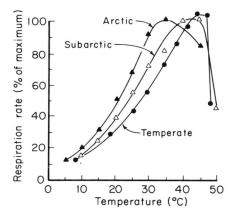

Figure 11   Generalized respiration–temperature curves of plants from different zones. Numbers of species: Arctic, 20; Subarctic, 18; Temperate, 12.

to a value that remains relatively constant during the most favorable part of the season. In autumn, some plants exhibit a new increase in respiration rate, which we attribute to bud formation (Fig. 13). The seasonal respiration pattern in Kola Peninsula plants is nearly the same as that on Wrangel Island. Investigations of the seasonal trend of environmental factors, especially temperature, show that the pattern of respiratory capacity is correlated more strongly with growth (Fig. 14) than with environment.

The maximum relative growth rates (RGR) that we measured were 50–150 mg g^{-1} d^{-1} (Table VI), values close to those found elsewhere in the Arctic (Chapin, 1983; Chapin and Shaver, 1985) and similar to those of temperate plants. Growth rate is highest at the beginning of the season but only for a short period: 7–10 days at Wrangel Island and 10–14 on the Kola Peninsula. Afterward, the aboveground part of the plant grows very slowly. Different

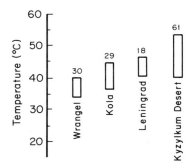

Figure 12   Critical temperatures for respiration of plants from different regions. Values above rectangles show the number of species investigated.

Table V Differentiation of Plant Species According to Critical Temperature of Leaf Respiration

Wrangel Island species

	Temperature (°C)			
	34–35	36–37	38	40
	Primula tschuktschorum	*Astragalus umbellatus*	*Polygonum bistorta*	*Papaver polare*
	Lagotis minor	*Potentilla emarginata*	*Oxygraphis glacialis*	*Rhodiola borealis*
	Nardosmia frigida	*Parrya nudicaulis*	*Oxyria digyna*	*Dryas punctata*
	Oxytropis tschukotica		*Alopecurus alpinus*	*Ranunculus sulphureus*
	Caltha arctica			
	Salix glauca			
	Cardamine pratensis			

Kola Peninsula species

	Temperature (°C)		
	35–38	40	41–44	45
	Cardamine bellidifolia	*Oxygraphis vulgaris*	*Saussurea alpina*	*Ranunculus acris*
	Oxyria digyna	*Salix reticulata*	*Primula elatior*	*Rhodiola borealis*
	Caltha palustris	*Oxytropis sordida*	*Cardamine pratensis*	*Trollius europaeus*
		Papaver lapponicum	*Astragalus frigidus*	
		Polygonum viviparum	*Cicerbita alpina*	
		Tussilago farfara	*Dryas octopetala*	
			Polemonium coeruleum	
			Betula nana	
			Poa alpina	

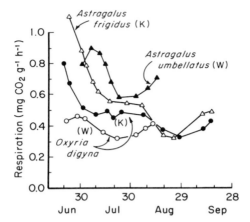

Figure 13   Seasonal leaf respiration patterns in plants from Wrangel Island (W) and the Kola Peninsula (K).

plant parts of some species do not grow simultaneously (Chap. 9). Thus *Primula tschuktschorum* has several periods of growth: after a short initial growth period, root growth stops, and only the aboveground plant parts increase their biomass; when this increase becomes very slow, the roots begin to grow again (Fig. 15). In general, the relative growth rate of roots is less than that of the aboveground parts.

Oxyria digyna had similar relative growth rates ($100–160$ mg g^{-1} d^{-1}) when measured at Wrangel Island, Kola Peninsula, and Leningrad. But because the growing season differed in length at the three sites, plant vigor was very different. Wrangel Island plants had 3–4 leaves, whereas those grown in Leningrad had 7–10 leaves and produced new shoots.

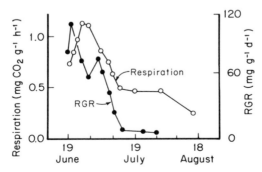

Figure 14   Seasonal patterns of leaf respiration rate at 15° C and shoot relative growth rate (RGR) of shoots in *Primula elatior* from Kola Peninsula.

Table VI   The Maximum Relative Growth Rates of
Wrangel Island Species

	Relative growth rates ($mg\ g^{-1}\ d^{-1}$)	
	Leaves	Whole plant
Primula tschuktschorum	112	92
Lagotis minor	125	50
Oxygraphis glacialis	60	50
Oxyria digyna		148
Cardamine pratensis		91

For a more detailed analysis of the interrelations of growth and respiration, we separated total respiration into its functional components: growth (or construction) respiration and maintenance respiration. These components can be expressed as (1) the percent of total respiration rate and (2) coefficients of growth and maintenance respiration. These coefficients indicate the amount of assimilate that must be respired to provide the energy needed to synthesize one gram of biomass, or to perform all "maintenance" functions in one gram of biomass (i.e., protein turnover, osmoregulation and compartmentation). If the chemical composition of the resulting biomass is known, the theoretical coefficient of growth respiration can be calculated (Penning de Vries *et al.*, 1974). Comparison of such calculated data and those gathered experimentally can show the efficiency of growth processes. Both coefficients can also be compared with those of temperate plants (Ivanova and Semikhatova, 1984; Semikhatova, 1990).

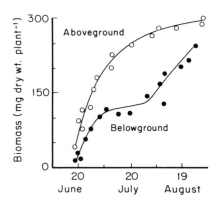

Figure 15   Seasonal patterns of above- and belowground biomass of *Primula tschuktschorum* on Wrangel Island.

Table VII Measured Coefficients of Growth
Respiration in Plants from Wrangel Island and
the Kola Peninsula[a]

	Wrangel Island	Kola Peninsula
Primula tschuktschorum	0.40	
P. elatior		0.25
Oxygraphis glacialis	0.40	
Lagotis minor	0.27	
Oxytropis sordida		0.25
Polygonum viviparum		0.20
Oxyria digyna	0.37	0.26

[a]For comparison, the theoretical coefficient for temperate
plants is 0.23–0.37 (Penning de Vries *et al.*, 1974).

Growth coefficients measured on Wrangel Island and Kola Peninsula plants
were close to the theoretical values calculated by Penning de Vries *et al.*, 1974;
Table VII), indicating that synthetic processes in arctic plants operate at maximum efficiency (Semikhatova, 1990). By contrast, the coefficients of maintenance respiration are much higher than theoretical values and those of
temperate-region plants when compared at the same temperature (Fig. 16),
suggesting that the cost of maintenance processes is higher in the north
(Lechowicz *et al.*, 1980).

We also compared the maintenance-respiration coefficients of mature
leaves in arctic and temperate-region plants at the mean temperatures of their
corresponding habitats (Semikhatova *et al.*, 1979). In such comparisons, the
ranges of the coefficients for arctic and temperate-region plants overlap (see

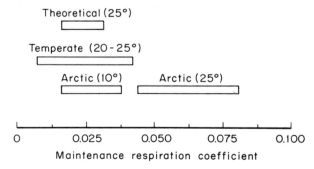

Figure 16 Ranges of maintenance respiration coefficients (g carbohydrate g^{-1} d^{-1}) in arctic (Wrangel Island) and temperate-region plants. For comparison, the ranges for arctic plants
are shown at 25° C and at ambient temperature (10° C); see text.

Table VIII   Quantity and Size of Mitochondria in Temperate-Zone and Arctic Plants[a]

	Mitochondria (% of value for temperate plants)[b]		Length (μm)		Width (μm)	
	Temperate	Arctic	Temperate	Arctic	Temperate	Arctic
Poa pratensis	100	300	0.90	0.80	0.53	0.45
P. annua	100	—	0.87	—	0.52	—
P. alpina	—	220	—	0.98	—	0.55
Astragalus cicer	100	—	0.99	—	0.62	—
A. richardsonii	—	220	—	1.36	—	0.74

[a]Data from Miroslavov and Bubolo (1980).
[b]Number of mitochondria per cross-sectional area of a cell, expressed as percentage of the value in the temperate species.

Fig. 16), suggesting that the high respiration rates of arctic plants are mainly a consequence of higher maintenance costs and opening the question of the mechanisms responsible for the high rates.

Arctic plants have high maintenance-respiration rates for several reasons: First, they have more mitochondria per cell than do leaves of temperate plants (Miroslavov and Bubolo, 1980; Table VIII). Arctic plants also have a lower critical temperature of respiration than do temperate plants. According to Alexandrov (1977), the critical temperature of a physiological process is inversely correlated with protein flexibility. High protein flexibility would enable the enzymes of arctic plants to work at low temperatures at the same rate as the enzymes of plants in temperate environments, but this is achieved at the cost of high protein turnover. If this hypothesis is correct, the protein turnover rate in arctic plants should be proportionally higher than in temperate plants. Since maintenance respiration depends on rates of protein turnover, the high maintenance-respiration rate of arctic plants is consistent with Alexandrov's idea. Consequently, we suggest that a difference in molecular structure of enzyme proteins is partially responsible for the high respiration rate of arctic plants.

IV. Summary and Conclusions

Arctic plants must adapt their main metabolic processes to low temperature, and the mechanisms of this adaptation are apparently the same for photosynthesis and respiration: increased number of structural units (chloroplasts and mitochondria) that compensate for the low-temperature effect and an increase in protein flexibility, which enables the enzymes to operate rapidly even at low temperature.

It seems that arctic plants can readily use these mechanisms, for there are consistent differences both within a season and between sites in the optimum temperature for gas exchange. Such physiological adjustment comes at a high carbon cost, however—not only the cost of new structures formed but also the respiratory cost, that is, a large amount of assimilate respired to provide energy for synthetic and maintenance processes. Serious restrictions like these explain not only incomplete metabolic acclimation to low temperature but also the fact that some species do not change the temperature relations of their metabolism.

Acclimation to light conditions is obviously more complicated, as it requires changes not only of enzymes, but of the whole light-absorbing system. Acclimation to low light requires a high chlorophyll content. Chlorophyll synthesis is costly, however, because it needs a complicated enzyme system and formation of special proteins that make up the chlorophyll–protein complex. These costs include not only carbon molecules but nitrogen as well.

Another way of increasing absorption at low light is to make thicker leaves, which many arctic plants do. But this is also costly, again requiring both carbon and nitrogen. In this regard, the comparatively high carotenoid content of arctic plants is noteworthy. Carotenoid synthesis is cheaper than that of chlorophyll, and carotenoids can absorb light. We speculate that with low chlorophyll contents stemming from low nitrogen supply, carotenoids can partially support photosynthesis in arctic plants. If such speculation is correct, it follows that northern plants are more limited by nitrogen than by the amount of assimilate.

Last, we suggest some answers to the riddle of why the growth of arctic plants ceases at the mildest time of year. Studying the functional components of respiration in the whole plant shows that over the growing season maintenance respiration "evicts" growth respiration by taking precedence in the usage of assimilates. Because maintenance respiration in arctic plants is high, this "eviction" occurs early in the season. We suggest that these relationships between respiratory components play a regulatory role; indeed, they link together the amount of available assimilate, respiration, growth, the length of the growing season, and photosynthesis. This regulatory picture is certainly not complete. Allocation and reserve formation ought to be taken into account, and more data on respiration are needed. The relationships between growth and respiration should therefore be important areas of future research in the Arctic.

Our physiological data permit us to speculate on the response of arctic plants and plant communities to climatic change. Interspecific differences in plant responses to temperature and light are significant in this respect. Acclimation to changes in temperature would be comparatively "easier" for arctic plants than acclimation to new light conditions. Although the characteristics of these responses strictly apply only to the species that were studied, they enable us to predict patterns of acclimation to higher (or lower) temperature and light.

The possible increase in ambient temperatures will not greatly increase photosynthetic rates because CO_2 output in the light will also be greater, and arctic plants have no special features to increase CO_2 reassimilation as do C_3 desert plants. If ambient temperature stabilizes at higher levels, the cost of maintenance respiration will decrease, and the growth of plants will be prolonged. While the climate is changing, however, this maintenance cost could increase, for the process of acclimation itself requires energy. Finally, the presence in the same community of species responding differently to environmental changes will alter community composition.

References

Alexandrov, V. J. (1977). "Cells, Molecules, and Temperature: Conformational Flexibility of Macromolecules and Ecological Adaptation." Springer-Verlag, Berlin.

Berry, I., and Björkman, O. (1980). Photosynthetic response and adaptation to temperature in higher plants. *Annu. Rev. Plant Physiol.* **3**, 491–529.

Billings, W. D., and Mooney, H. A. (1968). The ecology of arctic and alpine plants. *Biol. Rev.* **43**, 481–529.

Bykov, O. D. (1962). Analysis of the kinetics of gas exchange of illuminated plants (in Russian with English summary). *Physiol. Rast.* **9**, 326–334.

Chapin, F. S., III (1983). Direct and indirect effects of temperature on arctic plants. *Polar Biol.* **2**, 47–52.

Chapin, F. S., III, and Shaver, G. R. (1985). Arctic. *In* "Physiological Ecology of North American Plant Communities" (B. F. Chabot and H. A. Mooney, eds.), pp. 16–40. Chapman and Hall, New York.

Gerasimenko, T. V. (1973). Dependence of photosynthesis in tundra plants of Wrangel Island (in Russian with English summary). *Bot. Zh.* **58**, 493–504.

Gerasimenko, T. V., and Shvetsova, W. M. (1989). Main results of ecophysiological investigations of photosynthesis in the Arctic (in Russian). *In* "Ecophysiological Investigations of Plant Photosynthesis and Respiration" (O. A. Semikhatova, ed.), pp. 65–114. Nauka, Leningrad.

Gerasimenko, T. V., and Zalensky, O. V. (1973). Diurnal and seasonal dynamics of photosynthesis in plants of Wrangel Island (in Russian with English summary). *Bot. Zh.* **58**, 1655–1666.

Gerasimenko, T. V., and Zalensky, O. V. (1982). The characteristic features of assimilation of plants of tundra ecosystems (in Russian). *In* "Space Structure of Ecosystems" (Y. P. Kojevnikow, ed.), pp. 128–143. USSR Geographic Society Press, Leningrad.

Gerasimenko, T. V., Popova, I. A., Alexandrova, N. M., and Gagen, T. K. (1988). Chlorophyll content and photosynthesis of Wrangel Island plants in the course of their growth (in Russian with English summary). *Bot. Zh.* **73**, 1085–1103.

Gerasimenko, T. V., Popova, I. A., and Alexandrova, N. M. (1989). On the characterization of the photosynthetic apparatus and photosynthesis in plants of the arctic tundra (Wrangel Island) (in Russian with English summary). *Bot. Zh.* **74**, 669–679.

Ivanova, T. I., and Semikhatova, O. A. (1984). Respiration of arctic plants and its adaptive features (in Russian). *In* "The Adaptation of Organisms to the High North Environments" (J. L. Martin *et al.*, eds.), pp. 59–64. Academy of Sciences of Estonia, Tallinn.

Ivanova, T. I., and Vaskovsky, M. D. (1976). Respiration of plants of Wrangel Island (in Russian with English summary). *Bot. Zh.* **61**, 324–331.

Ivanova, T. I., Lokteva, T. N., and Shukhtina, H. G. (1983). Respiration of Kola plants (in Russian). *Bot. Zh.* **68**, 1637–1643.

Ivanova, T. I., Vaskovsky, M. D., and Vladimirov, V. K. (1985). Seasonal changes in respiration of herbaceous plants from Wrangel Island (in Russian). *Bot. Zh.* **70**, 1675–1682.

Ivanova, T. I., Semikhatova, O. A., Judina, O. S., and Leina, G. D. (1989). The effect of temperature on the respiration of plants from different zones (in Russian). *In* "Ecophysiological Investigations of Plant Photosynthesis and Respiration" (O. A. Semikhatova, ed.), pp. 140–167. Nauka, Leningrad.

Kislyuk, I. M., and Sheinina, G. A. (1985). The investigation of photosynthesis in herbaceous plants of arctic and taiga zones (with the aid of polynomial regression method) (in Russian with English summary). *Bot. Zh.* **70**, 169–179.

Kostytschew, S., Tschesnokow, W., and Bazyrina, K. (1930). Investigation of diurnal rates of photosynthesis on the beach of the Arctic Ocean (in German). *Planta* **11**, 160–169.

Lechowicz, M. J., Hellen, L. E., and Simon, J.-P. (1980). Latitudinal trends in the responses of growth respiration and maintenance respiration to temperature in a beach pea, *Lathyrus japonicus. Can. J. Bot.* **58**, 1521–1524.

Ledyaikina, N. A., and Gerasimenko, T. V. (1983). CO₂ gas exchange of Chibin plants (in Russian). *In* "Ecophysiological Investigations of Photosynthesis and Water Regime in the Field" (R. K. Saljaev, ed.), pp. 56–64. Siberia Dep., Academy of Sciences of the USSR, Irkutsk.

Lubimenko, V. N. (1916). Conversions of plastid pigments in living plant tissues (in Russian). *Proc. Acad. Sci. St. Petersburg* **33**, 1–274.

Lukyanova, L. M., Lokteva, T. N., and Bulytscheva, T. M. (1986). "Gas Exchange and Pigment Structure of Plants of the Kola Subarctic" (in Russian) (V. L. Vosnesensky, ed.). Kola Dep., Academy of Sciences of the USSR, Apatit.

Mayo, J. M., Despain, D. E., and van Zinderen Bakker, E. M., Jr. (1973). CO₂ assimilation by *Dryas integrifolia* on Devon Island, Northwest Territories. *Can. J. Bot.* **51**, 581–588.

Miroslavov, E. A., and Bubolo, L. S. (1980). The ultrastructure of the cells of leaf chlorenchyma in some arctic plants (in Russian with English summary). *Bot. Zh.* **65**, 1523–1529.

Miroslavov, E. A., and Jakovleva, O. V. (1983). The structure of mesophyll chloroplasts of some northern species (in Russian with English summary). *Ekologiya (Sverdl.)* **6**, 23–29.

Penning de Vries, F. W. T., Brunsting, A., and Van Laar, H. H. (1974). Product requirements and efficiency of biosynthesis: A quantitative approach. *J. Theoret. Biol.* **45**, 339–377.

Piankov, W. I. (1980). Analysis of the temperature response of photosynthesis of related species of arctic and temperate plants (in Russian with English summary). *Ekologiya (Sverdl.)* **3**, 37–41.

Popova, I. A., Maslova, T. G., and Popova, O. F. (1989). The characteristic features of the pigment apparatus of plants from different botanical–geographical zones (in Russian). *In* "Ecophysiological Investigations of Plant Photosynthesis and Respiration" (O. A. Semikhatova, ed.), pp. 115–139. Nauka, Leningrad.

Scholander, P., and Kanwisher, J. (1959). Latitudinal effect on respiration in some northern plants. *Plant Physiol.* **34**, 574–576.

Semikhatova, O. A. (1990). "Energetics of Respiration in Plants at Normal Conditions and under Ecological Stress" (in Russian). 48th Timiryasev Lecture. Nauka, Leningrad.

Semikhatova, O. A., Ivanova, T. I., and Golovko, T. (1979). Maintenance respiration of arctic plants (in Russian with English summary). *Physiol. Rast.* **26**, 1092–1102.

Semikhatova, O. A., and Shukhtina, H. G. (1973). The respiration rate of several species of plants of Heith Island (Franz Josef Land) (in Russian with English summary). *Bot. Zh.* **58**, 1816–1819.

Shvetzova, V. M. (1970). Dependence of photosynthesis on temperature in some plants of Western Taimyr. *Bot. Zh.* **55**, 701–705.

Stocker, O. (1935). Assimilation and respiration of West Java Island tropic trees (in German). *Planta* **24**, 402–445.

Tieszen, L. L. (1973). Photosynthesis and respiration in arctic tundra grasses: Field light intensity and temperature responses. *Arct. Alp. Res.* **5**, 239–251.

Tieszen, L. L. (1975). CO₂ exchange in the Alaskan arctic tundra: Seasonal changes in the rate of photosynthesis of four species. *Photosynthetica* **9**, 376–390.

Vosnesensky, V. L. (1989). CO₂ gas exchange and its components (in Russian). *In* "Ecophysiological Investigations of Plant Photosynthesis and Respiration (O. A. Semikhatova, ed.), pp. 65–114. Nauka, Leningrad.

Vosnesensky, V. L., Ledyaikina, N. A., and Achmedov, A. A. (1984). CO_2 gas exchange of desert plants from southeast Karakum (in Russian with English summary). *Bot. Zh.* **69**, 24–32.

Zalensky, O. V. (1977). "Ecophysiological Aspects of Photosynthesis Investigations" (in Russian). 37th Timiryasev lecture. Nauka, Leningrad.

9

Phenology, Resource Allocation, and Growth of Arctic Vascular Plants

Gaius R. Shaver and Jochen Kummerow

I. Introduction

Plant growth in the Arctic is constrained spatially, temporally, climatically, and nutritionally. One might expect that, taken together, all these constraints would strongly limit the diversity of plant growth patterns, or growth "strategies," that are possible in such an environment. Yet considerable diversity in plant growth pattern exists, both within and among vegetation types, despite the relatively small number of species in the arctic flora (Bliss, 1981; Andreev and Aleksandrova, 1981; Matveyeva, 1988; Chap. 5). In this chapter we will describe some of the major patterns of plant growth and their controls in the Arctic, focusing on vegetative and reproductive phenology, biomass allocation, and storage and internal recycling of essential growth resources.

The extremes of the arctic environment also suggest that most or all species should be highly responsive to changes in environmental variables such as temperature, light, or nutrient availability. In fact, however, arctic plants vary widely in their responsiveness to environmental changes. In the second part of this chapter, we will briefly review the growth responses of arctic plants to changes in their environment and discuss the relationships between growth rates of individual organs and the productivity of whole plants and vegetation.

Phenology, allocation, and growth of arctic plants have been relatively well studied in the past, and a number of excellent reviews are available (Warren-Wilson, 1966; Bliss, 1962, 1971; Billings and Mooney, 1968; Bliss *et al.*, 1973; Lewis and Callaghan, 1976). Extensive compilations and summaries of data from a wide range of arctic sites are available in the various IBP (International Biological Programme) volumes published in the 1970s and early 1980s (e.g., Wielgolaski, 1975; Bliss, 1977; Tieszen, 1978; Bliss *et al.*, 1981). Rather than attempting another complete review of this subject, we will focus on a few points that are of particular relevance to the problem of climatic change in the Arctic and on more recent research.

The diversity of arctic plant growth patterns and their controls has considerable implications for the effects of global climate change on the structure and productivity of arctic vegetation. Phenology, biomass allocation, and growth rates are all sensitive to climatic variables such as temperature, but sensitivity varies among species. Thus it is difficult to extrapolate from process-level studies of individual species and organs to complete vegetation. The problem of how to scale up from process studies has not yet been solved, but fortunately, in the Arctic we can accompany simulation modeling with whole-ecosystem experiments to provide better validation of predictions than in many other biomes.

II. Phenology, Allocation, and Storage

A. Phenology

The total length of the arctic growing season is rarely more than 100 days and may be less than 50 days in the high Arctic or in areas of late snowmelt (Miller, 1982b; Bliss, 1977; Chapin and Shaver, 1985a). During this short time the entire year's production of leaves, stems, roots, and reproductive parts must take place. Yet despite the short time available, considerable variation exists in the timing of individual growth events, or phenophases, both within and among species (Sørenson, 1941; Bliss, 1956; Wielgolaski and Karenlampi, 1975; Svoboda, 1977; Webber, 1978; Murray and Miller, 1982). Within a single site, this variation is due largely to microclimatic differences among species in the location of their perennating buds and to intrinsic differences in allocation patterns that affect the timing of growth requirements. The timing of the growth of above- versus belowground organs also varies because the optimal conditions for growth of these organs occur at different times, with the highest air temperatures and light intensities in late June and July, while most of the soil is still frozen solid (Chapin and Shaver, 1985a). After it finally thaws, the soil may not freeze again until late September or October, a month or more after average air temperatures go below freezing, and aboveground growth stops.

Arctic plants differ from their temperate-zone counterparts mainly in the rapidity with which phenological events, such as leaf expansion, are completed (Fig. 1) Such rapid growth is usually supported by draw-down of carbon and nutrient storage reserves as well as concurrent uptake (discussed below and in Chap. 16); storage of nutrients is particularly important if aboveground growth occurs when roots are still frozen (Chapin *et al.*, 1975; Shaver and Chapin, 1986). In other cases, however, arctic plants take much longer than temperate-zone plants to complete some aspects of their growth; reproductive growth, for example, is a multiyear process in many species in which most of the differentiation of flower primordia takes place in buds formed at least one year before anthesis (Sørenson, 1941).

The timing of leaf growth often varies among the major plant growth forms (Fig. 1). In tussock tundra at Toolik Lake, Alaska, for example, the two major sedge species (*Eriophorum vaginatum* and *Carex bigelowii*) begin leaf growth first, probably because their intercalary meristems lie just at the soil–moss—tussock surface, in the warmest microenvironment at that time. The dense, erect tussock form of *E. vaginatum*, on which *C. bigelowii* also grows, provides an especially warm and favorable microclimate, emerging from snow cover earlier than the lower intertussock areas (Chapin *et al.*, 1979). Erect deciduous shrubs (such as *Salix* and *Betula* species) begin leaf expansion about two weeks later. Prostrate deciduous species and forbs such as *Rubus chamaemorus* or *Petasites frigidus*—which have apical meristems 1–10 cm below the soil

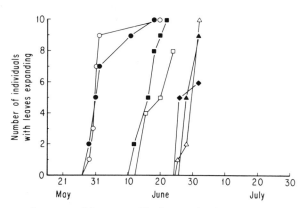

Figure 1 Seasonal pattern of the onset of leaf expansion in common or dominant species of tussock tundra near Toolik Lake, Alaska. Data represent the numbers of individual tillers or branches with leaves expanding or elongating, out of 10 individuals marked at the beginning of the season. Where the number of individuals failed to reach 10 for a given species, the remaining tillers or branches failed to grow at all. Circles represent the graminoid species *Eriophorum vaginatum* (filled) and *Carex bigelowii* (open); squares represent the deciduous shrubs *Salix pulchra* (filled) and *Betula nana* (open); filled diamonds represent *Rubus chamaemorus*; triangles represent the evergreen shrubs *Ledum palustre* (filled) and *Vaccinium vitis-idea* (open). [Unpublished data of G. R. Shaver and F. S. Chapin III.]

surface, often in the lower intertussock spaces—are delayed still further. One might expect early establishment of leaf area to be important in deciduous species because the loss of only a few days' photosynthesis represents a significant portion of the potential seasonal total. Evergreen shrubs, on the other hand, start the season with a significant amount of overwintering leaf mass, which regreens even before the deciduous leaves are fully expanded. Evergreen leaf buds do not begin to open until 3–4 weeks after graminoid leaf elongation begins and after the erect deciduous shrubs have nearly completed their leaf expansion.

These seasonal patterns of leaf growth also vary considerably among sites, especially in relation to climatic and microclimatic variation along latitudinal and topographic gradients (Sørenson, 1941; Bliss, 1956; Billings, 1973; Murray and Miller, 1982). Where the growing season is extremely short, as in the High Arctic or beneath late-melting snowbanks, there may be little discernible difference in the timing of growth events between species or growth forms. On the other hand, variation in the onset of growth among individuals of a single species inhabiting a range of microenvironments may vary by two weeks or more (Wielgolaski and Karenlampi, 1975; Murray and Miller, 1982). The major factor controlling the onset of growth among sites, and also among years, is the timing of snowmelt and above-freezing temperatures.

Root elongation begins as soon as the soil is thawed and elongation is possible; depending on the depth of individual root tips, this occurs anytime from 1 to 6 weeks after average air temperatures rise above freezing and the snow cover melts (Shaver and Billings, 1975; Kummerow *et al.*, 1983). Although roots elongate faster at higher temperatures, there is no evidence that low temperatures prevent root growth as long as the soil is thawed (Shaver and Billings, 1977). The distribution of roots within the soil appears to be an intrinsic feature of each species' basic growth pattern rather than a reflection of any particular patterns or gradients in the soil environment (Shaver and Billings, 1975, 1977; Shaver and Cutler, 1979; Miller *et al.*, 1982).

The seasonal pattern of root growth is also related to overall plant growth demands and the inability of the plants to satisfy all demands simultaneously (Kummerow *et al.*, 1983). In tussock tundra, for example, evergreen species begin root growth simultaneously with leaf regreening, whereas in deciduous species the most intensive period of root growth is delayed until leaf expansion is well underway. Part of this difference between evergreen and deciduous species may be due to the evergreens' typically shallow rooting depth, so that their roots thaw earlier; part of the difference may also be due to the intense energy and nutrient requirements of early, rapid leaf growth in the deciduous species. Graminoid roots tend to grow according to schedules that are highly species-specific (Shaver and Billings, 1977). Root tips of *Eriophorum* species start the season near the surface and follow the progress of soil thaw downward, growing at the freeze–thaw interface until mid-August (Bliss, 1956). Roots of *Carex* species, which are concentrated at a depth of 10–25 cm, cannot grow until they are thawed out in late June or early July. Shallow-rooted grass species, such as *Dupontia fisheri,* often grow rapidly for a short while but stop growth by mid-July (perhaps because of a shift in allocation priorities toward seed production or storage for the winter).

Flowering and seed development occur throughout the growing season in arctic plants, but most species reach anthesis within 2–5 weeks after snowmelt (Sørenson, 1941; Bliss, 1956; Wielgolaski and Karenlampi, 1975; Svoboda, 1977; Webber, 1978; Murray and Miller, 1982). Some species, such as *Eriophorum vaginatum,* actually begin culm elongation before the snow is fully melted. *Salix* and *Betula* species also flower early, before vegetative buds open. The sequence of species flowering is in general related to time since snowmelt, which may vary by 2–4 weeks between years, or even longer where late-melting snowbanks occur. In virtually all arctic species, inflorescence buds are formed the year before flowering, so climate during the current year has no effect on the maximum potential number of inflorescences. Weather and herbivory during the winter may dramatically reduce the number of inflorescence buds available at the start of a growing season, and weather during the season may determine the success of those inflorescences in producing viable seed, but these variables have not yet been systematically studied in the Arctic (Chester and Shaver, 1982).

B. Allocation

Most of the live biomass of arctic vegetation occurs belowground but there is considerable variation among vegetation types (Kjelvik and Karenlampi, 1975; Webber, 1977; Bliss, 1977; Jonasson, 1982; Miller *et al.*, 1982). Aboveground: belowground biomass ratios range from about 1:1 or higher in polar deserts and semideserts to 0.05:1 or lower in wet sedge tundras (Bliss, 1981). In general, the lowest aboveground:belowground ratios occur in wet, herbaceous vegetation and the highest ratios in relatively well-drained shrubby vegetation.

Among species, belowground allocation also varies considerably (Table I). For example, in Alaskan tussock tundra, the two dominant sedge species have aboveground:belowground mass ratios of less than 0.2:1, compared with ratios of about 1:1 for the two dominant woody evergreens. Deciduous shrubs are intermediate, with ratios of 0.6:1 to 0.8:1. These ratios for arctic species are all below the normal ranges for species of Mediterranean-type ecosystems or tropical forests, for example, which fluctuate between 1.5 and 4.0 (Kummerow, 1981; Klinge, 1973).

Much of the belowground biomass of arctic plants is not roots but belowground stems or rhizomes, which function primarily as storage tissues and as a mechanism of clonal spread (Shaver and Cutler, 1979; Bliss, 1981). The mass of fine roots, which are actually responsible for most nutrient and water uptake, is even smaller (approximately 3–5% of belowground biomass; Kummerow, unpubl. data). A more useful index of relative allocation to nutrient uptake versus carbon uptake in plants is the leaf:fine root ratio; in tussock tundra this ranges from 0.6:1 in sedges to 3.4–5.5:1 in evergreens to 4.1–12.1:1 in deciduous shrubs (Kummerow, unpubl. data).

Table I Above- and Belowground Biomass of Individual Species in a Moist Tussock Tundra at Toolik Lake, Alaska[a]

	Total biomass ($g\ m^{-2}$)	Percentage of total biomass for each species							
		Above-ground	Below-ground	Stems above-ground	Leaves	Rhizomes and stems belowground	Old roots ($> 1\ yr$)	New roots	Leaf:new root ratio
V. vitis-idaea (n = 10)	53.4 ±3.79 S.E.	49.7	50.3	12.6	37.1	35.2	10.1	4.8	5.5
L. palustre (n = 10)	9.4 ±1.16 S.E.	53.4	46.6	37.7	15.8	42.5	2.1	2.5	3.4
C. bigelovii (n = 20)	86.0 ±13.29 S.E.	7.8	92.2	—	7.8	36.9	42.7	12.9	0.6
E. vaginatum (n = 12)	35.5 ±7.42 S.E.	16.0	84.0	—	39.8	33.8	—	25.8	0.6
B. nana (n = 20)	66.3 ±9.61 S.E.	38.2	61.8	31.2	7.0	53.0	7.1	1.7	4.1
S. pulchra (n = 20)	26.9 ±1.72 S.E.	43.6	56.4	31.5	12.1	52.9	2.8	1.0	12.1

[a] Unpublished data from J. Kummerow.

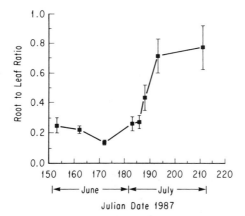

Figure 2 Seasonal changes in fine root:leaf mass ratio in *Betula nana* in tussock tundra near Toolik Lake, Alaska. [Unpublished data of J. Kummerow.]

Leaf:fine root ratios are not static but change seasonally as a result of the different patterns of leaf and root growth (Fig. 2). The precision of the fine-root mass measurements is not sufficient, however, to detect any significant year-to-year variation in relation to climate, either in fine root:leaf ratios or in the overall root:shoot or aboveground:belowground biomass ratios. In controlled environments, proportional allocation to roots in sedges tends to increase with increasing overall plant growth rate (Kummerow and Ellis, 1984). The root:shoot ratio in arctic grasses is relatively insensitive to nutrient availability, but field experiments with *E. vaginatum* have shown significant decreases in fine root:leaf ratio with the addition of fertilizer (Shaver *et al.*, 1986a).

Leaf:stem mass ratios also vary considerably among arctic plant species (Bliss, 1981), as does the proportion of the total stem mass that lies above ground (Fig. 3). In sedges and in evergreen and deciduous woody shrubs, leaf mass is usually only 10–25% of the total stem mass. In grasses and forbs, leaf mass often exceeds stem mass. For all growth forms, at least half of the total stem mass is belowground; in the graminoids and forbs this proportioning results from rhizomes actually spreading actively beneath the surface, whereas in the woody shrubs "belowground" stems usually start out aboveground and are later overgrown by mosses.

The proportional allocation of annual primary production differs considerably from the allocation of biomass because the various organs differ in their turnover rates. Thus, although leaves are only a small proportion of arctic plant biomass, annual leaf growth accounts for a large proportion of aboveground and total production (Fig. 4). Much less information is available on root production in the Arctic, but it usually at least equals aboveground production and may be as much as five or six times higher (Bliss, 1981; Kummerow *et al.*, 1987). In most previous studies, secondary stem production in

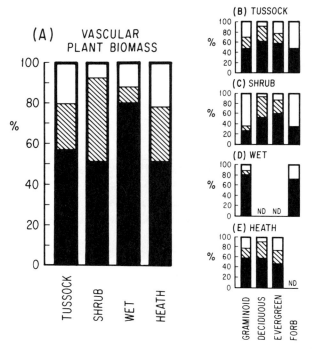

Figure 3 Proportional allocation of vascular plant biomass in four vegetation types near Toolik Lake, Alaska. *Tussock* indicates typical tussock tundra; *shrub* indicates deciduous shrub-dominated riparian shrub tundra; *wet* indicates wet sedge tundra; *heath* indicates a low lichen–heath tundra. In (A), biomass allocation for each vegetation type as a whole is presented, broken down into leaves (blank portions), aboveground stems (diagonal striping), and below-ground stems (filled portions). In (B) through (E), percentage allocation of the same biomass categories is shown within individual vascular plant growth forms at each site. ND, no data (biomass less than 0.1 g m^{-2}). [Reprinted from Shaver and Chapin (1991) with permission from the Ecological Society of America.]

arctic plants was considered negligible with respect to root, leaf, and apical stem production. Secondary stem production by shrubby species is significant in some deciduous communities, however, nearly equalling leaf production (Shaver, 1986).

Inflorescences and seeds account for only a small portion of either production or biomass in most arctic plants in most years (Chester and Shaver, 1982). Allocation to sexual reproduction varies greatly among years, however. This annual variation has not been well documented in terms of production or biomass, but inflorescence density (number of inflorescences per square meter), for example, varies by more than two orders of magnitude in *E. vaginatum* (Shaver *et al.*, 1986b). The climatic controls on this allocation are not well understood, despite the fact that the annual variation in flowering is uniform over broad regions.

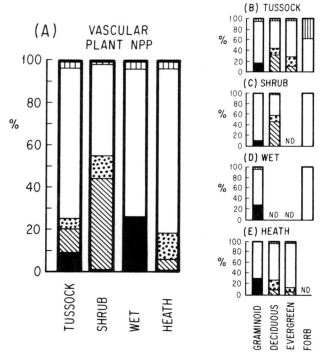

Figure 4   Proportional allocation of net primary production (NPP) by vascular plants in the same four vegetation types as in Fig. 3. In (A), total NPP (excluding roots) is presented. Inflorescence production is indicated by vertical stripes, leaf production by the blank portions of each bar, apical stem growth (current year's twigs) by the dotted portions, secondary stem growth by diagonal stripes, and belowground rhizome growth by the solid portions. In (B) through (E), the percentage allocation of the same tissue categories is shown within individual vascular plant growth forms at each site. ND, no data (NPP less than 0.1 g m⁻²). [Reprinted from Shaver and Chapin (1991) with permission from the Ecological Society of America.]

C. Storage

Arctic plants typically store large amounts of mobile carbon and nutrient reserves, but species differ greatly in the tissues where the reserves are stored, the chemical form of the reserves, and the seasonal patterns of their use. In many arctic (and other) species, these carbon and nutrient reserves are important sources of the substrates for growth when growth requirements cannot all be met concurrently by new uptake. Carbon and nutrient reserves are drawn down during periods of growth and replenished when growth has slowed or stopped. In other species, it is unclear why reserves, especially carbon reserves, are accumulated in such high concentrations.

Reserves are stored in all of the perennial plant organs (Chapin *et al.*, 1980; Chapin and Shaver, 1989a). In grasses and sedges, most of the storage is below ground, in rhizomes and leaf sheaths; in woody shrubs, the storage may be either above or below ground, in stems or old evergreen leaves. Deciduous shrubs store most of their reserves in old stems, and evergreens use old leaves (although Jonasson [1989] has shown that in some evergreen species, this form of storage may be less important than previously believed; see Chap. 16). The concentrations of storage reserves are usually lower in woody organs than in herbaceous organs like rhizomes, but because the total woody mass is large, the total storage is also large.

The chemical composition of organs that function primarily for storage (such as rhizomes) varies much more among species than does the chemistry of organs with more specific functions (such as green leaves) (Chapin and Shaver, 1988). Forbs and graminoids are especially variable in their storage chemistry. Carbon stores vary from predominately polysaccharides to predominately sugars, with polysaccharides usually predominating in lichens and mosses, and shrubs storing mostly sugars. Nitrogen is stored either as protein or as amino acids, with protein predominating in shrubs. Phosphorus is stored in a wide variety of forms, with no single fraction clearly predominating in any shrub species; in graminoids, either soluble organic phosphorus or inorganic phosphorus is sometimes accumulated in high concentrations.

The use of storage reserves during periods of rapid growth is reflected in decreased concentrations of these compounds in storage tissues (e.g., Chapin *et al.*, 1980; Chapin and Shaver, 1989a). Substrates that are strongly limiting to growth often show the most dramatic fluctuations in concentration. For example, in *Eriophorum vaginatum* at Toolik Lake, Alaska, the total pool size of available amino acids decreases to less than half its initial size during rapid, early-season leaf growth (Fig. 5). At the same time, the pool size of "storage," nonstructural carbohydrate actually increases slightly, suggesting that the carbon requirements for growth are more than adequately met by concurrent photosynthesis.

Another form of storage occurs in older organs that die during periods of rapid growth, but before they die they translocate much of their nitrogen and phosphorus content into new, rapidly growing tissues. Old leaves of evergreen shrubs are a particularly good example of this pattern (Reader, 1978; Shaver, 1981; Chapin and Shaver, 1989a; but see Jonasson, 1989, and Chap. 16), and some evidence indicates that rhizomes of *E. vaginatum* function similarly (Shaver *et al.*, 1986a). In graminoids with continuously overlapping leaf generations, as much as 90% of the nitrogen supply to new leaves may come from simultaneous senescence of old leaves (Jonasson and Chapin, 1985). In deciduous species this retranslocation occurs in the fall, when it is reflected in increased storage nitrogen and phosphorus concentrations in woody stems (Chapin *et al.*, 1980; Chapin and Shaver, 1988).

The large pools of storage reserves that are typical of arctic plants do not necessarily indicate an unusually strong dependence on storage in the Arctic relative to other parts of the world. When comparisons of carbon and nutri-

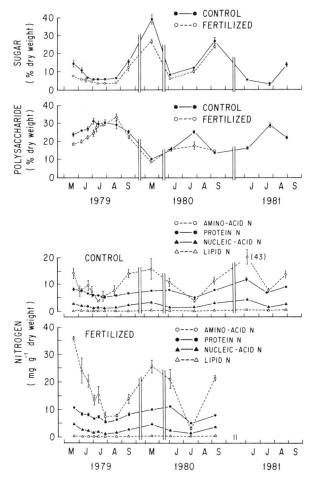

Figure 5 Seasonal changes in concentration of mobile carbon and nitrogen pools in rhizomes of *Eriophorum vaginatum* near Toolik Lake, Alaska. [Reprinted from Chapin *et al.* (1986) with permission from the British Ecological Society.]

ent storage are confined to close relatives of arctic species growing in the Temperate Zone, the arctic plants often do not differ with respect to the size of their storage pools or their dependence on storage in support of rapid growth (Chapin and Shaver, 1989b). Thus it is not clear that dependence on storage is a particular adaptation to arctic conditions. On the other hand, the fact that arctic soils are frozen solid for several days to weeks after air temperatures go above freezing suggests that at least some storage of nitrogen and phosphorus must be necessary to take advantage of aboveground growing conditions early in the year.

It is uncertain how global climate change may affect the role of carbon and nutrient storage in support of arctic plant growth. If the permafrost melts and soils warm dramatically, uptake of nutrients early in the season may be possible, and dependence on storage could be reduced. If nutrient mineralization rates, and thus nutrient availability, increase, there may be a lesser advantage to highly efficient recovery of nutrients associated with tissue senescence. These changes could certainly affect the competitive balance controlling the species composition of arctic vegetation and may make it possible for other species to invade. The greatest uncertainties in our current knowledge have to do with the ability of arctic plants to shift from storage and recycling to concurrent uptake of carbon and nutrients in support of new growth.

III. Growth Rates and Productivity

Arctic plants are especially well adapted to growth at low temperatures, often having lower optimum temperatures for growth than plants from the Temperate Zone and less temperature sensitivity at temperatures below 10° C (Chapin, 1983, 1987). On the other hand, the temperatures at which arctic plants grow are rarely optimal, even for arctic species. Furthermore, arctic plants do not necessarily concentrate their growth, or even grow most rapidly, in the warmest microsites. For example, root growth of *Eriophorum* species is concentrated deep in the thawed layer of soil, at temperatures of 0–4° C (Bliss, 1956; Shaver and Billings, 1975; Kummerow *et al.*, 1987), despite the much warmer temperatures near the soil surface. The relative growth rates of arctic plants are fairly high when compared with temperate-zone species of similar growth form (Chapin and Shaver, 1985b; Chapin, 1987; Chap. 8), suggesting that differences in productivity between the Arctic and the Temperate Zone may be due as much to the short growing season and the low overwintering biomass aboveground in the Arctic as to low temperature per se.

Some arctic plants do possess features that ameliorate their thermal environment to some extent (Sørenson, 1941; Bliss, 1962; Billings and Mooney, 1968). The most important of these is probably their low stature and relatively smooth canopy structure. Low, compact "cushion plants," in particular, tend to maximize the thickness of the boundary layer of air around the leaves, reducing heat loss by convection. In contrast, some of the advantages of the erect, tussock growth form are probably early emergence from snow cover, increased thawing, and higher root temperatures in the central column of the tussock (Chapin *et al.*, 1979). Inflorescences of many species of arctic plants are heliotropic or have morphological features that increase their temperature, such as parabolically shaped flowers or dense pubescence.

Growth rates and productivity of arctic plants vary widely among sites, often in association with dramatic changes in the growth-form composition of the vegetation (e.g., Miller, 1982a,b; Muc and Bliss, 1977; Jonasson, 1982). The most rapidly growing and productive species are typically deciduous shrubs and grasses; the most slowly growing are usually ericaceous evergreens or sedges. This variation in growth rate, productivity, and vegetation composition is more strongly correlated with topography, snow distribution, and soil characteristics, especially nutrient availability, than with air or soil temperatures.

Several existing lines of evidence might be used to predict the effects of global climate change on plant growth rates and primary production. These include both laboratory and field experiments (e.g., Oberbauer *et al.*, 1986), long-term monitoring of single sites (e.g., Shaver *et al.*, 1986b), transplant and phytometer studies (e.g., Chapin and Chapin, 1981), and intersite comparisons in relation to environment (Wielgolaski *et al.*, 1981). Unfortunately, however, the results of these studies often lead to very different predictions about the effects of global climate change on growth of arctic plants.

For example, when grown in the laboratory, arctic plants respond strongly to variation in temperature, light intensity, and nutrient availability, especially when these variables are manipulated within the ranges experienced by plants in the field. Almost invariably, arctic plants grow faster at higher temperatures, higher light levels, and higher nitrogen and phosphorus availability. Yet plants actually growing in the field are much more variable in their responsiveness to analogous manipulations. In some factorial fertilizer experiments, growth of all species responds strongly and essentially identically (Shaver and Chapin, 1980), whereas in others, individual species show widely different responses to the same treatments (Fig. 6; Chapin and Shaver, 1985b). From the laboratory studies and the field experiments where all species respond strongly and uniformly, one might predict major changes in plant growth and primary production with little change in species composition, at least among the dominant species. Where species respond individualistically, major changes in species composition could occur, with little or no change in overall production or biomass because some species would increase while others would decrease.

Similarly, although primary production, plant biomass, and plant growth rates are often well correlated with climatic variables along environmental gradients (Wielgolaski *et al.*, 1981), it is often unclear whether the correlation is due to the effects of climate directly on plant growth, or whether growth is actually controlled by some other variable (such as nitrogen supply) that is also correlated with climate (Chapin, 1983, 1987). Within single species, it often appears that climatic correlation is due to genetically based, ecotypic variation in plant size and growth rates rather than direct control by temperature, nutrient availability, or some other climatic factor (Shaver *et al.*, 1979; Shaver *et al.*, 1986b; Fetcher and Shaver, 1990). Thus, predictions of the effects of climate change based on differences in plant growth along climatic gradients may be based on incorrect assumptions.

One aspect of arctic plant growth that is potentially critical in a changing climate is the interaction between potential growth rate and photoperiod. In growth chamber experiments, arctic plants normally require photoperiods close to 24 hours to flower or to grow normally. At shorter photoperiods,

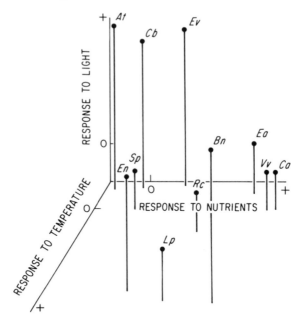

Figure 6   Growth response (as percent of maximum response shown by any species) of tussock and wet sedge tundra species to variations in nutrient, light, and air temperature regimes. Species are *Eriophorum vaginatum* (*Ev*), *E. angustifolium* (*Ea*), *Carex bigelowii* (*Cb*), *C. aquatilis* (*Ca*), *Betula nana* (*Bn*), *Salix pulchra* (*Sp*), *Rubus chamaemorus* (*Rc*), *Ledum palustre* (*Lp*), *Empetrum nigrum* (*En*), *Vaccinium vitis-idaea* (*Vv*), and *Aulacomnium turgidum* (*At*). [Reprinted from Chapin and Shaver (1985b) with permission from the Ecological Society of America.]

growth is reduced more than can be accounted for by the reduction in photoperiod. For example, root elongation rate in *Eriophorum angustifolium* is strongly related to temperature under controlled conditions with continuous light (Fig. 7; Shaver and Billings, 1977). Daily elongation rates with an 18-hour photoperiod are three-fourths the rates under continuous light, but when photoperiod decreases to 15 hours, there is no root elongation at all. The implications of this behavior are important because if global warming occurs, and temperatures favorable to plant growth extend earlier into spring and, especially, later into the fall, many arctic plants may not be able to take advantage of this extended growing season. (Photoperiod effects are likely to be less important when the growing season is extended into the spring than into the fall because in much of the Arctic, the photoperiod will still be almost 24 hours even if snowmelt begins a month or more earlier.) Growth of other species, like *Carex aquatilis*, is less affected by declining photoperiod, and thus a shift in species composition may result from a lengthening of the growing season even though root elongation in the two species has virtually identical temperature relationships under 24-hour photoperiods.

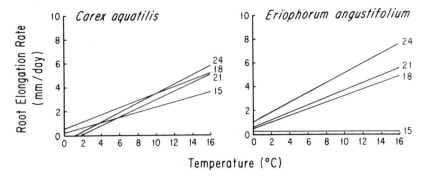

Figure 7 Root elongation rate in mm day^{-1} as a function of temperature and daily photoperiod (photoperiod in hours indicated at the right end of each regression line). [Adapted from Shaver and Billings (1977).]

A second important characteristic of arctic plants is that they often do not respond immediately to an improvement in growing conditions. For example, field fertilizer experiments usually show little or no growth response within the first year (Shaver *et al.*, 1986a). Instead, storage reserves are accumulated, and a dramatic growth increase happens in the second year. Flowering responses may take two or more years to develop. The reasons for this delayed response are not entirely understood, but they may be related to the time needed to initiate new meristematic sinks for the increased resource supply. The lack of a significant growth response in the first year is consistent with the hypothesis that apical meristem growth and leaf number are largely determined when buds are formed at the end of the previous year. The delayed response is important in the context of global climate change because, if species vary in the rates of their responsiveness to a change in resource availability, then the most responsive species should be the greatest winners or losers.

IV. Conclusions

One conclusion to be drawn from this review is that the direct effects of global climate change on arctic plants are likely to be strongly constrained by existing controls on phenology, allocation, and growth. One example of such an existing control is the induction of autumn senescence even when temperatures are favorable for growth, as discussed above. Another example is the generally low nutrient availability (despite large standing stocks of nutrients) in arctic soils, which restricts the plants' ability to sustain an initially large increase in growth when air temperatures are increased (Shaver *et al.*, 1986a). Some of these constraints, such as the controls on autumn senescence, have a strong genetic component to them that is unlikely to change except over very long time spans, as a result of genetic adaptation. Other constraints, such as nutrient supply from the soil, are influenced directly by climate change. The nature and importance of these latter constraints is thus likely to change over years to decades in response to climate, whereas the genetic constraints may take decades to millennia to change.

A second conclusion is that arctic plant species vary greatly in their controls on phenology, allocation, and growth. As a result, they should also vary in their responses to changes in temperature, CO_2 concentration, or precipitation regimes. Some of these controls may be related to growth form or phylogeny (e.g., graminoids do not produce woody stems, and most of their stem mass will always be below ground), but other controls are much more individualistic (such as the effect of day length on root growth of *Eriophorum* versus *Carex*). Because species respond individualistically, it is easier and more reliable to generalize about effects of climate change at the level of species-groups or growth forms with similar patterns of allocation, vegetative phenology, and storage (Shaver and Chapin, 1986).

Perhaps the most important issue in predicting responses of arctic plants to climate change is the relative importance of the direct climatic effects on plants versus indirect effects on other variables (such as nutrient availability) affecting phenology, allocation, or growth. Presumably the direct responses of plants (to a change in air temperature, for example) will be rapid initially, but further plant responses will be limited by the rate of change in these other variables. In the long term, changes in the vegetation itself will feed back on the direct responses to climatic change—for example, through changes in the chemical quality of litter input to the soil (Chap. 13) or through the effects of a change in canopy structure on soil temperatures and the thawing of permafrost. Before long-term predictions of plant response to climatic change can be made, we need a better understanding of indirect climatic effects and long-term feedbacks caused by changes in the vegetation itself.

V. Summary

This chapter describes the major patterns of growth, phenology, and carbon and nutrient allocation in arctic plants. One conclusion is that the direct effects of global climate change on arctic plants are likely to be strongly constrained by existing controls over plant growth, such as photoperiod requirements or low nutrient availability. Individual species also differ greatly in their responsiveness to changes in temperature, CO_2 concentration, or precipitation regime. Although it is now possible to predict short-term responses of arctic plants and vegetation to many kinds of climatic change, longer-term predictions require a greater understanding of feedback effects on element cycling, microclimate, and other ecosystem-level changes resulting from vegetation change.

References

Andreev, V. N., and Aleksandrova, V. D. (1981). Geobotanical division of the Soviet Arctic. *In* "Tundra Ecosystems: A Comparative Analysis" (L. C. Bliss, O. W. Heal, and J. J. Moore, eds.), pp. 25–37. Cambridge Univ. Press, Cambridge.

Billings, W. D. (1973). Arctic and alpine vegetation: Similarities, differences, and susceptibility to disturbance. *Bioscience* **23**, 697–704.

Billings, W. D., and Mooney, H. A. (1968). The ecology of arctic and alpine plants. *Biol. Rev. Camb. Philos. Soc.* **43**, 481–530.

Bliss, L. C. (1956). A comparison of plant development in microenvironments of arctic and alpine plants. *Ecol. Monogr.* **26**, 303–337.

Bliss, L. C. (1962). Adaptations of arctic and alpine plants to environmental conditions. *Arctic* **11**, 180–188.

Bliss, L. C. (1971). Arctic and alpine plant life cycles. *Annu. Rev. Ecol. Syst.* **2**, 405–438.

Bliss, L. C. (1977). "Truelove Lowland, Devon Island, Canada: A High Arctic Ecosystem." University of Alberta Press, Edmonton.

Bliss, L. C. (1981). North American and Scandinavian tundras and polar deserts. *In* "Tundra Ecosystems: A Comparative Analysis" (L. C. Bliss, O. W. Heal, and J. J. Moore, eds.), pp. 8–24. Cambridge Univ. Press, Cambridge.

Bliss, L. C., Courtin, G. M., Pattie, D. I., Riewe, R. R., Whitfield, D. W. A., and Widden, P. (1973). Arctic tundra ecosystems. *Annu. Rev. Ecol. Syst.* **4**, 359–399.

Bliss, L. C., Heal, O. W., and Moore, J. J., eds. (1981). "Tundra Ecosystems: A Comparative Analysis." Cambridge Univ. Press, Cambridge.

Chapin, F. S. (1983). Direct and indirect effects of temperature on arctic plants. *Polar Biol.* **2**, 47–52.

Chapin, F. S., III. (1987). Environmental controls over growth of tundra plants. *Ecol. Bull.* **38**, 69–76.

Chapin, F. S., III, and Chapin, M. C. (1981). Ecotypic differentiation of growth processes in *Carex aquatilis* along a latitudinal gradient. *Ecology* **62**, 1000–1009.

Chapin, F. S., III, and Shaver, G. R. (1985a). Arctic. *In* "Physiological Ecology of North American Plant Communities" (B. Chabot and H. A. Mooney, eds.), pp. 16–40. Chapman and Hall, London.

Chapin, F. S., III, and Shaver, G. R. (1985b). Individualistic growth response of tundra plant species to manipulation of light, temperature, and nutrients in a field experiment. *Ecology* **66**, 564–576.

Chapin, F. S., III, and Shaver, G. R. (1988). Differences in carbon and nutrient fractions among arctic growth forms. *Oecologia* **77**, 506–514.

Chapin, F. S., III, and Shaver, G. R. (1989a). Differences in growth and nutrient use among arctic plant growth forms. *Funct. Ecol.* **3**, 73–80.

Chapin, F. S., III, and Shaver, G. R. (1989b). Lack of latitudinal variations in graminoid storage reserves. *Ecology* **70**, 269–272.

Chapin, F. S., III, Van Cleve, K., and Tieszen, L. L. (1975). Seasonal nutrient dynamics of tundra vegetation at Barrow, Alaska. *Arct. Alp. Res.* **7**, 209–226.

Chapin, F. S., III, Van Cleve, K., and Chapin, M. C. (1979). Soil temperature and nutrient cycling in the tussock growth form of *Eriophorum vaginatum* L. *J. Ecol.* **67**, 169–189.

Chapin, F. S., Johnson, D. A., and McKendrick, J. D. (1980). Seasonal nutrient allocation patterns in various tundra plant life forms in northern Alaska: Implications for herbivory. *J. Ecol.* **68**, 189–209.

Chapin, F. S., III, Shaver, G. R., and Kedrowski, R. A. (1986). Environmental controls over carbon, nitrogen, and phosphorus chemical fractions in *Eriophorum vaginatum* L. in Alaskan tussock tundra. *J. Ecol.* **74**, 167–196.

Chester, A. L., and Shaver, G. R. (1982). Reproductive effort in cottongrass tussock tundra. *Holarct. Ecol.* **5**, 200–206.

Fetcher, N., and Shaver, G. R. (1990). Environmental sensitivity of ecotypes as a potential influence on primary productivity. *Am. Nat.* **136**, 126–131.

Jonasson, S. (1982). Organic matter and phytomass on three north Swedish tundra sites, and some connections with adjacent tundra areas. *Holarct. Ecol.* **5**, 367–375.

Jonasson, S. (1983). Nutrient content and dynamics in north Swedish shrub tundra areas. *Holarct. Ecol.* **6**, 295–304.

Jonasson, S. (1989). Implications of leaf longevity, leaf nutrient re-absorption and translocation for the resource economy of five evergreen plant species. *Oikos* **56**, 121–131.

Jonasson, S., and Chapin, F. S., III. (1985). Significance of sequential leaf development for nutrient balance of the cotton sedge, *Eriophorum vaginatum* L. *Oecologia* **67**, 511–518.

Kjelvik, S., and Kärenlampi, L. (1975). Plant biomass and primary production of Fennoscandian subarctic and subalpine forests and of alpine willow and heath ecosystems. *In* "Fennoscandian Tundra Ecosystems" (F. E. Wielgolaski, ed.), Part 1, Plants and Microorganisms, pp. 111–120. Springer-Verlag, Berlin.

Klinge, H. (1973). Root mass estimation in lowland tropical rainforest of central Amazonia, Brazil. 1. Fine root masses of a pale yellow latisol and a giant humus podzol. *Trop. Ecol.* **14**, 29–38.

Kummerow, J. (1981). Carbon allocation to root systems in Mediterranean evergreen sclerophylls. *In* "Components of Productivity of Mediterranean-Climate Regions" (N. S. Margaris and H. A. Mooney, eds.), pp. 115–120. Dr. W. Junk, The Hague.

Kummerow, J., and Ellis, B. (1984). Temperature effect on biomass production and root/shoot biomass ratios in two arctic sedges under controlled environmental conditions. *Can. J. Bot.* **62**, 2150–2153.

Kummerow, J., Ellis, B. A., Kummerow, S., and Chapin, F. S., III. (1983). Spring growth of shoots and roots in shrubs of an Alaskan muskeg. *Am. J. Bot.* **70**, 1509–1515.

Kummerow, J., Mills, J. N., Ellis, B. A., Hastings, S. J., and Kummerow, A. (1987). Downslope fertilizer movement in arctic tussock tundra. *Holarct. Ecol.* **10**, 312–319.

Lewis, M. C., and Callaghan, T. V. (1976). Tundra. *In* "Vegetation and the Atmosphere" (J. J. Monteith, ed.), pp. 399–433. Academic Press, London.

Matveyeva, N. V. (1988). The horizontal structure of tundra communities. *In* "Diversity and Pattern in Plant Communities" (H. J. During, M. J. A. Werger, and H. J. Willems, eds.), pp. 59–66. SPB Academic, The Hague.

Miller, P. C. (1982a). The availability and utilization of resources in tundra ecosystems. *Holarct. Ecol.* **5**, 81–220.

Miller, P. C. (1982b). Environmental and vegetational variation across a snow accumulation area in montane tundra in central Alaska. *Holarct. Ecol.* **5**, 117–124.

Miller, P. C., Mangan, R., and J. Kummerow, J. (1982). Vertical distribution of organic matter in eight vegetation types near Eagle Summit, Alaska. *Holarct. Ecol.* **5**, 117–124.

Muc, M., and Bliss, L. C. (1977). Plant communities of Truelove Lowland. *In* "Truelove Lowland, Devon Island, Canada: A High Arctic Ecosystem" (L. C. Bliss, ed.), pp. 143–154. Univ. of Alberta Press, Edmonton.

Murray, C., and Miller, P. C. (1982). Phenological observations of major plant growth forms and species in montane and *Eriophorum vaginatum* tussock tundra in central Alaska. *Holarct. Ecol.* **5**, 109–116.

Oberbauer, S. F., Sionit, N., Hastings, S. J., and Oechel, W. C. (1986). Effects of CO_2 enrichment on growth, photosynthesis, and nutrient concentration of Alaskan tundra plant species. *Can. J. Bot.* **64**, 2293–2298.

Reader, R. J. (1978). Contribution of overwintering leaves to the growth of three broad-leaved, evergreen shrubs belonging to the Ericaceae family. *Can. J. Bot.* **56**, 1248–1261.

Shaver, G. R. (1981). Mineral nutrition and leaf longevity in an evergreen shrub, *Ledum palustre* ssp. decumbens. *Oecologia* **49**, 362–365.

Shaver, G. R. (1986). Woody stem production in Alaskan tundra shrubs. *Ecology* **67**, 660–669.

Shaver, G. R., and Billings, W. D. (1975). Root production and root turnover in a wet tundra ecosystem, Barrow, Alaska. *Ecology* **56**, 401–410.

Shaver, G. R., and Billings, W. D. (1977). Effects of daylength and temperature on root elongation in tundra graminoids. *Oecologia* **28**, 57–65.

Shaver, G. R., and Chapin, F. S., III. (1980). Response to fertilization by various plant growth forms in an Alaskan tundra: Nutrient accumulation and growth. *Ecology* **61**, 662–675.

Shaver, G. R., and Chapin, F. S., III. (1986). Effect of fertilizer on production and biomass of tussock tundra, Alaska, U.S.A. *Arct. Alp. Res.* **18**, 261–268.

Shaver, G. R., and Chapin, F. S., III. (1991). Production: biomass relationships and element cycling in contrasting arctic vegetation types. *Ecol. Monogr.* **61**, 1–31.

Shaver, G. R., and Cutler, J. C. (1979). The vertical distribution of phytomass in cottongrass tussock tundra. *Arct. Alp. Res.* **11**, 335–342.

Shaver, G. R., Chapin, F. S., III, and Billings, W. D. (1979). Ecotypic differentiation in *Carex aquatilis* on ice-wedge polygons in the Alaskan coastal tundra. *J. Ecol.* **67**, 1025–1046.

Shaver, G. R., Chapin, F. S., III, and Gartner, B. L. (1986a). Factors limiting growth and biomass accumulation in *Eriophorum vaginatum* L. in Alaskan tussock tundra. *J. Ecol.* **74**, 257–278.

Shaver, G. R., Fetcher, N., and Chapin, F. S., III. (1986b). Growth and flowering in *Eriophorum vaginatum*: Annual and latitudinal variation. *Ecology* **67**, 1524–1525.

Sørenson, T. (1941). Temperature relations and phenology of northeast Greenland flowering plants. *Medd. Gronl.* **125**, 1–305.

Svoboda, J. (1977). Ecology and primary production of raised beach communities, Truelove Lowland. *In* "Truelove Lowland, Devon Island, Canada: A High Arctic Ecosystem" (L. C. Bliss, ed.), pp. 185–216. Univ. of Alberta Press, Edmonton.

Tieszen, L. L., ed. (1978). "Vegetation and Production Ecology of an Alaskan Arctic Tundra." Springer-Verlag, New York.

Warren-Wilson, J. (1966). An analysis of plant growth and its control in arctic environments. *Ann. Bot.* **30**, 383–482.

Webber, P. J. (1977). Belowground tundra research: A commentary. *Arct. Alp. Res.* **9**, 105–111.

Webber, P. J. (1978). Spatial and temporal variation of the vegetation and its production, Barrow, Alaska. *In* "Vegetation and Production Ecology of an Alaskan Arctic Tundra" (L. L. Tieszen, ed.), pp. 37–112. Springer-Verlag, New York.

Wielgolaski, F. E., ed. (1975). "Fennoscandian Tundra Ecosystems," Part 1, Plants and Microorganisms. Springer-Verlag, Berlin.

Wielgolaski, F. E., and Kärenlampi, L. (1975). Plant phenology of Fennoscandian tundra communities. *In* "Fennoscandian Tundra Ecosystems" (F. E. Wielgolaski, ed.), Part 1, Plants and Microorganisms, pp. 94–102. Springer-Verlag, Berlin.

Wielgolaski, F. E., Bliss, L. C., Svoboda, J. C., and Doyle, G. (1981). Primary production of tundra. *In* "Tundra Ecosystems: A Comparative Analysis" (L. C. Bliss, O. W. Heal, and J. J. Moore, eds.), pp. 187–226. Cambridge Univ. Press, Cambridge.

10

The Ecosystem Role of Poikilohydric Tundra Plants

J. D. Tenhunen, O. L. Lange, S. Hahn, R. Siegwolf, and S. F. Oberbauer

I. Introduction

Tundra vegetation is restricted by physical abrasion from windblown particles, frost drought during winter, the mechanical action of ice and snow, solifluction that may dislodge seedlings, a short growing season, and grazing (Sakai and Larcher, 1987). All these factors combine to compress the space plants can occupy to approximately 50 cm above and below the soil surface. Given this vertical restriction, plus low nutrient availability that limits the development of leaf area in phanerogams and frequent precipitation, poikilohydric mosses and lichens play a large role in production, nutrient sequestering, determination of soil thermal regime, and evaporation at the tundra surface.

Because they grow so slowly, lichens are a minor component in many ecosystems with a more favorable climate than that of the Arctic. The more extreme conditions become, the more successful lichens are as a group, increasing toward high latitudes both in percentage of cover and numbers of species. Mattick (1953) described this increase with a "lichen coefficient": the ratio of the number of lichen species to the number of vascular plant species. He demonstrated a continuous increase in this coefficient northward from Germany (50°N, lichen coefficient = 0.48) through Scandinavia (0.95 in Sweden) to Greenland (2.0) and Spitzbergen (78°N, 3.5). Comparable changes in the proportion of lichens also occur locally. Lichens tend to dominate where vascular species maintain much of their biomass below the surface, succumb to stress, or are unable to establish themselves (Bliss, 1971; Kappen, 1988). Thus one finds a variety of lichen heath types in rocky, well-drained, windswept locations that are subject to severe drying during summer and to the lowest temperatures during winter.

In less severe tundra habitats, where aboveground vascular plant growth is substantial, lichens are gradually replaced by faster-growing bryophytes, which can compete with the vascular plants (cf. Longton, 1982; Lee and La Roi, 1979). As water availability in the habitat increases, mosses, such as *Rhacomitrium lanuginosum*, that can tolerate desiccation and freezing increase in importance (Proctor, 1982). In still moister habitats, *Sphagnum* mosses become dominant (Steere, 1978), strongly influencing habitat characteristics and modifying biomass allocation and production in the vascular species that grow together with them (Luken *et al.*, 1985; for detailed summaries of the distribution of arctic lichens and bryophytes, see Kershaw 1985, Kappen 1988, Smith 1982, and Longton 1982, 1988).

II. The Ecosystem Role of Mosses and Lichens

Large pools of carbon and nutrients are found in the poikilohydric plant component of tundra ecosystems (Wielgolaski *et al.*, 1981; Shaver and Chapin, 1991), and significant aspects of ecosystem function therefore depend on the physiological response, turnover, and production of mosses and lichens. *Sphagnum* species, for example, may play a particularly important role in landscape evolution on mountain foothills of the Alaskan North Slope (Walker *et al.*, 1989). In tussock tundra and in moist habitats, *Sphagnum* mosses contribute a major portion of the living biomass (Hastings *et al.*, 1989) and presumably modify soil characteristics and the accumulation of ground ice via their insulative properties. The ground surface where lichens and mosses are found is an important interface controlling the transfer of energy and materials: it thaws first, it is in contact with meltwater, and it offers the warmest microenvironment for the longest growth period during the arctic summer.

The ecosystem-level relationships between *Sphagnum* moss biomass and factors directly affecting carbon fixation and growth and between moss biomass,

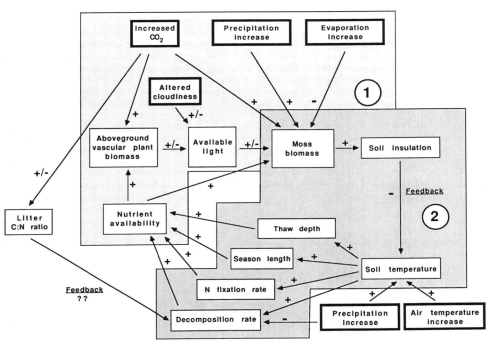

Figure 1 The ecosystem-level relationships between *Sphagnum* moss biomass and factors directly affecting carbon fixation and growth (box 1) and between moss biomass, soil temperature, and nutrient availability (box 2). Heavily framed boxes indicate driving variables in the physical environment that are predicted to change with changes in the climate of tundra regions. Arrow indicates an influence of one system variable on another; + indicates positive effect; − indicates negative effect; +/− indicates uncertain direction in response.

soil temperature, and factors affecting nutrient availability will be strongly affected by climate change (Fig. 1). Because of increased carbon fixation, moss production and biomass are likely to respond positively to increases in atmospheric CO_2 pressure and in hydration, which depends on the frequency of precipitation, the resulting water tables, and evaporation rates (Clymo and Hayward, 1982). In addition, moss production and biomass are strongly affected by incident light, which depends on the degree of development of the vascular plant community and on cloudiness. Initial increases in vascular plant biomass should promote *Sphagnum* growth, but the optimum light for *Sphagnum* growth depends on average hydration over the growing season and on individual species tolerances (Clymo and Hayward, 1982). The degree to which *Sphagnum* growth may be affected by modifications in light, CO_2, or hydration will depend on simultaneous changes in nutrient availability (Fig. 1: box 1), as demonstrated for tundra vascular plant species (Billings *et al.*, 1982, 1983, 1984), and on the growth potential of the species present.

In general, the increases in atmospheric CO_2, nutrient availability, and vascular plant biomass expected as the climate changes will lead to an increase in *Sphagnum* moss biomass. Increases in precipitation should also increase hydration of poikilohydric plants during the summer growing season, further increasing moss biomass. Depending on cloud cover effects, however, increased evapotranspiration in a warmer climate could lead to more drying than at present and thus a much different result. Soil moisture changes, which are correlated with water content changes in poikilohydric plants, are the least certain of climate projections. It has been suggested (see Chap. 2) that local conditions will lead to modest drying over the Canadian mainland but increased soil moisture in northern Alaska. Such differential hydration could lead to changes in the relative distribution of various tundra plant communities. Nevertheless, climate warming in the past has generally been associated with increases in *Sphagnum* accumulation (Walker *et al.*, 1989).

The thickness of the *Sphagnum* layer may be a critical factor controlling thaw depth and nutrient availability via its effects on soil insulation and temperature (Walker *et al.*, 1989; Fig. 1: box 2). In moist habitats, heat transfer to and from the permafrost and the seasonal course of permafrost thaw are determined by moss surface temperature, *Sphagnum* thermal properties, and the amount and frequency of precipitation (Murray *et al.*, 1989a). As illus-

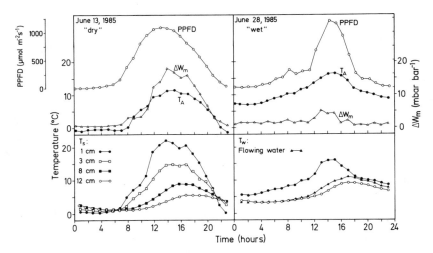

Figure 2 Typical microclimate and temperature gradients found during dry (left panels) and wet (right panels) weather in cushions of *Sphagnum* growing under *Salix* and *Betula* in a watertrack habitat in foothill tundra north of the Brooks Range, Alaska. Shown are daily courses for incident photosynthetic photon flux density (PPFD) on the vascular plant canopy, air temperature at 1 cm above the moss surface (T_A), vapor-pressure difference between moss and air (Δw_m), substrate temperatures at varying depths (T_s), and temperature of flowing water near the permafrost surface (T_w). T_s for June 28 at 3 and 8 cm depth was very close to flowing water temperature. [Adapted from Harley *et al.* (1989).]

trated in Figure 2 (left), temperature differences as large as 20° C may occur in the first 12 cm of *Sphagnum* cushions when they are dry (Harley *et al.*, 1989). Despite clear weather and a surface temperature above 22° C, the measured maximum temperature at 12 cm depth within the cushion was only 4.5° C, and permafrost remained only a few centimeters below this.

Increases in both air temperature and precipitation will tend to counteract *Sphagnum* insulation of the permafrost (see Fig. 1). Permafrost melting in tundra ecosystems after precipitation is triggered by heat accumulated diurnally in the discharging water, which is then transmitted directly to the permafrost surface (Fig. 2, right; Harley *et al.*, 1989). If there is any subsurface water flow, the temperature gradient in wet *Sphagnum* cushions is small, and heat is efficiently transferred. To predict the effects of climate change at the ecosystem level, it will be necessary to consider changes in the hydration state of *Sphagnum* and to use hydrology and energy-balance models of the type discussed by Kane *et al.* (Chap. 3). Changes in precipitation may be more important in controlling the thawing of permafrost than are the direct effects of changing surface temperature.

Sphagnum biomass may increase with climatic warming, providing more soil insulation, yet soil temperatures may not decrease if periods of dry weather shorten, and more water movement occurs (see Fig. 1: box 2). In this case, greater thaw depth, longer seasons, and any direct rise in soil temperature will promote decomposition, nitrogen fixation, and nutrient availability. These, in turn, will increase vascular plant and moss biomass. The competitive balance between mosses and vascular plants, and thus community structure, will be influenced by differential nutrient effects on growth but also by effects of equilibrium soil temperature on root growth, nutrient uptake by vascular plants, and movement of nutrients in the transpiration stream. Transpiration is coupled to the soil environment via limitations on stomatal conductance resulting from cold root temperatures (Dawson and Bliss, 1989).

In dry, exposed tundra habitats, lichens may play a role similar to that played by mosses in Figure 1. Well-developed fruticose lichen mats influence soil processes by cooling the soil (Kershaw and Field, 1975), acting as a mulch to slow soil drying (Kershaw and Rouse, 1971), and effectively absorbing large amounts of precipitation. Species composition is important because of the effects different pigments have on surface albedo. But since community structure and the size of carbon and nutrient pools in these habitats differ from those in moist locations, the relative importance of the effects depicted in Figure 1 may also differ.

III. Community Interactions and Poikilohydric Plants

A. Water Availability and Community Structure

Aboveground live biomass of vascular plants, leaf area index, and total chlorophyll content of tundra vegetation correlate strongly with the length of the growing season, growing season heat sum above 0° C, and moisture availability (Wielgolaski *et al.*, 1981). A changing arctic climate may well alter these same environmental factors (see Fig. 1). Water availability will be the most likely to influence the growth of poikilohydric plants and thus their role in controlling or contributing to ecosystem processes.

In response to climate change, the structure and function of tundra communities may shift along axes that are predictable from existing patterns, for example, a present-day "dry" tundra community may be replaced by a "wetter" one. The relative flux of carbon into vascular rather than poikilohydric plants is one comparative measure of the functional role played by these two ecosystem components. In considering the effects of climate change, we focus on present-day gradients in water availability, growth-form shifts in community composition, and the meaning of such shifts in terms of carbon flows through vascular or poikilohydric plants.

The relative proportions of photosynthetic biomass among the life-forms in tundra habitats change along with water availability (Fig. 3). In well-drained communities, such as upland lichen heaths, as much as 400 g m^{-2} dry mass of photosynthetically active lichens are found with a smaller evergreen shrub component. The same growth forms, although in smaller amounts, dominate rocky fell-fields. In upslope communities, such as *Cassiope*-dominated heath and dry shrub tundra, evergreen shrubs remain in large quantities, lichen biomass is reduced, and graminoid and deciduous shrubs contribute significantly to the mixed vegetation. Dry-adapted mosses are also important. In midslope intertrack and water-track areas and in well-drained riparian communities, a similar mixture of graminoid, deciduous shrub, and evergreen species occurs, but moss biomass becomes a major component (Hastings *et al.*, 1989; Murray *et al.*, 1989a). Finally, in waterlogged riparian areas, wet meadows, low-center polygons, and thaw lakes (Webber, 1978), moss biomass may be reduced, and the community is dominated by graminoids. Thus, along a gradient of increasing water availability, there is a transition from a lichen-dominated system to one dominated by *Sphagnum* and then to one from which poikilohydric plants may be eliminated.

Leaf area indices (LAI) of the vascular plants in all these tundra communities are low, ranging in most cases from 0.2 to 1.5 (Wielgolaski *et al.*, 1981), but leaf area index generally increases with increasing water availability. In dry upslope heath communities, it varies between 0 and 0.2. Along a gradient of water availability in the foothills tundra of the Philip Smith Mountains, Alaska, LAI increases to approximately 0.3–0.4 in dry shrub and tussock tundra communities, 0.4–0.8 in water tracks, and 0.8–3.2 in riparian situations (Tenhunen, unpubl. data). Along the same water-availability gradient, lichen groundcover was approximately 30% in heath vegetation types, with moss groundcover 10% or less. Lichens increased to 60% cover in stone stripe areas (see Walker *et al.*, 1989, for vegetation description), decreased to 8% in dry shrub tundra, and were generally negligible in moist communities. Moss cover increased to approximately 20% in stone stripe areas; to between 40 and

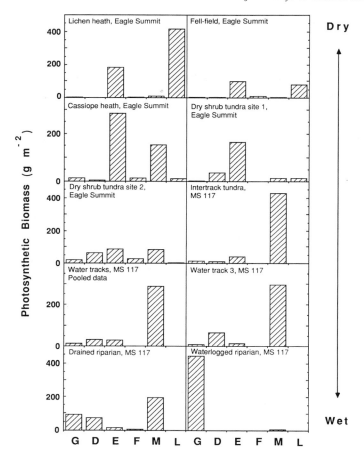

Figure 3 Distribution of photosynthetic biomass among life-forms in a variety of tundra plant communities at an upland tundra site north of Fairbanks, Alaska (Eagle Summit: Miller, 1982a,b; Miller *et al.*, 1982), and in foothill tundra north of the Brooks Range, Alaska (near material site 117 [MS 117] along the Dalton Highway; Tenhunen, unpubl. data). Water availability varies among the communities because of local topography. Upper panels illustrate the biomass for relatively dry communities, lower panels for relatively wet communities. "Water track 3" has a particularly well-developed *Salix* and *Betula* canopy; compare with "drained riparian," which it superficially resembles. "Intertrack tundra" refers to those tussock tundra areas found between water-track drainages (cf. Hastings *et al.*, 1989). G, graminoids; D, deciduous shrubs; E, evergreen shrubs; F, forbs; M, mosses; and L, lichens.

50% in dry shrub tundra, tussock tundra, and water tracks; and to more than 60% in lower-slope riparian communities. Despite low LAI, the vascular plant canopy significantly shades poikilohydric plants in moist habitats such as tussock tundra, water tracks, and drained riparian communities. Additional shading of mosses occurs in moist communities from the interception of radiation by woody stems, which have a projected area (stem area index) approximately equal to that of leaves, and in tussock tundra because of microtopographic variation.

B. Microclimate Gradients and Physiological Activity Periods

As community composition and structure change with increasing water availability, the poikilohydric plants in the system progressively experience less light, lower maximum temperatures during the growing season, and longer periods of optimal hydration (Table I). Heath communities are repeatedly moistened and dried, whereas downslope communities experience less fluctuation. Lichens of tundra heath vegetation show distinct diurnal activity patterns, taking up CO_2 after moistening by dew, rain, fog, mist, and melting snow, for example (Fig. 4). Since optimal hydration of arctic lichens often occurs during or immediately after cool, cloudy weather, light levels and temperatures may be low, or they may increase rapidly when gas-exchange processes are activated. Both light and temperature strongly influence patterns of diurnal carbon fixation until drying occurs. Nevertheless, as in leaves of arctic vascular plants, net photosynthesis in poikilohydric species saturates at a relatively low photosynthetic photon flux density (PPFD). This low saturation point apparently permits maximum photosynthesis in lichens despite cloudy weather.

For mosses such as *Sphagnum*, which are closely coupled to the water table by capillary systems, water content during dry periods decreases slowly from day to day as the soil water supply is depleted. For example, a decrease of 400% occurred in tussock tundra during a one-week warm, dry period (Murray *et al.*, 1989b). Diurnal variation in carbon exchange is determined primarily by light and temperature fluctuations. In a water track under a vascular plant canopy of *Salix pulchra* and *Betula nana*, diurnal temperature changes at the moss surface typically range from 0 to 20°C (Fig. 5, top). In tussock tundra, where the moss is fully exposed, surface temperatures are typically 0° C at night but nearly 30° C during the day (Harley *et al.*, 1989; Fig. 5, bottom).

Relatively high rates of photosynthesis are maintained over a broad temperature range in arctic lichens (Matthes and Feige, 1983; Kappen, 1988) and mosses (Harley *et al.*, 1989; Kallio and Heinonen, 1975). Temperature optima in most cases fall between 10 and 20° C, and carbon dioxide uptake often continues at temperatures considerably below 0° C (Lange, 1965; Lange and Kappen, 1972; Oechel and Sveinbjörnsson, 1978). Since net photosynthesis in *Sphagnum* is relatively insensitive to temperature between 20 and 30° C (Harley *et al.*, 1989), the differences between tussock tundra and water tracks in the diurnal course of carbon fixation, differences resulting from different temperature maxima, should be small. But observations of other moss species (Oechel and Sveinbjörnsson 1978)—many of which typically live in exposed situations—indicate that temperatures above 20° C strongly decrease carbon fixation. Thus, in drier vegetation, direct effects of high temperature on gas exchange in poikilohydric plants may be important in a changing climate.

Table I   Habitat Characteristics along a Water-Availability Gradient in Alaskan Tundra Vegetation[a]

	Lichen heath (dry)	Tussock tundra (moist)	Water track (wet)
Vascular plant LAI[b]	Very low 0.0–0.2	Low 0.3–0.4	Relatively high 0.4–0.8
Poikilohydric plant environment			
Light	High light but may be low during activity periods	High light	High light in early season Low light at midseason
Temperature	0–35° C Low temperature during activity periods	0–30° C	0–20° C
Moisture	Rapid fluctuations	Storage after rain	Maximum collection; longest storage
Belowground environment			
Temperature	Cold 1–8° C at 10 cm	Warm 0–15° C at 10 cm Decreases with dry moss	Cold 0–9° C at 10 cm Decreases with dry moss; increases after rain
Moisture	Aerobic	Aerobic	Periodically anaerobic Periodic high water table

[a] Data from the U.S. Department of Energy R4D site in the foothills north of the Brooks Range (near MS 117 along the Dalton Highway).
[b] LAI = leaf area index.

On the moss surface below vascular plant canopies, incident light levels are frequently below 50 μmol m^{-2} s^{-1} PPFD (see Fig. 5). Nevertheless, moss photosynthesis may still be light-saturated during most of the day (Oechel and Sveinbjörnsson, 1978). The ability of mosses to fix CO_2 at low light contributes to efficient solar-energy conversion under low radiation, even late in the season, and presumably between 2100 and 0400 hours during midseason. Various moss and lichen species can acclimate to prevailing light and temperatures, and this acclimation seasonally shifts the light compensation point and maximum photosynthetic capacity (Kershaw, 1985; Oechel and Sveinbjörnsson, 1978; Oechel and Van Cleve, 1986).

Low light saturation of photosynthesis in *Sphagnum* plays an important role in its day-to-day carbon balance. Maximum *Sphagnum* production occurs under the shade of *Salix* canopies in water-track drainages (Murray *et al.*, 1989a). On sunny days, positive CO_2 uptake may occur several centimeters deep in the moss mat, despite strong light attenuation. The euphotic zone, in which all but 1% of the incident light is absorbed, extends to a depth of 2 cm in *Sphagnum* cushions that develop under unshaded conditions and to 5 cm

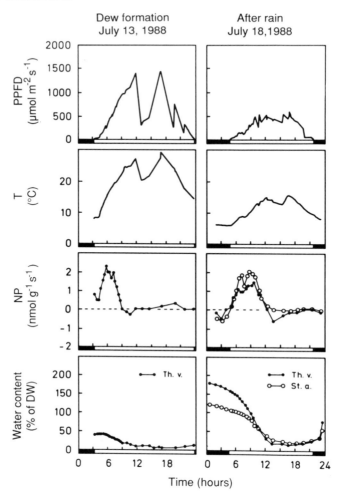

Figure 4   Observed diurnal courses of net photosynthesis (NP), water content, and prevailing environmental conditions for two lichen species (*Thamnolia vermicularis* [Th. v.] and *Stereocaulon alpinum* [St. a.]) growing in heath foothills tundra north of the Brooks Range, Alaska. Dark bar on abscissa indicates when PPFD on the lichens in situ was less than 3 μmol m^{-2} s^{-1}. Moistening of the lichens occurred on July 13 from dew formation during the low-light period (left). Moistening on July 18 occurred from rain, also during low light. Gas exchange was measured according to Lange *et al.* (1985).

in cushions that develop in shade (Clymo and Hayward, 1982). As incident light intensity decreases, a large portion of the moss canopy may begin to respire. How high photosynthetic rates remain in older moss tissues (i.e., whether lower strata in the moss canopy contribute to positive net carbon fixation) may depend on nutrient availability (Oechel and Van Cleve, 1986).

water track

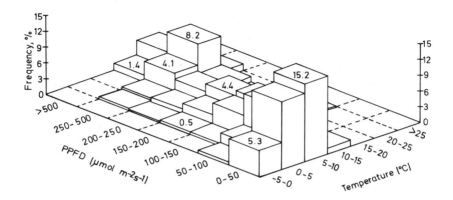

tussock tundra

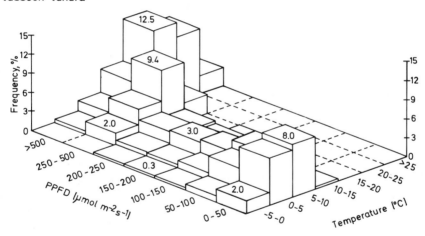

Figure 5 Light and temperature environment of *Sphagnum* growing in tussock tundra and water track drainages in foothills tundra north of the Brooks Range, Alaska (MS 117, Dalton Highway). Frequency represents the half-hour intervals in which combinations of average PPFD and moss surface temperature fell into each of 49 categories between June 15 and July 17, 1985. [Adapted from Harley *et al.* (1989).]

If climate change in the Arctic leads to replacement of present-day "dry" tundra communities by a slightly "wetter" type, then the change in the activity patterns of the poikilohydric plant component may be summarized with reference to Figure 5. Light and temperature microclimate at the ground surface would differ much as they do at present between exposed tussock tundra communities and water tracks with a developed vascular plant canopy. At present in exposed communities such as lichen heath, dry shrub tundra, and tussock tundra, surface temperatures between 15 and 25° C are common (Fig. 5, bottom). But in lichen-dominated vegetation, activity predominantly occurs at low light and low temperature, because hydration and drying occur most often in association with these conditions. As *Sphagnum* moss gains importance in exposed tussock tundra communities, high light and warm temperatures are common during activity periods, whereas water content remains relatively constant over days, depending on the depth of the water table. If moss and vascular plant biomass increased even more, so that a closed canopy developed, then water-track conditions would prevail (Fig. 5, top): water content would most likely increase and remain relatively constant during the day, average surface temperatures would decrease by as much as 5° C, and exposure to PPFD > 500 μmol m^{-2} s^{-1} would become uncommon.

C. Photosynthetic Capacity, Water Availability, and Atmospheric CO$_2$

As mosses, especially *Sphagnum*, replace lichens along water-availability gradients, changes that modify production potentials occur in photosynthetic capacity and in the amount of respiring tissue. As water availability rises, processes take place faster because of changes in photosynthetically active biomass, but also because of changes in innate physiological potentials. Maximum photosynthetic rates measured for a variety of lichen species in tundra heath vegetation from the foothills of the Philip Smith Mountains in northern Alaska (Table II; Hahn, unpubl, data) approximate those from nearby Anaktuvuk Pass (Moser and Nash, 1978; Moser *et al.*, 1983) and other arctic sites (Kappen, 1988). In general, the photosynthetic capacity of arctic species is lower than that of species from alpine areas to the south (cf. Kappen, 1988). Maximum carbon fixation rates reported for arctic moss species (Table II) generally exceed those of lichens on a dry weight basis but amount to only 20–30% of the maximum rates observed in leaves of tundra vascular plants. Photosynthetic capacity decreases with age in arctic lichens (Nash *et al.*, 1980) and mosses (Oechel and Van Cleve, 1986).

Photosynthetic capacity is directly influenced by tissue water content and atmospheric CO$_2$ pressure, factors that will be modified by climate change (see Fig. 1). In *Sphagnum*, the maximum rate of potential carbon fixation at CO$_2$ saturation decreases as water content decreases (Fig. 6a). Apparent carboxylation efficiency (the initial slope of the CO$_2$ response curve) stays constant between water contents of 700 and 3000% of dry weight but also decreases at lower water contents. These decreases are assumed to occur because the *Sphagnum* protoplasts shrink gradually as water is withdrawn from photosynthetically active cells (Kaiser, 1987), although the changes in maximum rate at high water content are not easily explained.

Other mosses and arctic lichens respond similarly (Tenhunen, unpubl. data), but the slope and saturation levels of response curves such as those shown in Figure 6 are species-dependent. Consequently, near maximum rates

Table II   Maximum Photosynthetic Rates of Common Lichens and Mosses of the Arctic Tundra at High Light Intensity, Optimal Temperature, and Optimal Water Content

Lichens[a]			Mosses		
Species	NP_{max}[b]	WC_{opt}[c]	Species	NP_{max}	WC_{opt}
Dactylina arctica	8.0	170	*Pogonatum alpinum*[d]	27.8	300
Stereocaulon alpinum	5.8	150	*Sphagnum squarrosum*[e]	18.9	700
Peltigera aphthosa	4.6	220	*Sphagnum angustifolium*[e]	18.3	700
Lobaria linita	3.8	250	*Sphagnum girgensohnii*[f]	18.3	—
Cetraria cucullata	3.7	300	*Polytrichum commune*[g]	18.3	100
Thamnolia vermicularis	3.1	100	*Calliergon sarmentosum*[d]	17.0	500
Masonhalea richardsonii	2.2	50	*Hylocomium splendens*[f]	13.9	425
Alectoria ochroleuca	2.0	60	*Dicranum fuscescens*[h]	9.5	300
Cetraria nivalis	2.0	250	*Pleurozium schreberi*[i]	7.6	680
Cetraria islandica	2.0	200	*Rhacomitrium lanuginosum*[j]	3.8	400[k]
Asahinea chrysantha	1.6	100			
Sphaerophorus globosus	0.8	100			

[a] From S. Hahn, unpublished field studies in lichen heath and xeric *Dryas* heath at the MS 117 research site, Dalton Highway, of the U.S. Dept. of Energy R4D Program.
[b] Maximum net photosynthesis in nmol g^{-1} s^{-1}.
[c] Optimal water content (% of dry weight) = the lowest water content at which >95% of maximum photosynthesis still takes place.
[d] Oechel and Collins (1976)
[e] Murray *et al.* (1989a)
[f] Stålfelt (1937)
[g] Oechel and Sveinbjörnsson (1978)
[h] Hicklenton and Oechel (1976)
[i] Skre and Oechel (1981)
[j] Kallio and Heinonen (1975)
[k] Tallis (1959)

of net photosynthesis may be observed over a broad range of tissue water content at the present-day atmospheric CO_2 pressure of 350 μbar. At very high water content, films of liquid water create diffusion resistances that reduce CO_2 availability and decrease net carbon fixation (Lange and Tenhunen, 1981). As water content decreases below the critical values indicated in Table II, net photosynthesis at present-day atmospheric CO_2 pressure rapidly decreases until the moisture compensation point is reached. This moisture compensation point depends on the balance between photosynthesis and respiration rates, which is affected by a variety of environmental and physiological controls on metabolism. Photosynthetic processes are inactivated at higher water content than is dark respiration; thus CO_2 loss may occur in the light despite high incident PPFD (see Fig. 4).

The activity of lichens and mosses may determine whether the long-term net CO_2 flux in some tundra ecosystems is positive or negative (Kjelvik *et al.*, 1975). Present-day physiological response (see Fig. 6) suggests that increased atmospheric CO_2 pressure will increase net carbon fixation by poikilohydric plants. The response will depend on water content, with greater increases in

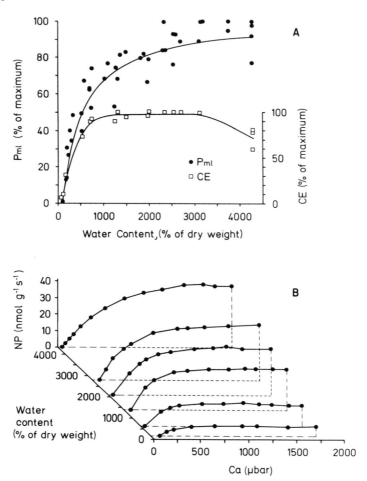

Figure 6   (A) Estimates of the CO_2 and light-saturated rate of net photosynthesis (Pml) and the initial slope of the carbon dioxide response curve (CE, carboxylation efficiency) at 15° C as affected by tissue water content in *Sphagnum palustre*. Values are derived from curves shown in Fig. 6B. (B) Net photosynthesis (NP) response of *Sphagnum palustre* from foothills tundra north of the Brooks Range, Alaska, to air CO_2 level (Ca) and water content. Ca dependencies were measured in the laboratory on a sequence of days during drying (each response curve on a separate day). Incident PPFD (200 μmol m^{-2} s^{-1}) was close to saturation, but was kept low to avoid photoinhibition. Air temperature was 15° C.

net fixation at high water content. An increase in atmospheric CO_2 pressure should also shift the moisture compensation point and reduce net CO_2 losses associated with frequent wetting or relatively low tissue water content. In *Cetraria cucullata*, for example, Lechowicz (1981) observed that respiratory CO_2 losses, which may occur in response to rewetting (Farrar and Smith, 1976;

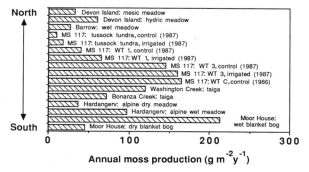

Figure 7 Comparison of estimated moss community annual production in a series of northern-latitude sites. Production values at Devon Island in the Canadian High Arctic are reported by Pakarinen and Vitt (1977); at Barrow, Alaska by Rastorfer (1978); at Washington Creek and Bonanza Creek in central Alaska by Oechel and Van Cleve (1986); at Hardangerv, Norway by Wielgolaski and Kjelvik (1973); and at Moor House, northern England, by Smith and Forrest (1978). Moss production at the series of foothill tundra sites north of the Brooks Range, Alaska (MS 117), reported by Murray *et al.* (1989b), was primarily *Sphagnum*, as at Moor House.

Eckardt *et al.*, 1982) and at the end of wetting cycles, exceeded CO_2 gain during a 15-day period. Kärenlampi *et al.* (1975) demonstrated that measured growth of *Cetraria nivalis* and *Cladonia alpestris* was correlated with the number of hours that thallus moisture content was elevated. Increased carbon fixation at high atmospheric CO_2 pressure and longer hydration periods should increase lichen productivity (estimated at 1.5–3.5 g dry weight m^{-2} year^{-1} at Devon Island).

D. Water, Light Microclimate, and Production

Moss production varies widely in northern-latitude sites, depending on topography and climate (Fig. 7). High water availability, delivery of nutrients in flowing water, and stable low pH are important for supporting maximum *Sphagnum* growth (Clymo and Hayward, 1982). At most tundra and taiga sites, annual moss production is estimated to be less than 100 g dry weight m^{-2} year^{-1}. The highest values of moss production for taiga and tundra sites (about 150 g dry weight m^{-2} year^{-1}) are reported for *Sphagnum*-dominated water tracks on tussock tundra slopes (Murray *et al.*, 1989a). In shady water-track habitats, shoot elongation of *Sphagnum* mosses is comparable to the maximum reported for *Sphagnum* at other sites (e.g., in northern England), despite the short growing season. Production at wet sites below vascular plant canopies exceeds that in open, relatively dry tussock tundra. Intermediate habitats exhibit intermediate productivity.

Although production varies greatly with water availability, short-term increases in water availability (irrigation) do not cause large increases in production, even with increased moss nutrient content. Murray *et al.*, (1989a) concluded that total production in tussock tundra is limited in part by the presence of many slow-growing individuals. The highly variable elongation rates of *Sphagnum* that occur in water tracks in response to precipitation suggest that these are rapidly growing species sensitive to the depth of the water table (Clymo and Hayward, 1982). In well-developed water tracks, irrigation does not dramatically affect production because the natural water supply is often sufficient to support maximum growth rates. We conclude that increases in water availability that may occur with climate change would affect overall production via changes in the moss community and selection of species with higher growth potentials. Furthermore, increased water availability would increase leaf area indices and shading by vascular plants, which would in turn stimulate *Sphagnum* growth.

Clymo and Hayward (1982) examined the effects of variation in light and water availability on the growth of several *Sphagnum* species and found that all species grew maximally at less than full sunlight. High light intensity appears to limit growth because of photoinhibition or photooxidative processes and may be the most important limitation on moss production in tundra ecosystems. Murray *et al.* (1989a) found negative correlations between *Sphagnum* growth rate in water-track habitats and incident solar irradiance. Experimental clearing of the vascular plant canopy in water tracks resulted in moss production that was only 18% of the control value under shaded conditions (Fig. 8). When shade was provided in relatively open riparian areas, cutting midday light on the moss surface to that typically found under well-developed water-track plant canopies, moss production increased by 290%. Experimental shading in tussock tundra increased moss production by 325%.

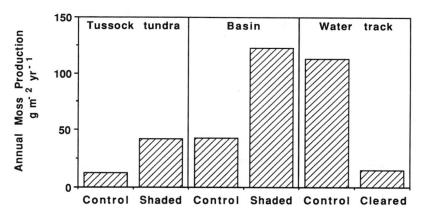

Figure 8   Effects on the annual production estimate for *Sphagnum* of experimental shading in open riparian areas (basin) and in tussock tundra and of vascular plant canopy removal in water tracks. Measurements in the basin and water tracks were taken in 1988 and in tussock tundra in 1989. For a description of methods see Murray *et al.* (1989b).

Artificial shading may simultaneously improve hydration of the capitula and reduce moss surface temperature. Nevertheless, chlorophyll is severely bleached, and photosynthetic capacity falls, in moss cushions exposed to high light (Harley *et al.*, 1989; Kallio and Valanne, 1975). Even in water tracks, production may be significantly limited by high light, inhibiting the development of full photosynthetic capacity until the vascular plant canopy provides shade near the end of June. Moss shaded by screens set out over the winter becomes green and active at least two weeks earlier than moss growing under ambient conditions.

E. Nutrient Availability and Production

Sphagnum growth is promoted by nutrients delivered in flowing water; it is especially responsive to additional phosphate, which is removed from solution to very low levels in *Sphagnum* mats (Gorham, 1956; Clymo and Hayward, 1982). Growth is inhibited by high Ca^{2+} and by high pH (Clymo, 1973). The high concentration of cation-exchange sites in cell walls contributes to a stable low soil pH. The degree to which acidification occurs depends on the rate new exchange sites are produced, the concentration of solutes in the soil, and the rate solutes circulate through the *Sphagnum* mats (Clymo and Hayward, 1982). Live *Sphagnum* affects decomposition in tundra ecosystems by maintaining acid conditions, and dead *Sphagnum* resists decay (Clymo, 1965; Martin and Holding, 1978). The nutrient-uptake and cation-exchange properties described for *Sphagnum* apply to other mosses as well, with *Pleurozium schreberi* and *Polytrichum commune* exhibiting more than double the exchange capacity of *Sphagnum* but lower production (Oechel and Van Cleve, 1986).

If climate change produces wetter soils, and thus greater moss biomass, the nutrient pools within poikilohydric plants will also enlarge. It is immaterial whether these pools build up in response to atmospheric inputs, increased weathering, or increased decomposition accompanying higher soil temperatures. What is important is that in the future tundra, mosses will play a role similar to the one they play at present in maintaining soil chemical properties and that the degree to which mosses can absorb and sequester nutrients within the ecosystem is limited by physical factors affecting total production, such as high light exposure, light attenuation in moss cushions, and water availability and CO_2 pressure as they affect net carbon exchange.

IV. Carbon Flows as an Indicator of the Ecosystem Role of Poikilohydric Plants

The complex effects of changing environmental conditions, phenology, and community structure on ecosystem carbon flows may be analyzed with the aid of a model simulating physiological response. Simulations from the computer model GAS-FLUX[1] demonstrate the sensitivity of net ecosystem carbon flux in tussock tundra to changes in the metabolic activities of vascular plants; *Sphagnum* moss growing below the vascular plant canopy; and the bulk soil, including roots (Fig. 9). The model simulates steady-state carbon flux between the atmosphere and each biomass compartment—leaves, moss, twigs and soil—for each hour and for prevailing microclimate conditions. Actual microclimate input for the simulations came from extensive field observations, and flux rates were validated by comparison with field measurements of leaf, moss, and soil CO_2 exchange.

For tussock tundra, the model indicates that the leaves of vascular plants fix considerably more carbon than mosses because they have a relatively high photosynthetic capacity and are exposed to high light. Lower light on overcast days reduces carbon fixation by leaves. Respiratory losses from twigs are unimportant, for most twigs of tundra species are small in diameter and fix enough carbon to compensate for their own respiration. As simulated, *Sphagnum* CO_2 exchange is strongly influenced by variation in light and temperature during the day but most strongly by long-term changes in soil water content. Predicted soil respiration decreases at both extremely high and extremely low water contents (Siegwolf, 1987; Oberbauer *et al.*, 1991). In the simulations, soil respiration is further limited by a high water table (Billings *et al.*, 1982) during wet weather (soil water content = 70% by volume), which reduces respiring soil volume. Although the model predicts net ecosystem CO_2 exchange (diurnal sum of all components shown in upper right of each panel in Fig. 9) to be positive at all soil water contents on clear days, it predicts carbon fixation by both leaves and *Sphagnum* to be reduced enough on overcast days so that daily integrated CO_2 flux may be negative.

These simulations for tussock tundra are particularly relevant because tussock tundra covers vast areas, and cloudiness, precipitation, evaporation, and soil water content in arctic regions are all predicted to change with global warming. Whether peat is being accumulated in tundra communities or carbon is being freed to the atmosphere depends on the frequency of situations like those in Figure 9 and on the transport of dissolved and particulate carbon to stream systems (cf. Chap. 7). Simulations can aid in evaluating the possible consequences of global climate change in tundra ecosystems (Chap. 20). If long-term water availability increased in the future, vascular plant leaf area indices and photosynthetic biomass would also increase. Poikilohydric plants

---

[1]GAS-FLUX is part of a General Arctic Simulator ecosystem model being constructed for the U.S. Department of Energy R4D program to examine potential impacts of energy development on arctic tundra. GAS-FLUX is specifically designed to calculate ecosystem net carbon dioxide exchange. It uses a canopy structure and light microclimate model (Caldwell *et al.*, 1986; Reynolds *et al.*, 1988) and mechanistically based subroutines for leaf photosynthesis, woody-stem gas exchange, and moss gas exchange that are sensitive to change in atmospheric CO_2 pressure (Tenhunen *et al.*, 1990). Soil respiration is based on correlation equations developed by Oberbauer *et al.* (1991).

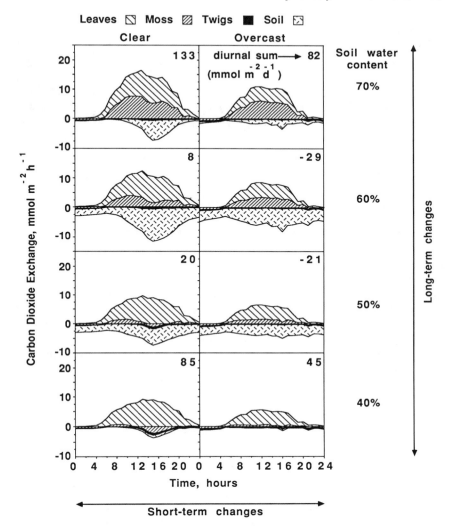

Figure 9 Simulated CO$_2$ gas exchange of tussock tundra vegetation at midseason. The fluxes for leaves, *Sphagnum* moss, twigs, and soil are presented additively, with carbon fixation by leaves and moss as positive fluxes and respiration of moss, twigs, and soil shown as negative. The left panels illustrate simulations for a clear day at midseason, the right panels for an overcast day. Soil moisture in the organic layer at 10 cm depth was assumed to vary from 70% by volume (top) to 40% by volume (bottom). These changes can occur slowly over a season (Kane and Hinzman, 1988), although water content rarely falls to 40% by volume. *Sphagnum* water contents were set at 1200, 1000, 700, and 300% of dry weight (top to bottom) based on the correlation of sample water contents with soil moisture measurements (Murray *et al.*, 1989a). The integrated total daily CO$_2$ flux (μmol m^{-2} d^{-1}) is shown in the upper right corner of each graph. Stratified sampling of biomass in several locations provided leaf area index (LAI = 0.22 m^2 m^{-2}) and stem area index (SAI = 0.17 m^2 m^{-2}).

would be exposed to less light because of increased shading, slightly lower temperatures, and longer periods of optimal hydration (see Table I). General gas-exchange characteristics of tundra ecosystems would change, possibly coming to resemble those found in present-day moist communities such as water tracks (Table III). The simulations suggest that if only CO_2 pressure increased, and community structure and other microclimate factors stayed constant, vascular plants would become more important in carbon fixation and water flux from their leaves would decrease (Table III: tussock tundra at 340 vs. 680 μbar atmospheric CO_2). Mosses would respond weakly to increased CO_2 pressure because of low water content. If the vegetation structure and microclimate changed simultaneously (Table III: water track), much more carbon would flow through the moss component of the ecosystem, and water loss from the moss surface would increase. Carbon fixation by the moss layer would increase from 8% of the total to 14% (relative to tussock tundra at 340 μbar CO_2). Water flux directly from the moss surface, rather than through vascular plant leaves, would increase from 15% to 42% as a result of moss hydration and stomatal closure. Large soil respiration flux is responsible for net carbon exchange remaining near zero in all three cases. Only when water tables become high does the model predict high values of net carbon fixation (Table III: riparian). In such situations, the proportion of carbon fixation by mosses and vascular plants approximates that found in present-day tussock tundra. Carbon flux through poikilohydric plants would thus increase because of increases in percent cover and high water tables, which improve hydration. Vascular plant carbon fixation would also increase because of high leaf area indices and the high carbon fixation potentials of dominant graminoid species.

V. Summary

Although predicted shifts in soil moisture associated with climate change are uncertain, we know enough about the physiological response of mosses and lichens to predict the altered role they might play in tundra ecosystems for given changes in climate. If soils become wetter during the summer growing season, tundra community characteristics may shift toward those found in present-day water-track and riparian habitats. These changes would modify light interception because of increases in and altered arrangement of photosynthetically active biomass. Physiological activity would also change as a result of greater hydration of the poikilohydric plants over longer periods and CO_2 effects on carbon fixation and stomatal conductance. The soil environment would differ, with important effects on nutrient cycling. The most important effect on poikilohydric plants would come from increased hydration. Faster-growing mosses would probably be favored, with growth potential, hydration, and prevailing light levels limiting production and, consequently, nutrient uptake. If soils become wetter in a future climate, present information suggests that poikilohydric plants will become more important in ecosystem

Table III Simulated Carbon Flows for Tundra Communities Estimated with the GAS-FLUX Model[a]

Vegetation (atmospheric CO_2 pressure)	Water table/soil condition	Carbon fixation (Leaves)	Carbon respired (Twigs)	Carbon fixation (Moss)	Carbon respired (Soil)	Net carbon balance	Canopy water loss	Moss water loss
Tussock tundra (340 μbar CO_2)	15 cm drained	152	−5	13	−133	27	33	6
Tussock tundra (680 μbar CO_2)	15 cm drained	175	−5	14	−133	52	19	6
Water track (680 μbar CO_2)	10 cm moist, aerated	238	−19	38	−261	−3	26	19
Riparian (680 μbar CO_2)	5 cm wet	502	−22	55	−126	409	56	28

[a] GAS-FLUX accounts for light interception by the plant community, CO_2 pressure effects on carboxylation and stomatal conductance, and hydration effects on moss and soil CO_2 (mmol m^{-2} d^{-1}) and water (mol m^{-2} d^{-1}) exchange (see Fig. 9).

carbon fixation and in controlling water loss from the soil to the atmosphere. The effect of increased moss cover on soil temperatures, and thus on decomposition and nutrient cycling, will depend on topography and moss distribution affecting how heat is transferred to the permafrost. Regardless of which direction soil water content shifts in the future, changes in the biomass of poikilohydric plants at the interface between the ground and the atmosphere will critically influence ecosystem function in arctic tundra habitats.

Acknowledgments

This work was supported by U.S. Department of Energy R4D Research Grant DEFG03-84ER60250 and by the Deutsche Forschungsgemeinschaft. We would like to thank A. Meyer, R. Gebauer, L. Balduman, and J. Lumianski for assistance in gathering field data. Assistance in data analysis and computer programming was provided by G. Radermacher and T. Liang.

References

Billings, W. D., Luken, J. O., Mortenson, D. A., and Peterson, K. M. (1982). Arctic tundra: A source or sink for atmospheric carbon dioxide in a changing environment? *Oecologia* 53, 7–11.

Billings, W. D., Luken, J. O., Mortenson, D. A., and Peterson, K. M. (1983). Increasing atmospheric carbon dioxide: Possible effects on arctic tundra. *Oecologia* 58, 286–289.

Billings, W. D., Peterson, K. M., Luken, J. O., and Mortenson, D. A. (1984). Interaction of increasing atmospheric carbon dioxide and soil nitrogen on the carbon balance of tundra microcosms. *Oecologia* 65, 26–29.

Bliss, L. C. (1971). Arctic and alpine plant life cycles. *Annu. Rev. Ecol. Syst.* 2, 405–438.

Caldwell, M. M., Meister, H. P., Tenhunen, J. D., and Lange, O. L. (1986). Canopy structure, light microclimate and leaf gas exchange of *Quercus coccifera* L. in a Portuguese macchia: Measurements in different canopy layers and simulations with a canopy model. *Trees* 1, 25–41.

Clymo, R. S. (1965). Experiments on breakdown of *Sphagnum* in two bogs. *J. Ecol.* 53, 747–758.

Clymo, R. S. (1973). The growth of *Sphagnum*: Some effects of environment. *J. Ecol.* 61, 849–869.

Clymo, R. S., and Hayward, P. M. (1982). The ecology of *Sphagnum*. *In* "Bryophyte Ecology" (A. J. E. Smith, ed.), pp. 229–289. Chapman and Hall, London.

Dawson, T. E., and Bliss, L. C. (1989). Intraspecific variation in the water relations of *Salix arctica*, an arctic–alpine dwarf willow. *Oecologia* 79, 322–331.

Eckardt, F. E., Heerfordt, L., Jørgensen, H. M., and Vaag, P. (1982). Photosynthetic production in Greenland as related to climate, plant cover, and grazing pressure. *Photosynthetica* 16, 71–100.

Farrar, J. F., and Smith, D. C. (1976). Physiological ecology of the lichen *Hypogymnia physodes*. IV: Carbon allocation at low temperatures. *New Phytol.* 81, 65–69.

Gorham, E. (1956). On the chemical composition of some waters from the Moor House National Nature Reserve *J. Ecol.* 44, 375–382.

Harley, P. C., Tenhunen, J. D., Murray, K. J., and Beyers, J. (1989). Irradiance and temperature effects on photosynthesis of tussock tundra *Sphagnum* mosses from the foothills of the Philip Smith Mountains, Alaska. *Oecologia.* 79, 251–259.

Hastings, S. J., Luchessa, S. A., Oechel, W. C., and Tenhunen, J. D. (1989). Standing biomass and production in water track drainages of the foothills of the Philip Smith Mountains, Alaska. *Holarct. Ecol.* 12, 304–311.

Hicklenton, P. R., and Oechel, W. C. (1977). The influence of light intensity and temperature on the field carbon dioxide exchange of *Dicranum fuscescens* in the subarctic. *Arct. Alp. Res.* 9, 407–419.

Kaiser, W. (1987). Effects of water deficit on photosynthetic capacity. *Physiol. Plant.* **71**, 142–149.

Kallio, P., and Heinonen, S. (1975). CO_2 exchange and growth of *Rhacomitrium lanuginosum* and *Dicranum elongatum. In* "Fennoscandian Tundra Ecosystems" (F. E. Wielgolaski, ed.), Part 1, Plants and Microorganisms, pp. 138–148. Springer-Verlag, Berlin.

Kallio, P., and Valanne, N. (1975). On the effect of continuous light on photosynthesis in mosses. *In* "Fennoscandian Tundra Ecosystems (F. E. Wielgolaski, ed.), Part 1, Plants and Microorganisms, pp. 149–162. Springer-Verlag, Berlin.

Kane, D. L., and Hinzman, L. D. (1988). Permafrost hydrology of a small arctic watershed. *In* "Permafrost: Fifth International Conference" (K. Senneset, ed.), pp. 590–595. Tapir Press, Trondheim, Norway.

Kappen, L. (1988). Ecophysiological relationships in different climatic regions. *In* "Handbook of Lichenology" (M. Galun, ed.), Vol. II, pp. 37–100. CRC Press, Boca Raton, Florida.

Kärenlampi, L., Tammisola, J., and Hurme, H. (1975). Weight increase of some lichens as related to carbon dioxide exchange and thallus moisture. *In* "Fennoscandian Tundra Ecosystems" (F. E. Wielgolaski ed.), Part 1, Plants and Microorganisms, pp. 135–137. Springer-Verlag, Berlin.

Kershaw, K. A. (1985). "Physiological Ecology of Lichens." Cambridge Univ. Press, Cambridge.

Kershaw, K. A., and Field, G. F. (1975). Studies on lichen-dominated systems. XV. The temperature and humidity profiles in a *Cladina alpestris* mat. *Can. J. Bot.* **53**, 2614–2620.

Kershaw, K. A., and Rouse, W. R. (1971). Studies on lichen-dominated systems. I. The water relations of *Cladonia alpestris* in spruce–lichen woodland in northern Ontario. *Can. J. Bot.* **49**, 1389–1399.

Kjelvik, S., Wielgolaski, F. E., and Jahren, A. (1975). Photosynthesis and respiration of plants studied by field techniques at Hardangervidda, Norway. *In* "Fennoscandian Tundra Ecosystems" (F. E. Wielgolaski, ed.), Part 1, Plants and Microorganisms, pp. 184–193. Springer-Verlag, Berlin.

Lange, O. L. (1965). Der CO_2-Gaswechsel von Flechten bei tiefen Temperaturen. *Planta* **64**, 1–19.

Lange, O. L., and Kappen, L. (1972). Photosynthesis of lichens from Antarctica. *In* "Antarctic Terrestrial Biology" (G. A. Llano, ed.). pp. 83–95. American Geophysical Union, Washington, DC.

Lange, O. L., and Tenhunen, J. D. (1981). Moisture content and CO_2 exchange of lichens. II. Depression of net photosynthesis in *Ramalina maciformis* at high water content is caused by increased thallus carbon dioxide diffusion resistance. *Oecologia* **51**, 426–429.

Lange, O. L., Tenhunen, J. D., Harley, P., and Walz, H. (1985). Methods for field measurements of CO_2 exchange: Diurnal courses of net photosynthesis capacity of lichens under Mediterranean climate conditions. *In* "Lichen Physiology and Cell Biology" (D. H. Brown ed.), pp. 23–39. Plenum, New York.

Lechowicz, M. J. (1981). The effects of climatic pattern on lichen productivity: *Cetraria cucullata* (Bell.) Ach. in the arctic tundra of northern Alaska. *Oecologia.* **50**, 210–216.

Lee, T. D., and La Roi, G. H. (1979). Gradient analysis of bryophytes in Jasper National Park, Alberta. *Can. J. Bot.* **57**, 914–925.

Longton, R. E. (1982). Bryophyte vegetation in polar regions. *In* "Bryophyte Ecology" (A. J. E. Smith, ed.), pp. 123–165. Chapman and Hall, London.

Longton, R. E. (1988). "The Biology of Polar Bryophytes and Lichens." Cambridge Univ. Press, Cambridge.

Luken, J. O., Billings, W. D., and Peterson, K. M. (1985). Succession and biomass allocation as controlled by *Sphagnum* in an Alaskan peatland. *Can. J. Bot.* **63**, 1500–1507.

Martin, N. J., and Holding, A. J. (1978). Nutrient availability and other factors limiting microbial activity in blanket peat. *In* "Production Ecology of British Moors and Montane Grasslands" (O. W. Heal and D. F. Perkins, eds.), pp. 113–135. Springer-Verlag, Berlin.

Matthes, U., and Feige, G. B. (1983). Ecophysiology of lichen symbioses. *In* "Encyclopedia of Plant Physiology, New Series" (O. L. Lange, P. S. Nobel, C. B. Osmond, and H. Ziegler, eds.), Vol. 12C, pp. 423–467. Springer-Verlag, Berlin.

Mattick, F. (1953). Lichenologische Notizen. I. Der Flechtenkoeffizient und seine Bedeutung für die Pflanzengeographie. *Ber. Dtsch. Bot. Ges.* **66**, 263–269.

Miller, P. C. (1982a). Environmental and vegetational variation across a snow accumulation area in montane tundra in central Alaska. *Holarct. Ecol.* **5**, 85–98.

Miller, P. C. (1982b). The availability and utilization of resources in tundra ecosystems. *Holarct. Ecol.* **5**, 81–220.

Miller, P. C., Mangan, R., and Kummerow, J. (1982). Vertical distribution of organic matter in eight vegetation types near Eagle Summit, Alaska. *Holarct. Ecol.* **5**, 117–124.

Moser, T. J., and Nash, T. H., III. (1978). Photosynthetic patterns of *Cetraria cucullata* (Bell.) Ach. at Anaktuvuk Pass, Alaska. *Oecologia.* **34**, 37–43.

Moser, T. J., Nash, T. H., III and Link, S. O. (1983). Diurnal gross photosynthetic patterns and potential seasonal CO_2 assimilation in *Cladonia stellaris* and *Cladonia rangiferina. Can. J. Bot.* **61**, 642–655.

Murray, K. J., Harley, P. C., Beyers, J., Walz, H., and Tenhunen, J. D. (1989a). Water content effects on photosynthetic response of *Sphagnum* mosses from the foothills of the Philip Smith Mountains, Alaska. *Oecologia.* **79**, 224–250.

Murray, K. J., Tenhunen, J. D., and Kummerow, J. (1989b). Limitations on moss growth and net primary production in tussock tundra areas of the foothills of the Philip Smith Mountains, Alaska. *Oecologia.* **80**, 256–262.

Nash, T. H., III, Moser, T. J., and Link, S. O. (1980). Nonrandom variation of gas exchange within arctic lichens. *Can. J. Bot.* **58**, 1181–1186.

Oberbauer, S. F., Tenhunen, J. D., and Reynolds, J. F. (1991). Environmental effects on CO_2 efflux from water track and tussock tundra in arctic Alaska, U. S. A. *Arct. Alp. Res.* **23**, 162–169.

Oechel, W. C., and Collins, N. J. (1976). Comparative CO_2 exchange patterns in mosses from two tundra habitats at Barrow, Alaska. *Can. J. Bot.* **54**, 1355–1369.

Oechel, W. C., and Sveinbjörnsson, B. (1978). Primary production processes in arctic bryophytes at Barrow, Alaska. *In* "Vegetation and Production Ecology of an Alaskan Arctic Tundra" (L. L. Tieszen, ed.), pp. 269–298. Springer-Verlag, New York.

Oechel, W. C., and Van Cleve, K. (1986). The role of bryophytes in nutrient cycling in the Taiga. *In* "Forest Ecosystems in the Alaskan Taiga" (K. Van Cleve, F. S. Chapin III, P. W. Flanagan, L. A. Viereck, and C. T. Dyrness, eds.), pp. 121–137. Springer-Verlag, New York.

Pakarinen, P., and Vitt, D. H. (1977). The ecology of bryophytes in Truelove Lowland, Devon Island. *In* "Truelove Lowland, Devon Island, Canada: A High Arctic Ecosystem" (L. C. Bliss, ed.), pp. 225–244. Univ. of Alberta Press, Edmonton.

Proctor, M. C. F. (1982). Physiological ecology: Water relations, light and temperature responses, carbon balance. *In* "Bryophyte Ecology" (A. J. E. Smith, ed.), pp. 333–381. Chapman and Hall, London.

Rastorfer, J. R. (1978). Composition and bryomass of the moss layers of two wet-tundra-meadow communities near Barrow, Alaska. *In* "Vegetation and Production Ecology of an Alaskan Arctic Tundra" (L. L. Tieszen, ed.), pp. 169–181. Springer-Verlag, New York.

Reynolds, J. F., Dougherty, R. L., Tenhunen, J. D., and Harley, P. C. (1988). PRECO: Plant Response to Elevated CO_2 Simulation Model. Parts I-III. Response of Vegetation to Carbon Dioxide #042. U. S. Dep. of Energy, Carbon Dioxide Research Division, Washington, DC.

Sakai, A., and Larcher, W. (1987). "Frost Survival of Plants" Springer-Verlag, Berlin.

Shaver, G. R., and Chapin, F. S., III (1991). Production:biomass relationships and element cycling in contrasting arctic vegetation types. *Ecol. Monogr.* **61**, 1–31.

Siegwolf, R. (1987). CO_2-Gaswechsel von *Rhodoendron ferrugineum* L. im Jahresgang an der alpinen Waldgrenze. Ph.D. diss., University of Innsbruck.

Skre, O., and Oechel, W. C. (1981). Moss functioning in different taiga ecosystems in interior Alaska. I. Seasonal, phenotypic, and drought effects on photosynthesis and response patterns. *Oecologia.* **48**, 50–59.

Smith, A. J. E. (1982). "Bryophyte Ecology." Chapman and Hall, London.

Smith, R. A. H., and Forrest, G. I. (1978). Field estimates of primary production. *In* "Production Ecology of British Moors and Montane Grasslands" (O. W. Heal and D. F. Perkins, eds.), pp. 17–37. Springer-Verlag, Berlin.

Stålfelt, M. G. (1937). Der Gasaustausch der Moose. *Planta* **27**, 30–60.

Steere, W. C. (1978). Floristics, phytogeography, and ecology of Arctic Alaskan bryophytes. *In* "Vegetation and Production Ecology of an Alaskan Arctic Tundra." (L. L. Tieszen, ed.), pp. 141–167. Springer-Verlag, New York.

Tallis, J. H. (1959). Studies in the biology and ecology of *Rhacomitrium lanuginosum* Brid. II. Growth, reproduction, and physiology. *J. Ecol.* **47**, 325–350.

Tenhunen, J. D., Sala Serra, A., Harley, P. C., Dougherty, R. L., and Reynolds, J. F. (1990). Factors influencing carbon fixation and water use by Mediterranean sclerophyll shrubs during summer drought. *Oecologia.* **82**, 381–393.

Walker, D. A., Binnian, E., Evans, B. E., Lederer, N. D., Nordstrand, E., and Webber, P. J. (1989). Terrain, vegetation, and landscape evolution of the R4D research site, Brooks Range Foothills, Alaska. *Holarct. Ecol.* **12**, 238–261.

Webber, P. J. (1978). Spatial and temporal variation of the vegetation and its production, Barrow, Alaska. *In* "Vegetation and Production Ecology of an Alaskan Arctic Tundra" (L. L. Tieszen, ed.), pp. 37–112. Springer-Verlag, New York.

Wielgolaski, F. E., and Kjelvik, S. (1973). Production of plants (vascular plants and cryptogams) in alpine tundra, Hardangervidda. *In* "Primary Production and Production Processes, Tundra Biome" (L. C. Bliss and F. E. Wielgolaski, eds.), pp. 75–86. Tundra Biome Steering Committee, Edmonton.

Wielgolaski, F. E., Bliss, L. C., Svoboda, J., and Doyle, J. (1981). Primary production of tundra. *In* "Tundra Ecosystems: A Comparative Analysis." (L. C. Bliss, O. W. Heal, and J. J. Moore, eds.), pp. 187–225. Cambridge Univ. Press, Cambridge.

11

Arctic Tree Line
in a Changing Climate

Bjartmar Sveinbjörnsson

I. Introduction

The boundary between the subarctic forest, or taiga, and the tundra marks a distinct separation between species as well as life-forms. Its location is ultimately determined by the climate, which affects ecosystem and physiological processes. A change in location would thus be an important indicator of changes in climatic conditions. The sharpness and regularity of the taiga—tundra boundary reflect both the topography and the history of the land surface. Fingerlike projections of forest and tall shrubland follow rivers into the tundra, and elsewhere, islands of taiga, such as lichen woodland, punctuate the tundra. Likewise, tundra islands lie within the taiga (Aleksandrova, 1980; Larsen, 1989). Beyond the forest border, a zone of scattered individual trees or small clusters of trees is sometimes found and referred to as the tree line. When apical dominance of the tree species is destroyed, or their natural growth form is otherwise affected, krummholz, coppices, or mats result, and a transition zone dominated by these growth habits is often named after them. A transect across the forest–tundra boundary may thus cut through forest border, tree line, krummholz, shrub tundra, and herbaceous tundra several times (Hustich, 1979). The number of tree genera at this circumpolar boundary is low. *Larix, Betula, Chosenia,* and *Populus* are deciduous; *Pinus* and *Picea* are evergreen (*Alnus* and *Salix* are considered shrubs).

Because of the obvious change in physiognomy and the economic importance of forest timber, the existence and position of the forest–tundra boundary, broadly referred to hereafter as tree line, has probably inspired more ecological questions than any other vegetation boundary. Most research has been conducted at alpine tree lines, however, because the circumpolar tundra–taiga boundary is remote and often diffuse (Tranquillini, 1979; Crawford, 1989). It is therefore often necessary, although questionable, to equate some aspects of the elevational and circumpolar tree lines.

II. Environmental Correlates of Tree Line

The location and structure of the edge communities along the tree line are often heavily influenced by humans, albeit sometimes indirectly. Logging, especially when combined with grazing, has strongly reduced the forest cover in continental Europe (Tranquillini, 1979). In Iceland, only 5% of the presettlement birch forest remains (Thorsteinsson, 1986), and the Orkney Islands have been treeless since the arrival of Neolithic humans 5000 years ago (Crawford, 1989). In the Torneträsk area of northern Sweden, the Lapps grazed reindeer intensively at tree line from the seventeenth century until around 1900 (Emanuelsson, 1987). Knowing the history of a particular tree line is important when one tries to unravel the causal relationships determining its location.

Circumpolar tree lines have fewer possible factors limiting tree performance than do midlatitude elevational tree lines, where ultraviolet radiation, low carbon dioxide partial pressure, and sharp precipitation gradients (Tranquillini, 1964) may all be important. In this respect, circumpolar tree lines are easier to study. Among the factors varying across the tundra–taiga gradient, foremost are solar radiation and temperature, which are, of course, closely interrelated.

Global solar radiation generally diminishes with increasing latitude and thus may correlate with tree-line position. In Canada (Hare and Hay, 1974; Hare and Ritchie, 1972), the circumpolar tree line receives about 75%, but only absorbs 60%, of the solar energy measured at the border between open woodland and closed forest. Such a radiation gradient may not be universal, however, for potential radiation varies along longitudinal tree lines, as in western Alaska. The location of different air masses (e.g., the arctic front) has been correlated with Canada's latitudinal tree line (Larsen, 1974).

As a rule, tree-line position is closely related to temperature; this is true for the circumpolar tree line (Köppen, 1936), northern and western Alaskan tree lines (Hopkins, 1959), and elevational tree lines (Dahl and Mork, 1959; Daubenmire, 1954). Several air temperature indices have been devised correlating with world tree-line positions. Köppen (1936) correlated the circumpolar tree line with a mean temperature of 10° C for the warmest month. Iversen (1944), in his distribution study of *Ilex*, included the mean temperature of the coldest month as well. Hopkins (1959) and Walter (1979) ascribed a minimum number of days (30 at tree line, according to Walter) with a mean daily temperature above 10° C as a requirement for tree existence. Skre (1972), and others cited by him, related average temperature of the four warmest months to alpine tree-line locations. A variety of degree-day or degree-month indices above different threshold values has been used (Hare, 1968). Secondary indices based on temperature, including nonlinear ones like potential evapotranspiration (Hare, 1950) and Norway spruce twig dark-respiration-rate equivalents (Skre, 1972), have also been devised. Significantly higher air temperatures correlate with circumpolar tree lines than with midlatitude alpine tree lines (Davitaja and Melnik, 1962, as cited in Tranquillini, 1979). One explanation for this finding is that greater flux densities of incident radiation at middle latitudes result in greater differences between tissue and air temperatures at midlatitude alpine tree lines than at high latitudes.

Soils along the tundra–taiga boundary are generally shallow and often underlain by permafrost, especially on poorly drained level ground or on north-facing slopes. The active layer is deepest along waterways (Viereck, 1970). The zone of continuous permafrost, which correlates with the position of tree line in parts of Canada, dips far into the larch forests of Siberia (Larsen, 1974). Soil temperatures are generally very low, and frost heaving and subsidence produces palsa bogs, "drunken forests," and other such phenomena.

Taiga soils are podzolized where the trees are evergreen or where evergreen heath species dominate (Tedrow, 1970; Werren, 1979), but no distinct qualitative soil border seems to be associated with the taiga–tundra boundary (Tedrow, 1970). Podzol distribution was shown to be unrelated to alpine tree-line position in Norway (Dahl, 1956, as cited in Tedrow, 1970). Soil nutrient conditions across the tundra–taiga border are poorly known. Nitrogen fixation is strongly temperature-dependent, more so than photosynthesis (Kallio, 1974), and so nitrogen input may decrease poleward. The active layer becomes more shallow through the taiga to the tundra (Rieger, 1974), and consequently, the potential nutritive capital diminishes; however, this situation is highly variable. In the taiga proper, the depth of the active layer varies greatly, depending on topography and successional stage (Viereck *et al.*, 1986).

III. Physiological Processes

The environmental correlates discussed in Section II do not provide mechanistic models for overall tree performance at tree line. In developing such models, it is important to realize that factors responsible for tree performance are interrelated (Larcher, 1980). Tree growth, for example, is related to reproductive effort and success (Werren, 1979). Therefore, understanding environmental limitations on growth provides a basis for explaining general tree performance. Reduced tree growth toward tree line has been attributed to insufficient photosynthesis, insufficient or excessive respiration, reduced positive carbon balance, increased tissue loss, and some vaguely defined soil processes.

A. Growth and Growth Rates

Incremental growth, be it elongation of roots or shoots or width of tree rings, decreases toward tree line with latitude and across topographical exposure gradients in northern lichen woodlands (Auger, 1974; Larsen, 1974; Scott *et al.*, 1987b; Werren, 1979) (Fig. 1). At tree line these growth increments are small and show little variation, whereas farther away the variation, as well as the maximum increment, increases (Larsen, 1974; Werren, 1979). Root elongation is highly temperature-sensitive, and reduction in aboveground tree growth on cold sites may be related to poor root growth (Chapin, 1986).

The decrease in incremental growth toward tree line is not due to lower relative growth rates. Auger (1974), studying larch (*Larix laricina*) in the Schefferville area of northern Quebec, showed that relative growth rates in a larch stand at tree line were similar to those in a lower stand. The growing season, however, began 10 days later at tree line, and its shorter duration limited total seasonal growth.

B. Photosynthesis and Water Relations

In contrast to the reduction in instantaneous rates of photosynthesis with altitude for high-elevation tree lines at middle latitudes (Benecke, 1972), no differences in maximum photosynthetic rates have been found between elevations for subarctic black spruce (*Picea mariana*), larch, or mountain birch (*Betula pubescens*) (Auger, 1974; Sveinbjörnsson, 1981; Vowinckel, 1975). Furthermore, the maximum photosynthetic rates of these trees are within the range of the same and related species growing farther south.

Low summer air temperatures are not likely to directly reduce net photosynthesis differentially toward tree line. Ledig *et al.* (1977) demonstrated a latitudinal variation in the effect of temperature pretreatment on net photosynthesis in pitch pine (*Pinus rigida*), finding that the northernmost (Quebec) population was less sensitive to low temperature than southern populations. This reduced sensitivity at high latitudes may apply to tree-line species in general. In any case, the optimum temperature range for net photosynthesis (within which the rate of net photosynthesis equals or exceeds 85% of maximum) is very broad for black spruce (7–24° C: Vowinckel *et al.*, 1975; Fig. 2) and for white spruce (*Picea glauca*) seedlings (7–23° C: Goldstein, 1981). In contrast to more temperate trees (Slatyer and Ferrar, 1977), these species show no evidence of major seasonal or acclimatization shifts (Gold-

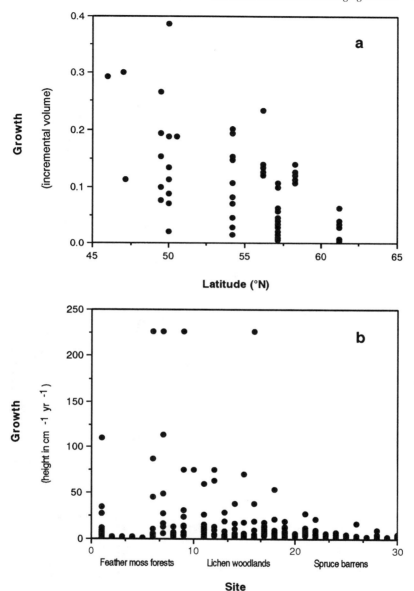

Figure 1   Spruce growth as related to latitude (a) and topographical exposure (b). Both gradients are from Canada. In (a) growth is a relative value based on diameter, height, and age; in (b) growth is based on the difference in ring number between the base of the tree and breast height. Note the decreased growth and decreased variation in growth toward tree line. [Modified from (a) Larsen (1974) and (b) Werren (1979).]

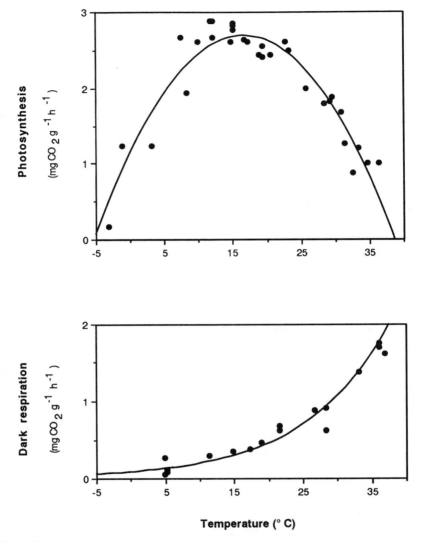

Figure 2   Temperature relations of net photosynthesis and dark respiration in black spruce (*Picea mariana*) near Schefferville, Quebec [Data from Vowinckel *et al.* (1975).]

stein, 1981). The optimum temperature range for net photosynthesis in deciduous trees is narrower and shows some seasonal change (Auger, 1974; Sveinbjörnsson, unpublished data). The optimum temperatures are again closely coupled to prevailing air temperatures; for example, in mountain birch at tree line, the optimum range in late fall is 5–17° C.

Water relations at the circumpolar tree line have not been much studied, at least partly because water stress is not evident in established trees (Black and Bliss, 1980). In the eastern American Subarctic, the presence of a continuous lichen cover significantly reduces evaporation, and water stress is therefore probably only sporadic (Cowles, 1984). Black and Bliss (1980) reported that black spruce trees in Canada's Mackenzie Valley showed no detrimental water stress, whereas, Goldstein (1981), measuring water potential and climatic conditions for white spruce at tree line in Alaska's Brooks Range, concluded that vapor-pressure deficits significantly reduced photosynthesis, although less at tree line than below it. His conclusions rested on a model of photosynthesis developed from lab measurements of seedlings grown in Washington. It is possible that seedlings at tree line are more prone to water stress than established trees, for Black and Bliss (1980) concluded that few microsites were suitable for seedling survival because of heat and water stress.

C. Respiration

Researchers disagree over the relationship between respiration and growth at tree line (Dahl and Mork, 1959; Tranquillini, 1959). Is respiration excessive, depleting resources that could be used for growth, or insufficient and unable to provide the energy required for growth? In subarctic trees, leaf dark respiration rate does not change with distance from tree line (Sveinbjörnsson, 1981; Vowinckel *et al.*, 1975) as it does in mountain ash (*Sorbus aucuparia*) in Scotland, where respiration rate increases with elevation (Crawford, 1989). The twig dark respiration rate in the subarctic varies between 10 and 25% of maximum net photosynthetic rates (Auger, 1974; Sveinbjörnsson, 1981; Vowinckel *et al.*, 1975). Models predicting instantaneous branch respiration and growth from average air temperatures in the Norwegian mountains (Skre, 1972; Skre, 1979) have shown correspondence with tree-line position. Vowinckel (1975) reports that only about 5% of photosynthate produced yearly by twigs is lost through twig night respiration during the season. Twigs, however, account for only a portion of the respiring tree; the respiration of other plant parts may consume substantial proportions of photosynthate. The role of plant respiration may be more important than recognized if the proportion of nonassimilating tree parts increases toward tree line.

D. Annual Carbon Balance

Studying CO_2 exchange in aboveground parts of *Pinus cembra* at the forest limit in the Austrian Alps, Tranquillini (1964) attributed reduced growth to a low annual carbon balance and implied that above the tree line, photosynthate balance is insufficient for growth. Vowinckel (1975), using weather-station data and a model of photosynthesis in black spruce twigs, calculated seasonal photosynthesis for several locations along the Canadian arctic tree line. Assuming that assimilatory–respiratory proportions stay constant and that the scant weather-station data were representative, he found that the length of the assimilation season was critical; up to 24% of the yearly total photosynthesis in black spruce twigs took place in May and October. In contrast, Haukioja *et al.* (1985) found no correlation between the timing of budbreak and seasonal growth of mountain birch in northernmost Finnish Lapland, indicating that the length of the photosynthetic season is unimportant for their growth. Moreover, stimulation of growth in arctic plants in response to increased CO_2 has been shown to be short lived (Tissue and Oechel, 1987). The importance of annual carbon balance in determining tree line thus remains tentative.

E. Tissue Loss

Winter tissue damage and loss at the circumpolar tree line can be substantial. At tree line in Finnish Lapland, trees often break from the weight of encasing snow and ice (Eurola and Huttonen, 1984). Werren (1979) compared rates of spruce tissue gain and loss in lowland feather moss forests with those in lichen woodlands and on barrens at Schefferville, northern Quebec, and found that the trees from the barrens grew less and lost more tissue than those from the two lower habitats. Especially important for future growth was the loss of green tissue needed for storage and photosynthetic functions. Needle loss is generally thought to be related to drying, partly caused by decreased cuticular thickness toward tree line (Hansen-Bristow, 1986; Tranquillini, 1979). Browsing by animals and snow abrasion may result in loss of a season's growth. Winter tissue loss also depends partly on summer conditions. For example, more summer growth would hasten maturation from juvenile to adult tree and thus confer resistance to winter grazing (Bryant *et al.*, 1983); it might also raise terminal buds out of the zone of snow abrasion (Wardle, 1965), ensuring survival and continued growth.

F. Plant Nutrient Relations

Several observations suggest that soil temperature may be more limiting for trees than air temperature, at least for deciduous species. Poplar (*Populus balsamifera*) trees grow around hot springs far beyond the tree line in the Brooks Range (Viereck, 1979). The fact that soil temperature is a better predictor than air temperature of budbreak at circumpolar and middle-latitude tree lines (Hansen-Bristow, 1986; Scott *et al.*, 1987b), of mid- to late-season photosynthesis at a midlatitude elevational tree line (DeLucia and Smith, 1987), and growth at a circumpolar tree line (Scott *et al.*, Fig. 3) suggests greater limitation below ground than above ground. The soil-related influence on growth and reproduction at elevational tree lines (Walter and Medina, 1969; Wardle, 1968) and arctic tree lines (Marr, 1948; Werren, 1979) may result from direct effects of soil temperature on tree function or from other, indirect effects, such as limiting soil nutrient availability. Direct effects include the impacts of temperature on root respiration (Lawrence and Oechel, 1983), water relations (Goldstein, 1981), nutrient uptake and hormone production by roots (Chapin, 1983), and root growth (Chapin, 1986). Nutrient limitation caused

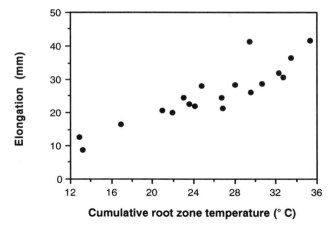

Figure 3   White spruce (*Picea glauca*) growth as related to soil temperature across the tree-line near Churchill, Manitoba. The temperatures are sums of repeated measurements. [Modified from Scott *et al.* (1987b).]

by the influence of low soil temperature on nitrogen mineralization would be an example of an indirect effect. Supporting the importance of indirect effects, Chapin *et al.* (1983) found that seedlings of all deciduous taiga tree species in Fairbanks, except alder, grow better at higher than lower phosphate concentrations, whereas temperature, in the ecologically meaningful range, had less effect on phosphate uptake rates and root growth (Chapin, 1986).

Because soil and leaf nitrogen concentrations increase with altitude (Ehrhardt, 1961; Körner, 1989) (Fig. 4), soil nitrogen is not considered limiting at elevational tree lines at middle latitudes (Tranquillini, 1979). A similar increase in foliar nitrogen concentration toward tree line at high latitudes has also been shown in mountain birch (Sveinbjörnsson *et al.*, 1991; Fig. 5) and Norway spruce (Marchenko and Karlov, 1962). But in plants native to infertile soils, nutrient deficiency does not usually show up as a reduction in nutrient concentrations but as a reduction in growth matched to nutrient availability (Chapin, 1980).

If tree-line trees are nutrient-limited because of low soil temperature, then one would expect reduced growth rather than reduced foliar nutrient concentrations. Tree growth, then, should respond to fertilization more strongly at tree line than away from it. Such a response has been demonstrated in mountain birch in Swedish Lapland (Sveinbjörnsson *et al.*, 1991; Fig. 6). Trees fertilized with nitrogen and nitrogen, phosphorus, and potassium showed an increasing growth response with increasing elevation, indicating that nutrient availability decreased with increasing elevation.

A functional relationship has thus been established between soil temperature and growth of a deciduous tree species at tree line. In evergreen trees,

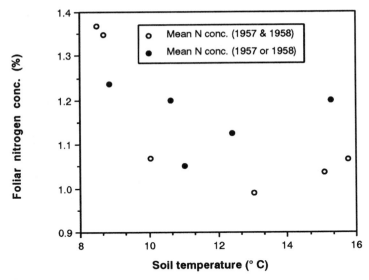

Figure 4   Needle nitrogen concentration of Norway spruce (*Picea abies*) as related to soil temperature along an elevation gradient in the Alps. Soil temperature was assessed by enzymatic conversion of standard organic substrate incubated in the soils. [Modified from Ehrhardt (1961).]

however, the mechanisms limiting growth are still unknown. Taiga black and white spruce seedlings do not increase growth in response to increased phosphate concentration (Chapin, 1983). Moreover, the generalization that evergreen trees and forests are more nutrient-efficient than deciduous ones argues either for direct effects of soil temperature on evergreen conifers, or at least for indirect effects other than primary-nutrient limitation. Such influences may act through plant carbon balance, other nutrient limitations, or plant root or mycorrhizal hormone production.

IV. Soil Processes

Soil mineralization rates are determined by temperature, moisture, and the chemical and physical quality of the plant litter (Flanagan and Van Cleve, 1983; Chap. 13), and all three factors may be important at tree line. For example, at different elevations in Swedish Lapland, the rates of soil respiration, soil nitrogen mineralization, and decomposition of birch litter are much lower in samples from tree line than in samples from forest limit or valley sites under comparable conditions (Davis and Sveinbjörnsson, unpubl. manuscript). A drop in litter quality can also come about partly from a change in vegetation composition where an increasing proportion of species produce low-quality litter. Nitrogen mineralization rates in French heath soils, for example, were

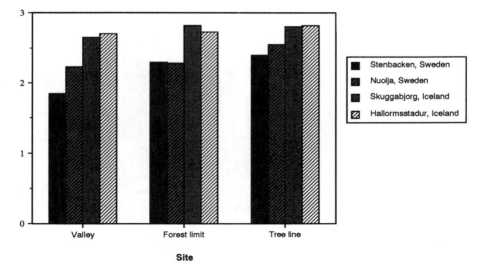

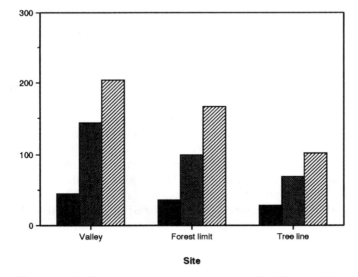

Figure 5   Leaf nitrogen concentration and annual height growth in mountain birch (*Betula pubescens* Ehr.) in northern Sweden (Stenbacken and Nuolja) and northern Iceland (Skuggabjörg and Hallormsstadur). Note that both growth and nitrogen concentrations are higher in Icelandic than Swedish trees and that at all sites, growth decreases with elevation while nitrogen concentration increases.

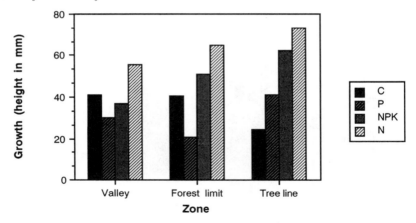

Figure 6 Mountain birch (*Betula pubescens*) height growth in different elevational zones in Swedish Lapland (Stenbacken) two years after fertilizer application. Note the large response to nitrogen fertilizer. [Reprinted from Sveinbjörnsson *et al.* (1991) with permission from Blackwell Scientific.]

reduced by the addition of foliage from some of the heath species (Goma-Tchimbakla and Roze, 1985). Larsen (1974) points out that the importance of heath species in the boreal forest increases toward tree line. Black spruce seedlings in Newfoundland were inhibited by forest-floor substrates (Mallik and Newton, 1988), and the same phenomenon has been described for an alpine tree line in France (André *et al.*, 1987). Thus, the species composition of ground cover may strongly influence soil nutrient availability and seedling establishment.

Gradients in tree mycorrhizal infection through the taiga–tundra boundary, analogous to those along temperate gradients of decreased nitrogen availability, have not been studied. Along temperate gradients, increasing concentrations of foliar nitrogen and phosphorus correlate with increasing infection (Boerner, 1986). Conversely, Haselwandter (1979) found that, with increasing elevation, a decrease in mycorrhizal infection in ericaceous shrubs correlated with decreased growth and size. At tree line, then, reduced tree growth might result either from decreased nutrient uptake partly brought about by reduced infection or, if mycorrhizal infection is substantial, from an increased proportion of assimilate going to maintain an increased biomass of mycorrhizal fungi to compensate for the reduced availability of soil nutrients.

V. Life History

The absence of trees may be caused by factors operating at any phase in their life cycle. Seed production, seed quality, seed banks, and the conditions for germination and survival of seedlings may all lead to the absence of trees (Kullman, 1979; Scott *et al.*, 1987a). For example, cones of tree-line spruces are few, often deformed or damaged, and frequently lost in winter storms (Werren, 1979). Many tree-line seeds are empty and thus not viable (Black and Bliss, 1980). Diminishing seed rain along elevational gradients has been found for mountain birch in central and northern Sweden (Kullman, 1984). Some of the tree-line tree species, such as birch, shed seeds primarily in the winter, and although the seeds generally do not travel far from the source tree, they tend to be collected with snow in depressions, as along waterways (Nordell, pers. comm.). Thus, taiga extensions into the tundra along sheltered depressions might be partly due to a greater seed collection in such areas.

The number of safe sites for germination and seedling establishment decreases with general environmental severity. Black and Bliss (1980) found that black spruce seed germination was greatest between 17° and 28° C, and seedling survival was possible in only very few microsites in the forest–tundra of the Mackenzie Valley of northern Canada. The microsites where the seeds germinated were hummock sides and troughs. There are apparently few windows in time when black spruce seeds have enough warmth or moisture to germinate in these two microsites. Seedlings in the trough sites had less summer mortality, because they suffered less water and temperature stress, but greater winter mortality from frost heaving. In central Sweden, Kullman (1984, 1986) found that even though seed germination was less at tree line than in the forest below, it did not limit mountain birch establishment, whereas seedling mortality did. In the forest, seedling mortality was caused by summer drought, above the forest by frost heaving and unknown factors. Older saplings transplanted into these habitats survived a broader range of conditions. Werren (1979) and Cowles (1984) state that black spruce and white spruce seedlings of the lichen woodlands near Schefferville, Quebec, are limited to desiccation cracks in the continuous lichen mat on the woodland floor.

The location of tree line is not necessarily in harmony with present climate (Kullman, 1979; Scott *et al.*, 1987a). The forest may persist under adverse climatic conditions without sexual reproduction through layering of branches (Black and Bliss, 1980; Werren, 1979). Conversely, in many areas, tree lines advance only slowly despite improved conditions (Sonesson and Hoogesteger, 1981). Insect outbreaks and fires can cause forests to disappear, either temporarily (on the order of 100 years) or more permanently (on the order of 1000 years), depending on whether the forest before destruction was in phase with the current climate or a relict one (Black and Bliss, 1980).

VI. Future Scenarios

The above discussion indicates that the processes involved in reduced growth and decreased performance in trees at tree line are complex, uncertain, or unknown. Consequently, predictions of changes are dubious. As pointed out in Chapter 1, the degree of change is uncertain for many of the factors likely to have the greatest impact on physiological and ecosystem processes. Even if the change in environmental factors could be anticipated, the final effect might not be. For example, nutrient availability might increase because of increased summer soil temperatures. Although this in turn might by itself increase tree nutrient uptake and deciduous tree growth, increased winter temperatures could cause increased snow loads and delay the start of the growing season. The timing of the onset of the growing season is more critical to deciduous trees, with their photoperiodically controlled leaf loss, than to evergreen trees, and the relative advantage that deciduous trees might thus gain from increased soil nutrient availability would be lost. Thus, the evergreen *Pinus* and *Picea* may be expected to advance more than *Betula, Larix, Populus,* or *Chosenia*. If winter tissue losses are the most critical factors in evergreen tree growth, it is not clear that tree performance will improve at all in future climates.

Seed production may be expected to increase if growth improves, but competition for space with other species may limit reproductive success. Increased cover of shrubs, graminoids, and herbs may make tree seedling establishment even more difficult than today even if seeds germinate more often. Herbivores can add to the uncertainty. The moth *Oporinia autumnata*, for example, espisodically devastates midelevation mountain birch forests in northern Fennoscandia, whereas lower- and higher-elevation stands are spared because minimum winter temperatures there are lethal to the moth's eggs (Tenow, 1975). Depending on the dynamics of change, if an advancing tree line lags behind climatic change during a major insect outbreak, the forest may in fact recede. Similarly, increased tree production coupled with summer droughts can cause more frequent and more widespread forest fires, with similar consequences.

VII. Summary

A variety of environmental factors correlate with the location of tree line. Although instantaneous rates of net photosynthesis do not differ along gradients from forest to tree line, whole-tree carbon gain based on actual biomass proportioning and a shortened growing season may limit growth through reduced tissue production or carbohydrate usage. Increasing scarcity of nutrients brought about by low soil temperatures and low litter quality toward tree line is likely to limit deciduous trees and perhaps also evergreens. Tissue loss limits tree-sized individuals, especially among evergreens. More so than trees of the same species, seedlings and saplings are endangered by water stress, heat stress, or both, and by frost heaving in the soil. In addition, seedling establishment, seedling and sapling mycorrhizal formation, and growth of mature trees are all hampered by allelopathic effects of the ground cover, including cryptogams. Thus, the uncertainty of local climate changes, coupled with the complexity of the processes affecting tree-line performance, combine to make predictions of the future location of tree line highly speculative.

Acknowledgments

This work was supported by NSF grant BSR-8706510 and draws heavily on research conducted at the Abisko Scientific Research Station and the University of Alaska Anchorage. I thank the staff at both institutions, as well as my graduate students, for their help.

References

Aleksandrova, V. D. (1980). Division of the Arctic into geobotanical areas. "The Arctic and Antarctic: Their Division into Geobotanical Areas." Cambridge Univ. Press, Cambridge.

André, J., Gensac, P., Pellissier, F., and Trosset, L. (1987). Regeneration of spruce populations at high altitude: Preliminary research on the role of allelopathy and mycorrhizal formation in the initial stages of development. (in French). *Rev. Ecol. Biol. Sol.* **24**, 301–310.

Auger, S. (1974). Growth and photosynthesis of *Larix laricina* (Du Roi) D. Koch in the subarctic at Schefferville, Québec. M.S. thesis, McGill University, Montréal.

Benecke, U. (1972). Growth, CO_2 exchange, and pigment content of several tree species after budbreak at various elevations (in German). *Angew. Bot.* **46**, 117–135.

Black, R. A., and Bliss, L. C. (1980). Reproductive ecology of *Picea mariana* (Mill) B. S. P., at tree line near Inuvik, Northwest Territories, Canada. *Ecol. Monogr.* **50**, 331–354.

Boerner, R. E. J. (1986). Seasonal dynamics, nutrient resorption, and mycorrhizal infection intensity of two perennial forest herbs. *Am. J. Bot.* **73**, 1249–1257.

Bryant, J. P., Chapin, F. S., III., and Klein, D. R. (1983). Carbon/nutrient balance of boreal plants in relation to vertebrate herbivory. *Oikos* **40**, 357–368.

Chapin, F. S., III. (1983). Adaptation of selected trees and grasses to low availability of phosphorus. *Plant Soil* **72**, 283–287.

Chapin, F. S., III. (1986). Controls over growth and nutrient use by taiga forest trees. *In* "Forest Ecosystems in the Alaskan Taiga" (K. Van Cleve, F. S. Chapin III, P. W. Flanagan, L. A. Viereck, and C. T. Dyrness, eds.), pp. 96–111. Springer-Verlag, New York.

Chapin, F. S. III., Tryon, P. R., and Van Cleve, K. (1983). Influence of phosphorus supply on the growth and biomass distribution of Alaskan taiga tree seedlings. *Can. J. For. Res.* **13**, 1092–1098.

Cowles, S. (1984). Effects of lichen ground cover on conifer growth in northern taiga. Ph. D. diss., McGill University, Montréal.

Crawford, R. M. M. (1989). "Studies in Plant Survival." Blackwell Scientific, Oxford.

Dahl, E. (1956). Mountain vegetation in South Norway and its relation to the environment. No. 3. Norwegian Academy of Sciences, I. Mathematics–Natural Sciences Section, Oslo.

Dahl, E., and Mork, E. (1959). On the relationships of temperature, respiration, and growth in spruce (*Picea abies* L. Karst.) (in Norwegian). *Medd. Nor. Skogforsøksves.* **16**, 81–93.

Daubenmire, R. F. (1954). Alpine timberlines in the Americas and their interpretation. *Butler Univ. Bot. Stud.* **11**, 119–136.

Davitaja, F. F., and Melnik, J. J. (1962). Radiative heating of the plants' active surface and the latitudinal and altitudinal limits of forest (in Russian). *Meteorol. Gidrol. Mosk.* **1962**, 3–9.

DeLucia, E. H., and Smith, W. K. (1987). Air and soil temperature limitations on photosynthesis in Engelmann spruce during summer. *Can. J. For. Res.* **17**, 527–533.

Ehrhardt, F. (1961). Studies on the influence of climate on the nitrogen mineralization of forest humus at various elevations in the Tirolean Alps (in German). *Forstwiss. Zentralbl.* **80**, 193–215.

Emanuelsson, U. (1987). Human influences on the vegetation in the Torneträsk area during the last three centuries. *Ecol. Bull.* **38**, 95–111.

Eurola, S., and Huttonen, A. (1984). Riisintunturi. *Oulanka Univ. Rep.* **5**, 65–67.

Flanagan, P. W., and Van Cleve, K. (1983). Nutrient cycling in relation to decomposition and organic matter quality in taiga ecosystems. *Can. J. For. Res.* **13**, 795–817.

Goldstein, G. H. (1981). Ecophysiological and demographic studies of white spruce (*Picea glauca* [Moench] Voss) at treeline in the central Brooks Range of Alaska. Ph.D. diss,. University of Washington, Seattle.

Goma-Tchimbakla, J., and Roze, F. (1985). Nitrogen mineralization in an acid heath soil. I. Study of the effect of litter in situ (in French). *Rev. Ecol. Biol. Sol* **22**, 281–290.

Hansen-Bristow, K. (1986). Influence of increasing elevation on growth characteristics at timberline. *Can. J. Bot.* **64**, 2517–2523.

Hare, F. K. (1950). Climate and zonal divisions of the boreal forest formation in eastern Canada. *Geogr. Rev.* **40**, 615–635.

Hare, F. K. (1968). The Arctic. *Q. J. R. Meteorol. Soc.* **94**, 439–459.

Hare, F. K., and Hay, J. E. (1974). The climate of Canada and Alaska. *In* "Climates of North America" (R. A. Bryson and F. K. Hare, eds.), pp. 49–192. Elsevier, Amsterdam.

Hare, F. K., and Ritchie, J. C. (1972). The boreal bioclimates. *Geogr. Rev.* **62**, 334–365.

Haselwandter, K. (1979). Mycorrhizal status of ericaceous plants in alpine and subalpine areas. *New Phytol.* **83**, 427–431.

Haukioja, E., Niemela, P., and Sirén, S. (1985). Foliage phenols and nitrogen in relation to growth, insect damage, and ability to recover after defoliation in the mountain birch *Betula pubescens* ssp. *tortuosa. Oecologia* **65**, 214–222.

Hopkins, D. M. (1959). Some characteristics of the climate in forest and tundra regions in Alaska. *Arctic* **12**, 214–220.

Hustich, I. (1979). Ecological concepts and biogeographical zonation in the north: The need for a generally accepted terminology. *Holarct. Ecol.* **2**, 208–217.

Iversen, J. (1944). *Viscum, Hedera,* and *Ilex* as climate indicators. *Geol. Foren. Stockh. Forh.* **66**, 463–483.

Kallio, P. (1974). Nitrogen fixation in subarctic lichens. *Oikos* **25**, 194–198.

Köppen, W. (1936). Das geografische System der Klimate. *In* "Handbuch der Klimatologie" (W. Köppen and R. Geiger eds.), Vol. IC, p. 44. Bornträger, Berlin.

Körner, C. (1989). The nutritional status of plants from high altitudes: A worldwide comparison. *Oecologia* **81**, 379–391.

Kullman, L. (1979). Change and stability in the altitude of the tree-limit in the southern Swedish Scandes, 1915–1975. *Acta Phytogeogr. Suec.* **65**, 1–121.

Kullman, L. (1984). Transplantation experiments with saplings of *Betula pubescens* ssp. *tortuosa* near the tree-limit in Central Sweden. *Holarct. Ecol.* **7**, 289–293.

Kullman, L. (1986). Demography of *Betula pubescens* ssp. *tortuosa* sown in contrasting habitats close to the birch tree-limit in Central Sweden. *Vegetatio* **65**, 13–20.

Larcher, W. L. (1980). "Physiological Plant Ecology," 2nd ed. Springer-Verlag, Berlin.

Larsen, J. A. (1974). The ecology of the northern continental forest border. *In* "Arctic and Alpine Environments" (J. D. Ives and R. G. Barry, eds.), pp. 254–276. Methuen, London.

Larsen, J. A. (1989). "The Northern Forest Border in Canada and Alaska: Biotic Communities and Ecological Relationships." Springer-Verlag, New York.

Lawrence, W. T., and Oechel, W. C. (1983). Effects of soil temperature on the carbon exchange of taiga seedlings. I. Root respiration. *Can. J. For. Res.* **13**, 840–849.

Ledig, T. F., Clark, J. C., and Drew, A. P. (1977). The effects of temperature treatment on photosynthesis of pitch pine from northern and southern latitudes. *Bot. Gaz.* **138**, 7–12.

Mallik, A. U., and Newton, P. F. (1988). Inhibition of black spruce seedling growth by forest-floor substrates of central Newfoundland. *For. Ecol. Manage.* **23**, 273–283.

Marchenko, A., and Karlov, Y. M. (1962). Mineral exchange in spruce forests of the northern taiga and forest tundra in the Archangelsk Oblast. *Soviet Soil Sci.* **7**, 722–734.

Marr, J. W. (1948). Ecology of the forest–tundra ecotone on the east coast of Hudson Bay. *Ecology* **18**, 119–144.

Rieger, S. (1974). Arctic soils. *In* "Arctic and Alpine Environments" (J. D. Ives and R. G. Barry, eds.), pp. 749–769. Methuen, London.

Scott, P. A., Hansell, R. I. C., and Fayle, D. C. F. (1987a). Establishment of white spruce populations and responses to climatic change at the treeline, Churchill, Manitoba, Canada. *Arct. Alp.
Res.* **19**, 45–51.

Scott, P. A., Bentley, C. V., Fayle, D. C. F., and Hansell, R. I. C. (1987b). Crown forms and shoot
elongation of white spruce at the treeline, Churchill, Manitoba, Canada. *Arct. Alp. Res.* **19**,
175–186.

Skre, O. (1972). High temperature demands for growth and development in Norway spruce
(*Picea abies* [L.] Karst.) in Scandinavia. *Medd. Norg. Landbrukshøgsk.* **51**, 1–29.

Skre, O. (1979). The regional distribution of vascular plants in Scandinavia with requirement for
high summer temperatures. *Norw. J. Bot.* **26**, 295–318.

Slatyer, R. O., and Ferrar, P. J. (1977). Altitudinal variation in the photosynthetic characteristics
of snow gum, *Eucalyptus pauciflora* Sieb. ex Spreng. II. Effects of growth temperature under
controlled conditions. *Aust. J. Bot.* **25**, 1–20.

Sonesson, M., and Hoogesteger, J. (1981). Recent tree-line dynamics (*Betula pubescens* Ehr. ssp.
tortuosa [Ledeb.] Nyman) in northern Sweden. *Nordicana* **47**, 47–54.

Sveinbjörnsson, B. (1981). Bioclimate and its effect on the carbon dioxide flux of mountain
birch (*Betula pubescens* Ehrh.) at its altitudinal tree-line in the Torneträsk area, northern Sweden. *Nordicana* **47**, 111–222.

Sveinbjörnsson, B., Nordell, O., and Kauhanen, H. (1991). Nutrient relations of mountain birch
growth at and below the elevational tree line in Swedish Lapland. *Funct. Ecol.* (in press).

Tedrow, J. C. F. (1970). Soils of subarctic regions. *In* "Ecology of the Subarctic Regions," pp.
189–205. UNESCO, Paris.

Tenow, O. (1975). Topographical dependence of an outbreak of *Oporinia autumnata* Bkh. (*Lep.
Geometridae*) in a mountain birch forest in northern Sweden. *Zoon* **3**, 85–110.

Thorsteinsson, I. (1986). The effect of grazing on the stability and development of northern
rangelands: A case study of Iceland. *In* "Grazing Research at Northern Latitudes." (O. Gudmundsson, ed.), pp. 37–43. Plenum, New York.

Tissue, D. T., and Oechel, W. C. (1987). Response of *Eriophorum vaginatum* to elevated CO_2 and
temperature in the Alaskan tussock tundra. *Ecology* **68**, 401–410.

Tranquillini, W. (1959). Dry matter production of pine, *Pinus cembra* L. at the forest limit during
a year (in German). *Planta* **54**, 107–151.

Tranquillini, W. (1964). The physiology of plants at high altitudes. *Annu. Rev. Plant Physiol.* **15**,
345–362.

Tranquillini, W. (1979). "Physiological Ecology of the Alpine Timberline: Tree Existence at High
Altitudes with Special References to the European Alps." Springer-Verlag, New York.

Viereck, L. A. (1970). Forest succession and soil development adjacent to the Chena River in interior Alaska. *Arct. Alp. Res.* **2**, 1–26.

Viereck, L. A. (1979). Characteristics of treeline plant communities in Alaska. *Holarct. Ecol.* **2**,
228–238.

Viereck, L. A., Van Cleve, K., and Dyrness, C. T. (1986). Forest ecosystem distribution in the
taiga environment. *In* "Forest Ecosystems in the Alaskan Taiga" (K. Van Cleve, F. S. Chapin
III, P. W. Flanagan, L. A. Viereck, and C. T. Dyrness, eds.), pp. 22–43. Springer-Verlag, New
York.

Vowinckel, T. (1975). The effect of climate on the photosynthesis of *Picea mariana* at the subarctic tree line. Ph.D. diss. McGill University, Montreal.

Vowinckel, T., Oechel, W. C., and Boll, W. G. (1975). The effect of climate on the photosynthesis of *Picea mariana* at the subarctic tree line. 1. Field measurements. *Can. J. Bot.* **53**, 604–620.

Walter, H. (1979). "Vegetation of the Earth and Ecological Systems of the Geo-biosphere," 3rd
ed. Springer-Verlag, Berlin.

Walter, H., and Medina, E. (1969). Soil temperature as the determining factor in the formation
of subalpine and alpine zones in the Venezuelan Andes (in German). *Ber. Dtsch. Bot. Ges.* **82**,
275–281.

Wardle, P. (1965). A comparison of alpine timberlines in New Zealand and North America. *N. Z. J. Bot.* **3**, 113–135.

Wardle, P. (1968). Engleman spruce (*Picea engelmannii* Engel.) at its upper limits on the Front Range, Colorado. *Ecology* **49**, 483–495.

Werren, G. L. (1979). Winter stress in subarctic spruce associations: A Schefferville case study. M.S. thesis, McGill University, Montreal.

Part III

Water and Nutrient Balance

12

Water Relations of Arctic Vascular Plants

Steven F. Oberbauer and Todd E. Dawson

I. The Importance of Water Stress

The arctic region encompasses a tremendous variety of microhabitats and with them, moisture regimes. Plant communities range from permanently wet coastal and shrub tundra in the Low Arctic to polar semidesert and barrens of the High Arctic (Bliss, 1988). Nevertheless, some generalizations about the importance of water as a controlling factor on vegetation structure and function are possible. In most arctic locations, including wet habitats such as coastal tundras, the environmental factor most closely correlated with vegetation type is soil moisture (Bliss, 1977; Webber, 1978; Walker *et al.*, 1989). Nevertheless, the connection between soil moisture and vegetation characteristics differs substantially among arctic habitats. In the High Arctic (Bliss, 1977; Bliss *et al.*, 1984) and at high elevations, water is clearly a limiting factor and important determinant of vegetation structure, productivity, and composition. In contrast, for much of the Low Arctic in North America and eastern Eurasia, vegetation structure and productivity are probably not directly controlled by soil moisture but rather by factors correlated with or affected by soil moisture, such as nutrient availability, thaw depth, soil aeration, redox potential, and pH (de Molenaar, 1987). Consequently, clarification of the role that water availability plays in controlling the productivity and vegetation structure for moist arctic systems has been difficult (Chapin and Shaver, 1985a). For tussock, shrub, and coastal tundra, water stress does not appear to strongly limit productivity (Stoner and Miller, 1975; Oberbauer and Miller, 1979; Miller, 1982; Chapin and Shaver, 1985a,b). Nevertheless, species differences in terms of water relations have been reported for these communities (Stoner and Miller, 1975; Kedrowski, 1976; Oberbauer and Miller, 1979, 1981, 1982a; Matthes-Sears *et al.*, 1988; Dawson and Bliss, 1989a,b), suggesting a past and a potential role for limitation of productivity by water stress.

In nearly all tundra environments, plants are susceptible to moisture stress during seedling establishment. The moisture requirement for the germination of arctic plants appears to be high (Bliss, 1958; Billings and Mooney, 1968; Oberbauer and Miller, 1982b; Karlin and Bliss, 1983). Furthermore, the seeds of most species are small, lacking the reserves to produce deep roots that would be less sensitive to the fluctuations of moisture content in the soil surface layers. Seedling establishment may be the primary limitation to colonization in polar desert and semideserts (Bliss, 1988).

Adult as well as juvenile plants of all arctic areas are subject to damage or death from desiccation during winter as a result of both wind abrasion and cuticular water loss while water uptake is restricted. Damage by winter desiccation is particularly severe in exposed, windy habitats.

II. Unique Aspects of the Arctic Environment

The arctic environment presents the water balance of plants with some unusual conditions not encountered in temperate or tropical regions. For example, 24-hour daylight during the growing season prevents complete nighttime equilibration of tissue water potential with soil water potential. Furthermore, arctic plants are exposed to large temperature differences between their roots and shoots; roots are frequently subject to soil temperatures at or near freezing, whereas shoots attain moderate to warm temperatures that can be as high as 20° C higher than air temperature (Warren-Wilson, 1957; Mølgaard, 1982).

In much of the Arctic, permafrost prevents water percolation to deeper soil layers. Instead, moisture is retained in the rooting zone, where it is available for plant uptake or is lost as runoff. Permafrost also provides a steady, but limited, supply of water as frozen soils melt, particularly in moist sites. Furthermore, because a high proportion of the yearly precipitation arrives as snowpack and because humid conditions often prevail during the growing season, the annual ratio of evapotranspiration (*ET*) to precipitation (*P*) in arctic areas is usually low.

Consequently, plant communities may be found in the Arctic under what might be considered atypical moisture regimes. For example, in Central Alaska, yearly precipitation as low as 180 mm can support forest vegetation (Woodward, 1987). Wetlands are found in areas where the balance between precipitation and potential evapotranspiration (*PET*) would normally be unfavorable for their development (Ford and Bedford, 1987). On the other hand, differences between *P* and *PET* suggest that tundra areas of northern Alaska and eastern Eurasia should support greater leaf areas indices (LAI) than are actually found. With the exception of the Low Arctic, LAI for most vegetation types worldwide can be accurately predicted from a simple model using *P* and *PET* (Woodward, 1987). The overestimate of LAI for the Low Arctic apparently arises out of limitations on vegetation imposed by low air and soil temperatures (Woodward, 1987).

III. Factors Influencing Plant Water Relations

Although the physiologically active period for arctic plants is relatively short, considerable spatial and seasonal variation occurs in the above- and belowground factors affecting plant water status (Oberbauer and Miller, 1982a; Chapin and Shaver, 1985a; Billings, 1987a; Dawson and Bliss, 1989a,b). Many of these factors follow south-to-north latitudinal or low-to-high elevational gradients. The interactive relationships between these environmental factors and plant water relations are diagrammatically presented in Figure 1 and considered below.

A. Above- and Belowground Factors

Soil moisture, primary among the determinants of plant water status, is highly variable on multiple spatial scales. On the microscale, moisture gradients from the troughs to the center of high-center polygons and across the two to three meters from wet meadows to beach ridges have been associated with different patterns of water relations (Stoner and Miller, 1975; Dawson and Bliss, 1989a). Frost boils and stone stripes within shrub tundra are often vegetated by drier-site species. On the mesoscale, soil moisture decreases characteristically along slopes from riparian zones in valley bottoms up to dry fell-fields on ridgetops (Billings, 1987a). On the macroscale, precipitation and soil moisture decrease with increasing latitude.

Soil moisture varies seasonally as well. In the Low Arctic, upper soil layers often begin the growing season saturated by meltwater. Summer precipitation and water released during soil thaw frequently, but not always, maintain high soil moisture levels throughout the remainder of the growing season. For much of the High Arctic, soils are saturated following snowmelt but dry severely as the season progresses as a result of small snowpacks, low summer

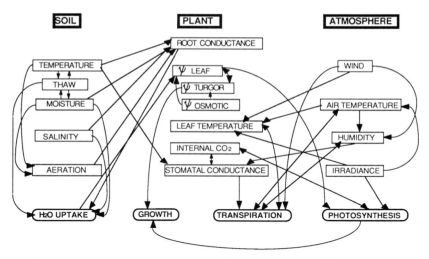

Figure 1   Interactions of arctic plant water relations with above- and below-ground environmental factors and related ecosystem processes.

precipitation, poor water retention, and exposed soils. Only below snowbanks and glaciers or adjacent to lakes does soil moisture remain high long enough to allow development of lusher vegetation.

Other important belowground factors include soil temperature, soil anoxia, and soil salinity. Soil temperature varies spatially in a manner similar to, and is influenced by, soil moisture and, of course, seasonally. Soil anoxia occurs in many arctic wetlands, and disruption of water relations by anoxia may prove to be an extremely important controlling factor for the distribution of species, although altered water balance is just one of many plant stress responses to oxygen deprivation of the roots. Soil salinity is locally important for plant water balance in some coastal arctic habitats (Dawson and Bliss, 1987; Bliss, 1988; Grulke and Bliss, 1988). Relatively few species have met the additional physiological challenge that soil salinity imposes and colonized the arctic coastal salt marsh habitat (Bliss, 1988).

Aboveground factors important for water relations include solar radiation, air temperature, atmospheric humidity, and wind velocity. Solar radiation, like soil moisture, decreases with increasing latitude. At many arctic sites, the latitudinal decrease in incident radiation is compounded by extensive cloud cover during the growing season. Air temperature closely follows the pattern of incident radiation. Low-arctic sites may have as many as 1400 degree-days (degree-days > 0° C; Bliss, 1988), whereas high-arctic sites have been reported with as few as 108 degree-days for a growing season (Grulke and Bliss, 1988). Humidities can be high throughout most of the growing season, particularly in coastal regions where low clouds are frequent. During warm sunny periods, however, the combination of low humidity and high leaf temperatures may produce substantial water vapor gradients between air and leaf (Δw). High winds are common during the growing season throughout much of the Arctic, particularly in the High Arctic and on exposed ridges and slopes.

B. Physiological Factors

Water status as measured by leaf water potential (ψ_L) differs considerably among arctic vascular plants, depending on microsite and growth form (Tables I and II). Plants of the wet sedge and tussock tundra rarely experience ψ_L as low as -2.0 MPa (Stoner and Miller, 1975; Oberbauer and Miller, 1979, 1981; Miller *et al.*, 1980; Chapin and Shaver, 1985a), whereas those plants inhabiting exposed ridge crests, polar semideserts and deserts, or fell-field communities have been reported to have ψ_L that range from -0.5 to -5.0 MPa (Teeri, 1973; Courtin and Mayo, 1975; Johnson and Caldwell, 1976; Addison, 1977; Mayo *et al.*, 1977; Oberbauer and Miller, 1979; Addison and Bliss, 1984; Grulke and Bliss, 1988; Dawson and Bliss, 1989a).

One aspect of the water relations of arctic plants that has not been well studied is root hydraulic conductance and its response to temperature, which in a warming Arctic may have important implications. At Barrow, Alaska, Stoner and Miller (1975) estimated that root conductances of species were not affected by low soil temperatures. Nevertheless, root conductances are likely to be limited by low soil temperatures in some species (Passioura, 1988).

Stomatal conductances (g) of arctic plants differ with habitat and growth form (see Tables I and II), although comparisons of growth form are frequently complicated by environmental differences between sites. At one tussock-tundra site in northern Alaska, graminoids and deciduous shrubs had higher g than forbs and evergreens (Table III; Oberbauer and Oechel, 1989).

Stomatal conductances of arctic plants are responsive to low soil temperatures and high Δw although the generality of these responses is uncertain. Dawson and Bliss (1989a) found that g was limited by low soil temperatures, but the effect was probably not mediated through water potential (lower uptake, Fig. 2). In contrast, Grulke and Bliss (1988) found that g decreased with increased soil temperatures. Limitation of g by high Δw has also been demonstrated for high-arctic species (Grulke and Bliss, 1988; Dawson and Bliss, 1989a,b), but little work has been done in sites where soil moisture remains high, such as tussock and coastal tundra (Johnson and Caldwell, 1976). Tieszen (1978) reported midday depressions in photosynthesis at Barrow, Alaska, which suggest such a response may be operating there. Gebauer *et al.* (1989), however, found no evidence for such a response in the tussock tundra near Toolik Lake, Alaska.

Wind velocity interacts strongly with Δw and air, leaf, and soil temperatures to affect leaf transpiration. At low wind velocities, Δw can be enhanced by increasing wind velocity, although at increasingly high wind velocity, the leaf and ground boundary layers may be removed, resulting in lower Δw, air, leaf, and soil temperatures (Fig. 3; Warren-Wilson, 1959; Bliss 1962). Higher

Table I Summary of Some Water Relations Characteristics of Arctic Vascular Plant Species from Wet to Mesic Habitats[a]

Species	Leaf water potential (MPa) Range	Mean	Stomatal conductance (mol m⁻² s⁻¹)	Osmotic adjustment/ cell wall elasticity	Reference
Graminoids					
Carex aquatilis	−0.80 to −4.4	−2.0	0.019–0.16	+	Johnson & Caldwell (1976)
Carex stans	−0.04 to −4.0	−1.4	0.015–0.06	−	Addison (1977)
Carex ursina	−0.25 to −1.0	−0.8	n.d.	+	Dawson & Bliss (1987); Dawson (unpubl.)
Alopecurus alpinus	−0.45 to −1.5	−1.1	n.d.	−	Dawson & Bliss (1987); Dawson (unpubl.)
Arctophila fulva	−0.02 to −1.5	−0.9	0.012–0.20	n.d.	Stoner & Miller (1975)
Dupontia fisheri	−0.80 to −3.9	−1.5	0.031–0.15	−	Johnson & Caldwell (1976); Dawson & Bliss (1987)
Eriophorum angustifolium	−0.30 to −0.7	−0.5	n.d.	+	Stoner & Miller (1975)
Puccinellia phryganodes	−0.21 to −0.7	−0.4	n.d.	+	Dawson & Bliss (1987); Dawson (unpubl.)
Forbs					
Cochleria officinalis ssp. *groenlandica*	−0.45 to −1.3	−0.7	n.d.	+	Dawson & Bliss (1987); Dawson (unpubl.)
Potentilla hyparctica	−0.03 to −1.2	−0.6	0.010–0.17	n.d.	Stoner & Miller (1975)
Deciduous Shrubs					
Salix arctica	−0.03 to −1.1	−0.8	0.011–0.25	+ + +	Dawson & Bliss (1989a,b)
Salix pulchra	−0.03 to −1.1	−0.7	0.008–0.10	n.d.	Stoner & Miller (1975)
Salix reticulata	−0.05 to −1.6	−1.1	0.035–0.20	+ + +	Dawson (unpubl.)
Evergreen Shrubs					
Vaccinium vitis-idaea	−0.01 to −2.2	−1.6	n.d.	n.d.	Oberbauer & Miller (1979)

[a] n.d. = no data; + = tissue elasticity high; − = tissue elasticity low; + + = osmotic adjustment; + + + = both high tissue elasticity and osmotic adjustment.

264

Table II Summary of Some Water Relations Characteristics of Arctic Vascular Plant Species from Mesic to Dry Habitats[a]

Species	Leaf water potential (MPa) Range	Mean	Stomatal conductance (mol m^{-2} s^{-1})	Osmotic adjustment/ cell wall elasticity	Reference
Graminoids					
Luzula confusa	−0.03 to −1.0	−0.5	0.050–0.15	−	Addison & Bliss (1984)
Phippsia algida	−0.05 to −4.9	−3.3	0.050–0.45	n.d.	Grulke & Bliss (1988)
Puccinellia vaginata	−0.50 to −3.0	−0.6	0.050–0.43	+	Grulke and Bliss (1988)
Deciduous Shrubs					
Salix arctica	−0.03 to −1.7	−1.1	0.020–0.41	+ + +	Dawson & Bliss (1989a,b)
Salix pulchra	−0.03 to −1.1	−0.6	0.015–0.21	n.d.	Stoner & Miller (1975)
Evergreen Shrubs					
Cassiope tetragona	−0.27 to −2.5	−1.8	0.024–0.30	+ + +	Dawson (unpubl.)
Dryas integrifolia	−1.10 to −4.2	−3.1	0.019–0.10	+	Mayo *et al.* (1977); Dawson (unpubl.)
Vaccinium vitis-idaea	−0.01 to −2.2	−1.6	n.d.	n.d.	Oberbauer & Miller (1979)

[a]n.d. = no data; + = tissue elasticity high; − = tissue elasticity low; + + = osmotic adjustment; + + + = both high tissue elasticity and osmotic adjustment.

265

Table III   Mean Maximum Stomatal Conductances
(g, mol m^{-2} s^{-1}) for Different Growth Forms along
an Arctic Tundra Slope in the Foothills of the Philip
Smith Mountains, Alaska

Growth form	g	Number of species
Deciduous shrubs	0.27	7
Graminoids	0.26	5
Perennial forbs	0.21	2
Evergreen shrubs	0.18	4

wind velocity and lower temperatures also reduce transpiration rates by reducing stomatal conductance (Grulke and Bliss, 1988; Dawson and Bliss, 1989a,b; Fig. 4).

Arctic plants of mesic to dry habits have been shown to acclimate to periodic or seasonal drought by osmotic adjustment, maintaining relatively elastic cell walls, or both (Johnson and Caldwell, 1976; Grulke and Bliss, 1988; Dawson and Bliss, 1989a,b). These tissue attributes have been suggested (Johnson and Caldwell, 1976; Grulke and Bliss, 1988) or shown (Dawson and Bliss, 1989a,b) to have a marked influence on turgor maintenance and the ability of some species to maintain higher stomatal conductances to both CO_2 (Johnson and Caldwell, 1976) and water vapor (Dawson and Bliss, 1989b).

The response of arctic plant water relations to changes in the environment is site-dependent. Plants of wet to mesic habitats are subjected to low soil and root temperatures that may increase root resistance to water uptake, and their stomata tend· be insensitive to increased Δw when soils are moist. With periodic soil moisture stress and increased wind velocity, however, the stomatal conductance of moist-site species declines much more quickly than that of most dry-site inhabitants (Johnson and Caldwell, 1976). The bulk tissue elastic modulus is often higher (low elasticity) for wet-site species than those plants of drier sites, and osmotic adjustment is rarely seen. Greater soil moisture and nutrient availability, lower wind velocity, and higher humidity allow for greater net primary production (150–300 g m^{-2} yr^{-1}; Bliss, 1981; Chapin and Shaver, 1985a) than in dry-site species despite low soil temperatures. In contrast, plants of drier sites are exposed to warmer soil and root temperatures, yet these sites often experience desiccating conditions from higher wind velocities and periodic soil moisture deficits. These conditions impose marked constraints on dry-site inhabitants; the ψ_L and g of most species closely follow seasonal and diurnal changes in soil water potential and Δw. As water stress develops in these sites, the ability to osmotically adjust or develop higher tissue elastic properties allows for turgor maintenance and continued gas exchange. Despite these adaptations, arctic vascular plants of drier sites

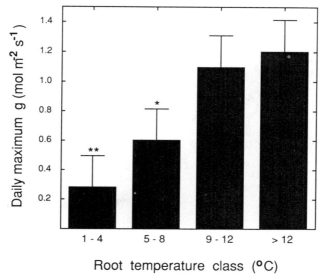

Figure 2   Daily maximum stomatal conductance (*g*) as a function of root temperature class in the deciduous shrub, *Salix arctica*. Values marked with asterisks are significantly different at $p = 0.01$** and $p = 0.05$* (Student's *t*-test). The data were derived from both field- and laboratory-based gas-exchange experiments. [Adapted from Dawson and Bliss (1989a,b).]

have lower net primary production ($5–50 \ g \ m^{-2} \ yr^{-1}$) than plants of wet to mesic habitats.

IV. Interactions between Water Relations and Whole-Plant Function

Water relations affect whole-plant function via numerous direct and indirect linkages, including some feedback and feedforward relationships (see Fig. 1). Water relations are linked to whole-plant growth via cell extension (Bradford and Hsiao, 1982); water-stressed plants have reduced leaf expansion (Dale, 1988). Thus the amount of leaf area, hence photosynthetic potential, is regulated by water status. In the Arctic, particularly the High Arctic, the situation is complicated because plants have little opportunity for nighttime recovery with soil water potential. Consequently, osmotic and therefore leaf water potentials are low, even under conditions of high soil water potentials. In the high-arctic grass species *Phippsia algida* (Sol). R. Br. and *Puccinellia vaginata* (Lge.) Fern. & Weath., for example, leaf water potentials during a week-long period near mid-season never attained values above -0.9 and -1.2 MPa, respectively, despite soil water potentials greater than -0.1 MPa (Grulke and Bliss, 1988). The effect this incomplete equilibrium with soil moisture has on leaf expansion is unclear.

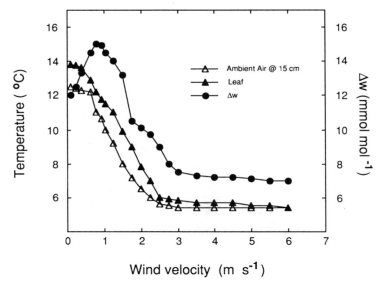

Figure 3   The ambient air (at 15 cm above the surface) and leaf (4 species) temperature, and leaf-to-air vapor-pressure gradient (Δw) as a function of increasing wind velocity on a dry beach ridge, Truelove Lowland, Devon Island, Canada. These data are means taken from weekly averages during the growing seasons (June–August) of 1983 and 1984. Standard errors are not shown. [Data from Dawson (1987) and Dawson and Bliss (1989a).]

Because of the difficulty in establishing the moisture levels required for extension growth, little data exist on this subject for the dominant species of different tundra communities. Oberbauer and Miller (1982a) found that in central Alaska, species from wet sites had higher water requirements for growth than those from dry sites, but they suggested that growth under natural conditions was not water-limited in the species tested. In sites where water stress is now common, increased canopy development as a result of greater leaf expansion can be expected with any improvement in water status.

The effect of water stress on carbon gain is well known and has been repeatedly documented (Warren-Wilson, 1966). Among arctic species, including those of moist sites, substantial differences have been found in the relationships between leaf water potential and stomatal conductance, thus presumably in photosynthesis (Stoner and Miller, 1975; Kedrowski, 1976; Grulke and Bliss, 1988). In an extreme case, Grulke and Bliss (1988) found that one of two high-arctic grass species showed no stomatal control with loss of turgor, suggesting decoupling between photosynthesis and g at low water potentials. These species differences could provide the basis for species separation in a warmer arctic.

Photosynthesis is also indirectly affected by g via the modulation of leaf temperature by transpiration, though such effects are likely to be small for arctic species because their photosynthetic responses to temperature are fairly broad (Tieszen, 1973, 1978; Limbach *et al.*, 1982; Chap. 8). Furthermore, leaves

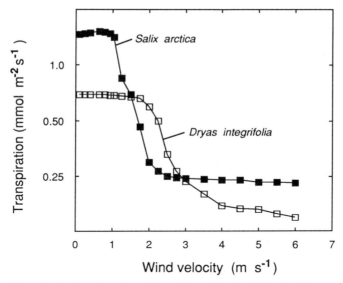

Figure 4 Transpiration as a function of increasing wind velocity in a deciduous (*Salix arctica*) and evergreen (*Dryas integrifolia*) shrub inhabiting a dry beach ridge, Truelove Lowland, Devon Island, Canada. [Data from Dawson (unpubl.).]

of most arctic species are small, and leaf temperatures should be only weakly affected by changes in transpiration.

Unlike limitation of carbon gain, mortality of established plants caused by water stress during the growing season is rare, at least in undisturbed habitats of the Low Arctic. This is not the case in the High Arctic. Grulke and Bliss (1988) reported substantial drought-induced mortality in two grass species, *Phippsia algida* and *Puccinellia vaginata*, on King Christian Island; during dry years, *Phippsia* mortality was especially high.

Desiccation mortality and damage may occur during the winter throughout the Arctic. Snow cover is extremely important in protecting plants against freezing, desiccation, and the abrasion of winter winds (Sørenson, 1941; Billings and Mooney, 1968; Bliss, 1988), as can easily be seen by the conformity of shrubs to the microtopography. Unfortunately, very little is known about plant water losses while roots are frozen (Chapin and Shaver, 1985a). Presumably, the potential for damage is compounded if plants are not fully hydrated at freeze-up. Because of the importance and potential of changes in winter snow cover and duration, this aspect of plant water relations could be a key determinant for the composition of future vegetation communities.

Undoubtedly, alleviation of water-related limitations of photosynthesis will result in increased plant carbon gain. Nevertheless, the general perception for tussock and coastal tundra is that the species are not carbon-limited, but rather carbon-sink-limited (Chapin, 1980; Chapin and Shaver, 1985a,b; Tissue and Oechel, 1987). Consequently, in moist tundra an easing of water limitations may produce minimal increases in productivity. In contrast, polar barrens and semideserts are more likely to experience increases in productivity with alleviation of water deficiency or stomatal limitation.

Even so, despite their desiccated appearance, the productivity of many dry sites may also be limited by nutrient availability, perhaps even more than by water. At a fell-field site in central Alaska, McGraw (1985) found that productivity was unaffected by irrigation but strongly affected by nutrient addition. Miller (1982) concluded that the large difference in vegetation type and productivity along a tundra mesotopographic gradient from fell-field to sedge–moss was related to nutrient, rather than water or temperature, regimes. In barren gravel areas of the Low Arctic, lush graminoid growth can be induced within two growing seasons with the application of nitrogen–phosphorus–potassium fertilizer (Oberbauer, pers. obs.). Even in the high-arctic polar barrens, nitrogen and phosphorus limit plant growth (Grulke and Bliss, 1985; Bliss, 1988).

Nutrient availability and water are tightly linked in arctic environments. In areas of standing water, anoxia limits the availability of some nutrients. On tundra slopes, nutrients move naturally with runoff (Kummerow *et al.*, 1987). In areas with flowing water, nutrient availability and water are further interrelated in that the thermal energy of the flowing water increases the depth of thaw in frozen soil and thus the volume of active soil available for root exploration. Increased soil temperatures also increase soil carbon and nutrient pools by increasing decomposition, mineralization, and perhaps nitrogen fixation in soil-dwelling microbes and algae (Billings, 1987b; Chap. 14). Flowing water also increases nutrient availability by overcoming diffusion limits by means of the mass flow of nutrients past roots (Karlsson, 1985; Chapin *et al.*, 1988). In the High Arctic, nitrogen availability is linked to moisture availability via support of freeliving nitrogen-fixing cyanobacteria (Henry and Svoboda, 1986; Bliss 1988; Chap. 14).

Furthermore, water and nutrient stresses produce similar leaf morphological characteristics, including scleromorphy, folded and rolled leaves, and small cell sizes (Sørenson, 1941; Small, 1972). Plants under nutrient stress may have altered water relations as well as altered nutrient-use patterns (Radin and Ackerson, 1981). Because tundra plants are frequently nutrient-limited by nitrogen and phosphorus (Ulrich and Gersper, 1978; Shaver and Chapin, 1980; McKendrick, 1987), observed differences in water relations between species and sites may result in part from nutrition.

V. Scaling Up from Leaf to Canopy Processes

A theoretical framework exists for predicting water fluxes from plant canopies under nonlimiting moisture regimes (Penman, 1948; Montieth, 1965), conditions that are satisfied for much of the Low Arctic during the growing season. The importance of leaf physiological processes such as stomatal opening for ecosystem transpiration fluxes is not entirely obvious, however, and depends on the coupling between the canopy and the atmosphere. Apparently, under many circumstances this coupling is weak (McNaughton and Jarvis, 1983; Jarvis and McNaughton, 1986), which explains the frequent result that evapotranspiration from vegetation can be predicted without knowing stomatal conductance (Jarvis and McNaughton, 1986; but see Pau U and Gao, 1988). The coupling between plant canopy and atmosphere in turn depends on local climate conditions and canopy structure. In smooth canopies, such as that of coastal and shrub tundra, the coupling will likely be weak, and g will have little impact on the overall ecosystem water efflux. In sparse vegetation and under conditions of strong water limitation, the coupling between plant canopy and atmosphere will be greater, and g will have a greater influence on water loss from the plant canopy.

VI. Water Relations, Global Climate Change, and Ecosystem Processes

Plant responses to changes in temperature, precipitation, wind, and snow cover that are projected to accompany the increase in atmospheric CO_2 and trace gases will depend heavily on the magnitude and seasonal timing of these changes. The influence of water relations on ecosystem processes in the face of global change will also depend on the importance of nutrient limitation in the regions under consideration, a factor that will not remain constant. These changes are expected to have multiple effects on plant water relations, including some that may be counteracting (Martin *et al.*, 1989; Rosenberg *et al.*, 1989, 1990) Furthermore, because of the tight linkage between soil moisture and soil nutrients, any change that alters the ecosystem moisture regime will affect the nutrient regime.

The increase in atmospheric CO_2 concentration will potentially affect tundra ecosystems via plant water relations both directly and indirectly. Reductions of g in direct response to higher atmospheric CO_2 concentrations should in theory increase the ratio of carbon gained to water transpired, or water-use efficiency. Crop species have shown a fairly uniform reduction in g of about 34% with a doubling of CO_2 (Cure, 1985). For native and crop species, this increase in water-use efficiency has been associated with improved water status and growth (Oechel and Strain, 1985). Surprisingly, little information now exists on the response of the stomata of arctic plants to CO_2 concentration, although it is likely that most species show such a response. Consequently, where soil moisture or atmospheric humidity limits carbon gain, productivity, and survivorship, such as in the High Arctic (Bliss *et al.*, 1984; Svoboda and Henry, 1987) and in disturbed areas (talus slopes, gravel pads), vegetation should expand into what are currently marginal areas. This effect should be particularly important where seedling establishment is the primary limitation to the development of vegetative cover. The extent to which this expansion occurs will depend on the accompanying changes in temperature and rainfall and the importance of nutrient limitation at the site.

In areas where water limitation is comparatively low, such as coastal and shrub tundra, increases in water-use efficiency may not have a large impact on productivity but could reduce system transpiration as a result of direct effects of CO_2 on stomatal conductance. The importance of this reduction for ecosystem water fluxes depends both on the proportion of ecosystem evapotranspiration attributable to vascular plants and on how well the plant canopy is coupled to the atmosphere (Jarvis and McNaughton, 1986). Martin *et al.* (1989) and Rosenberg *et al.* (1989) suggest, based on simulations of forest and agricultural systems, that the stomatal response to elevated atmospheric CO_2 will significantly moderate the increase in evapotranspiration that might result from increased temperature alone.

In tundra systems, evapotranspiration accounts for a substantial proportion of the annual water balance. Kane *et al.* (1991; Chap. 3) found that *ET* varied from 34 to 66% of the annual water balance in an arctic Alaskan watershed. Few attempts have been made to determine the proportion of *ET* attributable to vascular plants in tundra. At Barrow, Alaska, transpiration from vascular plants was estimated to constitute only 15% of *ET*, the other 85% coming from mosses (Miller *et al.*, 1976). Stuart *et al.* (1982) estimated that 38% of *ET* from tussock tundra results from vascular plant transpiration. In both of these ecosystems, however, the canopy smoothness and climatic conditions suggest leaf and atmosphere should be poorly coupled and that system transpiration will change little with increased CO_2 concentration.

At the same time that higher CO_2 concentrations should reduce stomatal conductance, higher plant productivity could increase the transpiring surface area, as has been reported for some crop species (Cure, 1985). Large increases in leaf area, however, are not anticipated for wet and shrub tundra. Studies of coastal tundra (Billings *et al.*, 1982, 1983, 1984) and tussock tundra (Tissue and Oechel, 1987; Grulke *et al.*, 1990) suggest that because of nutrient limitation, increases in CO_2 will not result in substantially greater production of leaf area. For example, under enriched CO_2 concentrations, field-grown plants of *Eriophorum vaginatum* rapidly lowered their photosynthesis to rates similar to those of plants growing at ambient CO_2 levels, suggesting that factors other than carbon are limiting productivity in this species (Tissue and Oechel, 1987). In any event, increased transpiring surface area, as in the case of decreased *g*, may have little impact on system water fluxes because in moist systems, vapor-pressure deficit from soil to atmosphere determines evapotranspiration (Jarvis and McNaughton, 1986).

Increased temperatures are certain to affect arctic plant water relations, although it is unclear whether the balance of these changes will be positive or negative. Increased temperatures will lead to higher evaporation rates in both wet and dry habitats as a result of greater Δw (Nielson *et al.*, 1989; Rosenberg *et al.*, 1990) and therefore will tend to decrease overall soil moisture. In wet to mesic habitats, however, increased temperatures may increase moisture availability by freeing water that would normally be held in the permafrost (Peterson *et al.*, 1984). Such increased water availability would be short-term if precipitation fails to recharge the soil. In dry habitats, higher temperatures should increase soil- and atmospherically induced water stress, depending on the rainfall regime, although reduced *g* in response to Δw and CO_2 concentrations may counteract this effect to some extent. Whether water loss and increases in Δw are sufficient to reduce carbon gain depends on the relative changes in CO_2, temperature, precipitation, humidity, and their timing.

Increases in precipitation will have both direct and indirect effects on arctic ecosystems. Precipitation effects in moist and shrub tundra will be predominantly indirect via alteration of the nutrient regime, rather than direct effects on water status. It is questionable whether the additional moisture alone will have any positive effect on growth. Indeed, Peterson *et al.* (1984) found reduced growth for *Carex aquatilis* and *Dupontia fisheri* with a higher water table. Increased runoff should support the spread of riparian vegetation in downslope areas as well as an increase in *Sphagnum* growth in colluvial areas (Miller, 1983). For polar semideserts and deserts, however, increased precipitation will have a direct, positive effect on plant productivity, the magnitude of which will depend on nutrient availability and the accompanying increases in temperature. Nitrogen limitation may be alleviated somewhat by the increased precipitation because increased moisture should support greater growth of freeliving nitrogen-fixing cyanobacteria (Billings, 1987b). Seedling establishment should be improved in dry high-arctic sites, particularly in combination with increased temperature. The two together should substantially improve the growth potential of the High Arctic and lead to increased vegetative cover and invasion by species with greater growth rates.

These scenarios will be contingent on how fast, and in what form, the global change comes about. Many arctic species have very broad physiological response characteristics (Tieszen, 1973, 1978; Limbach *et al.*, 1982). A 1–5° C change may in fact be very much within the range of tolerances of the arctic flora and thus cause little change in the productivity of current communities. Nevertheless, faster-growing southern species whose current northern limits are determined by cold tolerance will have the opportunity to move northward into competition with the slow-growing arctic flora of today. The findings of Woodward (1987) would suggest that given the ratio of precipitation to potential evapotranspiration in many low-arctic areas, large increases in leaf area index should result as temperature limitation is ameliorated. For dry high-arctic sites, it seems that understanding the direction of change in *P:PET* ratios of future climates is fundamental to predicting changes in vegetation structure, composition, and productivity.

VII. Arctic Plant Water Relations and the Biosphere

The influence of arctic plant water relations on the global biosphere operates predominantly via effects on the fluxes of atmospheric greenhouse gases: methane, carbon dioxide, and water vapor. Plant water relations are a consideration for methane fluxes in that considerable biogenic methane enters the atmosphere through plants rooted in anaerobic soils. Arctic wetlands contribute a significant proportion of annual biogenic methane emissions (Sebacher *et al.*, 1986; Whalen and Reeburgh, 1988). Because methane fluxes through plants appear to be driven by pressure gradients (Dacey, 1981; Sebacher *et al.*, 1985), stomata do not exert a direct control on these fluxes (Nouchi *et al.*, 1990). Insofar as the pressure gradients are influenced by leaf heating and stomatal opening, however, plant water relations may play a role in methane fluxes.

The contribution of CO_2 to the atmosphere by arctic ecosystems is currently unresolved because of difficulties in estimating the balance between decomposition and photosynthesis (see Chap. 4 and 7). The role of vascular-plant water relations in this balance lies in regulating CO_2 uptake as affected by factors discussed in Section V,B. If water limitation in barrens and semidesert areas were removed by additional precipitation, then considerably greater plant productivity and carbon storage would occur in these areas. Colonization of barren arctic areas may ultimately lead to reduced albedo in these areas and therefore affect global energy balance (Lashof, 1989).

For the next century, the potential global influence of the Low Arctic is considerably greater than that of the High Arctic, where production and carbon storage are currently minimal. Over the longer term, however, increased plant production and carbon storage over the vast area of the High Arctic as a result of global warming and reduced moisture stress could significantly affect global energy balance and biogeochemical cycles.

VIII. Summary

Vegetation composition and structure in the Arctic are strongly correlated with soil moisture. For many maritime or wet meadow arctic sites, however, vegetation structure and productivity are not controlled directly by soil moisture but rather indirectly via factors correlated with soil moisture, such as nutrient availability, soil aeration, depth of thaw, soil pH, and others. All sites probably experience some limitation of carbon uptake by water stress, but for most areas in the Low Arctic, nutrient limitation is the primary control over landscape physiognomy. Direct control of vegetation form and productivity by moisture stress does occur in some sites, particularly high-elevation tundras and in the barrens and semideserts of the High Arctic.

Indirect effects of global climate changes on arctic ecosystems will likely be greater than the direct consequences that these changes have on plant water relations. Only in polar desert and semidesert areas will the direct effects of increased water-use efficiency be important in system function. If nutrient availability increases as a result of increased decomposition, the importance of water as a primary control over productivity may increase.

The effects of increased temperature and CO_2 may partially offset each other; increased temperature will tend to increase vapor pressure deficits and transpiration, but increased CO_2 will tend to decrease transpiration via a direct effect on the stomata.

Because of the greater storage of carbon and carbon fixation potential, low-arctic vegetations are more likely than high-arctic vegetations to have immediate feedback effects on the global biosphere, primarily via release of the greenhouse gases methane and CO_2. Though the High Arctic extends over a considerably greater area, its low biomass and productivity will limit its potential to affect the biosphere in the short term. As the climate in the High Arctic improves, however, its productivity, and therefore its potential to affect global biogeochemical cycles and energy balance, will increase.

References

Addison, P. A. (1977). Studies on evapotranspiration and energy budgets on the Truelove Lowland. *In* "Truelove Lowland, Devon Island, Canada: A High Arctic Ecosystem" (L. C. Bliss, ed.), pp. 281–300. University of Alberta Press, Edmonton.

Addison, P. A., and Bliss, L. C. (1984). Adaptations of *Luzula confusa* to the polar semi-desert environment. *Arctic* **37**, 121–132.

Billings, W. D. (1987a). Constraints to plant growth, reproduction, and establishment in arctic environments. *Arct. Alp. Res.* **19**, 357–365.

Billings, W. D. (1987b). Carbon balance of Alaskan tundra and taiga ecosystems: Past, present and future. *Quat. Sci. Rev.* **6**, 165–177.

Billings, W. D., and Mooney, H. A. (1968). The ecology of arctic and alpine plants. *Biol. Rev. Camb. Philos. Soc.* **43**, 481–529.

Billings, W. D., Luken, J. O., Mortenson, D. A., and Peterson, K. M. (1982). Arctic tundra: Source or sink for atmospheric carbon dioxide in a changing environment? *Oecologia* **53**, 7–11.

Billings, W. D., Luken, J. O., Mortenson, D. A., and Peterson, K. M. (1983). Increasing carbon dioxide: Possible effects on arctic tundra. *Oecologia* **58**, 286–289.

Billings, W. D., Peterson, K. M., Luken, J. O., and Mortenson, D. A. (1984). Interaction of increasing atmospheric carbon dioxide and soil nitrogen on the carbon balance of tundra microcosms. *Oecologia* **65**, 26–29.

Bliss, L. C. (1958). Seed germination in arctic and alpine species. *Arctic* **11**, 180-188.

Bliss, L. C. (1962). Adaptations of arctic and alpine plants to environmental conditions. *Arctic* **15**, 117–144.

Bliss, L. C. (1977). General summary: Truelove Lowland ecosystem. *In* "Truelove Lowland, Devon Island, Canada: A High Arctic Ecosystem" (L. C. Bliss, ed.), pp. 657–675. University of Alberta Press, Edmonton.

Bliss, L. C. (1981). North American and Scandinavian tundras and polar deserts. *In* "Tundra Ecosystems: A Comparative Analysis" (L. C. Bliss, O. W. Heal, and J. J. Moore, eds.), pp 8-24. Cambridge Univ. Press, Cambridge.

Bliss, L. C. (1988). Arctic tundra and polar desert biome. *In* North American Terrestrial Vegetation" (M. G. Barbour and W. D. Billings, eds.), pp. 1–32. Cambridge Univ. Press, New York.

Bliss, L. C., Svoboda, J., and Bliss, D. I. (1984). Polar deserts, their plant cover and plant production in the Canadian High Arctic. *Holarct. Ecol.* **7**, 305–324.

Bradford, K. J., and Hsiao, T. C. (1982). Physiological responses to moderate water stress. *In* "Encyclopedia of Plant Physiology, New Series" (O. L. Lange, P. S. Nobel, C. B. Osmond, and H. Ziegler, eds.), Vol. 12B, pp. 263–324. Springer-Verlag, Berlin.

Chapin, F. S., III. (1980). The mineral nutrition of wild plants. *Annu. Rev. Ecol. Syst.* **11**, 233–260.

Chapin, F. S., III, and Shaver, G. R. (1985a). Arctic. *In* Physiological Ecology of North American Plant Communities (B. F. Chabot and H. A. Mooney, eds.), pp. 16–40. Chapman and Hall, New York.

Chapin, F. S., III, and Shaver, G. R. (1985b). Individualistic growth response of tundra plant species to environmental manipulations in the field. *Ecology* **66**, 564–576.

Chapin, F. S., III, Fetcher, N., Kielland, K., Everett, K., and Linkins, A. E. (1988). Productivity and nutrient cycling of Alaskan tundra: Enhancement by flowing soil water. *Ecology* **69**, 693–702.

Courtin, G. M., and Mayo, J. M. (1975). Arctic and alpine plant water relations. *In* "Physiological Adaptations to the Environment" (F. J. Vernberg, ed.), pp. 201–224. Intext Educational, New York.

Cure, J. D. (1985). Carbon dioxide doubling responses: A crop survey. *In* "Direct Effects of Increasing Carbon Dioxide on Vegetation" (B. R. Strain and J. D. Cure, eds.), pp. 99-116. DOE/ER-0238. Natl. Tech. Info. Serv., Springfield, Virginia.

Dacey, J. W. H. (1981). Pressurized ventilation in the yellow waterlily. *Ecology* **62**, 1137–1147.

Dale, J. E. (1988). The control of leaf expansion. *Annu. Rev. Plant Physiol. Plant Mol. Biol.* **39**, 267–295.

Dawson, T. E., and Bliss, L. C. (1987). Species patterns, edaphic characteristics, and plant water potentials in a high-arctic brackish marsh. *Can. J. Bot.* **65**, 863–868.

Dawson, T. E., and Bliss, L. C. (1989a). Intraspecific variation in the water relations of *Salix arctica*, an arctic–alpine dwarf willow. *Oecologia* **79**, 322–331.

Dawson, T. E., and Bliss, L. C. (1989b). Patterns of water use and tissue water relations in the dioecious shrub, *Salix arctica*: The physiological basis for habitat partitioning between the sexes. *Oecologia* **79**, 332–343.

de Molenaar, J. G. (1987). An ecohydrological approach to floral and vegetational patterns in arctic landscape ecology. *Arct. Alp. Res.* **19**, 414–424.

Ford, J., and Bedford, B. L. (1987). The hydrology of Alaskan wetlands, USA: A review. *Arct. Alp. Res.* **19**, 209–229.

Gebauer, R. L. E., Tenhunen, J. D., and Oberbauer, S. (1989). Diurnal patterns of leaf gas exchange of tundra plants growing in the foothills of the Philip Smith Mountains, Alaska. *Bull. Ecol. Soc. Am.* **70**, 120.

Grulke, N. E., and Bliss, L. C. (1985). Environmental control of the prostrate growth form in two high arctic grasses. *Holarct Ecol.* **8**, 204–210.

Grulke, N. E., and Bliss, L. C. (1988). Comparative life history characteristics of two high arctic grasses, Northwest Territories. *Ecology* **69**, 484–496.

Grulke, N. E., Riechers, G. H., Oechel, W. C., Helm, U., and Jaegar, C. J. (1990). Carbon balance in tussock tundra under ambient and elevated CO_2 *Oecologia* **83**, 485–494.

Henry, G. H. R., and Svoboda, J. (1986). Dinitrogen fixation (acetylene reduction) in high arctic sedge meadow communities. *Arct. Alp. Res.* **18**, 181–187.

Jarvis, P. G., and McNaughton, K. G. (1986). Stomatal control of transpiration: Scaling up from leaf to region. *Adv. Ecol. Res.* **14**, 1–47.

Johnson, D. A., and Caldwell, M. M. (1976). Water potential components, stomatal function, and liquid phase water transport resistance of four arctic and alpine species in relation to moisture stress. *Physiol. Plant.* **36**, 271–278.

Kane, D. L., Gieck, R. E., and Hinzman, L. D. (1991). Evapotranspiration from a small Alaskan watershed. *Nordic Hydrol.* **21**, 253–272.

Karlin, E. F., and Bliss, L. C. (1983). Germination ecology of *Ledum groenlandicum* and *Ledum palustre* ssp. *decumbens*. Arct. Alp. Res. **15**, 397–404.

Karlsson, P. S. (1985). Effects of water and mineral nutrient supply on a deciduous and an evergreen dwarf shrub: *Vaccinium uliginosum* L. and *V. vitis-idaea* L. *Holarct. Ecol.* **8**, 1–8.

Kedrowski, R. A. (1976). Plant water relations in the arctic tundra near Meade River, Alaska. M.S. thesis, San Diego State University, San Diego, California.

Kummerow, J., Mills, J. N., Ellis, B. A., Hastings, S. J., and Kummerow, A. (1987). Downslope fertilizer movement in arctic tussock tundra. *Holarct. Ecol.* **10**, 312–319.

Lashof, D. A. (1989). The dynamic greenhouse: Feedback processes that may influence future concentrations of atmospheric trace gases and climatic change. *Clim. Change* **14**, 213–242.

Limbach, W. E., Oechel, W. C., and Lowell, W. (1982). Photosynthetic and respiratory responses to temperature and light of three Alaskan tundra growth forms. *Holarct. Ecol.* **5**, 150–157.

McGraw, J. B. (1985). Experimental Ecology of *Dryas octopetala* ecotypes. III. Environmental factors and plant growth. *Arct. Alp. Res.* **17**, 229-239.

McKendrick, J. D. (1987). Plant succession on disturbed sites, North Slope, Alaska, U.S.A. *Arct. Alp. Res.* **19**, 554–565.

McNaughton, K. G., and Jarvis, P. G. (1983). Predicting effects of vegetation changes on transpiration and evaporation. *In* "Water Deficits and Plant Growth" (T. T. Kozlowski, ed.), Vol. 7, pp. 1–47. Academic Press, New York.

Martin, P., Rosenberg, N. J., and McKenney, M. S. (1989). Sensitivity of evapotranspiration in a wheat field, a forest, and a grassland to changes in climate and direct effects of carbon dioxide. *Clim. Change* **14**, 117-151.

Matthes-Sears, U., Matthes-Sears, W., Hastings, S. J., and Oechel, W. C. (1988). Biomass production, photosynthesis, and tissue nutrient concentrations of two dwarf deciduous shrub species along a natural slope in arctic Alaska. *Arct. Alp. Res.* **20**, 342–351.

Mayo, J. M., Hartgerink, A. P., Despain, D. G., Thompson, R. G., van Zinderen Bakker, E. M., Jr., and Nelson, S. D. (1977). Gas exchange studies of *Carex* and *Dryas*, Truelove lowland. *In* "Truelove Lowland, Devon Island, Canada: A High Arctic Ecosystem" (L. C. Bliss, ed.), pp. 265–280. University of Alberta Press, Edmonton.

Miller, P. C. (1982). Environmental and vegetational variation across a snow accumulation area in montane tundra in central Alaska. *Holarct. Ecol.* **5**, 85–98.

Miller, P. C. (1983). Plant and soil water storage in arctic and boreal forest ecosystems. *In* "Variations in the Global Water Budget" (A. Street-Perrot, M. Beran, and R. Ratcliffe, eds.), pp. 185–196. D. Reidel, Dordrecht, Netherlands.

Miller, P. C., Stoner, W. A., and Tieszen, L. L. (1976). A model of stand photosynthesis for the wet meadow tundra at Barrow, Alaska. *Ecology* **57**, 411–430.

Miller, P. C., Webber, P. J., Oechel, W. C., and Tieszen, L. L. (1980). Biophysical processes and primary production. *In* "An Arctic Ecosystem: The Coastal Tundra at Barrow, Alaska" (J. Brown, P. C. Miller, L. L. Tieszen, and F. L. Bunnell, eds.), pp. 66-101. Dowden, Hutchison and Ross, Stroudsburg, Pennsylvania.

Mølgaard, P. (1982). Temperature observations in high arctic plants in relation to microclimate in the vegetation of Peary Land, North Greenland. *Arct. Alp. Res.* **14**, 105–115.

Monteith, J. L. (1965). Evaporation and environment. *In* "The State and Movement of Water in Living Organisms" (C. E. Fogg, ed.), pp. 205–234. Cambridge Univ. Press, Cambridge.

Neilson, R. P., King, G. A., DeVelice, R. L., Lenihan, J., Marks, D., Dolph, J., Campbell, B., and Glick, G. (1989). "Sensitivity of Ecological Landscapes and Regions to Global Climate Change." U.S. Environmental Protection Agency, Environmental Research Laboratory, Corvallis, Oregon.

Nouchi, I., Mariko, S., and Aoki, K. (1990). Mechanism of methane transport from the rhizosphere to the atmosphere through rice plants. *Plant Physiol.* **94**, 59–66.

Oberbauer, S., and Miller, P. C. (1979). Plant water relations in montane and tussock tundra vegetation types in Alaska. *Arct. Alp. Res.* **11**, 69-81.

Oberbauer, S., and Miller, P. C. (1981). Some aspects of plant water relations in Alaska arctic tundra species. *Arct. Alp. Res.* **13**, 205–218.

Oberbauer, S., and Miller, P. C. (1982a). Growth of Alaskan tundra plants in relation to water potential. *Holarct. Ecol.* **5**, 194–199.

Oberbauer, S., and Miller, P. C. (1982b). Effect of water potential on seed germination. *Holarct. Ecol.* **5**, 218–220.

Oberbauer, S. F., and Oechel, W. C. (1989). Maximum CO_2-assimilation rates of vascular plants along an Alaskan arctic tundra slope. *Holarct. Ecol.* **12**, 312–316.

Oechel, W. C., and Strain, B. R. (1985) Native species responses to increased atmospheric carbon dioxide concentration. *In* "Direct Effects of Increasing Carbon Dioxide on Vegetation" (B. R. Strain and J. D. Cure, eds.), pp. 117–154. DOE/ER-0238. Natl. Tech. Info. Serv., Springfield, Virginia.

Passioura, J. B. (1988). Water transport in and to roots. *Annu. Rev. Plant Physiol. Plant Mol. Biol.* **39**, 245–265.

Pau U, K. T., and Gao, W. (1988). Applications of solutions to non-linear energy budget equations. *Agric. For. Meteorol.* **43**, 121–145.

Penman, H. L. (1948). Natural evaporation from open water, bare soil, and grass. *Proc. Roy. Soc. Lond. Ser. A.* **193**, 120–146.

Peterson, K. M., Billings, W. D., and Reynolds, D. N. (1984). Influence of water table and atmospheric CO_2 concentration on the carbon balance of arctic tundra. *Arct. Alp. Res.* **16**, 331–335.

Radin, J. W., and Ackerson, R. C. (1981). Water relations of cotton plants under nitrogen deficiency. III. Stomatal conductance, photosynthesis, and abscisic acid accumulation during drought. *Plant Physiol.* **67**, 115–119.

Rosenberg, N. J., McKenney, M. S., and Martin, P. (1989). Evapotranspiration in a greenhouse-warmed world: A review and a simulation. *Agric. For. Meteorol.* **47**, 303–320.

Rosenberg, N. J., Kimball, B. A., Martin, P., and Cooper, C. F. (1990). From climate and CO_2 enrichment to evaporation. *In* "Climate Change and U. S. Water Resources" (P. E. Waggoner, ed.), pp. 151–175. Wiley, New York.

Sebacher, D. I., Hariss, R. C., and Bartlett, K. B. (1985) Methane emissions to the atmosphere through aquatic plants. *J. Environ. Qual.* **14**, 40-46.

Sebacher, D. I., Hariss, R. C., and Bartlett, K. B., Sebacher, S. M., and Grice, S. S. (1986). Atmospheric methane sources: Alaskan tundra bogs, an alpine fen, and a subarctic boreal marsh. *Tellus* **38**, 1–10.

Shaver, G. R., and Chapin, F. S. III. (1980). Response to fertilization by various plant growth forms in an Alaskan tundra: Nutrient accumulation and growth. *Ecology* **61**, 662–675.

Sørenson, T. (1941). Temperature relations and phenology of the northeast Greenland flowering plants. *Medd. Gronl.* **125**, 1–305.

Small, E. (1972). Photosynthetic rates in relation to nitrogen recycling as an adaptation to nutrient deficiency. *Can. J. Bot.* **50**, 2227–2233.

Stoner, W. A., and Miller, P. C. (1975). Water relations of plant species in the wet coastal tundra at Barrow, Alaska. *Arct. Alp. Res.* **7**, 109–124.

Stuart, L., Oberbauer, S., and Miller, P. C. (1982). Evapotranspiration measurements in *Eriophorum vaginatum* tussock tundra in Alaska. *Holarct. Ecol.* **5**, 145–149.

Svoboda, J., and Henry, G. H. R. (1987). Succession in marginal arctic environments. *Arct. Alp. Res.* **19**, 373–384.

Teeri, J. A. (1973). Polar desert adaptations of a high arctic plant species. *Science* **179**, 496–497.

Tieszen, L. L. (1973). Photosynthesis and respiration in arctic tundra grasses: Field light intensity and temperature responses. *Arct. Alp. Res.* **5**, 239–251.

Tieszen, L. L. (1978). Photosynthesis in the principal Barrow, Alaska, species: A summary of field and laboratory responses. *In* "Vegetation and Production Ecology of an Alaskan Arctic Tundra" (L. L. Tieszen, ed.), pp. 241–268. Springer-Verlag, New York.

Tissue, D. T., and Oechel, W. C. (1987). Response of *Eriophorum vaginatum* to elevated CO_2 and temperature in the Alaskan tussock tundra. *Ecology* **68**, 401–410.

Ulrich, A., and Gersper, P. L. (1978). Plant nutrient limitations of tundra plant growth. *In* "Vegetation and Production Ecology of an Alaskan Arctic Tundra" (L. L. Tieszen, ed.), pp. 457–481. Springer-Verlag, New York.

Walker, D. A., Binnian, E., Evans, B. M., Lederer, N. D., Nordstrand, E., and Webber, P. J. (1989).

Terrain, vegetation, and landscape evolution of the R4D research site, Brooks Range Foothills, Alaska. *Holarct. Ecol.* **12**, 238–261.

Warren-Wilson, J. (1957). Observations on the temperatures of arctic plants and their environment. *J. Ecol.* **45**, 499–531.

Warren-Wilson, J. (1959). Notes on wind and its effects in arctic-alpine vegetation. *J. Ecol.* **47**, 415–427.

Warren-Wilson, J. (1966). An analysis of plant growth and its control in arctic environments. *Ann. Bot.* **30**, 383–402.

Webber, P. J. (1978). Spatial and temporal variation of the vegetation and its production, Barrow, Alaska. *In* "Vegetation and Production Ecology of an Alaskan Arctic Tundra" (L. L. Tieszen, ed.), pp. 37–112. Springer-Verlag, New York.

Whalen, S. C., and Reeburgh, W. S. (1988). A methane flux time series for tundra environments. *Global Biogeochem. Cycles* **2**, 399–410.

Woodward, F. I. (1987). "Climate and Plant Distribution." Cambridge Univ. Press, Cambridge.

13

Microbial Processes and Plant Nutrient Availability in Arctic Soils

Knute J. Nadelhoffer, A. E. Giblin, G. R. Shaver, and A. E. Linkins

I. Introduction

Several characteristics of arctic soils influence microbial activity, nutrient mineralization, and nutrient availability to plants and will certainly figure prominently in changes in these processes in a warmer arctic climate. Arctic soils are generally overlain by a dense mat of organic matter and vegetation, wet for at least part of the year and permanently frozen at some depth (Tedrow, 1977; Reiger, 1983). These factors combine to lower summer soil temperatures, impede the progression and decrease the depth of seasonal thawing, and maintain relatively high soil moisture content. Cold, wet soil environments and short summers slow organic matter decomposition and nutrient mineralization and severely restrict nutrient availability to plants.

The accumulation of organic matter in arctic soils is determined largely by the combined effects of temperature and moisture on decomposition and primary production (French, 1974, 1977; Heal *et al.*, 1981). Because of climatic variations among arctic regions, the amounts of organic matter and

281

nutrients in tundra soils vary across broad geographic scales. Organic matter often accumulates at depth in permanently frozen peats in relatively wet arctic regions such as the coastal plain in northern Alaska, the Canadian Low Arctic, northern Fennoscandia, and northern Siberia (French, 1974; Bliss, 1981; Moore, 1981). Soils in colder and drier regions, including much of the Canadian High Arctic, northern coastal Greenland, and extreme northern Siberia, typically have thin or discontinuous organic mats (French, 1974; Andreev and Aleksandrova, 1981; Bliss, 1981).

Ecosystem structure often varies as much on local scales as across large regions. For example, a diversity of ecosystem types exists in the Truelove Lowlands on Devon Island in the Canadian High Arctic (Bliss, 1977; Babb and Whitfield, 1977). Although this site lies within a large semibarren to barren region, local microclimate and variations in geomorphology and drainage patterns have resulted in the development of distinct ecosystem types with predictable relationships among site moisture status, soil organic matter content, and vegetation type. Organic carbon in Truelove Lowland soils increases with moisture, from low amounts in well-drained beach-ridge ecosystems with cushion plant–lichen communities to high amounts in very poorly drained ecosystems with sedge–moss communities (Walker and Peters, 1977). Such an overall pattern— of organic carbon increasing with moisture from well- to poorly-drained ecosystems—also occurs elsewhere in the Arctic, such as in northern Siberia at Tareya, Ary-Mass, and Agapa (Chernov *et al.*, 1975; Norin and Ignatenko, 1975; Vassiljevskaya *et al.*, 1975); in Scandinavia at Hardangervidda and Stordalen (Hinneri *et al.*, 1975; Sonesson *et al.*, 1975); and in Alaska's coastal (Gersper *et al.*, 1980) and foothill tundra regions (Oechel, 1989; Shaver *et al.*, 1990; Giblin *et al.*, 1991).

In this chapter we compare and contrast decomposition and nutrient cycling in dry, moist, and wet tundra ecosystems (Fig. 1) and predict possible changes in the distribution of soil organic matter and the availability of nutrients in these ecosystems under a warmer arctic climate. Dry soils with thin organic mats (0–5 cm thick); relatively deep thaw depths; and lichen, cush-

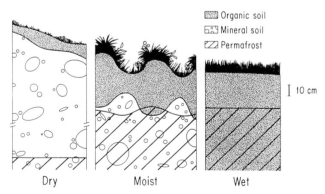

Figure 1   Generalized views of dry, moist, and wet tundra ecosystems. See text for details.

ion plant, heath, or dwarf shrub vegetation occur on ridges or steep slopes in hilly and alpine regions in the Low Arctic (Shaver *et al.*, 1990) and on well-drained uplands in the High Arctic (Bliss, 1977). Well-drained soils are less common in patterned ground regions with little relief, such as the Alaskan coastal plain, where more than 85% of soils are moist to poorly drained (Rydén, 1981). Moist soils with dense organic mats (5–40 cm thick), intermediate thaw depths, and diverse plant communities dominated by tussock-forming sedges occupy gently sloping land in much of the Low Arctic (Walker *et al.*, 1982; Oechel, 1989). Wet soils with thick peat accumulation (>40 cm, often below permafrost) and dominated by sedges and mosses occur in poorly to very poorly drained ecosystems in low-arctic regions and occasionally in the High Arctic (Everett *et al.*, 1981; Heal *et al.*, 1981).

This is a useful simplification because organic matter and moisture content are important determinants of soil temperature, thaw depth, cation exchange capacity, aeration, redox potential, and other properties affecting biological processes in soils (Gersper *et al.*, 1980). Decomposition rates and soil moisture balances will likely be affected by the warmer temperatures predicted for the Arctic (see Chap. 2, 3, 7). The resulting changes in soil organic matter, moisture, and microbial processes in arctic ecosystems will alter the amounts, seasonality, and forms of mineral nutrients available to plants. A warmer climate will likely have different overall effects on soil properties and on nutrient cycling in dry, moist, and wet arctic ecosystems.

II. Microbial and Soil Processes

Nutrient cycling and fertilization studies in arctic ecosystems show that plant growth is strongly limited by nutrient availability. Primary production is often nitrogen-limited, but phosphorus (especially on organic soils) or nitrogen and phosphorus together can also limit production (Tamm, 1954; Haag, 1974; McCown, 1978; McKendrick *et al.*, 1978, 1980; Ulrich and Gersper, 1978; Shaver and Chapin, 1980; Chapin and Shaver, 1985a).

Arctic ecosystems are generally conservative of nutrients, accumulating large amounts in soil organic matter pools with very long turnover times (Dowding *et al.*, 1981). Because of these characteristically slow turnover rates and, in some ecosystems, the gradual burial of organic matter in permafrost, nutrients become available to plants at very low rates. Long turnover times result from slow decomposition, which can become a bottleneck in nutrient cycling rates (Chapin *et al.*, 1980). Differences among ecosystem types in soil microclimate and decomposition may explain the inverse relationships between soil nutrient stocks and nutrient cycling rates or primary productivity as reported, for example, on Alaska's northern coast (Swift *et al.*, 1979: pp. 276–280) and in northern Sweden (Jonasson, 1983). Slow decomposition leads to greater accumulation of organic matter in soil and can lower nutrient mineralization rates, thereby decreasing primary productivity.

A. Decomposition

Although the decomposition of organic matter in arctic ecosystems is generally much slower than at lower latitudes, it is controlled by the same factors. Decomposition rates increase globally with actual evapotranspiration (AET) from warm temperate to arctic regions and decrease with the percentage of lignin in litter (Meentemeyer, 1978). Field studies comparing decomposition of common materials in tundra with forest ecosystems are consistent with this observation (e.g., Rosswall, 1974; Berg *et al.*, 1975). Models based on laboratory and field studies at a number of arctic International Biological Programme (IBP) sites have successfully used temperature, moisture content, and substrate quality to predict the decomposition of litter and soil organic matter (Bunnell and Dowding, 1974, Bunnell *et al.*, 1977a,b; Bunnell and Scoullar, 1981).

Temperature is the most important predictor of decomposition rates in arctic ecosystems (Flanagan and Veum, 1974; Heal *et al.*, 1981). Comparisons of arctic IBP studies suggest that, overall, decomposition rates increase by about 20% per year for every 1000 degree-days above 0° C (Heal and French, 1974). Microbial respiration in tundra litter and soil organic matter is measurable at −7° C and increases with temperature between about 5° and 20° to 30° C (Svensson *et al.*, 1975; Bunnell *et al.*, 1977a; Flanagan and Bunnell, 1980). There are, however, important functional differences in decomposition processes operating at low versus high ends of the range of typical summer field temperatures in arctic soils. Enzymatic degradation of cellulosics and simple organic compounds (e.g., water-soluble organic acids) both increase with temperature above about 10° C. Below 10° C, however, degradation of soluble organic compounds continues to vary with temperature, whereas cellulose decomposition is restricted, both by low numbers of active cellulolytic microbes (Christensen, 1974) and by greater thermodynamic constraints on cellulase activities as compared with enzymes that degrade organic acids (Linkins *et al.*, 1984). This probably contributes to the sharp increase in temperature sensitivity of microbial reactions in general in soils above 10° C (see Paul and Clark: Fig. 2.9).

Moisture is also a critical determinant of decomposition in arctic ecosystems. Decomposition is extremely slow and relatively independent of temperature when moisture content is less than 20% of dry mass. Above this level, temperature sensitivity generally increases until moisture content reaches 200% or more, with less moisture required for optimum decomposition as temperature increases (Bunnell and Tait, 1974; Flanagan and Veum, 1974; Heal *et al.*, 1981). Decreases in decomposition under highly saturated conditions probably result from the effects of poor aeration on the rates and pathways of microbial metabolism. The independent and interactive effects of moisture and temperature account for much of the variation in decomposer activity in the Arctic. For example, at the Stordalen IBP site in northern Sweden, these two factors accounted for 69% of the variation in carbon dioxide flux from different organic soils (Svensson, 1980). At the Hardangervidda IBP site in Norway,

65–80% of the seasonal variation in soil respiration within each of four ecosystem types could be accounted for by temperature alone, with moisture explaining much of the remaining variation in rates (Svensson, 1980).

The quality, or decomposability, of litter inputs to arctic soils varies among tissue types and plant growth forms. Studies done within individual arctic ecosystems or under controlled conditions show that deciduous shrub and graminoid leaf litters decompose faster than evergreen leaf litters, which in turn decompose faster than mosses, lichens, and woody stems (Heal and French, 1974; Rosswall *et al.*, 1975; Widden, 1977; Heal *et al.*, 1981). In the Arctic, as elsewhere, concentrations of secondary compounds such as lignin, phenols, and waxes are higher, and nutrient concentrations are lower, in poor-quality litter (Van Cleve, 1974). These factors affect not only decomposition rates, but also nutrient dynamics in decomposing litter and nutrient availability in soils. For example, litter types with low nutrient and high lignin concentrations decompose more slowly and can also immobilize more nutrients per unit mass lost than litter with high nutrient and low lignin concentrations (Aber and Melillo, 1982; Melillo *et al.*, 1982). Low-quality litter input from plants to soils can therefore lower nutrient availability to plants by serving as stronger sinks for inorganic nutrients than high-quality litter.

Plant litter quality in arctic ecosystems can play a pivotal role in determining the amount and quality of organic matter that accumulates in soil because in arctic soils, where decomposition rates decrease dramatically with depth, small differences in decomposability at the surface can produce large differences in the proportion of litter that is transferred to depths, where decomposition is slowed by cold, wet conditions (Jones and Gore, 1978; Heal *et al.*, 1981). The relative amounts of primary production allocated above and below ground also influence organic matter accumulation. Root detritus is inserted directly into soil profiles, where conditions are colder and wetter than on the surface, whereas aboveground litter can decompose for several years before burial occurs.

Other edaphic factors, such as exchangeable nutrients, cation-exchange capacity, base saturation, porosity, bulk density, and depth of thaw, also influence decomposition rates. Their effects are less clear, however, because these factors are themselves strongly influenced by organic carbon content (Gersper *et al.*, 1980; Heal *et al.*, 1981), which is in turn determined by temperature, moisture, and plant litter quality (see above). Mineral soils can influence decomposition, especially if calcium, magnesium, or phosphate are weathered from parent rock and cycled through vegetation and organic soils. Mineral soil properties, however, are more important in dry tundra than in either moist or wet tundra, where mineral soils thaw either slightly for very short periods or not at all (Everett *et al.*, 1981).

B. Nitrogen Mineralization

In the Arctic, as elsewhere, most of the inorganic nitrogen available for plant uptake is supplied by the mineralization of organic matter. Precipitation inputs of nitrogen to tundra ecosystems are generally ≤ 0.03 g m^{-2} yr^{-1}—small relative to annual nitrogen uptake into vegetation (Barsdate and Alexander, 1975). Nitrogen fixation by rhizobial or actinorhizal symbionts in certain nodulated vascular plants (e.g., lupines, alders) or by symbiotic or freeliving blue-green algae (e.g., *Nostoc, Anabaena*) can supply large proportions of the nitrogen taken up by vegetation in some arctic ecosystems (see Chap. 14).

When soil microbes are active because of favorable physical conditions and sources of readily oxidizable organic carbon, ammonium (NH_4^+) is released from organic matter to soil solution by enzymatically mediated processes (Burns, 1978). Active microbes are generally better competitors for mineral N than plant roots because of their greater surface area and closer proximity to microsites with actively decomposing organic matter. When microbes are subjected to substrate (carbon) limitation, low temperatures, low availabilities of electron acceptors (oxygen, sulfate, etc.), or other stresses, they often release NH_4^+ to soil solution. Mineralized NH_4^+ has several potential fates, including adsorption to negatively charged particles, chemical immobilization in clay lattices or humus, reimmobilization into microbial biomass, uptake by plants, and oxidation to nitrate (NO_3^-) by nitrifying bacteria (Paul and Clark, 1989). Evidence that nitrification can occur in arctic soils is accumulating (see below). The occurrence of nitrification has important implications for plants in arctic ecosystems because species differ in nitrate uptake potentials and assimilation efficiencies (Smirnoff and Stewart, 1985). NO_3^- is also more easily leached downslope than NH_4^+ and can be reduced and exported from soils as N_2O or N_2.

The difference between the gross nitrogen flux from soil organic matter mineral forms (gross mineralization) and immobilization by microbes is net nitrogen mineralization. Net N mineralization can be positive (gross mineralization > immobilization) or negative (immobilization > gross mineralization). Net N mineralization is generally positive during a growing season or across a year, but rates can be very low or negative during the arctic midsummer (see below). Variations in N mineralization rates across growing seasons are determined largely by asynchronies in competitive success for mineral N between heterotrophic microbes and plant roots (Gorham *et al.*, 1979). Thus, the amount and quality of organic carbon in soils, soil microclimate, and microbial activity interact to determine the amounts, forms, and seasonality of nitrogen availability to plants in arctic ecosystems.

Field measures in different biomes indicate that N mineralization in arctic soils is very low relative to rates in soils of lower latitudes. Annual N mineralization ranges from about 0.1 to 0.6 g m^{-2} yr^{-1} in arctic ecosystems, from 1.5 to 20 g m^{-2} yr^{-1} in boreal and temperate ecosystems, and up to 90 g m^{-2} yr^{-1} in some very productive tropical forest sites (Fig. 2). As such, net N mineralization in temperate and tropical soils can be more than 200-fold greater than in arctic soils. Primary production also increases from high to low latitudes, ranging from 50 g m^{-2} yr^{-1} in arctic ecosystems, excluding polar desert (Wielgolaski *et al.*, 1981), to highs of 4000 g m^{-2} yr^{-1} in moist tropical forests

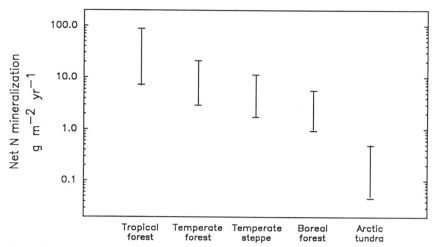

Figure 2   Comparative rates of net N mineralization in soils of several important ecosystem types based either on field measures of net N mineralization or on estimates of soil N turnover. Vertical bars indicate maximum and minimum rates reported for tropical forests (Ellenberg, 1971, 1977; Chandler, 1985; Vitousek and Denslow, 1986); temperate forests (Gosz, 1981; Melillo, 1981; Nadelhoffer *et al.*, 1983; Pastor *et al.*, 1984); temperate steppe (Ellenberg, 1971; Woodmansee *et al.*, 1981; Schimel *et al.*, 1985); boreal forest (Ellenberg, 1971; Van Cleve and Alexander, 1981); and arctic tundra (Rosswall and Granhall, 1980; Shaver *et al.*, 1990; Hart and Gunther, 1989; Giblin *et al.*, 1991). Rates within ecosystem types generally increase with soil moisture.

(Edwards *et al.*, 1981). These differences represent an 80-fold range in primary production globally. Because ratios of primary production to N mineralization are higher in arctic ecosystems than in many others, processes such as the direct uptake of amino acids from soils (Chap. 15), N fixation (Chap. 14), and N retranslocation within plants (Chapin and Shaver, 1985b; Chap. 9) are likely to be important in meeting nitrogen requirements for plant growth in arctic ecosystems.

Although net N mineralization is generally much lower in arctic soils than in soils of warmer regions, mineralization rates can vary among different arctic ecosystem types. For example, Hart and Gunther (1989) reported large differences in annual N mineralization measured in dry and moist tundra ecosystems at a high-elevation site in southern Alaska. In the foothills of Alaska's North Slope, fivefold differences in nitrogen mineralization were detected among dry, moist, and wet ecosystems using in situ soil incubations (Fig. 3). These differences were largely due to variations in the quality of soil organic matter and microclimate among ecosystems (Giblin *et al.*, 1991; Nadelhoffer *et al.*, 1991).

Seasonal variations in mineral nitrogen availability are also important in arctic soils. Many lines of evidence suggest that soil nutrients are in short supply during the peak growing season. Exchangeable NH_4^+ concentrations in

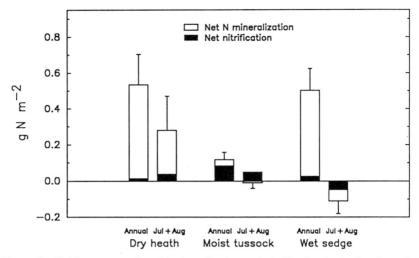

Figure 3   Field measures of net N mineralization and nitrification in dry heath, moist tussock, and wet meadow tundra ecosystems near the Sagavanirktok River in the northern foothills of the Brooks Range, Alaska. Annual values (means ± SE) are based on four years of in situ soil incubations, and July–August values are based on in situ soil incubations ($n = 7$) set out for seven weeks (9 July to 27 August) during the 1985 growing season. [Data from Giblin *et al.* (1991).]

organic soils are typically greater just after soil thaw or before fall freezing than in midseason (Gersper *et al.*, 1980). Net N mineralization in July (midseason) in a moist tussock tundra on Alaska's North Slope, estimated using in situ soil incubations, was just slightly greater than zero in a water-track site and was negative in an adjacent non-water-track site (Chapin *et al.*, 1988). In situ incubations in different ecosystems in the Sagavanirktok watershed reported by Giblin *et al.* (1991) showed that only half of the annual net N mineralization in moist tussock and dry heath ecosystems occurred in July–August 1985 and that mineralization was negative in a wet sedge ecosystem during this period (see Fig. 3). This pattern of relatively high N availability early and late in the growing season and low N availability in midsummer probably results from lower microbial demands when soil temperatures are cooler than when soils are warmer. Even if gross N mineralization is greatest in midsummer when soils are warmest, net N mineralization can be lowered by high rates of N immobilization by microbes.

Nitrification can occur in arctic soils even though cold, wet conditions are often not considered favorable for nitrifiers. Nitrate has been reported in tissues of arctic plants growing on stream banks and well-drained microsites and in arctic soil solutions and surface waters (Haag, 1974; Ulrich and Gersper, 1978). Direct evidence of nitrification in arctic soils also exists. For example, Chapin *et al.* (1988) found detectable nitrification during midsummer at a tussock tundra site on the Alaskan North Slope. In their study, nitrification was greater in tundra warmed by flowing soil water than in adjacent, cooler tundra. Giblin *et al.* (1991) measured nitrification in situ in a moist tussock tundra ecosystem on Alaska's North Slope but found no evidence for it in nearby wet sedge and dry heath ecosystems (see Fig. 3). Nitrification accounted for about half the annual net nitrogen mineralization in the moist tundra but, in contrast to ammonification, occurred mostly in midsummer. Because nitrification is more temperature-sensitive than mineralization (Paul and Clark, 1989), it may be largely restricted to the warmest part of the season in those arctic soils where it does occur. Nitrification rates could increase in warmer arctic soils. Such an increase has important implications for primary productivity and community composition in the Arctic, both because plants metabolize nitrate differently from ammonium and because species preferences for these forms differ (Smirnoff and Stewart, 1985).

C. Phosphorus Availability

Phosphorus is generally tightly conserved in arctic ecosystems, even when it is not the most limiting element (Chapin *et al.*, 1980). Soluble and exchangeable phosphate pools are typically very small and turn over rapidly. For example, in wet tundra soils at Barrow, Alaska, soluble phosphate must be replenished about 200 times annually to match plant uptake, as opposed to only about 10 times for mineral nitrogen (Bunnell *et al.*, 1975). Also, carbon:phosphorus ratios often increase with soil depth in arctic peats, whereas carbon:nitrogen ratios generally remain constant or decrease with depth (Malmer and Nihlgård, 1980; Giblin *et al.*, 1991). This difference suggests that strong biological demands for P in active soil layers minimize P burial in accumulating peats.

In contrast to nitrogen cycling, phosphorus cycling is controlled by physical–chemical as well as biological processes. For example, in many temperate soils, the chemical release of phosphate from weatherable minerals (e.g., apatite) is an important source of plant-available P (Williams and Walker, 1969). Weathering is less important in arctic ecosystems, however, because plant roots are isolated from mineral soil horizons, and low temperatures slow weathering reactions (Hill and Tedrow, 1961). In moist and wet tundra ecosystems, P input into organic surface soils is small, coming mainly from precipitation. Therefore, recycling of P from organic soils must supply nearly all of the P taken up annually by plants in these ecosystems (Shaver *et al.*, 1990). In dry tundra ecosystems, where organic horizons are thin, P availability depends more on properties of mineral soil horizons. Phosphorus availability can be high in young soils with large pools of weatherable P in primary minerals, but low in more intensively weathered soils, especially if phosphates are tightly held to secondary minerals such as aluminum or iron oxides (Sayers and Walker, 1969).

Seasonal variations in P availability are high in arctic soils. As with mineral N, soluble and exchangeable phosphate pools in organic horizons are greatest at snowmelt and decline as soil temperatures increase through the growing season (Dowding *et al.*, 1981). As a result, most P uptake by tundra plants probably occurs shortly after surface thawing in early summer, just before freezing in fall, or during other periods when microbial populations decline (Chapin and Bloom, 1976; Chapin *et al.*, 1978). Laboratory incubations of tundra soils also indicate that P availability is highest at temperatures just above freezing, when microbial demands for P are relatively low (Nadelhoffer *et al.*, 1991).

D. Controls on Carbon and Nutrient Cycles

Decomposition and nutrient cycling in many environments are linked to the relative amounts of soluble and insoluble carbon in litter and soils (Melillo *et al.*, 1982, 1989; Berg *et al.*, 1984; McClaugherty *et al.*, 1985). A simple but useful descriptor of these fractions is the lignocellulose index (LCI) as defined by Melillo *et al.* (1989). The LCI of organic material is defined as the ratio of acid-insoluble carbon to acid-soluble plus acid-insoluble carbon, or LCI = lignin/(cellulose + lignin). Plant litter typically enters soils with LCI values < 0.4 in temperate and arctic ecosystems. In temperate soils, LCI values gradually increase as acid-soluble (or "cellulose") C decomposes more rapidly than insoluble ("lignin") C. Eventually, the relative amounts of these carbon fractions stabilize in well-decomposed organic matter, and LCI values converge at about 0.7 to 0.8 in humus (Melillo *et al.*, 1989).

Comparisons of LCI values of soils at our Sag River sites on the North Slope with temperate soil values suggest that organic matter in arctic ecosystems is much less decomposed than in warmer ecosystems (Fig. 4). Furthermore, unlike soils in warmer climates, no increase in LCI with soil depth was found in either moist or wet tundra ecosystems, suggesting that decomposition rates decline rapidly with depth, temperature, and moisture in the soils at these sites. Soils of the dry heath ecosystem, which is warmer and drier and where organic matter stocks are lowest, had high LCIs relative to soils of the wetter ecosystems, suggesting that warmer and drier soils in even a modestly warmer arctic climate could lead to more complete oxidation of soil organic matter.

Variability in microbial respiration among laboratory incubations of different soils shows that the quality of organic matter can differ widely among arctic ecosystems (Fig. 5a). Respiration did not differ between 3 and 9° C within any soil from the Sagavanirktok River site (Nadelhoffer *et al.*, 1991). This somewhat surprising result is consistent with soil enzyme studies showing that below about 10° C, exocellulase activity can be limited by substrate rather than temperature in arctic soils (Linkins *et al.*, 1984).

Laboratory microcosm and soil enzyme studies both suggest that microbial activity is relatively temperature-insensitive across much of the 2–10° C range of growing-season temperatures typical of most arctic soils. The doubling of respiration that occurred between 9 and 15° C in soils from the Sagavanirktok sites (Fig. 5a) and enzyme studies (e.g., Linkins *et al.*, 1984), however, suggest that local or regional warming of surface soils above current upper limits could greatly increase carbon loss from arctic soils. Large losses in soil carbon

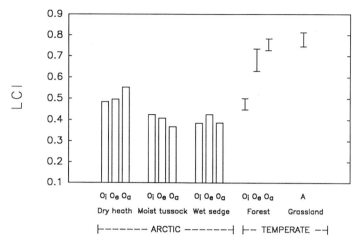

Figure 4 Lignocellulose indices (LCI) of organic matter in soils of some arctic and temperate ecosystems. The LCI is the ratio of acid-insoluble material to acid-soluble plus acid-insoluble material (or lignin/[lignin + cellulose]) in ash-free organic matter. LCIs of arctic soils were measured using methods of Melillo *et al.* (1989) on soil samples ($n = 2$) from the same ecosystems as in Figure 3. Ranges for temperate soils are from a summary by Melillo *et al.* (1989). Depths of organic horizons increase from O_i to O_a in tundra and forest soils. Organic matter at the soil surface in grasslands is generally mixed into mineral soils (A horizons).

could feed back in soils to cause warmer, drier conditions as the organic mat thins and thermal insulation declines.

Patterns of N mineralization in the laboratory were similar in many ways to respiration. As with respiration, N mineralization potentials differed greatly among the different ecosystem types (Fig. 5b). There was, however, a negative relationship between microbial respiration and net N mineralization across soils. Differences in N mineralized at 3 and 9° C did not occur in either moist tussock or wet sedge tundra soils. Differences in N mineralization among soils were greater than differences in microbial respiration (see Fig. 5b). Also, although respiration typically doubled between 9 and 15° C, N mineralization increased by up to tenfold. These results suggest that warming of soils a few degrees above current maximum growing-season values could increase net N mineralization rates in the field.

III. Soil Nitrogen and Phosphorus Cycling in a Warmer Arctic Climate

General circulation models (GCMs) predict increases of as much as 8–9° C in average annual temperatures at high latitudes (see Chap. 2). Even though the greatest temperature increases are likely to occur in winter, growing seasons

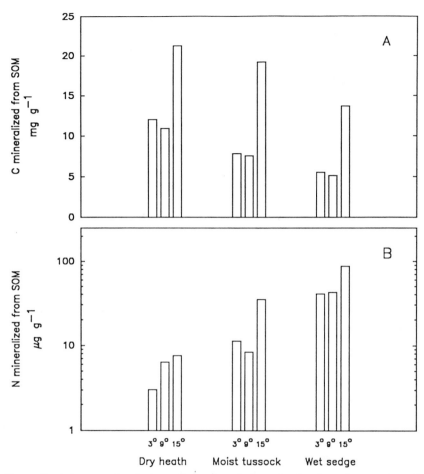

Figure 5   Laboratory incubations of organic soils from dry heath, moist tussock, and wet meadow tundra ecosystems near the Sagavanirktok River, Alaska. Soils were incubated for 13 weeks at −0.06 MPa moisture tension and either 3°, 9° or 15° C. Bars show (A) microbial respiration (CO_2-C production; $n = 3$) and (B) cumulative nitrogen mineralized ($NH_4-N + NO_3-N$ leached; $n = 6$) from soil organic matter (SOM). [Modified from Nadelhoffer *et al.* (1991).]

will likely be warmer as well. Increases in soil temperatures and thaw depths on Alaska's northern coastal plain during recent years suggest that gradual warming may now be occurring in the Arctic (Lachenbruch and Marshall, 1986). Studies summarized in this chapter suggest that even relatively modest increases in growing-season temperatures could have important implications for carbon and nutrient turnover in soils of tundra ecosystems and for plant processes affected by nutrient availability.

The cold, wet soils of many arctic ecosystems contain large amounts of potentially decomposable organic carbon that could be respired and lost under warmer or drier conditions. Increased soil respiration and carbon losses would likely increase rates of nutrient release from decomposing organic matter. Microbial activity and rates of element turnover in arctic soils should increase with soil temperatures. At present, temperatures between the surface layer of fresh litter and the freeze–thaw boundary generally decrease with depth from ≤10 to 0° C. Because microbial processes are less temperature-sensitive across this range than they are above 10° C, prolonged excursions of soil temperatures above ~10° C should increase carbon turnover and plant nutrient availability. In addition, the deeper thawing and longer growing seasons that accompany a warmer climate will increase both the volume of active soil and the length of time that microbes are active, thereby increasing annual carbon and nutrient mineralization.

Warmer conditions are also likely to change the seasonality and forms of plant nutrients available in arctic soils. If warming decreases soil organic carbon (especially soluble forms), nutrient immobilization by microbes could decrease, thereby increasing nutrient availability to plants during the peak growing season. The likelihood of nitrification in many soils will increase because of the direct effects of increases in temperature and oxygen and because higher N mineralization will increase the pools of substrate (NH_4^+) available to nitrifiers.

Along the dry-to-wet continuum of arctic ecosystems (see Fig. 1), soils of dry tundra ecosystems will probably be the least responsive to climatic warming. Soil temperatures, thaw depths, and nutrient availability should all increase somewhat with air temperature (Table I). Under current conditions,

Table I   Predicted Effects of a Warmer Climate on Critical Soil Properties and Processes in Dry, Moist, and Wet Arctic Ecosystems

	Tundra		
Characteristic	Dry heath	Moist tussock	Wet sedge
Thaw depth	+	+ + +	+ +
Soil respiration and C turnover	= to +	+ + +	= to + + + f(drainage)
Organic horizon thickness	= to −	− to − − −	− or + f(drainage)
Drainage	=	+ + to + + +	= to + + + f(local topography)
Soil moisture	=	− to − − −	= to − − −
N mineralization	= to +	+ + +	= to + +
Nitrification	More probable	More probable	More probable with better drainage
P availability	f(mineralogy)	+ + +	− to + + +

however, summer soil temperatures are high and thawing is deep relative to moister ecosystems. Thermal conductivity is greatest and summer temperatures are higher in soils of dry tundra because organic horizons are thin. Therefore, increases in soil respiration and decreases in stocks of organic matter would not feed back to further increase soil temperatures as strongly as in wetter tundra ecosystems. Relative increases in nitrogen mineralization rates would probably be less than in wetter ecosystems. Dramatic increases in nitrogen availability in dry tundra would probably require greater input via nitrogen fixation or precipitation. Changes in phosphorus availability are highly dependent on soil mineralogy. Phosphorus weathering and phosphorus availability could increase in dry tundra ecosystems on relatively young soils with large amounts of primary minerals. Large increases in phosphorus availability are unlikely on highly weathered mineral soil, however, where phosphorus is largely bound with iron and aluminum in secondary minerals (Giblin *et al.*, 1991).

Moist tussock tundra ecosystems have the greatest potential for changes in nutrient availability in a warmer climate (see Table I). The large stocks of poorly decomposed organic matter will likely decrease as soil respiration increases because of higher soil temperatures. Thinner organic mats would provide less thermal insulation in summer and would feed back to increase soil temperatures and depths of thaw. Thinner organic horizons, increased thaw depths, and soil warming would interact to increase hydraulic conductivity, drainage, and evaporative losses. Ratios of carbon to nitrogen are likely to decrease and rates of net nitrogen mineralization should increase in warmer, less saturated soils. This in turn would increase the amount of ammonium available as substrate for nitrifiers and increase the probability that nitrification will become an important process. Phosphorus availability will probably also increase as organic matter stocks decline and turn over more rapidly. Increases in phosphorus availability will also occur if mineral horizons thaw more deeply and plant roots become less isolated from mineral soils, especially where mineral horizons are not excessively weathered.

In wet tundra ecosystems, changes in soil microbial processes and nutrient availability will depend much on local topography and its effects on drainage (see Table I). Stocks of organic matter will probably remain high, or continue to accumulate where inputs of water from well-drained uplands are high, because microbial respiration and nutrient mineralization rates will not increase greatly with soil temperature and thaw depths, for soils will remain saturated and poorly aerated. In wet tundra soils that become better drained because of local perturbations (e.g., thawing of ice dams or diversion of water), soil responses may be similar to those in moist tundra ecosystems. Phosphorus availability should remain low where summer thawing does not extend to mineral soils. Increases in phosphorus availability are possible in wet tundra, however, where peats are thin enough to allow thawing and contact of roots with mineral soils.

IV. Summary

Soils of arctic ecosystems are overlain by mats of poorly decomposed organic matter, which increase in thickness along gradients of soil moisture. These mats, or organic horizons, contain most of the organic carbon and nutrients in arctic ecosystems. They inhibit drainage by decreasing hydraulic conductivity and impede the progression of soil warming during the growing season. Cold soil temperatures and short growing seasons restrict microbial activity, litter and humus decomposition, and nutrient mineralization. Thus, the availability of nutrients for plant uptake is generally low relative to the growth requirements of arctic plants.

Warmer growing seasons in the Arctic will likely increase soil microbial activity and the turnover rates of soil organic matter. As a result, in many arctic ecosystems, soil carbon stocks and the thickness of the surface organic mat are likely to decrease. Growing-season soil temperatures and thaw depths will increase because of warmer air temperatures and the lower insulation values of thinner organic mats. Amounts of mineral nutrients made available annually should increase with microbial activity and the turnover of soil carbon. Seasonal patterns of nutrient availability could also change, with more soil nutrients becoming available in midsummer. Nitrate availability will increase in many ecosystems if nitrification rates increase with ammonium availability and temperature. Increases in phosphorus availability will depend on the depths to which mineral soils thaw, on clay mineralogy, and on increases in the turnover of organic matter. The degree to which element cycling rates and plant nutrient availability will increase will vary among different arctic ecosystems, with the greatest and least changes occurring in moist tundra and dry tundra ecosystems, respectively. Wet tundra ecosystems also have a high potential for change, but changes in drainage patterns and soil moisture will be critical determinants in these ecosystems. Wet ecosystems that continue to receive and retain large amounts of water from upland areas will respond only slightly to warming, whereas soil microbial activity and nutrient availability could increase greatly in those that become better drained.

Acknowledgments

This work was supported by grants from the National Science Foundation, including NSF-BSR-8615191, NSF-BSR-8702328, and NSF-BSR-8806635.

References

Aber, J. D., and Melillo, J. M. (1982). Nitrogen immobilization in decaying hardwood leaf litter as a function of initial nitrogen and lignin content. *Can. J. Bot.* **60**, 2263–2269.

Andreev, V. N., and Aleksandrova, V. D. (1981). Geobotanical division of the Soviet Arctic. *In* "Tundra Ecosystems: A Comparative Analysis" (L. C. Bliss, O. W. Heal, and J. J. Moore, eds.), pp. 25–34. Cambridge Univ. Press, Cambridge.

Babb, T. A., and Whitfield, D. W. A. (1977). Mineral nutrient cycling and limitation of plant growth in the Truelove Lowland ecosystem. *In* "Truelove Lowland, Devon Island, Canada: A High Arctic Ecosystem" (L. C. Bliss, ed.), pp. 589–606. Univ. of Alberta Press, Edmonton.

Barsdate, K. J., and Alexander, V. (1975). The nitrogen balance of arctic tundra: Pathways, rates and environmental implications. *J. Environ. Qual.* **4**, 111–117.

Berg, B., Kärelampi, L., and Veum, A. K. (1975). Comparisons of decomposition rates measured by means of cellulose. *In* "Fennoscandian Tundra Ecosystems" (F. E. Wielgolaski, ed.), Part 1, Plants and Microorganisms, pp. 261–267. Springer-Verlag, New York.

Berg, B., Ekbohm, G., and McClaugherty, C. (1984). Lignin and holocellulose relations during long-term decomposition of some forest litters. Long-term decomposition in a Scots pine forest. IV. *Can. J. Bot.* **62**, 2540–2550.

Bliss, L. C. (1977). General summary: Truelove Lowland ecosystem. *In* "Truelove Lowland, Devon Island, Canada: A High Arctic Ecosystem" (L. C. Bliss, ed.), pp. 657–675. Univ. Alberta Press, Edmonton.

Bliss, L. C. (1981). North American and Scandinavian tundras and polar deserts. *In* "Tundra Ecosystems: A Comparative Analysis" (L. C. Bliss, O. W. Heal, and J. J. Moore, eds.), pp. 38–46. Cambridge Univ. Press, Cambridge.

Bunnell, F. L., and Dowding, P. (1974). ABISKO: A generalized decomposition model for comparisons between tundra sites. *In* "Soil Organisms and Decomposition in Tundra" (A. J. Holding, O. W. Heal, S. F. MacLean, Jr., and P. W. Flanagan, eds.), pp. 227–247. Tundra Biome Steering Committee, Stockholm.

Bunnell, F. L., and Scoullar, K. A. (1981). Between-site comparisons of carbon flux in tundra by using simulation models. *In* "Tundra Ecosystems: A Comparative Analysis" (L. C. Bliss, O. W. Heal, and J. J. Moore, eds.), pp. 685–715. Cambridge Univ. Press, Cambridge.

Bunnell, F. L., and Tait, D. E. N. (1974). Mathematical simulation models of decomposition processes. *In* "Soil Organisms and Decomposition in Tundra" (A. J. Holding, O. W. Heal, S. F. MacLean, Jr., and P. W. Flanagan, eds.), pp. 207–226. Tundra Biome Steering Committee, Stockholm.

Bunnell, F. L., MacLean, Jr., S. F., and Brown, J. (1975). Barrow, Alaska, USA. *In* "Structure and Function of Tundra Ecosystems" (T. Rosswall and O. W. Heal, eds.), pp. 73–124. Ecological Bulletins (Stockholm), Vol. 20.

Bunnell, F. L., Tait, D. E. N., Flanagan, P. W., and Van Cleve, K. (1977a). Microbial respiration and substrate weight loss. I. A general model of the influences of abiotic variables. *Soil Biol. and Biochem.* **9**, 33–40.

Bunnell, F. L., Tait, D. E. N., and Flanagan, P. W. (1977b). Microbial respiration and substrate weight loss. II. A model of the influences of chemical composition. *Soil Biol. Biochem.* **9**, 41–47.

Burns, R. G. (1978). "Soil Enzymes." Academic Press, New York.

Chandler, G. (1985). Mineralization and nitrification in three Malaysian forest soils. *Soil Biol. Biochem.* **17**, 347–353.

Chapin, F. S., III, and Bloom, A. J. (1976). Phosphate absorption: Adaptation of tundra graminoids to a low-temperature, low-phosphorus environment. *Oikos* **26**, 111–121.

Chapin, F. S., III, and Shaver, G. R. (1985a). Individualistic growth response of tundra plant species to manipulation of light, temperature, and nutrients in a field experiment. *Ecology* **66**, 564–576.

Chapin, F. S., III, and Shaver, G. R. (1985b). Arctic. *In* "Physiological Ecology of North American Plant Communities" (B. F. Chabot and H. A. Mooney, eds.), pp. 16–40. Chapman and Hall, London.

Chapin, F. S., III, Barsdate, R. J., and Barèl, D. (1978). Phosphorus cycling in Alaskan coastal tundra: A hypothesis for the regulation of nutrient cycling. *Oikos* **31**, 189–199.

Chapin, F. S., III, Miller, P. C., Billings, W. D., and Coyne, P. I. (1980). Carbon and nutrient budgets and their control in coastal tundra. *In* "An Arctic Ecosystem: The Coastal Tundra at Barrow, Alaska" (J. Brown, P. C. Miller, L. L. Tieszen, and F. L. Bunnell, eds.), pp. 458–482. Dowden, Hutchinson & Ross, Stroudsburg, Pennsylvania.

Chapin, F. S., III, Fetcher, N., Kielland, K., Everett, K. R., and Linkins, A. E. (1988). Productivity and nutrient cycling of Alaskan tundra enhanced by flowing soil water. *Ecology* **69**, 693–702.

Chernov, Yu. J., Dorogostaiskaya, E. V., Gerasimenko, T. V., Ignatenko, I. V., Matveyeva, N. V., Parinkina, O. M., Polozova, T. G., Romanova, E. N., Schamurin, V. F., Smirnova, N. V., Stepanova, I. V., Tomilin, B. A., Vinokurov, A. A., and Zalensky, O. V. (1975). Tareya, USSR. *In* "Structure and Function of Tundra Ecosystems" (T. Rosswall and O. W. Heal, eds.), pp. 159–181. Ecological Bulletins (Stockholm), Vol. 20.

Christensen, P. J. (1974). A microbiological study of some lake waters and sediments from the Mackenzie Valley with special reference to *Cytophagus. Arctic* **27**, 309–311.

Dowding, P., Chapin, F. S., III, Wielgolaski, F. E., and Kilfeather, P. (1981). Nutrients in tundra ecosystems. *In* "Tundra Ecosystems: A Comparative Analysis" (L. C. Bliss, O. W. Heal, and J. J. Moore, eds.), pp. 647–683. Cambridge Univ. Press, Cambridge.

Edwards, N. T., Shugart, H. H., McLaughlin, S. B., Harris, W. F., and Reichle, D. E. (1981). Carbon metabolism in terrestrial ecosystems. *In* "Dynamic Properties of Forest Ecosystems" (D. E. Reichle ed.), pp. 499–536. Cambridge Univ. Press, Cambridge.

Ellenberg, H. (1971). Nitrogen content, mineralization, and cycling. *In* "Productivity of Forest Ecosystems" (P. Duvigneaud, ed.), pp. 509–514. UNESCO, Paris.

Ellenberg, H. (1977). Stickstoff als Standortsfaktor, insbesondere für mitteleuropaische Pflanzengesellschaften. *Oecologia Plant.* **12**, 1–22.

Everett, K. R., Vassiljevskaya, V. D., Brown, J., and Walker, B. D. (1981). Tundra and analogous soils. *In* "Tundra Ecosystems: A Comparative Analysis" (L. C. Bliss, O. W. Heal, and J. J. Moore, eds.), pp. 139–179. Cambridge Univ. Press, Cambridge.

Flanagan, P. W., and Bunnell, F. L. (1980). Microflora activities and decomposition. *In* "An Arctic Ecosystem: The Coastal Tundra at Barrow, Alaska" (J. Brown, P. C. Miller, L. L. Tieszen, and F. L. Bunnell, eds.), pp. 291–334. Dowden, Hutchinson & Ross, Stroudsburg, Pennsylvania.

Flanagan, P. W., and Veum, A. K. (1974). Relationships between respiration, weight loss, temperature, and moisture in organic residues on tundra. *In* "Soil Organisms and Decomposition in Tundra" (A. J. Holding, O. W. Heal, S. F. MacLean, Jr., and P. W. Flanagan, eds.), pp. 249–277. Tundra Biome Steering Committee, Stockholm.

French, D. D. (1974). Classification of IBP tundra biome sites based on climate and soil properties. *In* "Soil Organisms and Decomposition in Tundra" (A. J. Holding, O. W. Heal, S. F. MacLean, Jr., and P. W. Flanagan, eds.), pp. 3–25. Tundra Biome Steering Committee, Stockholm.

French, D. D. (1977). Multivariate characteristics of IBP Tundra Biome site characteristics. *In* "Tundra Ecosystems: A Comparative Analysis" (L. C. Bliss, O. W. Heal, and J. J. Moore, eds.), pp. 47–75. Cambridge. Univ. Press, Cambridge.

Gersper, P. L., Alexander, V., Barkley, S. A., Barsdate, R. J., and Flint, P. S. (1980). The soils and their nutrients. *In* "An Arctic Ecosystem: The Coastal Tundra at Barrow, Alaska" (J. Brown, P. C. Miller, L. L. Tieszen, and F. L. Bunnell, eds.), pp. 219–254. Dowden, Hutchinson & Ross, Stroudsburg, Pennsylvania.

Giblin, A. E., Nadelhoffer, K. J., Shaver, G. R., Laundre J. A., and McKerrow, A. J. (1991). Biogeochemical diversity along a riverside toposequence in arctic Alaska. *Ecol. Monogr.* (in press).

Gorham, E., Vitousek, P. M., and Reiners, W. A. (1979). The regulation of chemical budgets over the course of terrestrial ecosystem succession. *Annu. Rev. Ecol. Syst.* **10**, 53–84.

Gosz, J. R. (1981). Nitrogen cycling in coniferous forests. *In* "Terrestrial Nitrogen Cycles" (F. E. Clark and T. Rosswall, eds.), pp. 405–426. Ecological Bulletins (Stockholm), Vol. 33.

Haag, R. W. (1974). Nutrient limitations to plant production in two tundra communities. *Can. J. Bot.* **52**, 103–116.

Hart, S. C., and Gunther, A. J. (1989). *In situ* estimates of annual net nitrogen mineralization and nitrification in a subarctic watershed. *Oecologia* **80**, 284–288.

Heal, O. W., and French, D. D. (1974). Decomposition of organic matter in tundra. *In* "Soil Organisms and Decomposition in Tundra" (A. J. Holding, O. W. Heal, S. F. MacLean, Jr., and P. W. Flanagan, eds.), pp. 279–309. Tundra Biome Steering Committee, Stockholm.

Heal, O. W., Flanagan, P. W., French, D. D., and MacLean, Jr., S. F. (1981). Decomposition and accumulation of organic matter in tundra. *In* "Tundra Ecosystems: A Comparative Analysis" (L. C. Bliss, O. W. Heal, and J. J. Moore, eds.), pp. 587–633. Cambridge Univ. Press, Cambridge.

Hill, D. E., and Tedrow, J. C. F. (1961). Weathering and soil formation in the arctic environment. *Amer. J. Sci.* **259**, 84–101.

Hinneri, S., Sonesson, M., and Veum, A. K., (1975). Soils of Fennoscandian IBP tundra ecosystems. *In* "Fennoscandian Tundra Ecosystems" (F. E. Wielgolaski, ed.), Part 1, Plants and Microorganisms, pp. 31–40. Springer-Verlag, Berlin.

Jonasson, S. (1983). Nutrient content and dynamics in north Swedish shrub tundra areas. *Holarct. Ecol.* **6**, 295–304.

Jones, H. E., and Gore, A. J. P. (1978). A simulation of production and decay in a blanket bog. *In* "Production Ecology of British Moors and Montane Grasslands" (O. W. Heal and D. F. Perkins, eds.), pp. 160–186. Springer-Verlag, Berlin.

Lachenbruch, A. H., and Marshall, B. V. (1986). Changing climate: Geothermal evidence from permafrost in the Alaskan Arctic. *Science* **234**, 689–696.

Linkins, A. E., Melillo, J. M., and Sinsabaugh, R. L. (1984). Factors affecting cellulase activity in terrestrial and aquatic ecosystems. *In* "Current Perspectives in Microbial Ecology" (M. J. Klug and C. A. Reddy, eds.), pp. 572–579. American Society for Microbiology, Washington, DC.

McClaugherty, C. A., Pastor, J., Aber, J. D., and Melillo, J. M. (1985). Forest litter decomposition in relation to soil nitrogen dynamics and litter quality. *Ecology* **66**, 266–275.

McCown, B. H. (1978). The interactions of organic nutrients, soil nitrogen, and plant growth and survival in the arctic environment. *In* "Vegetation and Production Ecology of an Alaskan Arctic Tundra" (L. L. Tieszen, ed.), pp. 435–456. Springer-Verlag, New York.

McKendrick, J. D., Ott, V. J., and Mitchell, G. A. (1978). Effects of nitrogen and phosphorus fertilization on the carbohydrate and nutrient levels in *Dupontia fisheri* and *Arctagrostis latifolia*. *In* "Vegetation and Production Ecology of an Alaskan Arctic Tundra" (L. L. Tieszen, ed.), pp. 565–578. Springer-Verlag, New York.

McKendrick, J. D., Batzli, G. O., Everett, K. R., and Swanson, J. C. (1980). Some effects of mammalian herbivores and fertilization on tundra soils and vegetation. *Arct. Alp. Res.* **12**, 565–578.

Malmer, N., and Nihlgård, B. (1980). Supply and transport of mineral nutrients in a subarctic mire. *In* "Ecology of a Subarctic Mire" (M. Sonesson, ed.), pp. 63–95. Ecological Bulletins (Stockholm), Vol. 30.

Meentemeyer, V. (1978). Macroclimate and lignin control of litter decomposition rates. *Ecology* **59**, 465–472.

Melillo, J. M. (1981). Nitrogen cycling in deciduous forests. *In* "Terrestrial Nitrogen Cycles" (F. E. Clark and T. Rosswall, eds.), pp. 427–442. Ecological Bulletins (Stockholm), Vol. 33.

Melillo, J. M., Aber, J. D., and Muratore, J. M. (1982). Nitrogen and lignin control of hardwood leaf litter decomposition dynamics. *Ecology* **63**, 621–626.

Melillo, J. M., Aber, J. D., Linkins, A. E., Ricca, A., Fry, B., and Nadelhoffer, K. J. (1989). Carbon and nitrogen dynamics along the decay continuum: Plant litter to soil organic matter. *In* "The Ecology of Arable Land" (M. Clarholm and L. Bergstrom, eds.), pp. 53–62. Kluwer Academic, Dordrecht.

Nadelhoffer, K. J., Aber, J. D., and Melillo, J. M. (1983). Leaf litter production and soil organic matter dynamics along a nitrogen availability gradient in southern Wisconsin. *Can. J. For. Res.* **13**, 12–21.

Nadelhoffer, K. J., Giblin, A. E., Shaver G. R., and Laundre, J. L. (1991). Effects of temperature and substrate quality on element mineralization in six arctic soils. *Ecology* **72**, 242–253.

Moore, J. J. (1981). Mires. *In* "Tundra Ecosystems: A Comparative Analysis" (L. C. Bliss, O. W. Heal, and J. J. Moore, eds.), pp. 35–37. Cambridge Univ. Press, Cambridge.

Norin, B. N., and Ignatenko, I. V. (1975). Ary-Mass, USSR. *In* "Structure and Function of Tundra Ecosystems" (T. Rosswall and O. W. Heal, eds.), pp. 183–191. Ecological Bulletins (Stockholm), Vol. 20.

Oechel, W. C. (1989). Nutrient and water flux in a small arctic watershed: An overview. *Holarct. Ecol.* **12**, 229–237.

Pastor, J., Aber, J. D., McClaugherty, C. A., and Melillo, J. M. (1984). Aboveground production and N and P cycling along a nitrogen availability gradient on Blackhawk Island, Wisconsin. *Ecology* **65**, 256–258.

Paul, E. A., and Clark, F. E. (1989). "Soil Microbiology and Biochemistry." Academic Press, New York.

Reiger, S. (1983). "The Genesis and Classification of Cold Soils." Academic Press, New York.

Rosswall, T. (1974). Cellulose decomposition studies on the tundra. *In* "Soil Organisms and Decomposition in Tundra" (A. J. Holding, O. W. Heal, S. F. MacLean, Jr., and P. W. Flanagan, eds.), pp. 325–340. Tundra Biome Steering Committee, Stockholm.

Rosswall, T., and Granhall, U. (1980). Nitrogen cycling in a subarctic ombrotrophic mire. *In* "Ecology of a Subarctic Mire" (M. Sonesson, ed.), pp. 209–234. Ecological Bulletins (Stockholm) Vol. 30.

Rosswall, T., Veum, A. K., and Kärenlampi, L. (1975). Plant litter decomposition at Fennoscandian tundra sites. *In* "Fennoscandian Tundra Ecosystems" (F. E. Wielgolaski, ed.), Part 1, Plants and Microorganisms, pp. 268–278. Springer-Verlag, Berlin.

Rydèn, B. E. (1981). Hydrology of northern tundra. *In* "Tundra Ecosystems: A Comparative Analysis" (L. C. Bliss, O. W. Heal, and J. J. Moore, eds.), pp. 115–137. Cambridge Univ. Press, Cambridge.

Sayers, J. K., and Walker, T. W. (1969). Phosphorus transformations in a chronosequence of soils developed on a windblown sand in New Zealand. I. Total and organic phosphorus. *J. Soil Sci.* **23**, 50–64

Schimel, D. S., Stillwell, M. A., and Woodmansee, R. G. (1985). Biogeochemistry of C, N, and P in a soil catena of the shortgrass steppe. *Ecology* **66**, 276–289.

Shaver, G. R., and Chapin, F. S., III. (1980). Response to fertilization by various plant growth forms in an Alaskan tundra: Nutrient accumulation and growth. *Ecology* **61**, 662–675.

Shaver, G. R., Nadelhoffer, K. J., and Giblin, A. E. (1990). Biogeochemical diversity and element transport in a heterogeneous landscape, the North Slope of Alaska. *In* "Quantitative Methods in Landscape Ecology" (M. Turner and R. Gardner, eds.), pp. 105–125. Springer-Verlag, New York.

Smirnoff, N., and Stewart, G. R. (1985). Nitrate assimilation and translocation by higher plants: Comparative physiology and ecological consequences. *Physiol. Plant.* **64**, 133–140.

Sonesson, M., Wielgolaski, F. E., and Kallio, P. (1975). Description of Fennoscandian tundra ecosystems. *In* "Fennoscandian Tundra Ecosystems" (F. E. Wielgolaski, ed.), Part 1, Plants and Microorganisms, pp. 3–28. Springer-Verlag, Berlin.

Svensson, B. H. (1980). Carbon dioxide and methane fluxes from the ombrotrophic parts of a subarctic mire. *In* "Ecology of a Subarctic Mire" (M. Sonesson, ed.), pp. 235–250. Ecological Bulletins (Stockholm), Vol. 30.

Svensson, B. H., Veum, A. K., and Kjelvik, S. (1975). Carbon losses from tundra soils. *In* "Fennoscandian Tundra Ecosystems" (F. E. Wielgolaski, ed.), Part 1, Plants and Microorganisms, pp. 279–286. Springer-Verlag, Berlin.

Swift, M. J., Heal, O. W., and Anderson, J. M. (1979). "Decomposition in Terrestrial Ecosystems." Univ. California Press, Berkeley.

Tamm, C. O. (1954). Some observations on the nutrient turnover in a bog community dominated by *Eriophorum vaginatum* L. *Oikos* **5**, 189–194.

Tedrow, J. C. F. (1977). "Soils of the Polar Landscapes." Rutgers Univ. Press, New Brunswick, New Jersey.

Ulrich, A., and Gersper, P. L. (1978). Plant nutrient limitations of tundra plant growth. *In* "Vegetation and Production Ecology of an Alaskan Arctic Tundra" (L. L. Tieszen, ed.), pp. 457–482. Springer-Verlag, New York.

Van Cleve, K. (1974). Organic matter quality in relation to decomposition. *In* "Soil Organisms and Decomposition in Tundra" (A. J. Holding, O. W. Heal, S. F. MacLean, Jr., and P. W. Flanagan, eds.), pp. 311–324. Tundra Biome Steering Committee, Stockholm.

Van Cleve, K., and Alexander, V. (1981). Nitrogen cycling in tundra and boreal ecosystems. *In* "Terrestrial Nitrogen Cycles" (F. E. Clark and T. Rosswall, eds.), pp. 375–404. Ecological Bulletins (Stockholm), Vol. 33.

Vassiljevskaya, V. D., Ivanov, V. V., Bogatyrev, L. G., Pospelova, E. B., Shalaeva, N. M., and Grishina, L. A. (1975). Abapa, USSR. *In* "Structure and Function of Tundra Ecosystems" (T. Rosswall and O. W. Heal, eds.), pp. 141–158. Ecological Bulletins (Stockholm), Vol. 20.

Vitousek, P. M., and Denslow, J. S. (1986). Nitrogen and phosphorus availability in treefall gaps of a lowland rainforest succession in Costa Rica, Central America. *Oecologia* **61**, 99–104.

Walker, B. D., and Peters, T. W. (1977). Soils of Truelove Lowland and Plateau. *In* "Truelove Lowland, Devon Island, Canada: A High Arctic Ecosystem" (L. C. Bliss, ed.), pp. 31–62. Univ. Alberta Press, Edmonton.

Walker, D. A., Acevedo, W., Everett, K. R., Gaydos, L., Brown, J., and Webber, P. J. (1982). "LANDSAT-assisted environmental mapping in the Arctic National Wildlife Refuge, Alaska." Rep. 82–87. U.S. Army Cold Regions Research and Engineering Laboratory, Hanover, New Hampshire.

Widden, P. W. (1977). Microbiology and decomposition on Truelove Lowland. *In* "Truelove Lowland, Devon Island, Canada: A High Arctic Ecosystem" (L. C. Bliss, ed.), pp. 505–530. Univ. Alberta Press, Edmonton.

Wielgolaski, F. E., Bliss, L. C., Svoboda, J., and Doyle, G. (1981). Primary production of tundra. *In* "Tundra Ecosystems: A Comparative Analysis" (L. C. Bliss, O. W. Heal, and J. J. Moore, eds.), pp. 187–225. Cambridge Univ. Press, Cambridge.

Williams, J. D. H., and Walker, T. W. (1969). Fractionation of phosphate in a maturity sequence of New Zealand basaltic soil profiles. *Soil Sci.* **107**, 213–219.

Woodmansee, R. G., Valis, I., and Mott, J. J. (1981). Grassland nitrogen. *In* "Terrestrial Nitrogen Cycles" (F. E. Clark and T. Rosswall, eds.), pp. 443–462. Ecological Bulletins (Stockholm), Vol. 33.

14

Nitrogen Fixation in Arctic Plant Communities

David M. Chapin and Caroline S. Bledsoe

I. Introduction

From an ecosystem perspective, rates of nitrogen (N_2) fixation are typically measured to determine how much nitrogen fixation contributes to an ecosystem and to assess the relative importance of control mechanisms regulating it. Since N_2 fixation is a physiological process subject to abiotic and biotic constraints, it serves as a useful model of how physiological processes at the organism level (in this case prokaryotes) can be scaled up to processes at the ecosystem level. Because N_2 fixation is critical to N cycling in the Arctic, understanding its spatial–temporal variation and environmental controls has broad significance in assessing the impact of climate change on nutrient cycling in arctic ecosystems.

Much of what we know about N_2 fixation in the Arctic began with the Tundra Biome studies of the International Biological Programme (IBP) (summarized in Alexander, 1974). In addition, a number of studies have been done in the Antarctic (e.g., Fogg and Stewart, 1968; Davey, 1983; Davey and Marchant, 1983) and in alpine tundra (e.g., Wojciechowski and Heimbrook, 1984; Fritz-Sheridan, 1988). The literature on arctic N_2 fixation provides an extremely valuable basis for comparing N_2 fixation processes among different locations in Alaska (Alexander and Schell, 1973; Alexander et al., 1978), Canada (Stutz and Bliss, 1975; Stutz, 1977; Jordan et al., 1978; Karagatzides et al. 1985; Henry and Svoboda, 1986), and the Scandinavian countries (Kallio and Kallio, 1975; Granhall and Lid-Torsvik, 1975). These investigations have shown that although estimated N_2 fixation rates (usually measured with the acetylene reduction method) are generally lower in the Arctic than in most temperate and tropical environments (Fig. 1), they are nonetheless the major source of N input to arctic ecosystems (Table I).

In this chapter, we summarize our current knowledge of N_2 fixation rates and their environmental controls in various arctic ecosystems. We then suggest how climate change may affect N_2 fixation rates and how these rate changes may alter other ecosystem processes.

II. Nitrogen Fixation Rates and Their Biological Importance in Arctic Ecosystems

Nitrogen is the mineral element required in the greatest amounts by plants in virtually all ecosystems (Paul and Clark, 1989), and N availability limits plant growth in many arctic ecosystems (Bliss, 1962; McKendrick et al., 1978). Nitrogen for plant growth is either released and recycled from soil organic matter (SOM) or it comes from precipitation, dry deposition, or fixation (Alexander, 1974). In general, fixation is responsible for about half of external N input, with precipitation accounting for most of the remainder (Table I). Dry deposition is relatively minor, although it may be underestimated because of measurement difficulties.

Decomposition and N mineralization are slow in the Arctic because soils are frozen much of the year, often cold (<5° C) even when thawed, and saturated in many areas (Chap. 13). Loss of fixed N in permanently frozen peat (i.e., in permafrost) and the slow release of mineral N from soil organic matter means that available N is chronically low in the Arctic, even in well-

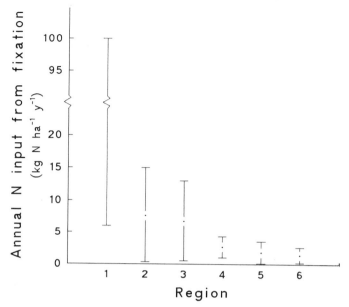

Figure 1   Range of N_2 fixation rates from various regions. Ranges shown are from the following sources: 1, tropical forests, Robertson and Rosswall (1986); 2, temperate forests, Waughman *et al.* (1981); 3, temperate arid and semi-arid, Skujins (1981); 4, temperate grasslands, Kapustka and Dubois (1987) and Woodmansee *et al.* (1981); 5, boreal forests, Van Cleve and Alexander (1981) and Alexander and Billington (1981); 6, arctic tundra, see Table I.

developed ecosystems, and fixation continues to be essential to replenish pools of cycled N.

As discussed below, N_2 fixation rates in the Arctic depend heavily on environmental conditions and are site specific. The following summary draws heavily on research in three areas: the Alaskan Arctic, the eastern Canadian Arctic, and the Fennoscandian tundra.

A. Nitrogen-Fixing Organisms

Most researchers studying N_2 fixation in the Arctic agree that the primary source of newly fixed N is cyanobacteria (also called blue-green algae), which live at the soil surface and are phototrophic. The principal genera of cyanobacteria in the Arctic are *Nostoc, Anabaena, Scytonema, Stigonema, Hapalosiphon, Tolypothrix,* and *Fischerella* (Alexander, 1974; Granhall and Lid-Torsvik, 1975; Leal Dickson, pers. comm.). *Chlorogloea, Gloeocapsa,* and *Calothrix* are also common, but because they lack heterocysts (specialized cells protecting the enzyme nitrogenase from oxygen), their importance as N_2 fixers is questionable. *Nostoc commune* is perhaps the most widespread species, predominating in many sites. Common N_2-fixing lichen genera found in the Arctic, most

Table I   Nitrogen Fixation Rates as Annual Ecosystem Input, Percentage of Total Annual Ecosystem
N Input, and Percentage of Total Annual N Uptake by Plants

Site	Habitat	N fixation rate (mg N m^{-2} yr^{-1})	% of input	% of uptake[a]
Barrow, USA	Sedge meadow	70[b]	70[a]	3.0
Devon Is., Canada	Beach ridge	19[c]	25[a]	8.6
	Sedge meadow	250	82[b]	10.6
Ellesmere Is., Canada	Sedge meadow	85[d]	73[d]	ND[g]
Kevo, Finland	Low alpine heath	136[e]	36[a]	ND[g]
Stordalen, Sweden	Mire	180[f]	45[a]	20.5
Hardangervidda, Norway	Lichen heath	56[f]	31[a]	2.7
	Wet meadow	94	43[b]	0.7
	Dry meadow	255	67[b]	3.4
Range		19–255	25–82	0.7–20.5
Mean		127	52.4	7.1

[a]Calculated using data from Van Cleve and Alexander (1981)
[b]Alexander *et al.* (1978)
[c]Stutz and Bliss (1975)
[d]Henry and Svoboda (1986)
[e]Alexander (1974)
[f]Granhall and Selander (1975)
[g]ND = not determined

often with *Nostoc* as the cyanobacterial symbiont, include *Nephroma, Stereo-caulon, Peltigera, Placopsis, Solorina,* and *Lobaria* (Kallio and Kallio, 1975; Stutz and Bliss, 1975; Waughman *et al.,* 1981).

The importance of anaerobic bacteria in arctic N_2 fixation is unclear. They appear to be major contributors of fixed N in some sites (Granhall and Lid-Torsvik, 1975) and are often abundant in the soil (Stutz, 1977), but generally, anaerobic bacteria may be of limited importance because of low soil temperatures and lack of a carbon substrate (Jordan *et al.,* 1978).

Symbiotic bacteria in association with legumes or other higher plants are not usually significant N_2 fixers in arctic environments, although they can be locally important. For example, *Astragalus alpinus* and *Lotus corniculatus* were active N_2 fixers at Hardangervidda, Norway, and Kevo, Finland (Granhall and Lid-Torsvik, 1975); *Astragalus alpinus* and two species of *Oxytropis* were found to fix N_2 and were the major sources of newly fixed N at Scarpa Lake in the Canadian Arctic (Karagatzides *et al.,* 1985). *Dryas octopetala* is known to be nodulated and fixes N_2 in some sites (Alexander *et al.,* 1978), but in others it is unnodulated (Granhall and Lid-Torsvik, 1975; Karagatzides *et al.,* 1985).

B. Rates of Nitrogen Fixation

Nitrogen fixation rates vary more than tenfold among arctic ecosystems (see Table I). Although much of this variability is due to site conditions, there are

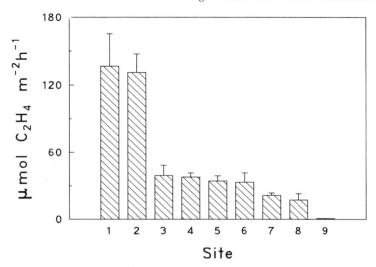

Figure 2   Spatial heterogeneity of acetylene reduction rates at Truelove Lowland, Devon Island, NWT, Canada. Measurements were made at 12 ± 1° C temperature and 300–400 μmol $m^{-2} s^{-1}$ PAR during the period 16 July to 2 August 1987. Data are means ($n = 5$); bars show standard errors. 1, *Puccinellia* salt marsh; 2, brackish pond margin; 3, aquatic algal mats; 4, willow-moss hummocks; 5, sedge meadow; 6, grass meadow; 7, frost boil; 8, freshwater pond margin; 9, beach ridge.

considerable differences among studies in methodology and in scaling up from short-term measurements to yearly input estimates, making comparisons of absolute rates among sites difficult.

At a given site, there can be substantial spatial heterogeneity in N_2 fixation (Fig. 2). At Devon Island, Canada, for example, mesic sites (e.g., sedge meadow and moss–herb hummocks) have considerably higher rates than dry beach ridges, indicating the importance of soil moisture in maintaining high N_2 fixation rates. Fixation rates in aquatic habitats, however, are not substantially higher than those in terrestrial sites, even in late season when soils have dried. This reflects, in part, the effectiveness of moss carpets in maintaining favorable moisture conditions. By far the highest rates of acetylene reduction occur in brackish environments, possibly a result of higher phosphorus concentrations. Very high rates of acetylene reduction in salt marsh communities have also been found at La Perouse Bay, Manitoba, by Bazely and Jefferies (1989), suggesting that in coastal arctic regions, salt marshes may generally be sites where N_2 fixation rates are high.

C. Importance of Nitrogen Fixation to Plant Growth and Ecosystem Processes

In discussing the importance of fixation as a source of N in arctic ecosystems, we must distinguish between N inputs and N demand. Although fixation is a major source of new N to the ecosystem, much of the N taken up by plants must come from the soil, since N_2 fixation rates are usually less than 10% of annual N uptake by plants (see Table I). Thus, release of N from decomposition processes must be the major source of N for plant growth, in addition to N retranslocated from storage tissues. Nonetheless, N fixed by cyanobacteria is released relatively rapidly through decomposition (C. Bledsoe, unpubl. data) and is rapidly taken up by plants (Alexander et al., 1978; Jones and Wilson, 1978), providing a source of readily available N. Since biologically fixed N is usually the major source of N input in arctic ecosystems, variation in rates has considerable long-term impact on plant growth and ultimately may be a major regulator of ecosystem productivity over a time scale of decades to centuries.

III. Environmental Controls

Since heterocystous cyanobacteria such as *Nostoc commune* are the major N_2-fixing organisms in the Arctic, investigations of the ecophysiology of N_2 fixation have focused on these organisms, either in a freeliving condition or as lichen symbionts. This section emphasizes cyanobacterial N_2 fixation but also includes information on bacterial fixation when available or relevant.

A. Temperature

Rates of nitrogen fixation, regardless of organism, are strongly affected by temperature and can increase threefold or more with a 10° C rise in temperature ($Q_{10} \geq 3$) (Alexander and Schell, 1973; Granhall and Lid-Torsvik, 1975). The temperature optimum for fixation in the Arctic varies from 15 to 25° C (Fig. 3; Alexander, 1974; Kallio and Kallio, 1978). Although this may seem surprisingly high in an arctic environment, soil surface temperatures in summer often exceed 15° C and thus provide a favorable thermal regime. The temperature optimum for *Nostoc commune* from temperate locations is around 20° C (e.g., Jones, 1974), suggesting that the physiology of *Nostoc* is similar over a wide range of latitudes and conditions. Lichens also appear to show little temperature adaptation, as temperature optima for two *Peltigera* species did not differ among lichens collected from arctic, subarctic, and boreal locations (Kallio and Kallio, 1978). In several studies, cyanobacteria have shown nitrogenase activity below 0° C (Englund and Meyerson, 1974; Davey and Marchant, 1983), and they readily resume N_2 fixation on thawing (e.g., Davey and Marchant, 1983; Fritz-Sheridan, 1988). Both traits have obvious adaptive value in tundra regions.

Temperature fluctuations, both diurnal and seasonal, strongly influence fixation rates. Although substantial seasonal variation is temperature-independent (Chapin et al., unpubl. manuscript), much of the variation results from changes in temperature over the growing season (e.g., Alexander et al., 1978).

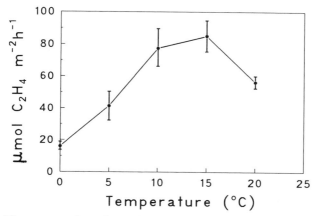

Figure 3 The response of acetylene reduction rates in moss–cyanobacterial associations (primarily with *Nostoc commune*) to varying temperature. Soil–moss cores from Truelove Lowland, Devon Island were frozen after collection, then thawed and cultured in growth chambers on a cycle of 12 hrs at 15° C, 12 hrs at 3° C, and 400 μmol m^{-2} s^{-1} PAR. Data are means; bars show standard errors.

B. Water

Moisture also appears to be one of the most important environmental factors controlling N$_2$ fixation rates in different arctic environments (Alexander, 1974). Correlation of N$_2$ fixation rate with soil moisture (Fig. 4; Billington and Alexander, 1978; Wojciechowski and Heimbrook, 1984) and with moisture content of *Nostoc* or lichen thalli (Kallio and Kallio, 1975; Scherer *et al.*, 1984; Coxson and Kershaw, 1983b; Fritz-Sheridan, 1988) provides strong evidence for the close dependence of N$_2$ fixation on moisture conditions. Much of the seasonal variation in N$_2$ fixation rates in freeliving cyanobacteria and lichens can be attributed to seasonal declines in soil moisture (Kallio and Kallio, 1975; Alexander *et al.*, 1978). Protection from desiccation is likely one of the major reasons cyanobacteria are so often associated with mosses, since moss mats are very effective in holding moisture and obtaining it from the soil by capillary action. Freeliving *Nostoc* thalli and lichen thalli have similar patterns of response to desiccation and rewetting. The mucilagenous matrix making up *Nostoc* thalli (sometimes several dm^2 in area) also retains moisture but probably less effectively than the fungal matrix of lichens.

Freeliving cyanobacteria and lichens have the ability to recover very rapidly from an extremely desiccated state and resume both photosynthesis and N$_2$ fixation (Henriksson and Simu, 1971; Coxson and Kershaw, 1983a; Scherer *et al.*, 1984). Adjustments during recovery occur both at the biochemical level (in the enzyme nitrogenase) and at the morphological level of cells and colonies (Potts and Bowman, 1985; Peat and Potts, 1987). The ability to tolerate desiccation is probably one of the major reasons that *Nostoc* spp. are

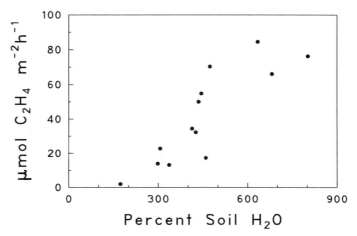

Figure 4   Acetylene reduction rates of soil–moss cores from two sedge meadow communities (Truelove Lowland, Devon Island) at varying levels of soil moisture. Cores were incubated within 12 hr after collection under conditions described for Figure 2. Data are means ($n = 5$) from 1987 and 1988 growing seasons.

ubiquitous from high to low latitudes and play such an important role in N_2 fixation in the Arctic.

Nitrogen fixation by freeliving bacteria can also be affected by soil moisture (Waughman *et al.*, 1981), probably because of the effects of soil drying on oxygen tension (see below) before actual desiccation becomes important. The water status of the host plant can affect N_2 fixation by symbiotic bacteria in higher plants (e.g., Kuo and Boersma, 1971), a relationship that should also be expected in arctic N_2 fixers.

C. Oxygen

Heterocysts protect nitrogenase from oxygen, but even in heterocystous cyanobacteria such as *Nostoc*, nitrogenase activity and oxygen tension are often closely associated (e.g., Alexander and Billington, 1986). Because soil oxygen tension is closely tied to moisture level, field measurements of N_2 fixation in relation to moisture may be confounded by the effects of oxygen. Although *Nostoc* is commonly epiphytic on mosses, and would thus seem to be growing primarily in an aerobic environment, the presence of methane in gas samples of moss–cyanobacterial associations suggests that oxygen tensions can be low in the field and may control some of the variation in rates of nitrogen fixation (Granhall and Selander, 1973; Alexander *et al.*, 1978).

Oxygen strongly influences fixation by freeliving soil bacteria, and N_2 fixation usually ceases when soils become aerobic (Jordan *et al.*, 1978). The oxygen environment of soil can be very heterogenous (Sexstone *et al.*, 1985), however, implying that even in unsaturated soils considerable N_2 fixation may take place in anaerobic microsites. In contrast to the high sensitivity of free-living bacterial fixation to oxygen, symbiotic bacterial fixation occurring within root nodules is effectively immune to oxygen effects.

D. Light

Nitrogenase activity in *Nostoc* is often light-dependent (Granhall and Lid-Tosvik, 1975; Alexander *et al.*, 1978; Basilier and Granhall, 1978). In other studies, however, nitrogenase activity shows little dependence on light (Alexander and Billington, 1986), or acetylene reduction continues in the dark for several hours (Coxson and Kershaw, 1983a; Fritz-Sheridan, 1988), indicating that nitrogenase can use stored energy. In some cyanobacteria, nitrogenase activity is fueled solely by respiratory energy, instead of by photosynthetically reduced ATP directly (Maryan *et al.*, 1986). Thus, there may be a considerable time lag before the N_2 fixation rate is affected by reduced light levels.

For *Nostoc commune*, photosynthetic rates saturate at a low photosynthetic flux density (<500 μmol m^{-2} s^{-1}) (Fig. 5; Coxson and Kershaw, 1983a; Fritz-Sheridan, 1988). This characteristic, combined with reliance on stored energy, continuous daylight when temperatures are warm, and reduced plant cover, suggests that light is not a major factor controlling fixation rates in polar environments.

E. Carbon Dioxide

Few data exist on responses of cyanobacteria to increased CO_2 levels. Nitrogen fixation rates in *Anabaenopsis circularis* rose tenfold when CO_2 levels were increased from ambient to 1% by volume (Fay, 1976), but such CO_2 levels are not ecologically realistic. It appears that HCO_3^- dissolved CO_2, or both may be important sources of carbon for cyanobacteria and that short-term internal accumulation of inorganic carbon may occur (Kaplan *et al.*, 1980). Since soil respiration probably strongly affects HCO_3^- and dissolved CO_2 concentrations in tundra surface water, photosynthetic rates (and thus N_2 fixation rates) may be closely linked to that ecosystem process. Some support for the importance of this relationship comes from stable-isotope measurements. Values of $\delta^{13}C$ from *Nostoc* from Devon Island are around $-31\%_{00}$ (D. Chapin, unpubl. data) compared with -16 to $-18\%_{00}$ for cyanobacteria from elsewhere (Behrens and Frishman, 1971; Calder and Parker, 1973). This difference suggests that *Nostoc* in these sites fixed CO_2 released by root respiration and the decomposition of vascular plants, which have a $\delta^{13}C$ of -25 to -28. Clearly, we need to know a great deal more about the sources of CO_2 for cyanobacterial photosynthesis, as well as the effects of increased CO_2 concentration on N_2 fixation in cyanobacteria.

Since the carbon nutrition of symbiotic bacteria depends on the CO_2-fixing abilities of their host plant, their N_2 fixation rates are likely to be indirectly affected by atmospheric CO_2 levels. We are not aware, however, of any studies that have shown this.

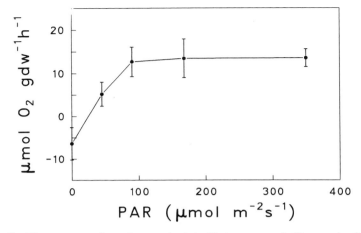

Figure 5   The response of net photosynthesis in *Nostoc commune* thalli to varying light intensity. Thalli collected from Truelove Lowland immediately before the experiment were incubated in BOD bottles at $13 \pm 1°$ C for 4 hr. Photosynthesis was measured as μmol O_2 evolved per gram dry weight (gdw) per hour, using a Clark type electrode [Source: Robert Lennihan (unpubl. data).]

F. Mineral Nutrients and pH

The availability of mineral nutrients, including N, P, Mo, Co, and Ca, affect N_2 fixation rates by freeliving cyanobacteria and lichens. Higher levels of N (as either NH_4^+ or NO_3^-) generally produce lower rates of N_2 fixation, as cyanobacteria avoid fixing atmospheric N when mineral N is available. Nitrogen fixation is positively correlated with most other nutrients: P is necessary for ATP to drive the N_2-fixing reactions; Mo and Co are nitrogenase enzyme cofactors; the function of Ca is unclear.

Experimental studies in the Arctic and Subarctic have shown that high NH_4^+ levels do inhibit N_2 fixation by *Nostoc* (C. Bledsoe, unpubl. data) and lichens (Kallio, 1978). There is also a weak negative correlation between soil NH_4^+ levels and N_2 fixation rate (Chapin *et al.*, unpubl. manuscript). The lack of a strong correlation suggests that field levels of mineral N do not reach a point where N_2 fixation rate is appreciably reduced. While the effect of ambient mineral N levels on N_2 fixation is an example of a classic negative feedback mechanism, its importance has yet to be demonstrated in the Arctic.

Increased rates of N_2 fixation rates in response to P fertilization have been shown in some arctic systems (Chapin *et al.*, unpubl. manuscript) but not in others (Alexander *et al.*, 1978). Correlations between P levels and N_2 fixation rates have been found in field studies (Basilier and Granhall, 1978; Chapin *et al.*, unpubl. manuscript), suggesting that P may play some role in regulating N_2 fixation rates in the Arctic. Very high rates of N_2 fixation in salt marsh systems (Bazely and Jefferies, 1989; Chapin *et al.*, unpubl. manuscript) may be due to high P inputs into sediments from marine sources. It has been hypothesized that levels of N_2 fixation are ultimately controlled by the availability of P (Gorham *et al.*, 1979), but the importance of P in regulating N_2 fixation rates in arctic systems remains unclear.

As trace elements, Mo and Co are often overlooked in ecosystem studies, but their importance in affecting N_2 fixation rates can be considerable (Alexander *et al.*, 1978). Adding Mo doubled N_2 fixation rates by *Nostoc* on Devon Island (C. Bledsoe, unpubl. data). Because the flux of these elements in arctic ecosystems is largely unknown, we cannot evaluate their role in regulating N_2 fixation there.

Cyanobacteria are sensitive to low pH. Fixation rates were very low in arctic and subarctic soils with pH <5 (Granhall and Selander, 1973; Basilier and Granhall, 1978), and cyanobacteria were largely absent from soils in Sweden with pH <5 (Granhall and Henriksson, 1969). Bacterial N_2 fixation in sub-boreal peats of Finland was also sensitive to a soil pH below 5 (Weber *et al.*, 1983).

G. Ecosystem Feedbacks

Although there are numerous linkages between rates of N_2 fixation and other ecosystem processes, we lack a quantitative understanding of these feedbacks. Using what we know about the ecophysiology of N_2 fixation, however, we can predict the directions of feedbacks that are likely to be ecologically significant.

Because levels of mineral N and P can directly affect N_2 fixation rates, N_2 fixation rates over long time scales (as in ecosystem succession) should be linked to the microbial and plant processes that regulate concentrations of soil N and P. How N and P levels and N_2 fixation rates change with succession in arctic ecosystems remains to be established.

Rates of N_2 fixation are also closely tied to ecosystem carbon balance. Carbon assimilation depends on available N, which is largely derived from biologically fixed N_2. Rates of N_2 fixation can in turn be strongly affected by whether carbon is lost from the soil or accumulates in peat. Carbon lost through respiration is likely to result in higher CO_2 availability and thus higher N_2 fixation rates for cyanobacteria. Carbon accumulation in peat can be a major controlling factor on N_2 fixation through its effect on soil water retention. For example, beach-ridge sites on Devon Island have significantly lower N_2 fixation rates than meadow habitats (see Fig. 2), primarily because they lack the accumulated organic matter necessary to hold moisture, and consequently they become very dry soon after snowmelt. Since O_2 levels are also largely determined by soil water levels, carbon accumulation will also affect N_2 fixation rates via that pathway.

IV. Climate Change, Nitrogen Fixation, and Arctic Ecosystem Processes

A. Direct and Indirect Effects of Climate Change on Fixation Rates

The activity of freeliving and, to a lesser extent, symbiotic cyanobacteria is very sensitive to ambient conditions. Although cyanobacteria do have the ability to survive extreme desiccation and marked temperature changes, N_2 fixation rates are likely to be strongly affected by climate-induced environmental change. Changes in temperature and moisture regime will have pronounced direct effects on N_2 fixation rates, and indirect effects will be associated with feedbacks from other processes, such as decomposition and nutrient cycling (Fig. 6).

1. Temperature

The higher temperatures associated with global climate change will exert a direct and straightforward effect on fixation rates, since N_2 fixation is so temperature-sensitive. Assuming a temperature response similar to that shown in Figure 2 and a current average surface temperature of 7° C (using 1987 data from Truelove Lowland, Devon Island), an increase in a mean summer temperature of 3° C at the soil surface will produce an approximately 45% increase in N_2 fixation rate.

2. Water and Oxygen

Because N_2 fixation rates on a spatial and seasonal basis are closely correlated with moisture, overall ecosystem fixation rates will be strongly affected by changes in the soil moisture regime. Manabe and Wetherald (1986) predicted that summer soil moisture will be reduced up to 20% in many areas of the circumpolar Arctic as a result of changing global climate. Alternatively, Maxwell (1987; see also Chap. 2 and 3) has hypothesized that moisture levels will increase. As discussed in Chapter 2, general circulation models (GCMs) disagree about the direction of change in soil moisture in the Arctic during summer. Considering the possibility of either an increase or decrease in summer soil moisture and using a linear regression model of soil moisture versus acetylene reduction rate derived from our field data at Devon Island [$nfix = 4.81 + 0.110(\% \text{ soil } H_2O)$, $R^2 = 0.54$], we predict that a 20% reduction or increase in soil moisture will result in a 15–20% reduction or increase in N_2 fixation rate. Given the heterogeneity in the distribution of moisture and the threshold effects of moisture level on N_2 fixation rate in *Nostoc* (Coxson and Kershaw, 1983a, Fritz-Sheridan, 1988), fixation rates will probably change much more than this amount in drier sites and lesser in wetter sites.

The presence of water is the primary mechanism by which oxygen concentration is reduced in soils. If soils are drier, oxygen tension will increase. Although the magnitude of this change is unknown and requires more detailed study, our estimate of moisture effects should account for changes in oxygen tensions.

3. Light

Some scenarios of climate change predict reduced cloud cover (Manabe and Wetherald, 1986), and others predict increases (Maxwell, 1987). Since cloud cover can reduce summer irradiance substantially in the Arctic, a decrease in summer cloud cover should have some positive effect on N_2 fixation and an increase should have a negative effect. Although changes in irradiance and

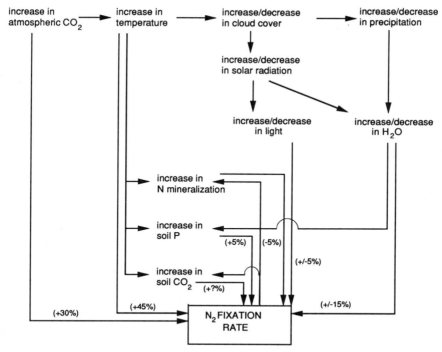

Figure 6 Schematic model showing how predicted climate change will alter the N_2 fixation rate in arctic ecosystems.

the response of N_2 fixation are not readily quantified, we expect the effects of light on N_2 fixation to be small relative to those of temperature and moisture and speculate that they will be +5% with decreased and −5% with increased cloud cover.

4. Carbon Dioxide

Increased atmospheric CO_2 will also have a direct effect on levels of CO_2 assimilation in cyanobacteria (see Chap. 7, 8, and 10). Since little work has been done on the influence of CO_2 levels on cyanobacterial growth rates or N_2 fixation, it is difficult to predict how much increasing CO_2 levels will increase fixation rates. The effects could be substantial, although limiting amounts of other factors such as P may constrain potential increases.

In addition to atmospheric-induced increases in CO_2 availability, increased soil respiration will probably result in higher bicarbonate and dissolved CO_2 concentrations in surface water in arctic ecosystems. Since freeliving cyanobacteria may derive much of their fixed CO_2 from these bicarbonate and dissolved CO_2 pools, they are likely to be affected by increases in CO_2 availability from both atmospheric and soil sources. Because the magnitude of increase in CO_2 availability and the response of cyanobacteria to CO_2 are so poorly known, our estimate of a CO_2 effect on N_2 fixation rate is largely speculation. Nevertheless, studies on aquatic and soil green algae indicate that a doubling of CO_2 in the organism's environment causes approximately a 30% increase in photosynthesis (Leal Dickson, pers. comm.), suggesting that there would be a concommitant 30% increase in N_2 fixation.

5. Mineral Nutrients

Changes in decomposition rates are likely to affect concentrations of mineral N and P in soil water (Chap. 13), another example where feedbacks from other ecosystem processes may be expected to alter N_2 fixation rates. With increasing decomposition rates, rates of N and P cycling should also increase, resulting in greater N and P availability for plant uptake. Increased mineral N could inhibit rates of N_2 fixation, but increased uptake by mosses and vascular plants associated with higher growth rates may reduce N concentrations in soil and surface water, eliminating this effect. In any case, we are still uncertain what concentrations of mineral N actually inhibit nitrogenase activity in the field. We speculate that effects of soil N concentration will be small in the short term, but over the long term, nitrogen turnover rates in arctic ecosystems may become faster than at present, resulting in a negative feedback on N_2 fixation rates as predicted by Gorham *et al.* (1979).

Changes in P availability may be more critical than changes in mineral N in affecting N_2 fixation rate. Since P has been shown to increase field rates of acetylene reduction, P is certainly limiting to fixation in some arctic ecosystems. If changes in decomposition and nutrient cycling rates as a result of climatic change generate greater P availability, N_2 fixation rates are likely to rise. To be able to predict the magnitude of this effect requires a good estimate of likely changes in P availability and the establishment of response curves of N_2 fixation to field P concentrations.

Effects of increased N and P levels with climate change are likely to be less important initially than effects of temperature, moisture, and CO_2. Although we lack quantitative data to make a confident prediction, we speculate that N_2 fixation rates will decrease by 5% because of changing N availability and increase by 5% because of changing P availability.

6. Net Effects of Climate Change on Nitrogen Fixation

If we add the effects resulting from predicted climate change, the net balance is a 65% increase in N_2 fixation under the dry and 85% increase under the wet scenario (Table II). Expected effects of temperature and CO_2 account for most of the increase, and changes in moisture take secondary importance (in contrast to the major role played by changes in moisture in other processes discussed in this book). The impact of changing light and soil N and P availability are expected to have only minor effects on N_2 fixation rates.

This estimate is most sensitive to our uncertainty of CO_2 effects, but we may be reasonably confident that N_2 fixation rates will increase by at least

Table II Estimated Effects of Climate Change on Nitrogen Fixation in Arctic Ecosystems[a]

	% change[b]		
	Low PPT	High PPT	Certainty of estimate[c]
Temperature	+45	+45	+++
Light	+5	-5	++
Water and O_2	-15	+15	+++
CO_2	+30	+30	+
N	+5	+5	+
P	-5	-5	+
Net balance	+65	+85	

[a]See text for basis of predictions.

[b]Two scenarios are shown: one with lower precipitation (PPT), lower cloud cover, and higher solar radiation; the other with higher precipitation, higher cloud cover, and lower solar radiation.

[c]Certainty of estimate: +++, high; ++, moderate; +, low.

half, and may even double, with predicted climatic change in the next 100 years.

B. Impact of Changing Nitrogen Fixation Rates on Arctic Ecosystems

1. Total N

The predicted increase in N_2 fixation rates will have a substantial effect on the amount of N input into arctic ecosystems. If N input from fixation is about 50% of total input (see Table I), the increase in N_2 fixation rates estimated here will result in about 25–50% more N input than the present rate.

Although the annual input of N from fixation is usually less than 10% of total uptake (Table I), arctic ecosystems with low N availability should be sensitive, even in the short term, to a doubling of this amount. Over the long term, increases in N should be cumulative, but arctic ecosystems may not necessarily be less N-deficient, depending on how other nitrogen cycling processes are affected by climatic change.

2. Nitrogen Cycling and Uptake

Since N fixed by cyanobacteria enters the nitrogen cycle relatively rapidly (Alexander *et al.*, 1978; Jones *et al.*, 1978), more fixed N will probably increase the rates of mineralization, nitrification, and denitrification. The magnitude of these increases must then be balanced with rate changes resulting from other effects of climate change on soil N transfers (Chap. 13).

Although N supply can limit decomposition rates, decomposition in arctic ecosystems is generally more limited by other factors, including substrate quality, low temperatures, and oxygen; consequently, relatively little N is immobilized by decomposers (Flanagan and Bunnell, 1980). Thus, we expect increased fixation to have little direct effect on decomposition.

Over long periods of time, increased N_2 fixation rates will increase the total N in arctic ecosystems. Plant response to increased N input, however, will be mediated by effects of climate change on other sources of available N and by direct effects of increased CO_2, temperature, water, and light on plant growth. If plant growth and soil organic matter increase because of these other factors, arctic ecosystems may remain N-limited despite increases in N input from fixation.

V. Summary

Nitrogen fixation by cyanobacteria is ubiquitous in arctic ecosystems, although in some areas anaerobic bacteria and nodulated higher plants may be important nitrogen fixers. Temperature and moisture probably exert the greatest environmental control on fixation rates in the Arctic, with levels of N, P, and light taking secondary importance. Current annual contributions of N by biological fixation average about half the total N input and supply about 10% of total annual uptake in arctic ecosystems.

We estimate that N_2 fixation rates will be one-and-a-half to two times current rates as a result of climate change over the next 50–100 years. The major effects will be via changes in temperature, moisture, and CO_2 levels. Our quantitative prediction for CO_2 effects, which is particularly weak, is probably the biggest source of error in this estimate. The magnitude of effects associated with changes in the availability of soil N and P is also uncertain. These gaps clearly indicate where further research is needed, especially in regard to links with other ecosystem processes.

This predicted increase in N_2 fixation rates will typically lead to a 25–50% increase in N input to arctic ecosystems. This greater input can have an immediate effect in nitrogen-deficient ecosystems but may not result in more nitrogen-rich arctic ecosystems over the long term because of feedbacks from other processes.

Almost all of our information about N_2 fixation comes from studies of short-term processes, making it difficult to extrapolate over large spatial and temporal scales. Considerable variation in methodology also makes cross-study comparisons problematic, and scaling problems in sampling ecosystem gas fluxes are not well understood. Nonetheless, fixation is one of the simplest physiological processes to scale up from the organism to the ecosystem, as incubations of soil–plant systems to measure N_2 fixation (e.g., acetylene reduction) are, in effect, performed on ecosystem microcosms. Thus, cyanobacterial N_2 fixation may prove to be particularly valuable in developing predictive models of ecosystem response to climate change.

Acknowledgments

We wish to thank L. C. Bliss, Terry Chapin, and Bob Jeffries for their critical and very helpful review of the manuscript. National Science Foundation grant DPP-85-20847 provided support for much of the unpublished research at Truelove Lowland, Devon Island, cited in this chapter.

References

Alexander, V. (1974). A synthesis of the IBP tundra biome circumpolar study of nitrogen fixation. *In* "Soil Organisms and Decomposition in Tundra" (A. J. Holding, O. W. Heal, S. F. MacLean, Jr., and P. W. Flanagan, eds.), pp. 109–121. Tundra Biome Steering Committee, Stockholm.

Alexander, V., and Billington, M. M. (1986). Nitrogen fixation in the Alaskan taiga. *In* "Forest Ecosystems in the Alaskan Taiga: A Synthesis of Structure and Function" (K. Van Cleve, F. S. Chapin III, P. W. Flanagan, L. A. Viereck, and C. T. Dyrness, eds.), 112–120. Springer-Verlag, New York.

Alexander, V., and Schell, D. M. (1973). Seasonal and spatial variation of nitrogen fixation in the Barrow, Alaska, tundra. *Arct. Alp. Res.* **5**, 77–88.

Alexander, V., Billington, M., and Schell, D. M. (1978). Nitrogen fixation in arctic and alpine tundra. *In* "Vegetation and Production Ecology of an Alaskan Tundra" (L. L. Tieszen, ed.), pp. 539–558. Springer-Verlag, New York.

Basilier, K., and Granhall, U. (1978). Nitrogen fixation in wet minerotrophic moss communities of a subarctic mire. *Oikos* **31**, 236–246.

Bazely, D. R., and Jefferies, R. L. (1989). Lesser snow geese and the nitrogen economy of a grazed salt marsh. *J. Ecol.* **77**, 24–34.

Behrens, E. W., and Frishman, S. A. (1971). Stable carbon isotopes in blue-green algal mats. *J. Geol.* **79**, 94–100.

Billington, M., and Alexander, V. 1978. Nitrogen fixation in a black spruce (*Picea mariana*) forest in Alaska. *In* "Environmental Role of Nitrogen-Fixing Blue-green Algae and Asymbiotic Bacteria" (U. Granhall, ed.), pp. 209–215. Ecological Bulletins (Stockholm), Vol. 26.

Bliss, L. C. (1962). Adaptations of arctic and alpine plants to environmental conditions. *Arctic* **15**, 117–144.

Calder, J. A., and Parker, P. L. (1973). Geochemical implication of induced changes in C^{13} fractionation by blue-green algae. *Geochim. Cosmochim. Acta* **37**, 133–140.

Coxson, D. S., and Kershaw, K. A. (1983a). Rehydration response of nitrogenase activity and carbon fixation in terrestrial *Nostoc commune* from *Stipa–Bouteloa* grassland. *Can. J. Bot.* **61**, 2658–2668.

Coxson, D. S., and Kershaw, K. A. (1983b). Nitrogenase activity during chinook snowmelt sequences by *Nostoc commune* in *Stipa–Bouteloa* grassland. *Can. J. Microbiol.* **29**, 938–944.

Davey, A. (1983). Effects of abiotic factors on nitrogen fixation by blue-green algae in Antarctica. *Polar Biol.* **2**, 95–100.

Davey, A., and Marchant, H. J. (1983). Seasonal variation in nitrogen fixation by *Nostoc commune* Vaucher at the Vestfold Hills, Antarctica. *Phycologia* **22**, 377–385.

Englund, B., and Meyerson, H. (1974). In situ measurement of nitrogen fixation at low temperatures. *Oikos* **25**, 283–287.

Fay, P. (1976). Factors influencing dark nitrogen fixation in a blue-green alga. *Appl. Environ. Microbiol.* **31**, 376–379.

Flanagan, P. W., and Bunnell, F. L. (1980). Microflora activities and decomposition. *In* "An Arctic Ecosystem: The Coastal Tundra at Barrow, Alaska" (J. Brown, P. C. Miller, L. L. Tieszen, and F. L. Bunnell, eds.), pp. 291–334. Dowden, Hutchinson & Ross, Stroudsburg, Pennsylvania.

Fogg, G. E., and Stewart, W. D. P. (1968). In situ determinations of biological nitrogen fixation in Antarctica. *Brit. Antarct. Bull.* **15**, 39–46.

Fritz-Sheridan, R. P. (1988). Physiological ecology of nitrogen fixing blue-green algal crusts in the upper-subalpine life zone. *J. Phycol.* **24**, 302–309.

Gorham, E., Vitousek, P., and Reiners, W. (1979). The regulation of chemical budgets over the course of terrestrial ecosystem succession. *Annu. Rev. Ecol. Syst.* **10**, 53–84.

Granhall, U., and Henriksson, E. (1969). Nitrogen-fixing blue-green algae in Swedish soils. *Oikos* **20**, 175–178.

Granhall, U., and Lid-Torsvik, V. (1975). Nitrogen fixation by bacteria and free-living blue-green algae in tundra areas. *In* "Fennoscandian Tundra Ecosystems" (F. E. Wielgolaski ed.), Part 1, Plants and Microorganisms, pp. 305–315. Springer-Verlag, New York.

Granhall, U., and Selander, H. (1973). Nitrogen fixation in a subarctic mire. *Oikos* **24**, 8–15.

Henriksson, E., and Simu, B. (1971). Nitrogen fixation by lichens. *Oikos* **22**, 119–121.

Henry, G. H. R., and Svoboda, J. (1986). Dinitrogen fixation (acetylene reduction) in high arctic sedge meadow communities. *Arct. Alp. Res.* **18**, 181–187.

Jones, K. (1974). Nitrogen fixation in a salt marsh. *J. Ecol.* **62**, 553–565.

Jones, K, and Wilson, R. E. (1978). The fate of nitrogen fixed by a free-living blue-green alga. *In* "Environmental Role of Nitrogen-Fixing Blue-green Algae and Asymbiotic Bacteria" (U. Granhall, ed.), pp. 158–163. Ecological Bulletins (Stockholm), Vol. 26.

Jordan, D. C., McNicol, P. J., and Marshall, M. R. (1978). Biological nitrogen fixation in the terrestrial environment of a high arctic ecosystem (Truelove Lowland, Devon Island, N.W.T.). *Can. J. Microbiol.* **24**, 643–649.

Kallio, S. (1978). On the effect of forest fertilizers on nitrogenase activity in two subarctic lichens. *In* "Environmental Role of Nitrogen-Fixing Blue-green Algae and Asymbiotic Bacteria" (U. Granhall, ed.), pp. 217–224. Ecological Bulletins (Stockholm), Vol. 26.

Kallio, P., and Kallio, S. (1978). Adaptation of nitrogen fixation to temperature in the *Peltigera apthosa*-group. *In* "Environmental Role of Nitrogen-Fixing Blue-green Algae and Asymbiotic Bacteria" (U. Granhall, ed.), pp. 225–233. Ecological Bulletins (Stockholm), Vol. 26.

Kallio, S., and Kallio, P. (1975). Nitrogen fixation in Lichens at Kevo, North-Finland. In "Fennoscandian Tundra Ecosystems" (F. E. Wielgolaski ed.), Part 1, Plants and Microorganisms, pp. 292–304. Springer-Verlag, New York.

Kaplan, A., Badger, M. R., and Berry, J. A. (1980). Photosynthesis and the intracellular inorganic carbon pool in the blue-green alga *Anabaena variabilis*: Response to external CO_2 concentration. *Planta* **149**, 219–226.

Kapustka, L. A., and DuBois, J. D. (1987). Dinitrogen fixation by cyanobacteria and associative rhizosphere bacteria in the Arapaho Prairie in the Sand Hills of Nebraska. *Am. J. Bot.* **74**, 107–113.

Karagatzides, J. D., Lewis, M. C., and Schulman, H. M. (1985). Nitrogen fixation in the high arctic tundra at Sarcpa Lake, Northwest Territories. *Can. J. Bot.* **63**, 974–979.

Kuo, T., and Boersma, L. (1971). Soil water suction and root temperature effects on nitrogen fixation in soybeans. *Agron. J.* **63**, 901–904.

Manabe, S., and Wetherald, R. T. (1986). Reduction in summer soil wetness induced by an increase in atmospheric carbon dioxide. *Science* **232**, 626–628.

Maryan, P. S., Eady, R. R., Chaplin, A. E., and Gallon, J. R. (1986). Nitrogen fixation by *Gloeothece* sp. PCC 6909: Respiration and not photosynthesis supports nitrogenase activity in the light. *J. Gen. Microbiol.* **132**, 789–796.

Maxwell, B. (1987). Atmospheric and climatic change in the Canadian Arctic: Causes, effects, and impacts. *Northern Perspect.* **15**, 2–6.

McKendrick, J. D., Ott, V. J., and Mitchell, G. A. (1978). Effects of nitrogen and phosphorus fertilization on carbohydrate and nutrient levels in *Dupontia fischeri* and *Arctagrostis latifolia*. *In* "Vegetation and Production Ecology of an Alaskan Arctic Tundra" (L. L. Tieszen, ed.), pp. 509–537. Springer-Verlag, New York.

Paul, E. A., and Clark, F. E. (1989). "Soil Microbiology and Biochemistry." Academic Press, New York.

Peat, A., and Potts, M. (1987). The ultrastructure of immobilized desiccated cells of the cyanobacterium *Nostoc commune* UTEX 584. *Fems Microbiol. Lett.* **43**, 223–227.

Potts, M., and Bowman, M. A. (1985). Sensitivity of *Nostoc commune* UTEX 584 (Cyanobacteria) to water stress. *Arch. Microbiol.* **141**, 51–56.

Robertson, G. P., and Rosswall, T. (1986). Nitrogen in West Africa: The regional cycle. *Ecol. Monogr.* **56**, 43–72.

Scherer, S., Ernst, A., Chen, T. W., and Boeger, P. (1984). Rewetting of drought-resistant blue-green algae: Time course of water uptake and reappearance of respiration, photosynthesis, and nitrogen fixation. *Oecologia* **62**, 18–423.

Sexstone, A. J., Parkin, T. B., and Tiedje, J. M. (1985). Temporal response of soil denitrification rates to rainfall and irrigation. *Soil Sci. Soc. Am. J.* **49**, 99–103.

Skujins, J. (1981). Nitrogen cycling in arid ecosystems. *In* "Terrestrial Nitrogen Cycles" (F. E. Clark and T. Rosswall eds.), pp. 477–491. Ecological Bulletins (Stockholm), Vol. 33.

Stutz, R. C. (1977). Biological nitrogen fixation in high arctic soils, Truelove Lowland. *In* "Truelove Lowland, Devon Island, Canada: A High Arctic Ecosystem" (L. C. Bliss, eds.), pp. 301–314. Univ. of Alberta Press, Edmonton.

Stutz, R. C., and Bliss, L. C. (1975). Nitrogen fixation in soils of Truelove Lowland, Devon Island, Northwest Territories. *Can. J. Bot.* **53**, 1387–1399.

Van Cleve, K., and Alexander, V. (1981). Nitrogen cycling in tundra and boreal ecosystems. *In* "Terrestrial Nitrogen Cycles" (F. E. Clark and T. Rosswall eds.), pp. 375–404. Ecological Bulletins (Stockholm), Vol. 33.

Waughman, G. J., French, R. J., and Jones, K. (1981). Nitrogen fixation in some terrestrial environments. *In* "Nitrogen Fixation" (W. J. Broughton, ed.), Vol. 1, pp. 135–192. Clarendon Press, Oxford.

Weber, A., Niemi, M., Sundman, V., and Skujins, J. (1983). Acetylene reduction (N_2 fixation) and endogenous ethylene release in sub-boreal soils and peats of Finland. *Oikos* **41**, 219–226.

Wojciechowski, M. F., and Heimbrook, M. E. (1984). Dinitrogen fixation in alpine tundra, Niwot Ridge, Front Range, Colorado, U.S.A. *Arct. Alp. Res.* **16**, 1–10.

Woodmansee, R. G., Vallis, I., and Mott, J. J. (1981). Grassland nitrogen. *In* "Terrestrial Nitrogen Cycles" (F. E. Clark and T. Rosswall eds.), pp. 443–462. Ecological Bulletins (Stockholm), Vol. 33.

15

Nutrient Absorption and Accumulation in Arctic Plants

Knut Kielland and F. Stuart Chapin III

I. Introduction

The Arctic is generally considered a temperature-limited system (Warren-Wilson, 1966; Billings and Mooney, 1968). Yet arctic species maintain high physiological activity at low temperatures, and the temperature optima of many physiological processes are often lower in arctic than in temperate species (Billings and Mooney, 1968). Furthermore, arctic species show low sensitivity to changes in temperature (Chapin, 1983a). Hence the direct effects of climatic warming on plant communities might be less important in the Arctic than originally expected. On the other hand, climatic warming may change the timing of snowmelt and freeze-up and, therefore, the length of the active growing season (Chap. 2). Moreover, some areas of the Arctic may become drier because of increased evapotranspiration (Chap. 3), and changes in soil moisture strongly influence the nutrient dynamics and carbon balance of natural ecosystems (Meentemeyer, 1978).

The availability of nutrients limits plant growth in the Arctic, as elsewhere (Chapin, 1987). Thus, the processes controlling nutrient supply to plants represent major controls over ecosystem function in arctic tundra. In this chapter we consider the known controls over plant nutrient uptake and attempt to predict the responses of arctic ecosystems to climate change in terms of plant nutrient acquisition and nutrient turnover.

II. Response of Tundra Plants to the Environment

A. Temperature

As in most species, absorption of phosphate and ammonium by tundra plants increases with increasing temperature (Chapin and Bloom, 1976; McCown, 1978). The effect of temperature on absorption rate may be greater for phosphate than for ammonium, however, as indicated by one study (Marion and Kummerow, 1990) that detected no temperature effect on ammonium uptake between 5° and 25° C in *Eriophorum vaginatum*, a dominant tundra sedge.

Tundra plants are similar to temperate species in having a temperature optimum for nutrient absorption of approximately 40° C (Chapin and Bloom, 1976). But nutrient absorption by tundra graminoids differs from that of their temperate counterparts in two respects. First, absorption of both ammonium and phosphate between 0.5° and 10° C is less temperature-sensitive in the arctic than in the temperate species that have been studied (Fig. 1; Chapin, 1983a; McCown, 1978). For example, tundra graminoids (e.g., *Carex aquatilis*, *Dupontia fisheri*, and *Eriophorum angustifolium*) at 1° C absorb phosphate at 20–60% of the rate at 20° C (Chapin and Bloom, 1976). Second, tundra species and ecotypes from polar regions have a higher phosphate absorption potential (i.e., maximum uptake rate per unit of root based on laboratory experiments) than do their temperate counterparts when the plants are grown and the potentials are measured under the same conditions (Chapin, 1974). Because of their high potential to absorb phosphate and their ability to take up ions at low temperatures, tundra species have phosphate absorption rates at 5° C that are similar to those of temperate species measured at 15° C.

A given genotype may acclimate to decreased root temperature by increasing its potential to absorb phosphate, and in some species this increase in uptake potential may fully compensate for the direct temperature effect on

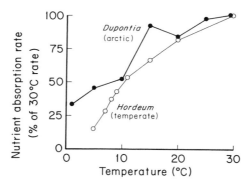

Figure 1 Temperature sensitivity of nutrient absorption exhibited by an arctic (*Dupontia fisherii*) and a temperate (*Hordeum vulgare*) species. [Reprinted from Chapin (1983) with permission from Springer-Verlag.]

absorption (Chapin, 1974). In sum, low temperature does limit phosphate and ammonium uptake by tundra graminoids, but this direct effect is relatively small because of the physiological adjustments that compensate for temperature effects.

B. Nutrient Availability

Nutrient uptake by tundra species increases sharply with increasing nutrient concentration over the range of concentrations typically found in soil solutions (Chapin, 1974; Chapin and Bloom, 1976). At high nutrient concentrations, uptake rates are much less sensitive to changes in external nutrient concentration.

The seasonal pattern of phosphate absorption by the tundra grass *Dupontia fisheri*, as estimated from seasonal measurements of uptake kinetics (Chapin and Bloom, 1976; Chapin *et al.*, 1978), correlated more closely with soil phosphate concentration ($r^2 = 0.92$) than with soil temperature ($r^2 = 0.004$). Thirty percent of the annual phosphate uptake was predicted to occur during the first two weeks after snowmelt, when phosphate concentrations in the soil solution were highest. (This period coincides with a time of high nutrient demand, however, see Section II,C.) Another 50% of the annual uptake likely occurred during the senescence of vegetation in the autumn, when the concentration of phosphate in the soil solution was again high (Chapin and Bloom, 1976). Simulations of ammonium uptake by *Eriophorum vaginatum* in the field, based on absorption experiments, suggest that the bulk of nitrogen is taken up early in the growing season (Marion and Kummerow, 1990). Thus, the time of greatest nutrient uptake by plants may not coincide with the period of highest air and soil temperatures but rather may occur in spring and autumn when soil solution concentrations are highest.

C. Plant Nutrient Status

Plants compensate for nutrient deprivation by increasing their absorption potential (Haynes and Goh, 1978). Thus, the requirement of a plant for nutrients exerts a strong effect on absorption rates. This general finding is supported by field studies in the Arctic. For example, large tillers of *Carex aquatilis* growing near the pond margin of ice-wedge polygons show higher phosphate uptake potential than do smaller tillers growing in the troughs and centers of polygons. Similarly, *Dupontia fisheri* shows the highest potential to absorb phosphate early in the season when the plants are growing most vigorously (Fig. 2; Chapin and Bloom, 1976). These observations have important implications for modeling nutrient flux through the ecosystem, because they suggest that plant requirements for nutrients may be as important as nutrient supply in determining uptake rate.

D. Type of Nitrogen Source

Ammonium, the predominant form of inorganic nitrogen in tundra soils, is consistently preferred over nitrate by northern plants (McCown, 1978; Chapin *et al.*, 1986). Tundra soils also contain high concentrations of soluble organic nitrogen, including free amino acids (Kielland, 1990). The concentrations of free amino acids generally exceed those of ammonium, perhaps because of the low rates of nitrogen mineralization in these soils (Marion and Miller, 1982; Kielland, 1990; Chap. 13). Free amino acids can serve as direct nitrogen sources for tundra species. For instance, *Eriophorum vaginatum* grows as well on glycine as on ammonium when the amino acid is the sole nitrogen source in hydroponic solutions. The growth rate of plants supplied with glycine was 30% greater than plants grown on nitrate, and uptake rates of ammonium and glycine were at least four fold greater than that of nitrate (Chapin, unpubl. data).

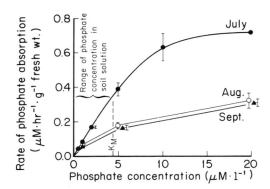

Figure 2 Seasonal change in phosphate absorption capacity in tundra graminoids. [Reprinted from Chapin and Bloom (1976) with permission from Munksgaard.]

Table I   Amino Acid to Ammonium Absorption Ratio in Major Vascular Growth Forms in Four Northern Alaska Tundra Communities

Growth form	Amino acid[a]		
	Glycine	Aspartic acid	Glutamic acid
Graminoids	0.76	0.10	0.04
	(0.09–1.46)	(0.04–0.20)	(0.01–0.09)
Deciduous shrubs	1.38	0.07	0.08
	(0.30–2.52)	(0.06–0.09)	(1.01–0.13)
Evergreen shrub[b]	4.31	0.17	0.18

[a] Numbers represent means and ranges. The ratio was calculated from the uptake rate at the seasonal average concentration of amino acids and ammonium in the soil. [Data from Kielland (1990).]
[b] Only one species (*Ledum decumbence*) was examined.

Roots of tundra species collected in the field also have a high potential to absorb free amino acids, particularly low molecular weight amino acids such as glycine (Table I). This indicates that free amino acids may represent an important source of nitrogen for arctic tundra species (Kielland, 1990) and that mineralization is not the only process by which plant-available nitrogen is generated in tundra ecosystems. This mechanism of nitrogen uptake may explain why annual nitrogen uptake estimated from biomass harvests exceeds nitrogen uptake estimated from simulations based on root growth, ammonium availability, and absorption kinetics (Marion and Kummerow, 1990).

III. Species and Growth-Form Differences

Arctic species vary considerably in their potential to absorb nutrients. For example, in a survey of 15 tundra species near Toolik Lake, Alaska, the potential to absorb phosphate varied about 20-fold (Kielland and Chapin, unpubl. data). Similar observations have been made among Alaskan taiga species (Chapin *et al.*, 1986). Within a given tundra ecosystem, deciduous shrubs, with their large annual nutrient requirement, have the highest potential to absorb phosphate, and evergreen shrubs the lowest; graminoids are intermediate (Fig. 3; Chapin and Tryon, 1982).

Moss species exhibit similarly great (10-fold) variation in rates of phosphate and ammonium uptake, and their capacity for nutrient uptake often exceeds that of vascular plants (Chapin *et al.*, 1987). Experiments conducted in the taiga with the same moss species that predominate in tundra show that mosses are efficient filters of nutrients from the atmosphere or from leaching of the vascular plant canopy and litter. For instance, mosses retained for more than a year about 90% of the nitrogen (Weber and Van Cleve, 1981) and phosphorus (Chapin *et al.*, 1987) applied to the forest floor. This observation suggests

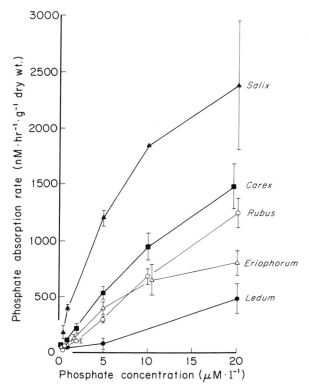

Figure 3 Growth-form differences in phosphate absorption capacity in selected tundra species. Absorption was measured at 10° C in excised roots of *Salix pulchra, Carex aquatilis, Rubus chamaemorus, Eriophorum vaginatum,* and *Ledum palustre.* Mean ± SE, *n* = 4. [Reprinted from Chapin and Tryon (1982) with permission from Munksgaard.]

that nutrient uptake by mosses plays a significant role in the nutrient dynamics of northern ecosystems and underscores the importance of internal recycling within plants as the main avenue of nutrient supply for vascular plant growth in these systems (Chap. 16).

IV. Uptake in the Field

A. Water

Annual nutrient uptake by tundra species shows a strong positive correlation with water availability, both within and among ecosystems (Chapin *et al.*, 1988). This correlation seems surprising because tundra soils are wet (60–90% water by mass) and may have low (−200 mV) redox potentials (Kielland, unpubl. data). The nitrogen and phosphorus entering plants directly as a result of transpirational water flux make up less than 5% of the amount actively absorbed (Chapin, 1988). Thus, the effect of water on nutrient acquisition must be largely indirect. High water flux might increase soil heat flux, but, as discussed in Section II,A, higher soil temperatures would have little direct effect on nutrient uptake. The major effect of high soil water is probably to increase nutrient flow to the root surface, thereby reducing the limits on uptake imposed by diffusion (Nye, 1977). For example, in one study of tussock tundra (Chapin *et al.*, 1988), mass flow of subsurface water was estimated to supply nitrogen and phosphorus to roots seven times more rapidly than could occur by diffusion. This facilitation of nutrient flow was of the right magnitude to explain the nine fold increase in annual nitrogen and phosphorus uptake by *Eriophorum vaginatum*, the species whose roots occupied the zone of major water movement.

B. Mineralization

As discussed in Section II,B, nutrient supply (see Chap. 13) is probably the single environmental factor that most strongly determines the rate of nutrient uptake by vegetation. The seasonality and constancy of nutrient supply have major influences on patterns of nutrient uptake by plants. Much of the nitrogen and phosphorus from mineralization and leaching of litter is released in early spring (Kielland, 1990; Chap. 13), coinciding with the time of greatest nutrient demand for plant growth (Chapin, 1980). This nutrient demand generally exceeds supply, and the difference is furnished largely from belowground storage reserves (Chap. 16). Nutrient uptake during the rest of the year mainly goes toward replenishing these reserves.

Periodic declines in soil microbial biomass, up to 75% in three weeks at Barrow (Miller and Laursen, 1974), can release substantial amounts of nutrients. For example, a 90% decrease in microbial biomass could satisfy the entire annual phosphorus requirement of the vascular vegetation (Chapin *et al.*, 1978). Subsequent periods of microbial population growth can immobilize nutrients and may account for the negative nitrogen mineralization rates observed in arctic tundra during midseason (e.g., Chapin *et al.*, 1988). The high nutrient uptake potential of arctic plants (Chapin, 1974) allows these species to capitalize on nutrient flushes to accumulate nutrients in excess of immediate requirements (luxury consumption). The pronounced luxury consumption of arctic plants observed in fertilization experiments (e.g., Shaver and Chapin, 1980) is consistent with this hypothesis.

C. Temperature

Although temperature has relatively minor direct effects on nutrient uptake, the indirect effects are substantial (Chapin, 1983a). Low temperatures in the Arctic result in a short growing season, shallow thaw depth, and reduced primary production. The presence of permafrost and frequently poor soil aeration exacerbate the low nutrient availability that largely controls the rate of nutrient gain in the field.

Raising the soil temperature fourfold (from 4 to 17° C) by means of heated pipes can increase the concentration of nitrogen in the soil solution (Chapin and Bloom, 1976). The increased nitrogen availability was associated with increased plant growth and a three- to sixfold increase in the potential of roots to absorb phosphate. The increased rate of phosphate uptake reflected an enhanced demand for phosphorus associated with the higher growth rate rather than increased phosphate availability.

D. Biotic Interactions

1. Mycorrhizae

Ecto- and endomycorrhizal associations are common in tundra plants (Miller, 1982), as in nearly all wild species. Nevertheless, sedge roots growing in saturated soils generally lack mycorrhizae (Hirrel *et al.*, 1978; Chapin *et al.*, 1988). Thus tundra species vary considerably in the types and occurrence of mycorrhizae.

Mycorrhizae enhance both the phosphorus (Tinker, 1978) and the nitrogen (Stribley and Read, 1974) status of plants, primarily by increasing the effective root surface area, rather than by exploiting otherwise unavailable soil pools (Gianinazzi-Pearson and Gianinazzi, 1983). Mycorrhizal tundra species have a higher ratio of amino acid to ammonium absorption than do nonmycorrhizal species, however, suggesting that mycorrhizae may increase the effectiveness with which tundra plants tap soil nitrogen pools (Kielland, 1990).

2. Herbivory

Grazing or browsing has substantial effects on nutrient uptake by arctic plants (Fig. 4). Productive, graminoid-dominated ecosystems that support high levels of grazing can respond with increasing plant root allocation, nutrient uptake, and compensatory growth (Wallace *et al.*, 1982; Cargill and Jefferies, 1984; Chapin and Slack, 1979; Archer and Tieszen, 1986). In contrast, infertile ecosystems such as the late-successional Alaskan black spruce forests (Bryant *et al.*, 1983) and evergreen heath tundra of northern Alaska support much lower levels of sustained herbivory. Grazing in infertile ecosystems can reduce nutrient uptake and plant growth (Davidson and Milthorpe, 1966; Clement *et al.*, 1978; Chapin and McNaughton, 1989).

3. Competition

In infertile soils, including most arctic soils, differences in absorptive root surface and the distribution of roots at different depths in the soil strongly influence competition for soil nutrients. Tundra species, even of the same growth form, may show 20-fold differences in belowground:aboveground biomass ratios (Shaver and Cutler, 1979), as well as large differences in root:shoot ratios, absolute root biomass, and fine root density (Kummerow *et al.*, 1983).

Shallow-rooted species, such as *Vaccinium vitis-idaea*, may have a competitive advantage over more deeply rooted species, such as *Salix pulchra*, in exploiting spring nutrient flushes, which leach down from the surface horizons of

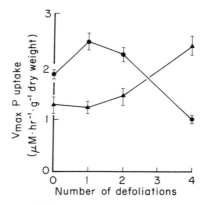

Figure 4 Effect of repeated defoliation at two-day intervals on phosphate absorption capacity in *Eriophorum vaginatum* (●) and *Carex aquatilis* (▲) on a dry weight basis. Mean ± variance, $n = 24$. [Reprinted from Chapin and Slack (1979) with permission from Springer-Verlag.]

the soil. Moreover, the timing of fine root initiation among various species at a site may differ by as much as a week (Kummerow *et al.*, 1983), which could greatly affect the capacity of some species to exploit spring nutrient flushes. Thus, spatial and temporal patterns of root growth interact with the root's absorptive functions to affect the outcome of competition for nutrients.

V. Role of Nutrient Uptake in Ecosystem Processes

A. Primary Productivity

IBP Tundra Biome studies demonstrated that net primary productivity varies more within a tundra site than among sites and correlates poorly with most climatic variables (Wielgolaski, 1975); this observation implies that soil conditions are major determinants of primary production in arctic tundra ecosystems. We suggest that soil factors governing nutrient availability probably drive biological production in tundra ecosystems.

In four structurally contrasting tundra ecosystems in northern Alaska, which differed 10-fold in aboveground biomass and annual production, the uptake of nitrogen and phosphorus as a proportion of the annual plant requirements for these nutrients was similar (Shaver and Chapin, 1991). Thus, large differences among ecosystems in the quantity of nutrients absorbed by plants is directly related to differences in production and rates of element cycling.

B. Nutrient Uptake and Turnover

Rapidly growing species that rely primarily on uptake to meet their annual nutrient requirement probably play a more important role in nutrient cycling than do slowly growing species that rely mainly on stored reserves. For example, understory shrubs in the Alaskan taiga replace about 40% of their aboveground nutrient pools annually, compared with 5% for the dominant tree species (Chapin, 1983b). In the understory shrubs, uptake accounts for approximately 65% of the annual nitrogen and phosphorus requirements, the remainder coming from stored reserves. These shrubs contribute 19–24% of the nitrogen and phosphorus cycled annually by vascular plants, even though they make up less than 3% of the vascular plant biomass. Thus, species differences in nutrient uptake can have important consequences for nutrient cycling.

VI. Climate Change and Plant Nutrient Absorption

A. Increased Temperature

Because nutrient absorption is more temperature-sensitive than is photosynthesis at present ambient temperatures (Chapin and Bloom, 1976; Miller *et al.*, 1976), temperature changes below ground will probably have a greater effect on plants than will temperature changes above ground. The increases in mineralization expected with climatic warming (see Chap. 13) could enhance nutrient uptake even more strongly. Manipulations of nutrient availability in the field suggest that species typical of nutrient-rich sites (e.g., deciduous shrubs and grasses) show a greater growth response to improved nutrient availability than species typical of nutrient-poor sites (e.g., evergreen shrubs like *Empetrum nigrum*; Haag, 1974; Shaver and Chapin, 1980; but see Chapin and Shaver, 1985). Thus, we may expect climatic warming to change the species composition of plant communities toward species with high nutrient requirements, high nutrient uptake potential, and rapid tissue turnover. The shift in species composition may come about largely through an expansion of species that are now uncommon in the community (Shaver and Chapin, 1980). The shift toward the predominance of grasses and deciduous shrubs will probably lead to increased productivity and more rapid nutrient cycling.

Deciduous shrubs and grasses have higher photosynthetic rates than present community dominants (Johnson and Tieszen, 1976). If they also have higher rates of nutrient uptake and higher concentrations of leaf nitrogen, these species will be particularly responsive to increased levels of atmospheric CO_2 (Oechel *et al.*, Chapter 7). Therefore, the increase in CO_2 and temperature-induced improved plant nutrient status could together increase primary productivity.

We expect that changes in plant species composition will be accompanied by changes in the community composition of soil organisms. Because nitrification tends to be more temperature-sensitive than ammonification (Haynes and Goh, 1978; Kielland, 1990), soil warming could increase nitrate availability and probably alter the relative abundance of soluble organic versus inorganic nitrogen. For example, soil heating experiments in wet coastal tundra at Barrow, Alaska, caused a 2.8-fold increase in soil-solution ammonium and a 4-fold increase in soil-solution nitrate, but no change in soluble or extractable phosphate (Chapin and Bloom, 1976). Because species differ strongly in their relative preference for different forms of soil nitrogen, these changes in the balance of soil nutrient pools will probably alter nitrogen uptake rates and also the competitive relationships among plants in ways that are now difficult to predict.

One of the greatest uncertainties is how changes in the length of the growing season with climatic warming will affect nutrient uptake and production. Computer simulations based on seasonal patterns of photosynthesis in the current climate suggest that a 10% increase in the length of the growing season caused by earlier snowmelt could increase annual carbon gain by 55% (Miller *et al.*, 1976). A longer growing season would also allow roots to absorb nutrients for a longer time, thus increasing the nutrient return per unit of root biomass. If, however, most nutrient uptake is associated with spring and autumn pulses of nutrient availability, then lengthening of the growing season might affect production considerably less than would be predicted on the basis of photosynthetic responses.

B. Decreased Moisture

Although predictions about climate change emphasize the effects of altered CO_2 and temperature on plants, changes in soil moisture could have an equal or greater effect on plant nutrient uptake and growth (Billings *et al.*, 1982). The projected changes in soil moisture are less certain than the expected soil warming, but in some regions of the Arctic, soils will probably be drier in the future than at present (Chap. 2, 3). The effects of soil drying on the availability and uptake of nitrogen could differ markedly from those for phosphorus (Fig. 5). In wet soils, increased aeration produced by a lowering of the water table would increase microbial respiration and nitrogen mineralization (Chap. 13), which would lead to increased nitrogen uptake and growth. Enhanced growth would in turn increase the demand for and absorption of phosphorus (Clarkson and Hanson, 1980). Because nitrogen is the nutrient that most frequently limits productivity in present tundra ecosystems, a drying of wet tundra soils would probably increase overall nitrogen uptake and plant production.

In contrast, further drying of well-aerated soils would have either no effect or a negative effect on mineralization and nutrient supply rates. Here, declines in ion mobility would probably reduce nitrogen uptake and net primary production.

Phosphorus availability responds quite differently to soil aeration than does nitrogen, because phosphate availability is governed primarily by redox-mediated precipitation reactions. Increased aeration of now-saturated soils might increase soil redox potentials, thereby reducing phosphate solubility (see Fig. 5). This could be particularly important in wet sedge tundra, which has high concentrations of iron (which precipitates out phosphate at redox potentials

Effects on N

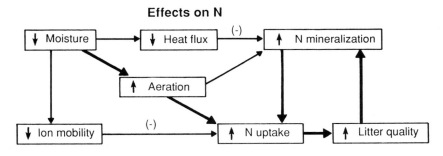

Effects on P

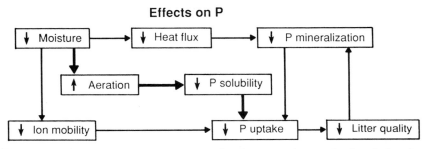

Figure 5   Differential effect of changes in soil moisture on nitrogen (top) and phosphorus (bottom) dynamics in arctic ecosystems. All effects are positive unless otherwise indicated (−).

above +100 mV). Thus, changes in soil moisture will lead to different results, depending on the initial moisture content of the soil and the nature of the limiting nutrient.

VII. Summary

Nutrient absorption in arctic plants is limited more by nutrient availability than by temperature. Thus, the indirect effects of increased temperature on factors such as soil nutrient availability and plant nutrient demand are most likely to exert the strongest control over changes in plant nutrient uptake. At the ecosystem level, increased soil nutrient supply can drive a change in community composition toward species with higher annual nutrient requirements, higher litter quality, and more rapid tissue turnover. Because soil processes, rather than plant physiology, impose the strongest limitations on nutrient absorption in the field, edaphic changes pertaining to increased nutrient movement in the soil matrix have major consequences for nutrient uptake.

From an ecophysiological perspective, plants are integrated systems of physiological processes for maintenance, growth, and reproduction. Thus, at the level of the individual plant, the effects of climate change on the process of nutrient absorption must be viewed in the context of the entire suite of plant physiological functions and the ways they respond to altered environmental conditions. The outcome of these feedbacks may be more important in determining changes in nutrient uptake than the direct effects of altered CO_2 concentration, temperature, and moisture in the environment.

References

Archer, S. R., and Tieszen, L. L. (1986). Plant response to defoliation: Hierarchical considerations. *In* "Grazing Research at Northern Latitudes" (O. Gudmundsson, ed.), pp. 45–59. Plenum, New York.

Billings, W. D., and Mooney, H. A. (1968). The ecology of arctic and alpine plants. *Biol. Rev. Camb. Philos. Soc.* **43**, 481–529.

Billings, W. D., Luken, J. O., Mortensen, D. A., and Peterson, K. M. (1982). Arctic tundra: A source or sink for atmospheric carbon dioxide in a changing environment. *Oecologia* **53**, 7–11.

Bryant, J. P., Chapin, F. S., III, and Klein, D. R. (1983). Carbon/nutrient balance of boreal plants in relation to vertebrate herbivory. *Oikos* **40**, 357–368.

Cargill, S. M., and Jefferies, R. L. (1984). The effects of grazing by lesser snow geese on the vegetation of a sub-arctic salt marsh. *J. App. Ecol.* **21**, 669–686.

Chapin, F. S., III. (1974). Morphological and physiological mechanisms of temperature compensation in phosphate absorption along a latitudinal gradient. *Ecology* **55**, 1180–1198.

Chapin, F. S., III. (1980). The mineral nutrition of wild plants. *Annu. Rev. Ecol. Syst.* **11**, 233–260.

Chapin, F. S., III. (1983a). Direct and indirect effects of temperature on arctic plants. *Polar Biology* **2**, 47–52.

Chapin, F. S., III. (1983b). Nitrogen and phosphorus nutrition and nutrient cycling by evergreen and deciduous understory shrubs in an Alaskan black spruce forest. *Can. J. For. Res.* **13**, 773–781.

Chapin, F. S., III. (1987). Environmental controls over growth of tundra plants. *Ecol. Bull.* **38**, 69–76.

Chapin, F. S., III. (1988). Ecological aspects of plant mineral nutrition. *Adv. Miner. Nutr.* **3**, 161–191.

Chapin, F. S., III, and Bloom, A. J. (1976). Phosphate absorption: Adaptation of tundra graminoids to a low-temperature, low-phosphorus environment. *Oikos* **26**, 111–121.

Chapin, F. S., III, and McNaughton, S. J. (1989). Lack of compensatory growth under phosphorus deficiency in grazing-adapted grasses from the Serengeti Plains. *Oecologia* **79**, 551–557.

Chapin, F. S., III, and Shaver, G. R. (1985). Individualistic growth response of tundra plant species to environmental manipulations in the field. *Ecology* **66**, 564–576.

Chapin, F. S., III, and Slack, M. (1979). Effect of defoliation upon root growth, phosphate absorption, and respiration in nutrient-limited tundra graminoids. *Oecologia* **42**, 67–79.

Chapin, F. S., III, and Tryon, P. R. (1982). Phosphate absorption and root respiration of different plant growth forms from northern Alaska. *Holarct. Ecol.* **5**, 164–171.

Chapin, F. S., III, Barsdate, R. J., and Barel, D. (1978). Phosphorus cycling in Alaskan coastal tundra: A hypothesis for the regulation of nutrient cycling. *Oikos* **31**, 189–199.

Chapin, F. S., III, Van Cleve, K., and Tryon, P. R. (1986). Relationship of ion absorption to growth rate in taiga trees. *Oecologia* **69**, 238–242.

Chapin, F. S., III, Oechel, W. C., Van Cleve, K., and Lawrence, W. (1987). The role of mosses in the phosphorus cycling of an Alaskan black spruce forest. *Oecologia* **74**, 310–315.

Chapin, F. S., III, Fetcher, N., Kielland, K., Everett, K. R., and Linkins, A. E. (1988). Productivity and nutrient cycling of Alaskan tundra: Enhancement by flowing soil water. *Ecology* **69**, 693–702.

Clarkson, D. T., and Hanson, J. B. (1980). The mineral nutrition of higher plants. *Annu. Rev. Plant Physiol.* **31**, 239–298.

Clement, C. R., Hopper, M. J., Jones, L. H. P., and Leafe, E. L. (1978). The uptake of nitrate by *Lolium perenne* from flowing nutrient solution. II. Effects of light, defoliation, and relationship to CO_2 flux. *J. Exp. Bot.* **29**, 1173–1183.

Davidson, J. L., and Milthorpe, F. L. (1966). The effect of defoliation on the carbon balance in *Dactylis glomerata. Ann. Bot.* **30**, 185–198.

Gianinazzi-Pearson, V., and Gianinazzi, S. (1983). The physiology of vesicular–arbuscular mychorrhizal roots. *Plant Soil* **71**, 197–209.

Haag, R. W. (1974). Nutrient limitations to plant production in two tundra communities. *Can. J. Bot.* **52**, 103–116.

Haynes, R. J., and Goh, K. M. (1978). Ammonium and nitrate nutrition of plants. *Biol. Rev. Camb. Philos. Soc.* **53**, 465–510.

Hirrel, M. C., Mehravaran, H., and Gerdemann, J. W. (1978). Vesicular–arbuscular mycorrhizae in the Chenopodiaceae and Cruciferae: Do they occur? *Can. J. Bot.* **56**, 2813–2817.

Johnson, D. A., and Tieszen, L. L. (1976). Aboveground biomass allocation, leaf growth, and photosynthesis patterns in tundra plant forms in arctic Alaska. *Oecologia* **24**, 159—173.

Kielland, K. (1990). Processes controlling nitrogen release and turnover in arctic tundra. Ph.D. diss., University of Alaska, Fairbanks.

Kummerow, J., Ellis, B. A., Kummerow, S., and Chapin, F. S., III. (1983). Spring growth of shoots and roots in shrubs of an Alaskan muskeg. *Am. J. Bot.* **70**, 1509–1515.

McCown, B. H. (1978). The interaction of organic nutrients, soil nitrogen, and soil temperature and plant growth and survival in the arctic environment. *In* "Vegetation and Production Ecology of an Alaskan Arctic Tundra" (L. L. Tieszen, ed.), pp. 435–456. Springer-Verlag, New York.

Marion, G. M., and Miller, P. C. (1982). Nitrogen mineralization in a tussock tundra soil. *Arct. Alp. Res.* **14**, 287–293.

Marion, G. M., and Kummerow, J. (1990). Ammonium uptake by field-grown *Eriophorum vaginatum* under laboratory and simulated field conditions. *Holarct. Ecol.* **13**, 50–55.

Meentemeyer, V. (1978). Macroclimate and lignin control over litter decomposition rates. *Ecology* **59**, 465–472.

Miller, O. K., Jr. (1982). Mycorrhizae, mycorrhizal fungi, and fungal biomass in subalpine tundra near Eagle Summit, Alaska. *Holarct. Ecol.* **5**, 125–134.

Miller, O. K., Jr., and Laursen, G. A. (1974). Belowground fungal biomass on U.S. tundra biome sites at Barrow, Alaska. *In* "Soil Organisms and Decomposition in Tundra" (A. J. Holding, S. F. MacLean, Jr., and P. W. Flanagan, eds.), pp. 151–158. Tundra Biome Steering Committee, Stockholm.

Miller, M. C., Stoner, W. A., and Tieszen, L. L. (1976). A model of stand photosynthesis for the wet meadow at Barrow, Alaska. *Ecology* **57**, 411–430.

Nye, P. H. (1977). The rate-limiting step in plant nutrient absorption from soil. *Soil Sci.* **123**, 292–297.

Shaver, G. R., and Cutler, J. D. (1979). The vertical distribution of live vascular phytomass in cottongrass tussock tundra. *Arct. Alp. Res.* **11**, 335–342.

Shaver, G. R., and Chapin, F. S., III. (1980). Response to fertilization by various plant growth forms in the Alaskan tundra: Nutrient accumulation and growth. *Ecology* **61**, 662–675.

Shaver, G. R., and Chapin, F. S., III. (1991). Production:biomass relationships and element cycling in contrasting arctic vegetation types. *Ecol. Monogr.* **61**, 1–31.

Stribley, D. P., and Read, D. J. (1974). The biology of mycorrhiza in the Ericaceae. IV. The effect of mycorrhizal infection on uptake of ^{15}N from labelled soil by *Vaccinium marcocarpon* Ait. *New Phytol.* **73**, 1149–1163.

Tinker, P. B. (1978). Effect of vesicular arbuscular mycorrhizas on plant nutrition and plant growth. *Physiol. Veg.* **16**, 743–751.

Wallace, L. L., McNaughton, S. J., and Coughenour, M. B. (1982). The effects of clipping and fertilization on nitrogen nutrition and allocation by mycorrhizal and nonmyccorhizal *Panicum coloratum* L., a C_4 grass. *Oecologia* **54**, 68–71.

Warren-Wilson, J. (1966). An analysis of plant growth and its control in arctic environments. *Ann. Bot.* **30**, 383–402.

Weber, M. G., and Van Cleve, K. (1981). Nitrogen dynamics in the forest floor of interior Alaska black spruce ecosystems. *Can. J. For. Res.* **11**, 743–751.

Wielgolaski, F. E. (1975). Primary productivity of alpine meadow communities. *In* "Fennoscandian Tundra Ecosystems" (F. E. Wielgolaski, ed.), Part 1, Plants and Microorganisms, pp. 121–128. Springer-Verlag, Berlin.

16

Nutrient Use and Nutrient Cycling in Northern Ecosystems

Frank Berendse and Sven Jonasson

I. Introduction

In many natural ecosystems, nutrient supply is the most important factor limiting plant growth (Elberse *et al.*, 1983; Tilman, 1984). In tundra ecosystems, many environmental factors seem to restrict primary production (Chap. 9), but even under the extreme conditions prevailing in these ecosystems, the addition of nitrogen or phosphorus always increases plant growth (Chapin and Shaver, 1985). Nutrient supply, therefore, has a major impact on plant species composition in both temperate (Vermeer and Berendse, 1983; Elberse *et al.*, 1983) and arctic (Chapin and Shaver, 1985) ecosystems.

Global environmental changes, such as the increase in atmospheric deposition of nitrogen and sulfur compounds, the increase in atmospheric carbon dioxide concentrations, and the rise in mean temperature will have important effects on nutrient cycles and may lead to changes in primary production and species composition of plant communities. Central to our analysis of the effects these changes will have on the dynamics of tundra ecosystems is a treatment of (1) the internal recycling of nutrients within plants; (2) the loss of organic matter and nutrients from plants to the soil and the consequences of this loss for long-term plant growth; and (3) the effects of this flow of organic matter and nutrients on soil fertility.

II. Storage and Recirculation of Nutrients or Additional Uptake?

A. Relationship to the Tundra Environment

The use of stored and recirculated nutrients allows plants to subsist on previously acquired nutrients, minimizing the need for nutrient uptake and making growth less dependent on the availability of soil nutrients (Ryan and Bormann, 1982; Schulze and Chapin, 1987). This characteristic is probably an important ecological attribute of tundra plants, as soil nutrient availability is low (Haag, 1974; McKendrick *et al.*, 1978), the growing season is short, and soils are often frozen at the time shoot growth begins.

Remobilization of nutrients from storage organs can occur even at subzero temperatures (Chapin *et al.*, 1986), permitting early-season canopy growth to be independent of uptake (Jonasson and Chapin, 1985). The use of stored nutrients therefore assures an early start to growth even if soil nutrients are relatively unavailable, whereas uptake during the season serves to compensate for litter losses and may support late-season growth (Chapin *et al.*, 1980, 1986). Stored reserves protect against years of suboptimal nutrient availability as well as against heterogeneity in soil nutrient resources. The tundra is a highly patchy environment consisting of numerous microenvironments. Recycling in rhizomatous and stoloniferous species, which are common life-forms in such environments, allows ramets to cross unfavorable microsites subsidized by nutrients from the remainder of the clone (Callaghan, 1988).

B. Growth Support

Growth support from stored nutrients has been demonstrated in many tundra plants. In *Dupontia fisheri* and *Arctagrostis latifolia*, two arctic grasses, 56 and 67%, respectively, of peak-season nitrogen in rhizomes plus shoots and 67 and 86% of peak-season phosphorus occurred as stored reserves in the early spring (McKendrick *et al.*, 1978). Tillers of the sedge *Eriophorum vaginatum*, with attached belowground stems and roots, that were grown in bags without access to soil nutrients, accumulated as much leaf biomass during one growing season as unbagged tillers (Jonasson and Chapin, 1985). Stored

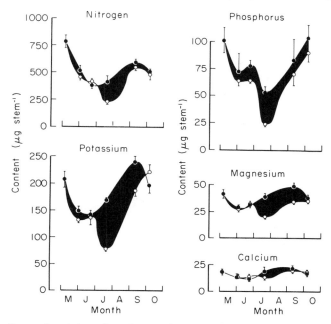

Figure 1   Seasonal variation of nutrient pools in stembases of *Eriophorum vaginatum* grown with (●) and without (○) access to soil nutrients [Reprinted from Jonasson and Chapin (1985) with permission from Springer-Verlag.]

nutrients were transported from belowground stems, the principal storage organ, regardless of whether the tillers had had access to soil nutrients during the two months after snowmelt. After canopy senescence, the content of most nutrients in the belowground stems was much higher than during the summer (Fig. 1). Nutrient content in the belowground stems of bagged tillers was only slightly lower than that of unbagged tillers, indicating a proportionally smaller contribution from new nutrient uptake than from retranslocation of old reserves. Aboveground growth of *Carex aquatilis* shows a similar high demand for stored nutrients (Shaver *et al.*, 1979).

In a study of deciduous shrubs, 14% of the total plant nitrogen and phosphorus was translocated from stems and large roots into leaves during three weeks following snowmelt (Chapin *et al.*, 1980). The elemental contents in stems and roots increased slowly thereafter. The storage organs were replenished by late autumn, presumably through a combination of root uptake and retranslocation from senescing leaves.

Jonasson *et al.* (1986) found that the concentrations of phloem-mobile nutrients (nitrogen, phosphorus, potassium) in leaves of a number of vascular plant species were negatively correlated with accumulated degree-days. The negative correlation between tissue nutrient concentrations and annual temperature was presumably caused by increased photosynthesis in warm and sunny years. This increase led to increased growth, accumulation of structural carbohydrates, increased allocation of carbon to tannin synthesis, and a dilution of concentrations of the mobile nutrient elements, which were retained in the plants from previous years. Less mobile elements, such as calcium and magnesium, did not show this trend; neither did any nutrients in mosses, which may lack storage organs. Hence, assimilation of carbon and the supply of poorly mobile nutrients are more strongly regulated by the current year's conditions for acquisition than is the supply of mobile nutrients, provided that they can be drawn from stored reserves.

C. Nutrient Resorption from Leaves

Deciduous shrubs export nutrients at the end of each growing season from leaves to stems and roots. Tundra evergreen plants remobilize nutrients before leaf senescence, which occurs throughout the whole growing season (Chapin *et al.*, 1980; Jonasson, 1989). Forbs export nutrients from senescing leaves to belowground systems, often to specialized storage organs such as thick rhizomes, rootstocks, or bulbs. Graminoids transport nutrients to belowground stems or swollen leaf bases (Webber and May, 1977; Chapin *et al.*, 1980, 1986).

Nutrient resorption from senescing leaves ranges from zero to more than 90% of the peak-season N, P, and K pools, although in most cases it falls between 40 and 60% (Table I). These values are comparable to those reported for other perennials (Small, 1972; Chapin and Kedrowski, 1983).

Nutrient withdrawal varies widely not only among species, but also within the same species growing at different sites (Jonasson, 1983). For example, the percentage removal of nutrients from *Betula nana* leaves (see Table I) differed appreciably within a distance of a few hundred meters.

The data in Table I do not indicate any pronounced differences in the efficiency of resorption among life-forms (Chapin and Shaver, 1989), although evergreen species may be less efficient than the others (Jonasson, 1989). Within a life-form, researchers disagree as to whether individuals growing in infertile sites are more efficient (Small, 1972) or similar in resorption efficiency (Chapin and Kedrowski, 1983) compared with individuals from fertile sites.

Nutrient resorption in graminoids in particular is often underestimated because plants with sequential leaf development transfer nutrients between successive leaf generations (Hopkinson, 1964, 1966), and this transfer is masked if retranslocation is calculated from data on the entire canopy. The transfer between leaves is generally high in graminoids, and the export starts even before leaves are fully expanded (Chapin, 1978; Mutoh and Nakamura, 1978). From a functional standpoint, the transfer between leaves in plants with sequential leaf production may be at least as important as the transport between canopy and belowground storage organs.

Table I   Nutrient Retranslocation from Leaves of Arctic Tundra Plants

Species	% Retranslocation			Measured as	Reference[a]
	N	P	K		
Deciduous shrubs					
Arctostaphylos alpina	51	21	—	Pool	Chapin and Shaver (1989)
Betula nana	55	53	35	Pool	Chapin *et al.* (1980)
Betula nana	64	56	53	Conc	Jonasson (1983)
Betula nana	64	57	62	Conc	Jonasson (1983)
Betula nana	65	71	73	Conc	Jonasson (1983)
Betula nana	52	43	—	Pool	Chapin and Shaver (1989)
Salix pulchra	48–56	64	53	Pool	Chapin *et al.* (1980)
Salix pulchra	68	70	—	Pool	Chapin and Shaver (1989)
Vaccinium uliginosum	69	20	60	Pool	Jonasson (unpubl. data)
Vaccinium myrtillus	64	35	26	Pool	Jonasson (unpubl. data)
Evergreen shrubs					
Ledum palustre	53	53	—	Pool	Chapin and Shaver (1989)
Empetrum nigrum	74	75	—	Pool	Chapin and Shaver (1989)
Empetrum hermaphroditum	42	28	44	Pool	Jonasson (1989)
Rhododendron lapponicum	69	56	3	Pool	Jonasson (1989)
Dryas octopetala	39	0	2	Pool	Jonasson (1989)
Vaccinium vitis-idaea	44	45	34	Pool	Jonasson (unpubl. data)
Lycopodium annotinum	63	64	90	Pool	Callaghan (1980)
Lycopodium annotinum	—	86	—	Pool	Headley *et al.* (1985)
Forbs					
Polygonum bistorta	62	76	—	Pool	Chapin and Shaver (1989)
Petasites frigidus	81	88	—	Pool	Chapin and Shaver (1989)
Epilobium latifolium	27	38	—	Pool	Chapin and Shaver (1989)
Graminoids[a]					
Eriophorum vaginatum	29	—	—	Pool	Chapin *et al.* (1980)
Eriophorum vaginatum	43	61	48	Pool	Chapin *et al.* (1975)
Eriophorum vaginatum	78	92	91	Pool	Jonasson and Chapin (1985)
Carex aquatilis	53	55	64	Pool	Chapin *et al.* (1975)
Dupontia fisheri	21	44	25	Pool	Chapin *et al.* (1975)

[a] All data for graminoids, except those of Jonasson and Chapin (1985), are calculated for the entire canopy. Those of Jonasson and Chapin (1985) are recalculated on a single-leaf basis.

Evergreen shrubs may also transfer nutrients directly from old to young leaves, based on evidence of the lack of pronounced storage in stems and roots (Chapin *et al.*, 1980) and the coincidence of leaf senescence with new leaf growth (Reader, 1978; Shaver, 1981). Jonasson (1989) showed, however, that similar amounts of nutrients were imported by new leaves of several tundra evergreens (Table II), even if the old leaves were removed just before senescence. Since it is questionable whether any appreciable amount of nutrients can be drawn from stems and roots to leaves of evergreens, they may be more dependent on nutrient uptake for leaf development than previously thought (cf. Chapin and Shaver, 1989).

Table II Mass and Nutrient Pool Sizes of New Leaves on Defoliated and Undefoliated Control Shoots of Alpine and Boreal Evergreens[a]

Species	Mass (mg)		N		P		K	
	Control	Defoliated	Control	Defoliated	Control	Defoliated	Control	Defoliated
Rhododendron lapponicum[b]	12.8	11.5	291	273	28.9	27.1	109.0	93.0
Empetrum hermaphroditum[b]	8.0	7.8	103	98	10.6	11.9	54.3	62.0
Ledum palustre[c]	9.0	6.9**	155	126	11.9	9.9	51.8	43.2

[a] Defoliation of old leaves was done when new leaf buds opened (*R. lapponicum* and *E. hermaphroditum*) or 1–1.5 months before budbreak (*L. palustre*); sampling of new leaves was done in August or September. Significant difference between treatments indicated by **; $p < 0.01$. [Data from Jonasson (1989).]

[b] Nutrient (N, K, P) pool sizes expressed as µg per shoot.

[c] Nutrient pool sizes in µg per leaf.

D. Nutrient Absorption and Conservation

Tundra plants primarily minimize nutrient losses rather than use resources for investments in structures enhancing uptake. Headley *et al.* (1985) found that the stoloniferous evergreen vascular cryptogram *Lycopodium annotinum* took up only 25% of the amount of phosphorus that was used for annual growth, whereas the rest was supplied from old and dying parts of the clone. Indeed, *L. annotinum* lacks pronounced nutrient storage organs (Headley *et al.*, 1985) and uses reserves from old tissues as the main source of nutrients. This recycling necessitates a minimal total biomass allocation to the roots of only 5% (Callaghan, 1988).

The total withdrawal of nutrients such as phosphorus from single leaves of *Eriophorum vaginatum* can exceed 90% of the leaves' peak content (see Table I), suggesting that approximately 10% of the annual demand needs to be absorbed to maintain the biomass. This estimate has been confirmed by labeling tussocks with ^{32}P in early spring as the tussock thawed (Jonasson and Chapin, 1991). Much of the uptake occurred within a few days after application. Uptake after the initial flush contributed only a minor part of the phosphorus pool in the tillers; most of the added P was rapidly immobilized in the soil. Much of the phosphorus that was taken up accumulated in the roots. Part of this root P was slowly redistributed to stores in leaf sheaths and stem bases during the growing season. Only a small part of the absorbed P went to leaf blades, indicating that current uptake contributed little to support canopy growth.

These figures confirm earlier findings of a very high dependence on recirculation of nutrients for aboveground growth of *E. vaginatum* (Goodman and Perkins, 1959; Jonasson and Chapin, 1985). Similar patterns of movement have been shown for phosphorus in *Lycopodium annotinum* (Headley *et al.*, 1985) and nitrogen in *Carex bigelowii* (Jonsdottir and Callaghan, 1989).

Despite the incomplete knowledge of nutrient losses and nutrient use in tundra plants, available data on internal nutrient circulation and annual nutrient supply for new growth can be used for a preliminary assessment of the dependence on stored versus absorbed nutrients for annual production.

Calculating the amounts of nutrients needed for annual canopy development of dominant species in two adjacent tundra communities (Table III) indicates that retranslocated nutrients supply 30 and 60% of the nutrients incorporated into the canopy of shrub and tussock tundra, respectively. Hence, the annual uptake demand is reduced to 70 and 40%, respectively, of the total nutrient requirement for canopy development.

The figure for shrub tundra is similar to previously computed values for northern hardwood forests (Ryan and Bormann, 1982). The low uptake requirement in tussock tundra reflects efficient nutrient transfer between leaves in *E. vaginatum,* allowing for their repeated use during a single growing season. The differences between the plant communities are a function of different abilities of the dominant species to retranslocate nutrients. The uptake demand (per m^2) for the evergreen shrubs exceeds that of *E. vaginatum* by 2.5

Table III   Demand for N and P from New Uptake for Annual Leaf Production of Dominant Growth Forms at Two Alaskan Tundra Sites[a]

Species	Leaf production[b] (g m^{-2})	N (g m^{-2})		P (g m^{-2})	
		Pool size[b]	Estimated uptake	Pool size[b]	Estimated uptake
Tussock tundra					
Eriophorum vaginatum	19.5	0.38	0.06	0.04	0.006
Betula nana	6.0	0.14	0.07	0.02	0.01
Ledum palustre	9.6	0.16	0.16	0.02	0.02
Shrub tundra					
Betula nana	13.0	0.24	0.12	0.02	0.01
Salix pulchra	18.8	0.36	0.18	0.02	0.01
Ledum palustre	18.3	0.29	0.29	0.03	0.03

[a]Calculated from data on leaf nutrient retranslocation. Calculations are based on annual withdrawal of 85% N and P from leaves of *E. vaginatum* and 50% withdrawal from deciduous shrubs. The retained amounts are used for new leaf development. Because of the lack of data on nutrient supply from stores to new leaves of evergreen shrubs, it is assumed that their canopy development relies entirely on new uptake (Jonasson, 1989; Chapin and Shaver, 1989).
[b]Data from Stoner *et al.* (1982).

times, even though they produce only half as much leaf biomass. Similarly, although the leaf production of deciduous shrubs is less than one-third that of *E. vaginatum*, the shrubs demand higher uptake of N and P because of differences in retranslocation, even though they recirculate about half the canopy nutrient pool.

The higher uptake demand of the shrub tundra than the tussock tundra (see Table 3) may be offset by higher mycorrhizal infection in the shrubs. Mycorrhizal infections enhance nitrogen and phosphorus supply to the host plants because the mycorrhizal hyphae improve exploitation of the soil (Chap. 15). Absorbed nitrogen and phosphorus are transported to the host or accumulate in the fungal sheaths for later release and transport, thereby creating an external storage pool for the host (France and Reid, 1983).

III. Nutrient Losses from the Plant

A. The Relative Nutrient Requirement

Adaptation of plants to nutrient-poor environments may involve two different evolutionary responses: maximizing the acquisition or minimizing the loss of nutrients. There has been some controversy about the importance of these two responses. Tilman (1988) emphasized the significance of maximizing nutrient absorption by increasing allocation of carbon and nitrogen to roots, whereas Grime (1979) and Chapin (1980) stressed mechanisms enabling plants to conserve absorbed nutrients. Because there have been no studies in tundra that fully document nutrient loss, we draw largely from the results of studies in tundra-like European heaths to discuss nutrient loss and nutrient cycling. A comparison of the characteristics of plants growing in nutrient-poor and nutrient-rich soils has shown that plants tolerate nutrient-poor environments primarily by minimizing nutrient losses (Berendse and Elberse, 1989, 1990).

To measure and compare nutrient uptake by different plant species in natural environments, we introduced the concept of relative nutrient requirement (Berendse, 1985; Berendse *et al.*, 1987b) Nutrient requirement is defined as the loss of nutrients from an individual or population during a given time interval (e.g., mg N m^{-2} yr^{-1}). This amount must be absorbed to maintain or replace existing biomass. Relative nutrient requirement (L) is defined as the nutrient requirement per unit of biomass, or the amount of a specific nutrient needed to maintain or replace each unit of biomass during a given time period (mg N g^{-1} dry weight yr^{-1}).

The most complete information on relative nutrient requirements comes from a comparison of two plant species dominating wet tundra–like heathlands in the Netherlands: *Erica tetralix*, a prostrate evergreen dwarf shrub, and *Molinia caerulea*, a deciduous perennial grass. *Erica* dominates nutrient-poor sites but is replaced by *Molinia* when nutrient availability increases because of humus accumulation or atmospheric deposition of nitrogenous compounds (Berendse and Aerts, 1984; Berendse *et al.*, 1987a). Similar vegetation changes have been observed in long-term fertilization studies in arctic tundra. Around animal carcasses (McKendrick *et. al.*, 1980; Bliss, 1977) and in plots that have received nutrient additions for a decade or more (Shaver and Chapin, 1986), grasses have increased in abundance, and evergreen shrubs have declined.

Both *Erica* and *Molinia* annually lose an unexpectedly large proportion of their nutrient pool (Table IV). The two species have different strategies to restrict nutrient losses. *Molinia* is a perennial grass that dies off above ground at the end of the growing season but withdraws most of the absorbed nutrients from dying leaves and culms. About 60% of the leaf nitrogen is stored in basal internodes and roots and is reallocated at the beginning of the new growing season. *Erica* is a woody evergreen and retains its nutrients as a consequence of the relatively long lifespan of its stems and leaves. Only about 40% of the nitrogen in the leaves is withdrawn before abscission. Despite these differences, the percentage of nitrogen lost from whole *Molinia* plants is two to three times greater than the loss from *Erica* (see Table IV), resulting in a relative nitrogen requirement in *Molinia* that is two to three times greater as well.

B. Relative Growth Rate and Relative Nutrient Requirement

What consequences do different relative nutrient requirements have for relative growth rates? Relative growth rates (RGR) of *Molinia* and *Erica* are about equal under nutrient-poor conditions (0.32 and 0.36 per month, respectively), whereas the RGR of *Molinia* is much higher than that of *Erica* when

Table IV　Percentage of Nitrogen Lost through Litter Production and the Relative Nitrogen Requirements of *Erica tetralix* and *Molinia caerulea*[a]

	Erica		*Molinia*	
	1982	1983	1982	1983
Percentage loss (yr^{-1})				
Aboveground	19	27	46	63
Whole plant	22–32	—	64–100	—
Relative nitrogen requirement ($mg\ N\ g^{-1}\ dw\ yr^{-1}$)				
Aboveground	2.0	2.6	6.0	7.5
Whole plant	2.3–3.4	—	7.4–11.7	—

[a] Percentage losses are expressed with respect to the amounts of nitrogen in the aboveground and total biomass at the end of the growing season. [After Berendse and Elberse (1989) with permission from SPB Academic Publishing.]

nutrients are ample (0.85 and 0.37 per month, respectively; Boot, unpubl. data). *Molinia* is able to respond more rapidly than *Erica* to increased nutrient availability by investing more carbohydrates and nutrients in photosynthetic tissues (Table V). These differences reflect the use of carbon and nitrogen for stem production and secondary stem growth in *Erica* and are not caused by differences between the species in the allocation to roots. Because of the low mortality of stems (see Table V), investment of nitrogen in woody tissues reduces the plant's relative nitrogen requirement. In contrast, despite efficient retranslocation, investing nitrogen in photosynthetic tissues increases the relative nitrogen requirement because these tissues are short-lived.

The chemical composition of plant tissues is a second factor that may affect both RGR and the relative nitrogen requirement of plants. The longer lifespan of *Erica* leaves and stems is linked to their relatively high lignin content. The biosynthesis of lignin requires more glucose than the synthesis of other carbohydrates such as cellulose (Penning de Vries *et al.*, 1974). Differences in the costs of biosynthesis may lead to differences in potential RGR between species that show equal rates of carbon assimilation. The higher lignin concentration in leaves and stems of *Erica* signifies that *Erica* requires more glucose per unit biomass production than *Molinia* (see Table 5). This difference may be expected to lead to a potential RGR in *Molinia* that is 25% greater than that in *Erica*. It is not clear, however, whether such differences also occur between plant species in tundra ecosystems. Chapin (1898) could not find any substantial difference in biosynthetic cost among a large number of tundra species or among growth forms.

C. Interspecific Competition and Nutrient Availability

To study the effects of increased nutrient availability on competition between *Erica* and *Molinia*, a competition experiment was carried out at different nitro-

Table V   Allocation of Nitrogen, Lifespan, Lignin
Concentration, and Costs of Biosynthesis of Tissues in
Erica tetralix and *Molinia caerulea*[a]

	Erica	*Molinia*
N allocation (% absorbed N)		
Photosynthetic tissues	15	29
Roots	65	60
Stems	13	0
Lifespan (years)		
Leaves	1.3	0.35
Stems	5.7	—
Lignin concentration (mass %)		
Leaves	33	24
Stems	46	—
Costs of biosynthesis		
(g glucose/g dry matter)		
Shoot	1.80	1.41
Roots	1.69	1.40

[a]After Berendse and Elberse (1989) with permission from SPB
Academic Publishing.

gen availabilities (Berendse and Aerts, 1984). In this experiment, the relative
competition coefficient (De Wit, 1960) was measured, based on the gross
absorption of nutrients during one growing season (Berendse *et al.*, 1987b).
If the relative competition coefficient of *Molinia* with respect to *Erica* (k_{me}) is
higher than unity, *Molinia* is the superior competitor for nutrients available
in the soil; if k_{me} is less than unity, *Erica* is superior. The relative competition
coefficient of *Molinia* (k_{me}) was 1.1 under unfertilized conditions and increased to 5 with increasing nitrogen supply. This increase in k_{me} is caused by
the much stronger response to fertilization in *Molinia* than in *Erica*, as
described in Section III, B. The results of this experiment show why *Molinia*
rapidly replaces *Erica* after an increase in nutrient availability, but they do not
explain why *Erica* is able to remain dominant in nutrient-poor environments.

To address this problem, we developed a simple model of competition
between two perennial populations with different relative nutrient requirements, including nutrient absorption as well as nutrient loss (Berendse, 1985;
Berendse *et al.*, 1987b). In competition between a species with a high relative
nutrient requirement, such as *Molinia*, and a species with a lower relative nutrient requirement, such as *Erica*, the first species will become dominant if

$$k_{me} > L_m/L_e \tag{1}$$

and the second species will become dominant if

$$k_{me} < L_m/L_e \tag{2}$$

where L_m and L_e are the relative nutrient requirements of *Molinia* and *Erica*, respectively. The relative nutrient requirement of *Molinia* is higher than that of *Erica* (L_m/L_e = 1.5–3). Under nutrient-poor conditions, *Molinia* wins in terms of nutrient uptake but ultimately loses in competition with *Erica*, for k_{me} is still below the critical limit ($1 < k_{me} < 3$). After an increase in nutrient supply, the relative competition coefficient exceeds this critical limit, and *Molinia* replaces *Erica* as the dominant species. To predict the outcome of competition between perennial plants, we must thus take into account possible differences in their relative nutrient requirements.

Similar conclusions were drawn by McGraw and Chapin (1989) on the growth of *Eriophorum vaginatum* and *E. scheuchzeri*. They showed that *E. vaginatum* increased its growth less when the nutrient supply increased, and suffered more from competition at high nutrient levels, than *E. scheuchzeri*. *E. scheuchzeri* is a rhizomatous, "foraging" species (Callaghan, 1988) growing on more fertile sites than *E. vaginatum*. It has a less pronounced internal recirculation of nutrients and probably has a higher relative nutrient requirement (Marks and Chapin, 1988).

These results suggest that species like *Erica* and *E. vaginatum* are more "successful" in nutrient-poor environments because they are more frugal with the nutrients they acquire. After an increase in nutrient supply, plant properties that lead to a high potential growth rate become more favorable than the ability to retain nutrients.

IV. Impact of Dominant Plant Species on Nutrient Cycles

In the Arctic, higher CO_2 levels in the atmosphere will probably lead to increases in air temperature, in the depth of the active soil layer, and in the length of the period during which the upper part of the soil is thawed (see Chap. 2, 3). Hence a major indirect effect of increased CO_2 levels will be increased rates of nutrient supply through mineralization (Chap. 13, 15). The changes in species composition that have occurred in the Netherlands during the last decades may provide some insights into changes that might occur in arctic ecosystems with an expected increase in nutrient supply.

Increased annual nitrogen input in the Netherlands has led to an accelerated accumulation of soil organic matter, which in turn has led to a rapid increase over earlier times in rates of nutrient mineralization (Berendse, 1990). This change has resulted in the rapid expansion of *Molinia* in Dutch heathlands. Although it is well known that nutrient availability has a major impact on species composition, it has not been as well appreciated that dominant plant species may in turn have an important effect on the quantities of available nutrients. Studies of nitrogen budgets in both temperate (Berendse *et al.*, 1987a) and arctic nutrient-poor ecosystems (Rosswall and Granhall, 1980) indicate that 70–90% of annual nitrogen uptake by the plant community comes from the decomposition of dead organic material. Different plant species produce different quantities of dead organic material (Staaf and Berg, 1981). Moreover, the chemical composition of the litter produced by a range of plant species can vary widely, resulting in different rates of decomposition (Berg and Staaf, 1980; Melillo *et al.*, 1982) and, hence, different rates of nutrient release.

In long-term decomposition experiments in the field, much higher mass losses were measured in *Molinia* litter than in *Erica* litter, probably because of the latter's higher lignin concentrations (Berendse *et al.*, 1989). Thus, in communities dominated by *Molinia*, the flow of carbon and nitrogen via litter production to the soil is greater than in communities dominated by dwarf shrubs. Carbon losses from the soil as a result of microbial respiration are also higher in *Molinia*-dominated vegetation. In the short term, higher rates of mass loss will lead to increased rates of nitrogen mineralization (Berg and Staaf, 1981). In the long term, however, higher decomposition rates result in less organic matter accumulating in the soil, which is likely to lead to lower overall rates of nitrogen mineralization. Hence, at first sight it is uncertain what will be the long-term effects on nutrient supply of the production of easily degradable litter.

The results of a model simulating secondary succession in heathlands with both *Erica* and *Molinia* (Berendse, 1988) show the different effects that *Erica* and *Molinia* can have on the nitrogen cycle (Fig. 2). The model assumed that secondary succession starts with a mineral substrate containing little organic matter. After succession starts, the amount of organic matter in the soil increases, resulting after a few years in increased rates of nitrogen mineralization. Because of this increase in nitrogen supply, the relative competition coefficient k_{me} increases, and after about 35 years it exceeds the ratio between the relative nitrogen requirements of the two species. Before this time, *Erica* can maintain itself as the dominant species, but afterward, *Molinia* gradually replaces *Erica*. Because of this increase in *Molinia*, after about 50 years the increase in nitrogen mineralization is accelerated, leading to a further increase in the relative competition coefficient. There is, therefore, a positive feedback through the mutual positive effects that nitrogen supply and the presence of *Molinia* have on each other. This positive feedback probably explains the rapid replacement in the Netherlands of the former *Erica*-dominated communities by the grass *Molinia*. In the next section, we discuss the possible implications of such positive feedbacks for the future of arctic ecosystems.

V. Climate Change and Nutrient Cycles in Tundra Ecosystems

In analyzing the possible effects of climate change on the nitrogen cycle, one must distinguish between the effects of elevated CO_2 alone and those of global warming. Elevated CO_2 levels will increase carbon assimilation by plants (Chap. 7). An increase in photosynthetic rate may—even in nutrient-poor ecosystems—lead to an increase in plant production (Pearcy and Björkman,

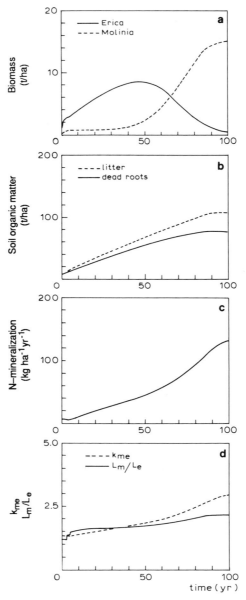

Figure 2   The course of secondary succession after turf removal in a wet heathland as simulated with the model NUCOM2 (Berendse, 1988). Total biomass of *Erica* and *Molinia* (a); amount of dead roots and amount of dead roots + aboveground litter (b); annual nitrogen mineralization (c), and relative competition coefficient of *Molinia* with respect to *Erica* (k_{me}) and the ratio between the relative nitrogen requirement of *Molinia* and that of *Erica* (L_m/L_e).

1983). Such an increase will probably be associated with a decrease in the nitrogen concentration of plant tissues. This predicted effect is similar to the temporary decline in the nutrient concentrations of tundra plants during warm and sunny years when conditions for photosynthesis and growth are favorable (Jonasson *et al.*, 1986). Other possible effects involve increased allocation to belowground plant parts and increased exudation of carbohydrates from roots (Oechel and Strain, 1985). Tannin levels in plant tissues are also likely to rise as carbon assimilation increases (Jonasson *et al.*, 1986; Laine and Henttonen, 1987). This may reduce the forage quality of arctic plants because tannins act as grazer-deterrent substances (Feeny, 1970).

We introduced a reduction in the percentage of nitrogen in plant biomass and an increase in the proportional allocation of this element to roots into a model for nitrogen and carbon cycling in tundra ecosystems. This model was similar to the one used in Section IV, but we adjusted parameter values to tundra conditions. The accumulation of soil organic matter because of water-saturated or frozen soil conditions was taken into account. The model was used to simulate the effects of increased CO_2 on equilibrium levels of plant biomass, nitrogen mineralization, and soil carbon accumulation. Both the nitrogen concentration of plant tissues and the proportional increase in allocation to roots increased 10 and 25%, respectively (Table VI). Standing biomass decreased because of the increase in the relative investment in roots, which have higher turnover rates than aboveground biomass. Net nitrogen

Table VI Effects of Elevated CO_2 Levels and of Raised Air Temperature and Extended Growing Period on Total Standing Biomass, Including Roots, Nitrogen Mineralization, and Soil Carbon Accumulation

Parameter change	Biomass (g dw m^{-2})	N mineralization (g m^{-2} yr^{-1})	Soil C accumulation (g m^{-2} yr^{-1})
Elevated CO_2 levels[a]			
0%	713 (100[c])	0.9 (100)	39 (100)
10%	647 (91)	0.7 (78)	43 (110)
25%	558 (78)	0.5 (56)	50 (128)
Elevated CO_2 levels + raised temperatures[b]			
0%	647 (100)	0.7 (100)	43 (100)
10%	668 (103)	0.8 (114)	42 (98)
25%	698 (108)	0.9 (129)	41 (95)
50%	749 (116)	1.0 (143)	40 (93)
100%	853 (132)	1.3 (186)	38 (88)

[a] Elevated CO_2 levels were simulated by reducing nitrogen concentrations in plant tissues and increasing the proportional allocation of nitrogen and carbon to roots by 0%, 10% or 25%.
[b] Raised air temperatures and longer growing periods are supposed to have a major, positive impact on annual microbial respiration. This effect is simulated by increasing relative decomposition rates, which measure annual microbial respiration, by 0%, 10%, 25%, 50% and 100%. In all simulations it was supposed that CO_2 levels had increased (cf. 10% row under elevated CO_2 levels).
[c] Percentages of control given in parentheses.

mineralization was reduced by increased microbial immobilization of nitrogen because of a lower nitrogen concentration in dead plant tissues and by reduced microbial respiration since dead roots decompose more slowly than aboveground litter. The accumulation of carbon in the soil was accelerated by the slower decay of dead roots.

The effects of global warming on the tundra nitrogen cycle contrast with those of elevated CO_2. Effects of elevated air temperatures and extended growing periods were simulated by increasing relative decomposition constants, which measure the annual percentage of mass loss from soil organic matter caused by microbial respiration (Chap. 13). These effects were calculated for conditions of elevated CO_2. Small increments (up to 25%) in microbial activity just compensate for the reductions caused by elevated CO_2 in total standing biomass and the annual amount of nitrogen mineralized (see Table VI). Greater increments in activity lead to increased mineralization rates, which result in increased plant biomass and a slower accumulation of soil organic carbon. But the positive effects of elevated CO_2 levels on soil carbon accumulation are much stronger. A 100% increase in the relative decomposition rates is just enough to compensate for the positive effects of a 10% increment in allocation to roots and a 10% decline in plant nitrogen concentration.

From these modeling exercises, it appears that possibly climate changes will increase the supply of nutrients and increase the rate of soil carbon accumulation. These changes are likely to be modified, however, by interactions between the plants and their environment. Observed responses in both heathland and tundra (Chapin and Shaver, 1985; McGraw and Chapin, 1989) suggest that some species will respond strongly to an increase in nutrient supply at the expense of other species. Those with low relative nutrient requirements will probably decline, whereas fast-growing species with high relative nutrient requirements will expand. These species may be those now occurring on relatively fertile tundra sites, or they may come from more southern latitudes. For example, species with a strategy similar to that of *Eriophorum scheuchzeri* will gain in competitive ability relative to those with a strategy like that of *E. vaginatum*. Vegetation types with low nutrient uptake demands (e.g., tussock tundra) will be more sensitive to the predicted changes than shrub tundra. If species with high relative nutrient requirements and high growth rates increase, they will affect both the quantity and the chemical composition of dead organic material. Such species will lose more litter, which will decompose faster. These plant characteristics may create a positive feedback, as in the Dutch heathlands. This phenomenon could result in an unexpectedly rapid change in the species composition of arctic ecosystems.

VI. Summary

Storage and internal recirculation of nutrients are important attributes enabling tundra plants to survive where the availability of nutrients is extremely low. Plant species retranslocate 40–90% of the peak leaf nutrient content to storage organs before leaf senescence. These and other plant features that restrict nutrient losses strongly reduce the demand for new nutrients. The relative nutrient requirement is a concept that enables us to measure and compare this uptake demand in different species. In the competition between perennial plant species, trade-offs exist between plant features minimizing nutrient losses and those leading to a high competitive ability under relatively nutrient-rich conditions. Plant species with a high relative nutrient requirement that produce readily decomposed litter appear to have a positive impact on the nutrient supply through mineralization. Elevated CO_2 levels and higher temperatures may increase nutrient supply. After an increase in nutrient supply, the positive impact on nutrient mineralization of plant species adapted to relatively nutrient-rich conditions can create a positive feedback, thereby causing an unexpectedly rapid change in the species composition of arctic ecosystems.

Acknowledgments

T. V. Callaghan, J. Goudriaan, and P. Ketner made valuable comments on parts of a first draft of this chapter. We thank also R. L. Jefferies and F. S. Chapin III for their many helpful comments on the different versions of the manuscript.

References

Berendse, F. (1985). The effect of grazing on the outcome of competition between plant populations with different nutrient requirements. *Oikos* **44**, 35–39.

Berendse, F. (1988). "Een simulatiemodel als hulpmiddel bij het beheer van vochtige heidevelden." CABO, Wageningen, Netherlands.

Berendse, F. (1990). Organic matter accumulation during secondary succession in heathland ecosystems. *J. Ecol.* **8**, 413–427.

Berendse, F., and Aerts, R. (1984). Competition between *Erica tetralix* L. and *Molinia caerulea* (L.) Moench as affected by the availability of nutrients. *Acta Oecol. Oecol. Plant.* **5**, 3–14.

Berendse, F., and Elberse, W. T. (1989). Competition and nutrient losses from the plant. *In* "Causes and Consequences of Variation in Growth Rate and Productivity of Higher Plants" (H. Lambers, M. L. Cambridge, H. Konings, and T. L. Pons, eds.), pp. 269–284. SPB Academic, The Hague.

Berendse, F., and Elberse, W. T. (1990). Competition and nutrient availability in heathland and grassland ecosystems. *In* "Perspectives on Plant Competition" (J. B. Grade and D. Tilman, eds.), pp. 93–116. Academic Press, San Diego, California.

Berendse, F., Beltman, B., Bobbink, R., Kwant, R., and Schmitz, M. (1987a). Primary production and nutrient availability in wet heathland ecosystems. *Acta Oecol. Oecol. Plant.* **8**, 265–279.

Berendse, F., Oudhof, H., and Bol, J. (1987b). A comparative study on nutrient cycling in wet heathland ecosystems. I. Litter production and nutrient losses from the plant. *Oecologia* **74**, 174–184.

Berendse, F., Bobbink, R., and Rouwenhorst, G. (1989). A comparative study on nutrient cycling in wet heathland ecosystems. II. Litter decomposition and nutrient mineralization. *Oecologia* **78**, 338–348.

Berg, B., and Staaf, H. (1980). Decomposition rate and chemical changes of Scots pine needle litter. II. Influence of chemical composition. *In* "Structure and Function of Northern Coniferous

Forests—An Ecosystem Study" (T. Persson, ed.), pp. 373–390. Ecological Bulletins (Stockholm), Vol. 32.

Berg, B., and Staaf, H. (1981). Leaching, accumulation, and release of nitrogen in decomposing forest litter. *In* "Terrestrial Nitrogen Cycles" (F. E. Clark and T. Rosswall, eds.), pp. 163–178. Ecological Bulletins (Stockholm), Vol. 33.

Bliss, L. C. (ed.) (1977). "Truelove Lowland, Devon Island, Canada: A High Arctic Ecosystem". Univ. of Alberta Press, Edmonton.

Callaghan, T. V. (1980). Age-related patterns of nutrient allocation in *Lycopodium annotinum* from Swedish Lapland: Strategies of growth and population dynamics of tundra plants. *Oikos* **35**, 373–386.

Callaghan, T. V. (1988). Physiological and demographic implications of modular construction in cold environments. *In* "Plant Population Ecology" (A. J. Davy, M. J. Hutchings, and A. R. Watkinson, eds.) Blackwell, Oxford.

Chapin, F. S., III. (1978). Phosphate uptake and nutrient utilization by Barrow tundra vegetation. *In* "Vegetation and Production Ecology of an Alaskan Arctic tundra" (L. L. Tieszen, ed.), pp. 483–507. Springer-Verlag, New York.

Chapin, F. S., III. (1980). The mineral nutrition of wild plants. *Annu. Rev. Ecol. Syst.* **11**, 233–260.

Chapin, F. S., III. (1989). The cost of tundra plant structures: Evaluation of concepts and currencies. *Am. Nat.* **133**, 1–19.

Chapin, F. S., III, and Kedrowski, R. A. (1983). Seasonal changes in nitrogen and phosphorus fractions and autumn retranslocation in evergreen and deciduous taiga trees. *Ecology* **64**, 376–391.

Chapin, F. S., III, and Shaver, G. R. (1985). Individualistic growth responses of tundra plant species to environmental manipulations in the field. *Ecology* **66**, 564–576.

Chapin, F. S., III, and Shaver, G. R. (1989). Differences in growth and nutrient use among arctic plant growth forms. *Funct. Ecol.* **3**, 73–80.

Chapin, F. S., III, Van Cleve, K., and Tieszen, L. L. (1975). Seasonal nutrient dynamics of tundra vegetation at Barrow, Alaska. *Arct. Alp. Res.* **7**, 209–226.

Chapin, F. S., III, Johnson, D. A., and McKendrick, J. D. (1980). Seasonal movement of nutrients in plants of different growth forms in Alaskan tundra ecosystems: Implications for herbivory. *J. Ecol.* **68**, 189–209.

Chapin, F. S., III, Shaver, G. R., and Kedrowski, R. A. (1986). Environmental controls over carbon, nitrogen, and phosphorus fractions in *Eriophorum vaginatum* in Alaskan tussock tundra. *J. Ecol.* **74**, 167–195.

De Wit, C. T. (1960). On competition. *Agric. Res. Rep.* **66**, 1–82.

Elberse, W. T., Van den Bergh, J. P., and Dirven, J. G. P. (1983). Effects of use and mineral supply on the botanical composition and yield of old grassland on heavy-clay soil. *Neth. J. Agric. Sci.* **31**, 63–88.

Feeny, P. P. (1970). Seasonal changes in oak leaf tannins and nutrients as a cause of spring feeding by winter moth caterpillars. *Ecology* **51**, 565–581.

France, C. R., and Reid, C. P. P. (1983). Interactions of nitrogen and carbon in the physiology of ectomycorrhiza. *Can. J. Bot.* **61**, 964–984.

Goodman, G. T., and Perkins, D. F. (1959). Mineral uptake and retention in cotton-grass (*Eriophorum vaginatum* L.) *Nature* **184**, 467–468.

Grime, J. P. (1979). "Plant Strategies and Vegetation Processes." Wiley, Chichester.

Haag, R. W. (1974). Nutrient limitations to plant productivity in two tundra communities. *Can. J. Bot.* **52**, 103–116.

Headley, A. D., Callaghan, T. V., and Lee, J. A. (1985). The phosphorus economy of the evergreen tundra plant *Lycopodium annotinum. Oikos* **45**, 235–245.

Jonasson, S. (1983). Nutrient content and dynamics in north Swedish tundra areas. *Holarct. Ecol.* **6**, 295–304.

Jonasson, S. (1989). Implications of leaf longevity, leaf nutrient reabsorption, and translocation for the resource economy of five evergreen species. *Oikos* **56**, 121–131.

Jonasson, S., and Chapin, F. S., III. (1985). Significance of sequential leaf development for nutrient balance of the cotton-sedge, *Eriophorum vaginatum* L. *Oecologia* **67**, 511–518.

Jonasson, S., and Chapin, F. S., III. (1991). Seasonal uptake and allocation of phosphorus in *Eriophorum vaginatum* L. measured by labelling with ^{32}P. *New Phytol.* **118**, 349–357.

Jonasson, S., Bryant, J. P., Chapin, F. S., III, and Andersson, M. (1986). Plant phenols and nutrients in relation to variations in climate and rodent grazing. *Am. Nat.* **128**, 394–408.

Jonsdottir, I. S., and Callaghan, T. V. (1990). The movement of nitrogen within interconnected tiller systems of *carex biegelowii* determined by ^{15}N and nitrate reductase assays. *New Phytol* **114**, 419–428.

Laine, K. M., and Henttonen, H. (1987). Phenolics/nitrogen ratios in the blueberry (*Vaccinium myrtillus*) in relation to temperature and microtine density in Finnish Lapland. *Oikos* **50**, 389–395.

Marks, A. F., and Chapin, F. S., III. (1988). Seasonal control over allocation to reproduction in a tussock-forming and a rhizomatous species of *Eriophorum* in central Alaska. *Oecologia* **78**, 27–34.

McGraw, J. B., and Chapin, F. S., III. (1989). Competitive ability and adaptation to fertile and infertile soils in two *Eriophorum* species. *Ecology* **70**, 736–749.

McKendrick, J. D., Ott, V. J., and Mitchell, G. A. (1978). Effects of nitrogen and phosphorus fertilization on carbohydrate and nutrient levels in *Dupontia fisheri* and *Arctagrostis latifolia*. *In* "Vegetation and Production Ecology of an Alaskan Arctic Tundra" (L. L. Tieszen, ed.), pp. 509–537. Springer-Verlag, New York.

McKendrick, J. D., Batzli, C. O., Everett, K. R., and Swanson, J. C. (1980). Some effects of mammalian herbivores and fertilization on tundra soils and vegetation. *Arct. Alp. Res.* **12**, 565–578.

Melillo, J. M., Aber, J. D., and Muratore, J. F. (1982). Nitrogen and lignin control of hardwood leaf litter decomposition dynamics. *Ecology* **63**, 621–626.

Mutoh, N., and Nakamura, T. (1978). An autecological study of nutrient economy of the plant. *In* "Ecophysiology of Photosynthetic productivity" (M. Monsi and T. Saeki, eds.), pp. 230–237. JIBP Synthesis, (Tokyo), Vol. 19.

Oechel, W. C., and Strain, B. R. (1985). Native species responses to increased atmospheric carbon dioxide concentration. *In* "Direct Effects of Increasing Carbon Dioxide on Vegetation" (B. R. Strain and J. D. Cure, eds.). DOE/ER-0238. Natl. Tech. Info. Serv., Springfield, Virginia.

Penning de Vries, F. W. T., Brunsting, A. H. M., and Van Laar, H. H. (1974). Products, requirements, and efficiency of biosynthesis: A quantitative approach. *J. Theor. Biol.* **45**, 399.

Pearcy, R. W., and Björkman, O. (1983). Physiological effects. *In* "CO_2 and Plants" (E. R. Lemon, ed.). Westview Press, Boulder, Colorado.

Reader, R. J. (1978). Contribution of overwintering leaves to the growth of three broad-leaved, evergreen shrubs belonging to the *Ericaceae* family. *Can. J. Bot.* **56**, 1248–1261.

Rosswall, T., and Granhall, U. (1980). Nitrogen cycling in a subarctic ombrotrophic more. *In* "Ecology of a Subarctic Mire" (M. Sonesson, ed.), pp. 209–234. Ecological Bulletin (Stockholm), Vol. 30.

Ryan, D. F., and Bormann, F. H. (1982). Nutrient resorption in northern hardwood forests. *Bioscience* **32**, 29–32.

Schulze, E.-D., and Chapin, F. S., III. (1987). Plant specialization to environments of different resource availability. *In* "Potentials and Limitations of ecosystem analysis" (E.-D. Schulze and H. Zwöfler, eds.), pp. 120–148. Springer-Verlag, Berlin.

Shaver, G. R. (1981). Mineral nutrition and leaf longevity in an evergreen shrub, *Ledum palustre* ssp. *decumbens*. *Oecologia* **49**, 362–365.

Shaver, G. R., and Chapin, F. S., III. (1986). Effects of fertilizer on production and biomass of tussock tundra, Alaska, U.S.A. *Arct. Alp. Res.* **18**, 261–268.

Shaver, G. R., Chapin, F. S., III, and Billings, W. D. (1979). Ecotypic differentiation in *Carex aquatilis* on ice-wedge polygons in the Alaskan coastal tundra. *J. Ecol.* **67**, 1025–1046.

Small, E. (1972). Photosynthetic rates in relation to nitrogen recycling as an adaptation to nutrient deficiency in peat bog plants. *Can. J. Bot.* **50**, 2227–2233.

Staaf, H., and Berg, B. (1981). Plant litter input to soil. *In* "Terrestrial Nitrogen Cycles" (F. E.

Clark and T. Rosswall, eds.), pp. 147–162. Ecological Bulletins (Stockholm), Vol. 33.

Stoner, W. A., Miller, P., and Miller, P. C. (1982). Seasonal dynamics and standing crops of biomass and nutrients in a subarctic tundra vegetation. *Holarct. Ecol.* **5**, 172–179.

Tilman, D. (1984). Plant dominance along an experimental nutrient gradient. *Ecology* **65**, 1445–1453.

Tilman, D. (1988). "Plant Strategies and the Structure and Dynamics of Plant Communities." Princeton Univ. Press, Princeton, New Jersey.

Vermeer, J. G., and Berendse, F. (1983). The relationship between nutrient availability, shoot biomass, and species richness in grassland and wetland communities. *Vegetatio* **53**, 121–126.

Webber, P. J., and May, D. E. (1977). The magnitude and distribution of belowground structures in the alpine tundra of Niwot Ridge, Colorado. *Arct. Alp. Res.* **9**, 157–172.

Part IV

Interactions

17

Response of Tundra Plant Populations to Climatic Change

James B. McGraw and Ned Fetcher

I. Introduction

In confronting global ecological issues, one immediately sees that processes understood at the physiological level must be linked to those measured at the ecosystem level. Less obvious is how population and community ecology contribute to understanding ecosystem-level change. In this chapter we point out some important connections of plant demography to physiological and ecosystem ecology and show that plant demography will play a key role in improving projections of the long-term consequences of global climate change in the arctic ecosystem.

II. Life Histories of Tundra Plants

As in most ecosystems, the life histories of all tundra plant species cannot be adequately characterized in a brief description. Nevertheless, several types of life histories are common among the dominant species in many arctic communities, and these are worth noting before beginning an explicit discussion of the importance of demography in response to climate change.

Tundra plants, although highly variable in growth form, are often able to spread vegetatively by production of more-or-less independent ramets that constitute the basic unit of modular construction. Growth of whole, genetically distinct individuals, or "genets," appears to be inherently slow in many species, although growth at this level has not often been studied in arctic plants in the field because genets are frequently difficult to distinguish. Morphological plasticity is high in many dominant arctic species, particularly for ecotypes or species found in relatively productive sites (Shaver *et al.*, 1986; McGraw, 1987; McGraw and Chapin, 1989; Fetcher and Shaver, 1990). In general, reproductive effort is lower in tundra plants than in their temperate counterparts even though the proportion of biomass allocated to seeds is not substantially lower (Chester and Shaver, 1982). Seedling establishment is rare or infrequent for most species in undisturbed tundra (McGraw and Shaver, 1982), but natural disturbances of many kinds occur (Billings, 1973; McGraw and Vavrek, 1989) so that opportunities for recruitment in are not uncommon. About one-half of all tundra species have a seed bank, which is the source of most recruitment in disturbed organic soils (see review by McGraw and Vavrek, 1989). It is not possible to generalize about the principal causes or timing of genet death in arctic plants, although genets of some species may well survive for hundreds or thousands of years. This idea has not been tested because it is difficult or impossible to age most species directly (see Mark *et al.*, 1985, for an example of indirect aging).

Great potential longevity, inability to estimate age directly, and inability to identify genets contribute to the difficulties of studying demography in arctic plant communities. One way around these difficulties comes from the realization that demography can be studied meaningfully at another level—that of the module (or plant part). A second out is to focus studies at certain critical stages of the life cycle (e.g., seed germination and establishment). Despite these useful research tactics, a great deal of ignorance concerning basic demography of all tundra plant species remains. Indeed, there are no simple life tables for genets of *any* tundra plant species.

III. Demography at the Modular Level: Implications for Productivity

Most plant species are constructed largely of repeated parts—meristems, stem segments, leaves, shoots, roots, and so on—that can be considered "populations" within the individual (White, 1979, 1984). As a plant develops, change in size occurs largely by the birth and death of these parts. For plants like vegetatively spreading arctic species, this concept provides a way to examine growth under field conditions. The demography of plant parts has not yet been applied to deriving ecosystem-level measures of response to the environment. Nevertheless, the studies that have used modular demography show that this linkage is worth pursuing.

One demographic approach is possible for plants in which parts such as leaves or shoots can be aged. This involves constructing a life table for leaves or shoots that summarizes birth and death rates as a function of age (Bazzaz and Harper, 1977). Growth over the short-term can be measured as the net change in shoot numbers. In addition, since the plant parts are typically tagged and censused, the timing of plant part loss and addition can be ascertained and an estimate made of turnover. This information is frequently lost or ignored in traditional harvest methods, where tissue losses between harvests may not be accounted for, resulting in potentially large errors in productivity estimates. Long-term projected plant growth can be estimated by recasting the life table into a Leslie matrix (module birth rates across the first row, survival rates along the subdiagonal), then finding the dominant eigenvalue (γ) of this matrix. γ is the finite rate of increase of the plant part population that would eventually be reached given the survival and fecundity patterns measured. The Leslie matrix can also be used to project the transient dynamics of the shoot population by iterative multiplication of the matrix by a state vector consisting of the number of shoots in each age class $[\eta_i(t)]$. Matrix methods are described in detail by Caswell (1989).

Fetcher and Shaver (1983) used the life-table and Leslie-matrix approach in studying the response of *Eriophorum vaginatum* to disturbance. They showed that age-specific birth and death rates of tillers increased in disturbed tundra relative to undisturbed tundra, although there was no net difference in γ between the two tiller populations (Fig. 1a,b). At the ecosystem level, this shift would have important consequences, with increased litter production and more rapid cycling of nutrients expected in the disturbed sites. Fertilization, in contrast (data not shown), increased tiller birth rates with no effect on survival (Fig. 1c,d), resulting in a rapid shoot population increase—a response expected from independent productivity measurements in fertilizer studies (Shaver and Chapin, 1980). Thus, Fetcher and Shaver (1983) were able to show, with a fine level of resolution, that the effects of fertilization were not equivalent to the effects of disturbance, although in one respect (tillering), they were similar. The researchers pointed out that the flexibility in tiller life histories may also be an important mechanism by which *E. vaginatum* maintains its dominance from early to mid- or late succession in mesic tundra (cf. Chap. 6).

Callaghan also studied population growth of *Carex bigelowii* tillers in a Norwegian lichen–heath community (Callaghan, 1976). He explicitly related tiller densities to *Carex* biomass within the community, thus connecting shoot demography to a property of the ecosystem. In this way, he demonstrated that physiological properties influencing tiller demography will in turn influence species biomass and productivity.

In many species shoot size varies considerably, and thus productivity or biomass may be different for shoot populations that differ in size structure,

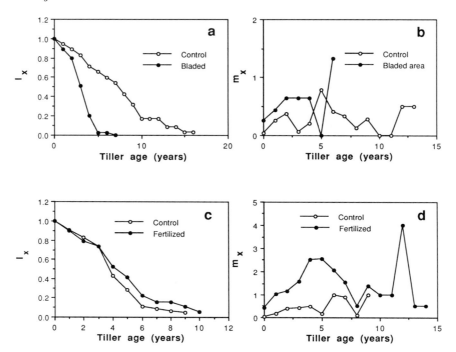

Figure 1   (a) Age-specific tiller survival (l_x) and (b) tiller fecundity (m_x) in terms of new tiller production in control and bladed plots at Eagle Creek, Alaska, and effect of fertilization on (c) age-specific tiller survival and (d) tiller fecundity at Toolik Lake, Alaska.

even if their density is the same. For such species, demographic models explicitly incorporating size structure have been used (McGraw and Antonovics, 1983b). In *Dryas octopetala*, shoot module growth and new shoot initiation were highly size-dependent (Fig. 2). A Lefkovitch-type matrix model (Lefkovitch, 1965) demonstrated an increase in shoot populations of two ecotypes growing in close proximity ($\gamma = 1.13$ and 1.17). A sensitivity analysis of this model showed that growth in number of shoots was very sensitive to the rate at which individual shoots grew, since large shoots produced offspring shoots at a much higher rate than small shoots. Thus, in size-structured shoot populations, growth of individual shoots, and the overall size structure of the population, may be as important at the ecosystem level (e.g., to productivity or biomass) as changes in shoot densities. Responses of aboveground production to fertilization over three years in *D. octopetala* were explained by variation in both shoot size and shoot population growth; neither was adequate by itself to explain variation in production (McGraw, 1985a).

The relative response to fertilization of shoot populations of coexisting deciduous (*Vaccinium uliginosum*) and evergreen (*Vaccinium vitis-idaea*) species was examined to test ideas about the significance of leaf longevity in nutrient-

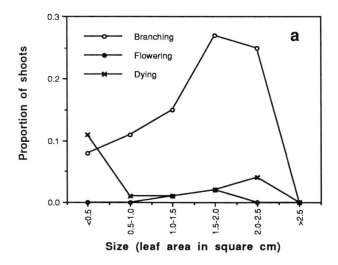

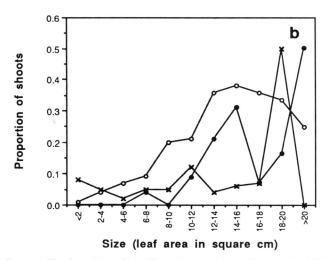

Figure 2   Size-specific shoot branching, flowering, and mortality rates in (a) fell-field and (b) snowbed *Dryas octopetala.*

poor environments (Chester and McGraw, 1983). The analysis showed that the two species shared a similar growth response to fertilization, contrary to expectation from other studies. Individual shoots of *V. uliginosum* increased in size with fertilization, but the rate of shoot production at a given size did not change (Fig. 3a). *V. vitis-idaea* responded by increasing shoot numbers, but its shoot size structure did not change (Fig. 3b). Without the demographic model to integrate the shoot dynamics and show that these two responses had similar overall effects, the results would have been difficult to interpret.

For some species, shoot dynamics may vary with both age and size. Law (1983) formulated a matrix population model that allows dual classification of individuals. Such a model was applied to shoot populations in an eastern deciduous forest shrub (*Rhododendron maximum*; McGraw, 1989). In this instance, the model incorporating both age and size projected a different growth rate than did models based on age or size alone because there was a significant interaction between age and size in determining the fate of individual shoots, suggesting that it may be important to incorporate both state variables in projecting growth of the shoot population. This approach has not been used on arctic plant species.

Demographic models of shoot populations allow projections from observations over some interval and are most useful in answering "What if . . . ?" questions about growth, assuming conditions remain the same. By themselves, they are not useful for attempting to *predict* what will happen. To make such models predictive requires that the dynamics of shoot populations be connected to the controls over shoot death, growth, and birth, that is, establishing explicit relationships between module dynamics and plant physiology. Although models have attempted to include modular demography (e.g., Lawrence *et al.*, 1978), little empirical research has been done to provide a sound basis for even the simplest model assumptions for tundra plant or any other nonagronomic species. Hence, this represents one of the largest gaps in scaling up from physiological responses to ecosystem responses.

Scaling up from studies of plant growth to studies of genet demography also involves modular demography since seed production, per square meter or per individual, is closely related to shoot density and shoot population size structure (McGraw and Wulff, 1983; McGraw and Antonovics, 1983b). Moreover, genet death occurs when the shoot population of an individual decreases to zero.

IV. Demography at the Individual Level: Implications for Ecosystem Change

Ecosystem characteristics in a given environment depend largely on the assembly of species existing there. Likewise, the ecosystem response to environmental change should depend on the particular group of species in the community (Chapin *et al.*, 1986). One of the critical controls over ecosystem change is potential change in species composition, particularly the immigration of new species. It would be a mistake to assume that species composition will remain unchanged in response to changing climate (Chap. 5, 7, 16). Yet changing species composition is also one of the most problematic determinants of ecosystem response because immigration of a new species will be a complex, often stochastic function of seed availability, disturbance frequency

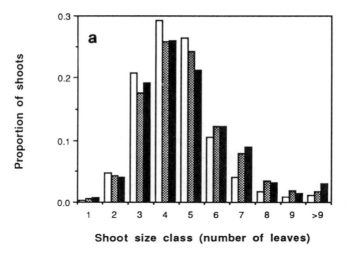

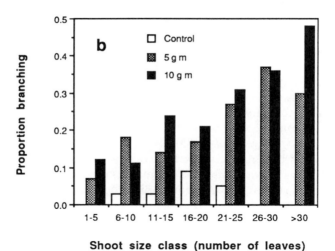

Figure 3 (a) Shift in size distribution of shoots caused by fertilization in *Vaccinium uliginosum* at Eagle Creek, Alaska, and (b) increase in rate of branching as a function of shoot size caused by fertilization of *V. vitis-idaea* at Eagle Creek, Alaska.

and intensity, and the ability of existing species in a community to outcompete invaders or vice versa. The question, "How rapidly will new species invade an existing community?" is crucial to projecting ecosystem change, and this question can only be addressed with demographic studies. Although few studies have examined the invasion of intact tundra communities by new species, several investigators have studied the process of germination and establishment. It seems reasonable to assume that invasion by vegetative growth from adjacent communities will be too slow to account for large-scale community change, although this could certainly occur on a local scale for some species. On a time-scale of thousands of years, the assumption that vegetative expansion will be of limited importance as an invasion mechanism may be faulty.

The persistent dogma concerning reproduction by seed in tundra is that (1) few species have intrinsic seed dormancy, and germination occurs opportunistically early in the growing season (Bliss, 1959; Amen, 1966), and (2) establishment of seedlings occurs only rarely because of the short growing season, low temperatures, and exposure to hazards such as needle ice or drought (Billings and Mooney, 1968). Studies of recruitment by seed in the field, however, have suggested that this dogma needs qualification.

Despite the fact that seeds of most tundra plant species have no intrinsic barriers to immediate germination, most tundra communities studied to date have dormant viable seed populations in the soil (16 of 18 communities; Table I; see review by McGraw and Vavrek, 1989). Seed bank densities tend to be highest in sites with intermediate moisture levels, although diversity is low in these seed banks. In dry sites, seed bank diversity is relatively high, but buried seed densities are low, possibly because of relatively thin soils. In wet sites, both diversity and density of seeds in the soil tend to be low.

The ubiquity of seed banks ensures a ready seed source when disturbance disrupts the soil in the Arctic, as long as the seed-containing soil is left on the site. In mesic sites such as cottongrass tussock tundra, where seedling establishment has been studied the most, recruitment in large disturbed areas is almost entirely from the seed bank (McGraw, 1980; Chester and Shaver, 1982; Gartner *et al.*, 1983; Shaver *et al.*, 1983). Rapid growth of seedlings on a cleared area can reestablish complete vegetative cover within a decade (Chapin and Chapin, 1980). Twenty years after the original disturbance, however, the abundance of species relative to adjacent undisturbed tundra remains strongly skewed in favor of the two most prevalent species in the seed bank (*Eriophorum vaginatum* and *Carex bigelowii*) (McGraw, pers. obs., 1990). The seed bank could reduce the probability of invasion of new species since, with a ready seed source, the opportunity for establishment in the relatively resource-rich conditions of an uncolonized soil is short-lived. Further decreasing the probability of successful invasion would be the relatively low seed rain of new colonizers relative to seeds from the existing tundra.

Surprisingly, the abundance of seedlings in undisturbed moist tundra is similar to that on disturbed sites (McGraw and Vavrek, 1989). Growth from seedling to adult size, however, is rare or absent for most species in the "closed" community (McGraw and Shaver, 1982; Gartner *et al.*, 1983). Establishment from seed may occur more frequently in undisturbed xeric tundra, where vegetative cover by plants is less (Wager, 1938; Soyrinki, 1938; Osburn, 1961; Bonde, 1968; Freedman *et al.*, 1982). In these sites, seedling establishment appears to occur preferentially within the canopy of established plants

Table I Seed Bank Sizes and Diversity in Xeric, Mesic, and Hydric Tundra Sites

Xeric sites		Mesic sites		Hydric sites	
Vegetation	Seed bank (seeds m^{-2})	Vegetation	Seed bank (seeds m^{-2})	Vegetation	Seed bank (seeds m^{-2})
Dryas fell-field[a]	130	Tussock tundra[d]	3367	Sedge meadow[g]	205
Cassiope heath[a]	55	Tussock tundra[c]	1173	Sedge meadow[f]	533
Dryas and heath[b]	85–3142	*Carex–Salix* tundra[f]	779	*Carex* moss[b]	80–2802
Lichen heath[c]	1769	Snowbed[b]	310–2717	Sedge grass[c]	10
Dryas heath[c]	0	Tussock tundra[c]	78	Sedge grass[c]	0
Alpine meadow[c]	521	*Carex* fen	1	Grass sedge[c]	462

[a]Freedman et al. (1982).

[b]Fox (1983).

[c]Archibold (1984).

[d]McGraw (1980).

[e]Gartner et al. (1983).

[f]Roach (1983).

[g]Leck (1980).

(Griggs, 1956; Urbanska and Schutz, 1986), although beyond a certain density, it is likely that competition would decrease the chance of establishment. Grulke and Bliss (1988, and unpubl. data) showed that in the High Arctic, establishment from seed is the predominant form of reproduction, and the canopy of neighbors does afford some protection to establishing plants (see also Chap. 6). It seems likely that with increasing exposure (and decreasing vegetative cover), the probabilities of establishment, and by extension the probability of a new species invading, would vary with plant density as shown in Figure 4. Unfortunately, too few data exist to allow us to attach a scale to the y-axis (probability of establishment). Therefore it is impossible to say whether establishment of new, invading species is more likely in dry, moist, or wet tundra. If this scheme is correct, however, it would suggest that some disruption of the vegetation that decreases plant density would be required more for successful invasion in moist or wet sites than in dry sites. The curves shown in Figure 4 should be viewed as hypotheses that could readily be subjected to experimental verification. Moreover, the figure suggests that the degree of change in the plant community due to invasion of new species will depend on whether disturbance frequency (e.g., any thermokarst-related disturbances) is altered by climate change.

V. Ecological Genetic Variation, Plasticity, and Ecosystem Change

Just as ecosystem response to climate change will depend on species composition, so a particular species' response will depend on the amount of genetic variability with respect to the ability to succeed under those changing conditions and the ability of existing genotypes to respond in a plastic manner to environmental change. Tundra plant species appear to have ample reservoirs of genetic variability and plasticity. Both factors should confer some "resis-

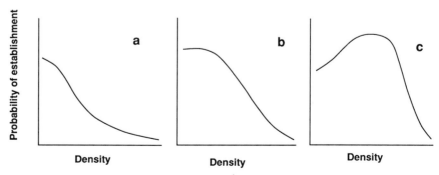

Figure 4   Hypothetical response of seedling establishment to density (which may be lowered by disturbance) in (a) hydric, (b) mesic, and (c) xeric tundra sites.

tance" to climate change for a given species, and hence to the community containing that species.

A high level of genetic variability is suggested by the frequent reports of ecotypic variation in arctic species (e.g., Mooney and Billings, 1961; Teeri, 1972; Shaver *et al.*, 1979; Chapin and Chapin, 1981; Chapin and Oechel, 1983; McGraw and Antonovics, 1983a; Shaver *et al.*, 1986; Fetcher and Shaver, 1990). These studies are unusual in the degree to which physiological, as well as morphological, characters are considered as part of an ecotype's set of adaptations. The classic study of Mooney and Billings (1961) clearly demonstrates differentiation with respect to physiological traits in *Oxyria digyna* over a latitudinal gradient. Based on their findings, it seems unlikely that these different populations would respond in precisely the same way to global climate change. Using a common-environment study, Shaver *et al.* (1979) demonstrated differentiation of *Carex aquatilis* populations over distances of a few decimeters. As in *Oxyria*, these differences were physiological: nutrient uptake rates and nutrient concentrations differed in the four populations studied. Moreover, the four populations responded differently to varying phosphorus levels, suggesting that a climatic change influencing phosphorus availability would have different effects in the different populations.

Through interactions with other factors, even single-factor climatic changes will result in multifactorial environmental change at a particular site. Within arctic species, ecotypes often inhabit sites that differ in several ways over short gradients, such as across snowbeds (Teeri, 1972; McGraw and Antonovics, 1983a). Field reciprocal transplants combined with factor manipulations have demonstrated that one or two factors may account for most of the differential success of the ecotypes along the gradient. But these factors frequently influence several other factors, often indirectly through effects on other species. In *Dryas octopetala*, total productivity of fell-field and snowbed communities responded identically to nitrogen–phosophorus fertilization, yet the fell-field population of *Dryas* declined in productivity, while that of the snowbed population increased (Fig. 5; McGraw, 1985a). Experiments under controlled conditions confirmed that this difference was caused by differential response to light (not nutrients), which was less for *Dryas* in the field study because taller competitors grew more when fertilized. A competition experiment with the two ecotypes in growth chambers confirmed that the snowbed population was competitively superior at high nutrient levels because of the plants' below- and aboveground properties (McGraw, 1985b). In an ecosystem context, these data (1) reemphasize the conclusion that response to global change is likely to differ from various ecotypes within a species and (2) suggest that if climatic changes are large enough, ecotypes would be expected to shift their distributions. Such shifts could lead to the extinction of some ecotypes, whereas others could expand their range. Ultimately, the existence of ecotypic variation should make extinction of an entire species less likely, since at least some genotypes may exist that can cope with the environmental change.

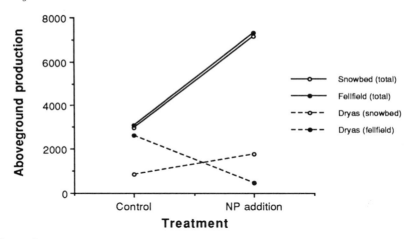

Figure 5 Similar effects of fertilization (N and P addition) on whole-community aboveground production in snowbed and fell-field sites, with contrasting response of *Dryas octopetala* ecotypes within these communities.

The ability of individuals to accommodate change by phenotypic plasticity should reduce the impact of climate change. The lifespan of tundra plants is likely to be long enough to buffer changes over decades or centuries in this way. It appears that tundra species or ecotypes occurring in resource-rich, productive environments are more plastic than their relatives in less productive environments (Shaver *et al.*, 1986; McGraw, 1987; McGraw and Chapin, 1989; Fetcher and Shaver, 1990). For *Dryas octopetala*, the greater plasticity of a snowbed ecotype plays an important role in allowing it to survive in that relatively resource-rich environment (McGraw, 1987).

Fetcher and Shaver (1990) explicitly examined the influence of ecotypic variation in plasticity on ecosystem productivity. They found that northern populations of *Eriophorum vaginatum* were less plastic than southern populations and speculated that productivity in the northern populations might therefore respond more slowly to climatic warming than in southern populations (Fig. 6). They recognized that, eventually, responsive and presumably more competitive individuals could replace less responsive individuals by immigration from southern populations or, more likely, by selection from within the population currently occupying the site. This selection process could take a long time, given the longevity (Mark *et al.*, 1985) and the low frequency of seedling recruitment of *Eriophorum* in undisturbed tundra (McGraw and Shaver, 1982).

The reservoir of genetic variability within populations, and their potential to evolve with respect to factors that are likely to change, may be critical to determining which species are successful and unsuccessful as the global climate shifts. Ultimately, these factors will affect ecosystem parameters such as productivity and biomass, since no two species respond in precisely the same

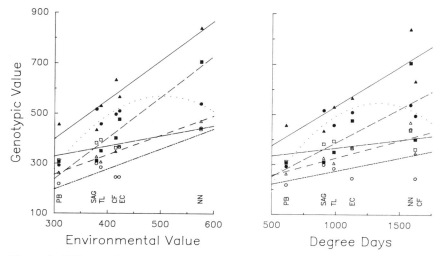

Figure 6   Difference in plasticity in response to the environment for *Eriophorum vaginatum* as shown in a reciprocal-transplant experiment. The genotypic value of tiller size index (length of the longest leaf times number of green leaves) is plotted against environmental value (determined by the mean performance at a site). Lines were fitted by least-squares regression. [© 1990 by the University of Chicago. Reprinted from Fetcher and Shaver (1990) with permission from the University of Chicago.]

way to perturbation (Chapin and Shaver, 1985). One study of within-population genetic variability examined variability at a certain time and changes over time using a "time-transect" provided by seeds buried under a solifluction lobe. Seed subpopulations of two species (*Carex bigelowii* and *Luzula parviflora*) from 0 to 300 years old were extracted from the buried organic horizon and aged by direct dating of seed coats (McGraw *et al.*, in press). In multigeneration, common-environment experiments using clonal replicates of each plant, substantial variation was found within the subpopulations, but the old and young subpopulations were phenotypically and genetically distinct as well (Vavrek *et al.*, 1991; Bennington *et al.*, 1991). Moreover, these plants appear to have evolved differences in ecological response to temperature and nutrients under controlled conditions (Figure 7a,b).

These ecological genetic analyses of buried seed pools have several implications for projecting the effects of global climate change. First, the buried seed populations have apparently recorded rapid evolutionary shifts in two species of arctic plants under natural conditions. Arctic populations are evolving at significant rates even on the century time-scale. These results are at once problematic and promising: They suggest on the one hand that arctic plants may be less affected by environmental change than predicted, if indeed they are capable of adaptive evolution in response to such change. On the

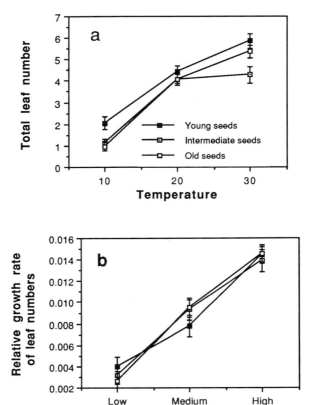

Figure 7   Differential response of different-aged buried seed subpopulations to environmental factors. (a) Response of *Carex bigelowii* to temperature under growth-chamber conditions; (b) response of *Luzula parviflora* to nutrient levels under greenhouse conditions. The oldest seeds are approximately 175 years old; the young seeds are about 0–20 years old.

other hand, this ability to evolve may mean that studies of the responses of present-day populations will inadequately characterize the response one or two centuries hence, because the population will comprise a different set of genotypes by that time. At the ecosystem level, it will be important to determine experimentally whether genotypes "matter," that is, whether the identity of a genotype significantly affects ecosystem function. At present, no such studies exist for arctic plants.

Reserves of genetic variability within seed banks could also furnish a source of genetic variability for evolutionary change. Since new recruits into the population will very likely come from the buried seed pool, such variability could be particularly important. Genetic variability among individuals in the aboveground population is important as well, since differential survival of adults and the genetic composition of the seed rain is determined by these individuals. Yet genetic variation in physiological and ecological traits within populations has received little attention from arctic ecologists. The study of such variation represents an urgent research need, for the first genetic response to climate change will probably be at this level.

VI. Summary and Conclusions

Common aspects of tundra plant life histories, such as vegetation or clonal propagation, a high level of plasticity, extreme longevity, and infrequent reproduction by seed, complicate demographic studies in the arctic. Nevertheless, a conceptual model integrating physiological, demographic, and ecosystem research can now be formed, which shows that demography is an important link in understanding the basis for ecosystem change (Fig. 8). Many of the consequences of different physiological responses to climate change of species within communities, populations within species, and genotypes within populations will show up in differential growth at the module population level. These differential responses will directly affect ecosystem productivity, biomass, net storage of carbon, and so on over tens to hundreds of years. At the same time, perhaps on a longer time scale (or perhaps not), differential recruitment and death of individuals among species, populations, or genotypes may also have direct consequences for ecosystem properties. At present, there is relatively strong evidence that such differential responses will be important for arctic ecosystems undergoing climatic change. But although it may be possible to predict that southern species, ecotypes, or genotypes will move northward (or up in elevation) in response to global warming, the existing demographic data give few clues concerning the rate of such movement, the predominant kinds of shifts that will occur (species versus ecotypic versus genotypic displacement), or its ecosystem level consequences. Unfortunately, very few studies have explicitly attempted to link demographic processes with ecosystem response to any perturbation. Questions such as, "In a given environment, is ecosystem functioning affected by the identity of the species (or ecotype or genotype) present?" need to be explicitly addressed with experimental studies. Questions concerning invasibility of communities or vegetational momentum are fundamentally demographic in nature.

In future research, demographic properties of tundra plants should be studied in detail in conjunction with physiological and ecosystem properties. Although purely demographic studies will be useful in other ways, with respect to research on the effects of climatic change, it is the links among demography, physiology, and ecosystem ecology that must be the focus of future experiments. Only with a better understanding of these relationships will we be able to speculate intelligently on the consequences of global climate change on any time scale.

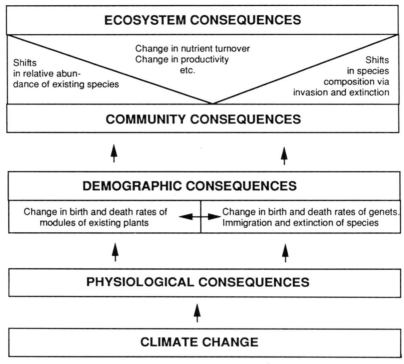

Figure 8   The role of demographic processes in determining ecosystem response to climatic change.

References

Amen, R. D. (1966). The extent and role of seed dormancy in alpine plants. *Q. Rev. Biol.* **41**, 271–281.

Archibold, O. W. (1984). A comparison of seed reserves in arctic, subarctic, and alpine soils. *Can. Field Nat.* **98**, 337–344.

Bazzaz, F. A., and Harper, J. L. (1977). Demographic analysis of the growth of *Linum usitatissimum*. *New Phytol.* **78**, 193–208.

Bennington, C. C., McGraw, J. B., and Vavrek, M. C. (1991). Ecological genetic variation in seed banks. II. Phenotypic and genetic differences between young and old subpopulations of *Luzula parviflora. J. Ecol.* (in press).

Billings, W. D. (1973). Arctic and alpine vegetations: Similarities, differences, and susceptibility to disturbance. *Bioscience* **23**, 697–704.

Billings, W. D., and Mooney, H. A. (1968). The ecology of arctic and alpine plants. *Biol. Rev. Camb. Philos. Soc.* **43**, 481–529.

Bliss, L. C. (1959). Seed germination in arctic and alpine species. *Arctic* **11**, 180–188.

Bonde, E. K. (1968). Survival of seeds of an alpine clover (*Trifolium nanum*). *Ecology* **49**, 1194–1195.

Callaghan, T. V. (1976). Growth and population dynamics of *Carex bigelowii* in an alpine environment. *Oikos* **27**, 402–413.

Caswell, H. (1989). "Matrix Population Models." Sinauer, Sunderland, Massachusetts.

Chapin, F. S., III, and Chapin, M. C. (1980). Revegetation of an arctic disturbed site by native tundra species. *J. Appl. Ecol.* **17**, 449–456.

Chapin, F. S., III, and Chapin, M. C. (1981). Ecotypic differentiation of growth processes in *Carex aquatilis* along latitudinal and local gradients. *Ecology* **62**, 1000–1009.

Chapin, F. S., III, and Oechel, W. C. (1983). Photosynthesis, respiration, and phosphate absorption by *Carex aquatilis* ecotypes along latitudinal and local environmental gradients. *Ecology* **64**, 743–751.

Chapin, F. S., III, and Shaver, G. R. (1985). Individualistic growth response of tundra plant species to manipulation of light, temperature, and nutrients in a field experiment. *Ecology* **66**, 564–576.

Chapin, F. S., III, Vitousek, P. M., and Van Cleve, K. (1986). The nature of nutrient limitation in plant communities. *Am. Nat.* **127**, 48–58.

Chester, A. L., and Shaver, G. R. (1982). Reproductive effort in cotton grass tussock tundra. *Holarct. Ecol.* **5**, 200–206.

Chester, A. L., and McGraw, J. B. (1983). Effects of nitrogen addition on the growth of *Vaccinium uliginosum* and *Vaccinium vitis-idaea. Can. J. Bot.* **61**, 2316–2322.

Fetcher, N., and Shaver, G. R. (1983). Life histories of tillers of *Eriophorum vaginatum* in relation to tundra disturbance. *J. Ecol.* **71**, 131–147.

Fetcher, N., and Shaver, G. R. (1990). Environmental sensitivity of ecotypes as a potential influence on primary productivity. *Am. Nat.* **136**, 126–131.

Fox, J. F. (1983). Germinable seed banks of interior Alaskan U.S.A. tundra. *Arct. Alp. Res.* **15**, 405–412.

Freedman, B., Hill, N., Svoboda, J., and Henry, G. (1982). Seed banks and seedling occurrence in a high arctic oasis at Alexandra Fjord, Ellesmere Island, Canada. *Can. J. Bot.* **60**, 2112–2118.

Gartner, B. L., Chapin, F. S., III, and Shaver, G. R. (1983). Demographic patterns of seedling establishment and growth of native graminoids in an Alaskan tundra disturbance. *J. Appl. Ecol.* **20**, 965–980.

Griggs, R. F. (1956). Competition and succession on a Rocky Mountain fellfield. *Ecology* **37**, 8–20.

Grulke, N. E., and Bliss, L. C. (1988). Comparative life-history characteristics of two high arctic grasses, Northwest Territories. *Ecology* **69**, 484–496.

Law, R. (1983). A model for the dynamics of a plant population containing individuals classified by age and size. *Ecology* **64**, 224–230.

Lawrence, B. A., Lewis, M. C., and Miller, P. C. (1978). A simulation model of population processes of arctic tundra graminoids. *In* "Vegetation and Production Ecology of an Alaskan Arctic Tundra" (L. L. Tieszen, ed.), pp. 599–619. Springer-Verlag, New York.

Leck, M. A. (1980). Germination in Barrow, Alaska, tundra soil cores. *Arct. Alp. Res.* **12**, 343–349.

Lefkovitch, L. P. (1965). The study of population growth in organisms grouped by stages. *Biometrics* **21**, 1–18.

McGraw, J. B. (1980). Seed bank size and distribution of seeds in cottongrass tussock tundra. *Can. J. Bot.* **58**, 1607–1611.

McGraw, J. B. (1985a). Experimental ecology of *Dryas octopetala* ecotypes. III. Environmental factors and plant growth. *Arct. Alp. Res.* **17**, 229–239.

McGraw, J. B. (1985b). Experimental ecology of *Dryas octopetala* ecotypes: Relative response to competitors. *New Phytol.* **100**, 233–241.

McGraw, J. B. (1987). Experimental ecology of *Dryas octopetala* ecotypes. IV. Fitness response to reciprocal transplanting in ecotypes with differing plasticity. *Oecologia* **73**, 465–468.

McGraw, J. B. (1989). Effects of age and size on life histories and population growth of *Rhododendron maximum* shoots. *Am. J. Bot.* **76**, 113–123.

McGraw, J. B., and Antonovics, J. (1983a). Experimental ecology of *Dryas octopetala* ecotypes. I. Ecotypic differentiation and life-cycle stages of selection. *J. Ecol.* **71**, 879–897.

McGraw, J. B., and Antonovics, J. (1983b). Experimental ecology of *Dryas octopetala* ecotypes. II. A demographic model of growth, branching, and fecundity. *J. Ecol.* **71**, 899–912.

McGraw, J. B., and Chapin, F. S. (1989). Competitive ability and adaptation to fertile and infertile soils in two *Eriophorum* species. *Ecology* **70**, 736–749.

McGraw, J. B., and Shaver, G. R. (1982). Seedling density and seedling survival in Alaskan cotton grass tussock tundra. *Holarct. Ecol.* **5**, 212–217.

McGraw, J. B., and Vavrek, M. C. (1989). The role of buried viable seeds in arctic and alpine plant communities. *In* "Ecology of Soil Seed Banks" (M. A. Leck, V. T. Parker, and R. L. Simpson, eds.), pp. 91–106. Academic Press, San Diego, California.

McGraw, J. B., and Wulff, R. D. (1983). The study of plant growth: A link between the physiological ecology and population biology of plants. *J. Theor. Biol.* **103**, 21–28.

McGraw, J. B., Vavrek, M. C., and Bennington, C. C. (1991). Ecological genetic variation in seed banks. I. Establishment of a time-transect. *J. Ecol.* (in press).

Mark, A. F., Fetcher, N., Shaver, G. R., and Chapin, F. S., III. (1985). Estimated ages of mature tussocks of *Eriophorum vaginatum* along a latitudinal gradient in central Alaska, U.S.A. *Arct. Alp. Res.* **17**, 1–5.

Mooney, H. A., and Billings, W. D. (1961). Comparative physiological ecology of arctic and alpine populations of *Oxyria digyna*. *Ecol. Monogr.* **31**, 1–29.

Osburn, W. S. (1961). Successional potential resulting from differential seedling establishment in alpine tundra stands. *Bull. Ecol. Soc. Am.* **42**, 146–147.

Roach, D. A. (1983). Buried seed and standing vegetation in two adjacent tundra habitats, northern Alaska. *Oecologia* **60**, 359–364.

Shaver, G. R., Chapin, F. S., III, and Billings, W. D. (1979). Ecotypic differentiation in *Carex aquatilis* on ice-wedge polygons in the Alaskan coastal tundra. *J. Ecol.* **67**, 1025–1046.

Shaver, G. R., and Chapin, F. S., III. (1980). Response to fertilization by various plant growth forms in an Alaskan tundra: Nutrient accumulation and growth. *Ecology* **61**, 662–675.

Shaver, G. R., Gartner, B. L., Chapin, F. S., III, and Linkins, A. E. (1983). Revegetation of arctic disturbed sites by native tundra plants. *In* "Permafrost: Fourth International Conference Proceedings," pp. 1133–1138. National Academy Press, Washington, DC.

Shaver, G. R., Fetcher, N., and Chapin, F. S., III. (1986). Growth and flowering in *Eriophorum vaginatum*: Annual and latitudinal variation. *Ecology* **67**, 1524–1535.

Soyrinki, N. (1938). Studien über die generative und vegetative vermehrung der samenpflanzen in der alpinen vegetation. Petsamo Lappland. I. Suomal. *Elain-ja Kasvit Seur van elain* **11**, 1–323.

Teeri, J. A. (1972). Microenvironmental adaptations of local populations of *Saxifraga oppositifolia* in the High Arctic. Ph.D. diss., Duke University, Durham, North Carolina.

Urbanska, K. M., and Schutz, M. (1986). Reproduction by seed in alpine plants and revegetation research above timberline. *Bot. Helv.* **96**, 43–60.

Vavrek, M. C., McGraw, J. B., and Bennington, C. C. (1991). Ecological genetic variation in seed banks. III. Phenotypic and ecological differentiation of young and old genotypes of *Carex bigelowii*. *J. Ecol.* (in press).

Wager, H. G. (1938). Growth and survival of plants in the Arctic. *J. Ecol.* **26**, 390–410.

White, J. (1979). The plant as a metapopulation. *Annu. Rev. Ecol. Syst.* **10**, 109–145.

White, J. (1984). Plant metamerism. *In* "Perspectives in Plant Population Ecology" (R. Dirzo and J. Sarukhan, eds.), pp. 15–47. Sinauer, Sunderland, Massachusetts.

18

Controls over Secondary Metabolite Production by Arctic Woody Plants

John P. Bryant and Paul B. Reichardt

I. Introduction

Secondary metabolites control herbivory on woody vegetation in arctic ecosystems (e.g., Bryant and Kuropat, 1980; Batzli, 1983; Kuropat, 1984; MacLean and Jensen, 1984) and may play an equally important role in controlling nutrient cycling through their effects on decomposition and nitrification (e.g., Handley, 1961; Swift *et al.*, 1979; Flanagan and Van Cleve, 1983; Rice and Pancholy, 1972, 1973, 1974). In this chapter we review what is known of the environmental controls over the production of secondary metabolites by woody plants and use this information to predict how these substances might influence the short- and long-term responses of arctic tundra ecosystems to climate change.

Arctic shrubs and boreal trees produce a variety of carbon-based secondary metabolites (phenolics and terpenes) in ecologically significant quantities. Nitrogen-containing secondary metabolites, on the other hand, have not been found in any significant quantities (e.g., Prudhome, 1983; Kuropat, 1984; Reichardt *et al.*, 1984, 1987, 1990a,b; Bryant *et al.*, 1985; Clausen *et al.*, 1986). For this reason, we consider only carbon-based secondary metabolites in this chapter.

II. Environmental Controls of Secondary Metabolite Production

A. Evolutionary Responses to Nutrient Availability and Disturbance

In the arctic tundra, high nutrient availability is associated with disturbances, such as wildfire (Chapin and Shaver, 1985a). Therefore, nutrient-demanding deciduous species (Chapin, 1980a), such as willow and shrub birch, have several physiological traits enabling them to rapidly replace destroyed above-ground parts: a high maximum potential growth rate (Chapin, 1980a), large belowground reserves of carbon and nutrients (Chapin, 1980b), the ability to resprout (Bryant and Chapin, 1986), a high nutrient absorption capacity (Chapin and Tryon, 1982), and a high photosynthetic rate (Small, 1972; Chapin, 1980b). These traits also permit a rapid regrowth after herbivory (Archer and Tieszen, 1980; Chapin, 1980b,c) and therefore minimize selection for antiherbivore defense (Bryant *et al.*, 1983; Coley *et al.*, 1985).

In contrast, slowly growing tundra evergreens, such as *Ledum palustre*, store more of their carbon and nutrient reserves in leaves and shoots eaten by herbivores and have a lower nutrient absorption capacity and a lower photosynthetic rate. Although these traits are advantageous in nutrient-deficient habitats (Chapin, 1980a), they are disadvantageous when herbivory occurs because they limit regrowth. Thus, tundra evergreens such as *L. palustre* (Reichardt *et al.*, 1990b) have been selected for very effective antiherbivore defenses that make the plants less palatable, less digestible, and more toxic than tundra deciduous species such as willow and shrub birth (e.g., Robus, 1981; Batzli, 1983; Kuropat, 1984; MacLean and Jensen, 1984).

B. Phenotypic Responses to Resource Availability

Insufficient mineral nutrition usually limits the growth of tundra plants more than it limits photosynthesis (Chapin, 1980a, 1991; Fig. 1). Thus, soluble carbohydrate in excess of growth demands accumulates in tissues of nutrient-stressed woody plants. This excess of carbohydrate favors increased production of carbon-based secondary metabolites (Bryant *et al.*, 1983). In contrast, good mineral nutrition increases growth, resulting in a demand for carbohydrate that reduces tissue carbohydrate concentrations (Shaver and Chapin, 1980), and this reduction in turn reduces the plant's potential to produce carbon-based secondary metabolites. Shade directly limits photosynthesis, thereby also reducing the availability of carbohydrate for secondary metabolite pro-

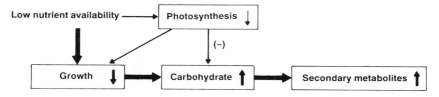

Figure 1 Effect of low nutrient availability on secondary metabolite production. Thickness of arrows indicates magnitude of effect. All effects and feedbacks are positive unless otherwise indicated (−).

duction (Bryant *et al.*, 1983). For example, nitrogen–phosphorus–potassium fertilizer, shade, and their combination reduced the concentration of condensed tannin in shoots of *S. alaxensis* (Bryant, 1987), a dominant willow on floodplains of arctic rivers (Viereck and Little, 1972).

The role of moisture stress in controlling plant secondary metabolism is poorly understood (Gershenzon, 1984). Moderate moisture stress is associated with increased chemical defense against insect attack (Lorio, 1986; Sharpe *et al.*, 1985), but drought is frequently associated with insect outbreak (Mattson and Haack, 1987).

C. Responses of Leaves to Herbivory

When vertebrate herbivores browse woody plants (Fig. 2; left), they prune shoots, thereby reducing the competition for nutrients among the remaining shoots (Moorby and Waring, 1963). Thus, concentrations of mobile nutrients such as nitrogen rise in the leaves of browsed plants (Chapin, 1980c). As a result of better nutrition, leaf growth increases (e.g., Danell and Huss-Danell, 1985; Haukioja *et al.*, 1990), creating a demand for carbohydrate (Chapin, 1980c) that limits the production of carbon-based secondary metabolites and fiber (Bryant *et al.* 1991b). Rising concentrations of nutrients and declining concentrations of secondary metabolites and fiber increase the value of leaves as food for herbivores so that leaf herbivory increases (Danell and Huss-Danell, 1985; du Toit *et al.*, 1990; Haukioja *et al.*, 1990).

In contrast, severe and repeated defoliation (Fig. 2; right) of rapidly growing deciduous species, such as occurs in insect outbreaks, removes leaves and the nutrients in them but not entire shoots, as does browsing by mammals. Thus, shoot numbers do not decline in the next year as they do in severely browsed plants. Defoliation also results in mortality of fine roots, reducing the plant's ability to absorb nutrients, and an increase in photosynthesis, which improves the plant's carbohydrate status (Prudhomme, 1982; Bassman and Dickmann, 1982a,b). Consequently, the availability of nutrients per leaf declines with each severe defoliation, and leaf growth becomes progressively more nutrient- than carbon-limited (Bryant *et al.*, 1988). In turn, nutrient limitation of leaf growth results in a surplus of carbohydrate that can be used for increased production of fiber and carbon-based secondary metabolites.

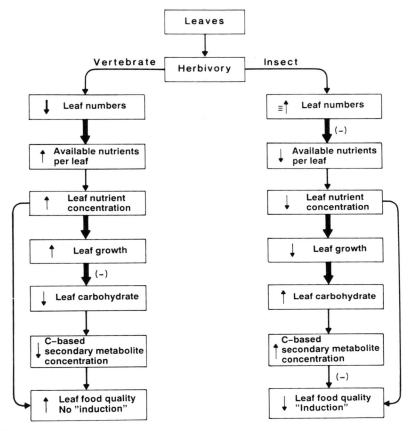

Figure 2   Effects of vertebrate and invertebrate herbivory on leaves of rapidly growing decid-uous arctic woody plants. Thickness of arrows indicates magnitude of effect. All effects and feed-backs are positive unless otherwise indicated ($-$).

These defoliation-induced changes in the chemistry of leaves of rapidly grow-ing deciduous species decrease the food value of leaves for herbivores (Bal-tensweiler *et al.*, 1977; Tuomi *et al.*, 1984; Bryant *et al.*, 1988).

Defoliation of slowly growing evergreens has the opposite effect on leaf chemistry and food value (Bryant *et al.*, 1988). In evergreens, defoliation leads to increases in leaf nitrogen and decreases in leaf carbohydrate (e.g., Ericsson *et al.*, 1980, 1985) so that concentrations of carbon-based secondary metabo-lites decrease (Bryant *et al.*, 1991a), thereby increasing the food value of leaves (Niemela *et al.*, 1984; Bryant *et al.*, 1991a).

Bryant *et al.* (1988) have suggested that these contrasting responses to defo-liation are consequences of herbivory-induced alteration of plant carbon—nutrient balance. Defoliation depletes the nutrient reserves of all woody species at approximately the same rate, but it reduces the carbon reserves of slowly growing evergreens more rapidly than those of rapidly growing deciduous species. Thus, defoliation is likely to induce nutrient stress in rapidly growing tundra deciduous species and carbon stress in slowly growing tundra evergreens. Nutrient stress results in reduced leaf nitrogen and increased production of fiber and carbon-based secondary metabolites (e.g., Baltensweiler, 1977; Tuomi *et al.*, 1984), chemical changes that reduce the food value of leaves for herbivores (Scriber and Slansky, 1981; Robbins, 1983). Carbon stress has the opposite effects on leaf chemistry and food value.

D. Age-Specific Selection for Defense by Vertebrate Herbivory

Throughout their ontogeny, woody plants undergo genetically programmed physiological changes that are evolutionary responses to age-specific selection pressures (Kozlowski, 1971). In arctic ecosystems, vertebrate browsing of the juvenile stage in winter is one such age-specific selection pressure (Bryant and Kuropat, 1980; Bryant *et al.*, 1983). The evolutionary response to winter browsing of the juvenile stage has been increased resistance to browsing by mammals and ptarmigan (e.g., Ryala, 1966; Bryant, 1981; Bryant *et al.*, 1983). Moreover, the juvenile stage of species and genotypes from arctic regions with high-amplitude hare cycles are more defended than those from regions with low-amplitude hare cycles or no hare cycle (Bryant *et al.*, 1989).

III. Responses of Secondary Metabolite Production to Climate Change

A. Direct Responses to Climate Change

Depending on the amount and physical properties of clouds, a doubling of CO_2 in the atmosphere is expected to increase the surface temperature in the Arctic by 3–8° K (Mitchell *et al.*, 1989; Chap. 2; Fig. 3; step 1). Chapin and Shaver (1985b) found that a temperature increase of this magnitude had variable effects on the growth of tundra vegetation. In an upland tussock tundra dominated by the graminoid *Eriophorum vaginatum*, two years of warming increased the growth of dominant canopy shrubs (*Betula nana; L. palustre*) but not the growth of *E. vaginatum* and less dominant graminoids, shrubs, and mosses. In a lowland wet sedge meadow, warming reduced the growth of one dominant species (*E. angustifolium*) and had no effect on the growth of the other (*Carex aquatilis*). The general conclusion to be drawn from Chapin and Shaver's experiment is that warming in the arctic tundra is likely to increase the growth of dominant canopy shrubs (Fig. 3: Step 2; see also Chap. 6, 17).

   The general insensitivity of arctic plant physiological processes to temperature indicates that the increased growth of canopy shrubs may well be an indirect response to an increase in the length of the growing season and increased nutrient availability and uptake (Chapin, 1983; Chapin and Shaver, 1985a). In accordance with principles outlined in Section II,B, an increase in

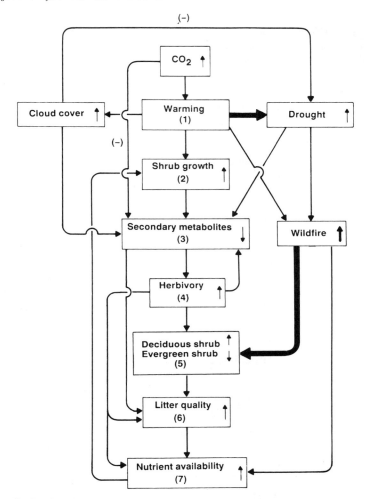

Figure 3   Predicted short-term effects of climate change on the function of arctic tundra ecosystems. Thickness of arrows indicates magnitude of effects. All effects and feedbacks are positive unless otherwise indicated (−).

growth because of better mineral nutrition should reduce carbohydrate reserves, resulting in decreased production of carbon-based secondary metabolites (Fig. 3: step 3) and therefore, increased herbivory (Fig. 3: step 4).

In the Arctic, cloud cover is expected to increase by 1–2% by the time CO_2 doubles (Mitchell *et al.*, 1989; Chap. 2). Although small, this increase in cloud cover might further reduce the antiherbivore defenses of woody plants (see Fig. 3). A variety of arctic plants of several growth forms respond to low light in cloudy summers by an increase in leaf nitrogen and a decrease in leaf phenolics, presumably because of carbon stress resulting from reduced photosynthesis (Jonasson *et al.*, 1986).

Although precipitation will increase slightly (Mitchell *et al.*, 1989; Chap. 2), warming is expected to produce increased drought in the Arctic (see Fig. 3) because of increased evapotranspiration (Chap. 2, 3). Moderate moisture stress might increase chemical defenses (Sharpe *et al.*, 1985; Lorio, 1986), but drought is more likely to decrease secondary metabolite production by woody plants and increase the availability of amino acids in leaves (Mattson and Haack, 1987), leading to increased herbivory by insects. Moreover, warming can also be expected to directly increase the rate at which insects grow (Ayres and MacLean, 1987a,b; Mattson and Haack, 1987). In combination, increased leaf food value and direct effects of warming on insect growth favor an increase in insect outbreaks (Mattson and Haack, 1987). Thus, it is likely that climate change in the Arctic will increase insect outbreaks (Singh, 1988).

B. Herbivory-Mediated Responses to Climate Change

In the Arctic, most herbivory of woody plants occurs in late spring and summer and involves deciduous species, especially willow, and to a lesser extent, birch (e.g., Batzli, 1983; Robus, 1981; White, 1983; Kuropat, 1984; MacLean and Jensen, 1984; Williams *et al.*, 1980). Browsing of deciduous species by vertebrates reduces concentrations of carbon-based secondary metabolites in leaves (see Fig. 2: left). In contrast, increased defoliation by insects in outbreaks can be expected to increase secondary metabolite concentrations (Fig. 2: right). Increased nutrient availability resulting from warming (Chap. 13) should, however, ameliorate the effects of increased insect defoliation (Tuomi *et al.*, 1984; Bryant *et al.*, 1988). Increased herbivory (either vertebrate or invertebrate) of slowly growing evergreens such as *L. palustre* would reduce secondary metabolite concentrations. Thus, the net effect of increases in herbivory will probably be a feedback that further reduces leaf secondary metabolite concentrations (see Fig. 3), unless the direct effect of increasing CO_2 on photosynthesis counters the effects of warming, cloud cover, drying, and increased herbivory on leaf secondary metabolite production. It is unlikely, however, that increased CO_2 will substantially alter photosynthetic rates (Chap. 7).

1. Responses to Wildfire and Herbivory

Warming and drying will probably greatly increase the frequency and intensity of wildfire (Davis, 1988; Singh, 1988; Fig. 3). Although it is unlikely that the fire return time will be shortened enough to favor the dominance of graminoids (Chapin and Van Cleve, 1981), it will probably increase the dominance of deciduous shrubs and decrease the dominance of evergreens.

Increased herbivory of evergreens is likely to increase the mortality of evergreen species more than that of deciduous species, further increasing dominance of deciduous shrubs (Chapin, 1980c). An increase in dominance of early successional deciduous shrubs, however, often brings an increase in the numbers of browsing mammals and therefore an increase in the intensity at which deciduous shrubs are browsed (Grange, 1949). This increased browsing drives succession toward dominance of the vegetation by late successional evergreens (Bryant and Chapin, 1986). Although browsing by mammals can be expected to favor the survival of evergreens, it is likely that the negative effect of increased wildfire on the survival of evergreens will greatly override the effects of browsing. Thus, the net effect of wildfire and herbivory on arctic tundra vegetation will be to decrease the abundance of woody evergreens, leading to an increase in the abundance of deciduous woody species such as willow and birch (Fig. 3: step 5). Interestingly, warming at the end of the Pleistocene and in the Holocene led to an increase in the abundance of willow and especially birch (Lamb and Edwards, 1988).

In comparison with willow and birch, arctic evergreens produce inherently slowly decomposing (low-quality) litter (Flanagan and Van Cleve, 1983; Pastor *et al.*, 1988). That is, evergreen litter has low nutrient concentrations and high concentrations of lignin and carbon-based secondary metabolites, which depress decomposition and nitrification (Handley, 1961; Swift *et al.*, 1979; Rice and Pancholy, 1972, 1973, 1974). If climate change increases the dominance of tundra vegetation by willow and birch, then litter quality will increase (Fig. 3: step 6). Increased litter quality will hasten decomposition and nitrification and therefore accelerate nutrient cycling, which will make more nutrients available to plants (Fig. 3: step 7). In turn, an increase in plant nutrition will increase growth, thereby increasing mammalian herbivory because of reduced plant defenses. This increase in herbivory will further accelerate nutrient cycling for two reasons. First, it will increase the nitrogen concentration of leaves while reducing their chemical defenses (Fig. 2: left). Increased leaf nitrogen and reduced leaf defense will be reflected in an increase in the nitrogen and secondary metabolite contents of leaf litter and, therefore, in its quality (Fig. 3: step 6). Second, increased herbivory will directly accelerate the rate at which nutrients become available to plants because of the increased return of nutrients to the soil in dung and urine (e.g., Schultz, 1968; Batzli, 1983).

An increase in wildfires will also increase the availability of nutrients to plants (Chapin and Van Cleve, 1981). The micronutrients in ash will increase the rate at which soil organic matter decomposes, especially in peat bogs, resulting in the rapid release of macronutrients such as nitrogen locked in peat (Silvola, *et al.*, 1985). Increased micro- and macronutrient availability because of fire will feed back positively to the growth of shrubs, further reducing the production of carbon-based secondary metabolites and producing even higher-quality litter. In addition, the increase in decomposition associated with ash from fire will necessarily be associated with an increase in soil respiration and, therefore, increased input of CO_2 into the atmosphere.

Invertebrates in arctic streams consume birch and willow leaf litter faster than the leaf litter of evergreens such as *L. palustre* (Irons, unpubl. data). Furthermore, either increased nutrient availability (Irons *et al.*, 1987) or increased browsing (Irons *et al.*, 1991) that enhances litter quality will increase the rate at which stream invertebrates process the leaf litter. Thus, climate change acting through plant secondary metabolism will affect not only the functioning of terrestrial ecosystems, but also that of stream ecosystems.

IV. Replacement of Tundra by Taiga

Vegetation models predict that climate change will profoundly alter the distribution of the earth's biomes (Davis, 1988; Pastor and Post, 1988), with arctic tundra being displaced northward by a mosaic of taiga forest and shrubland similar to the vegetation found at the current tree line (Rizzo, 1988). Less vegetation change is expected in high-arctic polar deserts because of the lack of soil development (Chap. 6).

Replacement of arctic tundra by taiga in periods of global warming is not unique in geologic time (Lamb and Edwards, 1988), but the rate at which vegetation change associated with the greenhouse effect is expected to occur is unique. Whereas post-Pleistocene and Holocene warming took place over hundreds to thousands of years, warming from the greenhouse effect is expected to occur within 100 years. Warming could thus occur so rapidly that the capacity of large-seeded southern woody species to migrate north will be exceeded (Davis, 1988; Chap. 5). Although this concern may be valid in mid-latitudes, it may not be a major impediment to the dispersal of taiga species into regions now occupied by tundra. Many taiga willows and balsam poplar (*Populus balsamifera*) already occur north of the tree line (Viereck and Little, 1972) and have light seeds adapted for rapid long-distance dispersal. Migration of the heavier-seeded birches, alders (*Alnus*), pines (*Pinus*), spruces (*Picea*), and larches (*Larix*) can follow valleys of north-flowing rivers, where seeds may be dispersed by water. Such rapid northward movement of trees along river valleys occurred in Europe at the end of the Pleistocene as the Fennoscandian Ice Sheet melted (Hoffman, 1983).

Nevertheless, seed dispersal is not the only factor that can influence the northward migration of taiga species. If the frequency and intensity of wildfires and insect outbreaks in evergreen forests increase dramatically (Singh, 1988), tree-line evergreens will be eliminated as a seed source, and tundra will ultimately become a woodland dominated by rapidly growing disturbance-adapted taiga species such as the birches and willows. Taiga willow and birch have more effective chemical defenses against browsing by mammals in winter than do tundra willow and birch, which have evolved in arctic regions where the ten-year hare cycle does not occur (Bryant *et al.*, 1989). If arctic mammals that rely on woody browse for their winter food are less able to detoxify the defenses of taiga willow and birch, as is indicated by the results of Bryant *et al.* (1989), they will starve in winter and be replaced by taiga-browsing mammals that have better detoxification abilities.

A further consequence of the invasion of tundra by taiga species will be a significant decline in the abundance of tundra graminoids. Loss of the graminoid resource will adversely affect populations of arctic grazers because these do not perform well on a diet of chemically defended woody plants (Batzli, 1983). Furthermore, *E. vaginatum* flowers are a critical resource for reindeer and caribou (*Rangifer tarandus*) and ptarmigan (*Lagopus*) at the time they produce young (Kuropat, 1984). A decrease in the abundance of *E. vaginatum* flowers would nutritionally stress reindeer, caribou, and ptarmigan at this critical period in their life cycles, greatly reducing their productivity (White, 1983). Such a reduction in the production of arctic grazers can be expected to have negative effects on populations of their predators. Furthermore, these changes will directly affect the indigenous peoples of the Arctic by reducing their food supply and the abundance of furbearers that they depend on for revenue.

V. Summary

In arctic ecosystems, secondary metabolites in woody plants reduce rates of herbivory, decomposition, and nitrification. Climate change will probably decrease the production of these substances. Climate change will probably also increase the frequency and intensity of wildfires, resulting in increased dominance of tundra vegetation by deciduous shrubs, which are comparatively poorly defended by secondary metabolites. The combined effect of an increase in deciduous shrubs in tundra vegetation and a decrease in the production of carbon-based secondary metabolites by shrubs will be to increase herbivory and nutrient cycling.

In the long term, arctic tundra will be displaced northward by a taiga woodland dominated by deciduous shrubs and trees that are better defended against browsing in winter by mammals than are tundra deciduous shrubs. Because taiga-browsing mammals can better detoxify secondary metabolites, they will likely replace tundra-browsing mammals. Furthermore, the abundance of graminoids will probably also decline, thereby leading to declines in tundra grazers and their predators. These changes will directly affect the indigenous peoples of the Arctic by reducing their food supply and the abundance of furbearers that they depend on for revenue.

Acknowledgments

This study was supported by NSF Grant BSR-870262 for long-term ecological research in the Alaskan taiga.

References

Archer, S., and Teiszen, L. L. (1980). Growth and physiological responses of tundra plants to defoliation. *Arct. Alp. Res.* **12**, 531–552.

Ayres, M. P., and MacLean, S. F., Jr. (1987a). Development of birch leaves and the growth energetics of *Epirrita autumnata* (Geometridae). *Ecology* **68**, 558–568.

Ayres, M. P., and MacLean, S. F., Jr. (1987b). Molt as a component of insect development *Galerucella sagittariae* (Chrysomelidae) and *Epirrita autumnata* (Geometridae). *Oikos* **48**, 273–279.

Baltensweiler, W., Benz, G., Bovey, P., and Delucchi, V. 1977. Dynamics of larch bud moth populations. *Annu. Rev. Entomol.* **22**, 79–100.

Bassmann, J. H., and Dickman, D. I. (1982a). Effects of defoliation in the developing leaf zone on young *Populus* × *euramericana* plants. I. Photosynthetic physiology, growth, and dry weight partitioning. *For. Sci.* **28**, 599–612.

Basmann, J. H., and Dickman, D. I. (1982b). Effects of defoliation in the developing leaf zone on young *Populus* × *euramericana* plants. II. Distribution of ^{14}C-photosynthate after defoliation. *For. Sci.* **31**, 358–366.

Batzli, G. O. (1983). Responses of arctic rodent populations to nutritional factors. *Oikos* **40**, 396–406.

Bryant, J. P. (1981). Phytochemical deterrence of snowshoe hare browsing by adventitious shoots of four Alaskan trees. *Science* **313**, 889–890.

Bryant, J. P. (1987). Feltleaf willow–snowshoe hare interactions: Plant carbon/nutrient balance and floodplain succession. *Ecology* **68**, 1319–1327.

Bryant, J. P., and Chapin, F. S., III. (1986). Browsing–woody plant interactions during boreal forest plant succession. *In* "Forest Ecosystems in the Alaskan Taiga" (K. Van Cleve, F. S. Chapin III, P. W. Flanagan, L. A. Viereck, and C. T. Dyrness, eds.), pp. 313–225. Springer-Verlag, New York.

Bryant, J. P., and Kuropat, P. J. (1980). Selection of winter forage by subarctic browsing vertebrates: The role of plant chemistry. *Annu. Rev. Ecol. Syst.* **11**, 261–285.

Bryant, J. P., Chapin, F. S., III, and Klein, D. R. (1983). Carbon/nutrient balance of boreal plants in relation to vertebrate herbivory. *Oikos* **40**, 357–368.

Bryant, J. P., Wieland, G. D., Clausen, T. P., and Kuropat, P. J. (1985). Interactions of snowshoe hares and feltleaf willow (*Salix alaxensis*) in Alaska. *Ecology* **66**, 1564–1573.

Bryant, J. P., Tuomi, J., and Niemala, P. (1988). Environmental constraint of constitutive and long-term inducible defenses in woody plants. *In* "Chemical Mediation of Coevolution" (K. Spencer, ed.), pp 376–389. Academic Press, New York.

Bryant, J. P., Tahvanainen, J., Sulkinoja, M., Julkunen-Titto, R., Reichardt, P., and Green, T. (1989). Biogeographic evidence for the evolution of chemical defense by boreal birch and willow against mammalian browsing. *Am. Nat.* **134**, 20–34.

Bryant, J. P., Heitkonig, I., Kuropat, P., and Owen-Smith, N. (1991a). Effects of severe defoliation on long-term resistance to insect attack and on leaf chemistry in six woody species of the southern African savanna. *Am. Nat.* **137**, 50–63.

Bryant, J. P., Danell, K., Provenza, F. P., Reichardt, P. B., and Clausen, T. P. (1991b). Effects of mammals browsing upon the chemistry of deciduous woody plants. *In* "Phytochemical Induction by Herbivores" (D. W. Tallamy and M. J. Raup, eds.). Wiley, New York (in press).

Chapin, F. S., III. (1980a). The mineral nutrition of wild plants. *Annu. Rev. Ecol. Syst.* **11**, 233–260.

Chapin, F. S., III. (1980b). Nutrient allocation and responses to defoliation in tundra plants. *Arct. Alp. Res.* **12**, 553–563.

Chapin, F. S., III. (1980c). Effect of clipping upon nutrient status and forage value of tundra plants in arctic Alaska. *In* "Proceedings of the Second International Reindeer/Caribou Symposium, Roros, Norway (E. Reïmers, E. Gaare, and S. Skjenneberg, eds.), pp. 19–25. Direktotatet for vilt og ferskvannsfisk, Trondheim.

Chapin, F. S., III. (1983). Direct and indirect effects of temperature on arctic plants. *Polar Biol.* **2**, 47–52.

Chapin, F. S., III. (1991). Integrated responses of plants to stress. *Bioscience* **41**, 29–36.

Chapin, F. S., III., and Shaver, G. R. (1985a). Arctic. *In* "Physiological Ecology of North American Plant Communities" (B. F. Chabot and H. A. Mooney, eds.), pp. 16–40. Chapman and Hall, New York.

Chapin, F. S., III., and Shaver, G. R. (1985b). Individualistic growth response of tundra plant species to environmental manipulations in the field. *Ecology* **66**, 564–576.

Chapin, F. S., III., and Tryon, P. R. (1982). Phosphate absorption and root respiration of different plant growth forms from northern Alaska. *Holarct. Ecol.* **5**, 164–171.

Chapin, F. S., III., and Van Cleve, K. (1981). Plant nutrient absorption and retention under differing fire regimes. *U.S. For. Serv. Gen. Tech. Rep. WO* **26**, 301–321.

Clausen, T. P., Bryant, J. P., and Reichardt, P. B. (1986). Defense of winter-dormant green alder against snowshoe hares. *J. Chem. Ecol.* **12**, 2117–2131.

Coley, P. D., Bryant, J. P., and Chapin, F. S., III. (1985). Resource availability and plant antiherbivore defense. *Science* **230**, 985–899.

Danell, K., and Huss-Danell, K. (1985). Feeding by insects and hares on birches earlier affected by moose browsing. *Oikos* **44**, 75–81.

Davis, M. B. (1988). Ecological systems and dynamics. *In* "Toward an Understanding of Global Change," pp. 69–105. National Academy Press, Washington, DC.

du Toit, J. T., Bryant, J. P., and Frisby, K. (1990). Regrowth and palatability of *Acacia* shoots following pruning by African savanna browsers. *Ecology* **71**, 149–154.

Ericsson, A., Larsson, S., and Tenow, O. (1980). Effects of early- and late-season defoliation on growth and carbohydrate dynamics in Scots pine. *J. Appl. Ecol.* **17**, 747–769.

Ericsson, A., Hellqvist, C., Langstrom, B., Larsson, S., and Tenow, O. (1985). Effects on growth of simulated and induced shoot pruning by *Tomicus piniperda* as related to carbohydrate and nitrogen dynamics in Scots pine. *J. Appl. Ecol.* **22**, 105–124.

Flanagan, P. W., and Van Cleve, K. (1983). Nutrient cycling in relation to decomposition and organic matter quality in taiga ecosystems. *Can. J. For. Res.* **13**, 795–817.

Gershenzon, J. (1984). Changes in levels of plant secondary metabolites under water and nutrient stress. *Rec. Adv. Phytochem.* **18**, 273–320.

Grange, W. B. (1949). "The Way to Game Abundance." Scribner's, New York.

Handley, W. R. C. (1961). Further evidence for the importance of residual leaf protein complexes in litter decomposition and the supply of leaf nitrogen for plant growth. *Plant Soil* **15**, 37–73.

Hoffman, G. W. 1983. "A Geography of Europe." Wiley, New York.

Haukioja, E., Ruohomaki, K., Senn, J., Soumela, J., and Walls, M. (1990). Consequences of herbivory in the mountain birch (*Betula pubescens* ssp. *tortuosa*): Importance of the functional organization of the tree. *Oecologia* **82**, 238–247.

Irons, J. G., Oswood, M. W., and Bryant, J. P. (1987). Choices of leaf feeding detritus by a shredding caddisfly: Effects of tree species and fertilization. *Hydrobiologia* **160**, 53–61.

Irons, J. G., Bryant, J. P., and Oswood, M. W. (1991). Effects of moose browsing on decomposition rates of birch leaf litter in a subarctic stream. *Can. J. Fish. Aquat. Sci.* **48**, 442–444.

Jonasson, S., Bryant, J. P., Chapin, F. S., III, and Anderson, M. (1986). Plant phenols and nutrients in relation to variations in climate and rodent grazing. *Am. Nat.* **128**, 394–408.

Kozlowski, T. T. (1971). "Growth and Development of Trees," Vol. 1. Academic Press, New York.

Kuropat, P. J. (1984). Foraging behavior of caribou on a calving ground in northwestern Alaska. M.S. thesis, University of Alaska, Fairbanks.

Lamb, H. F., and Edwards, M. E. (1988). The Arctic. *In* "Vegetation History" (B. Huntley and T. Webb, III, eds.), pp. 519–555. Kluwer Academic, Dordrecht.

Lorio, P. L., Jr. (1986). Growth-differentiation balance: A basis for understanding southern pine beetle–tree interactions. *For. Ecol. Manage.* **14**, 259–273.

MacLean, S. F., and Jensen, T. J. (1984). Food-plant selection by insect herbivores: The role of plant growth form. *Oikos* **47**, 211–221.

Mattson, W. J., and Haack, R. A. (1987). The role of drought in outbreaks of plant-eating insects. *Bioscience* **37**, 110–118.

Mitchell, J. F. B., Senior, C. A., and Ingram, W. J. (1989). CO_2: A missing feedback? *Nature* **341**, 132–134.

Moorby, J., and Waring, P. J. (1963). Aging in woody plants. *Ann. Bot.* **106**, 291–309.

Niemela, P., Tuomi, J., Mannila, R., and Ojala, P. (1984). The effect of previous damage on the quality of Scotch pine (*Pinus sylvestris*) foliage as food for diprionid sawflies. *Z. Agnew. Entomol.* **98**, 33–43.

Pastor, J., and Post, W. M. (1988). Response of northern forests to CO_2-induced climate change. *Nature* **334**, 55–58.

Pastor, J., Naimen, R. J., Dewey, B., and McInnes, P. (1988). Moose, microbes, and the boreal forest. *Bioscience* **38**, 770–777.

Prudhomme, T. I. (1982). The effect of defoliation history on photosynthetic rates in mountain birch. *Rep. Kevo Subarct. Res. Stn.* **18**, 5–9.

Prudhomme, T. I. (1983). Carbon allocation to antiherbivore compounds in a deciduous and evergreen arctic shrub species. *Oikos* **40**, 344–357.

Reichardt, P. B., Bryant, J. P., Clausen, T. P., and Wieland, G. D. (1984). Defense of winter-dormant Alaska paper birch against snowshoe hare. *Oecologia* **65**, 58–69.

Reichardt, P. B., Greene, T. P., and Chang, S. (1987). 3-O-malonylbetulafolientriol oxide I from *Betula nana* subsp. *exilis*. *Phytochemistry* **26**, 855–856.

Reichardt, P. B., Bryant, J. P., Mattes, R. R., Clausen, T. P., Chapin, F. S., III, and Meyer, M. (1990a). Winter defense of Alaskan balsam poplar against snowshoe hares. *J. Chem. Ecol.* **16**, 1941–1959.

Reichardt, P. B., Bryant, J. P., Anderson, B. J., Phillips, D., Chapin, F. S., III, Meyer, M., and Frisby, K. (1990b). Germacrone defends Labrador tea from browsing by snowshoe hares. *J. Chem. Ecol.* **16**, 1961–1970.

Rice, E. L., and Pancholy, S. K. (1972). Inhibition of nitrification by climax ecosystems. *Am. J. Bot.* **59**, 1033–1040.

Rice, E. L., and Pancholy, S. K. (1973). Inhibition of nitrification by climax ecosystems. II. Additional evidence and possible role of tannins. *Am. J. Bot.* **60**, 691–702.

Rice, E. L., and Pancholy, S. K. (1974). Inhibition of nitrification by climax ecosystems. III. Inhibitors other than tannins. *Am. J. Bot.* **61**, 691–702.

Rizzo, B. (1988). The sensitivity of Canada's ecosystems to climatic change. Canada Committee on Ecological Land Classification Newsletter 17, pp. 10–12. Supply and Services Canada, Ottawa.

Robbins, C. T. (1983). "Wildlife Nutrition." Academic Press, New York.

Robus, M. (1981). Foraging of muskoxen in arctic Alaska. M.S. thesis, University of Alaska, Fairbanks.

Ryala, P. (1966). On the choice of plant diet and feeding in trees of the willow grouse (*Lagopus lagopus*) and ptarmigan (*Lagopus mutus*) in winter (in Finnish and English summary). *Suom. Riista* **19**, 79–93.

Schultz, A. M. (1968). A study of an ecosystem: The arctic tundra. *In* "The Ecosystem Concept in Natural Resource Management" (G. M. Van Dyne, ed.), pp. 77–93. Academic Press, New York.

Scriber, J. M., and Slansky, F., Jr. (1981). The nutritional ecology of immature insects. *Annu. Rev. Entomol.* **26**, 183–211.

Sharpe, P. J. H., Wu, H. I., Cates, R. G., and Goeschl, J. D. (1985). Energetics of pine defense systems to bark beetle attack. *U.S. For. Serv. Gen. Tech. Rep. SO-56*, 206–223.

Silvola, J., Valijoki, J., and Aaltonen, H. (1985). Effect of draining and fertilization on soil respiration at three ameliorated peatland sites. *Acta For. Fenn.* **191**, 1–32.

Shaver, G. R., and Chapin, F. S., III. (1980). Response to fertilization by various plant growth forms in an Alaskan tundra: Nutrient accumulation and growth. *Ecology* **61**, 662–675.

Singh, T. (1988). Potential impacts of climate change on forestry. Canada Committee on Ecological Land Classification Newsletter 17, pp. 4–5. Supply and Services Canada, Ottawa.

Small, E. (1972). Photosynthetic rates in relation to nitrogen recycling as an adaptation to nutrient deficiency in peat bog plants. *Can. J. Bot.* **50**, 2227–2233.

Swift, M. J., Heal, O. W., and Anderson, J. M. (1979). "Decomposition in Terrestrial Ecosystems." Univ. of California Press, Berkeley.

Tuomi, J., Niemala, P., Haukioja, E., Siren, S., and Neuvonen, S. (1984). Nutrient stress: An explanation for plant anti-herbivore response to defoliation. *Oecologia* **61**, 208–210.

Viereck, L. A., and Little, E. J., Jr. (1972). "Alaska Trees and Shrubs." Agricultural Handbook 410. U.S. Government Printing Office, Washington, DC.

White, R. G. (1983). Foraging patterns and their multiplier effects on productivity of northern ungulates. *Oikos* **40**, 377–384.

Williams, J. B., Best, D., and Warford, C. (1980). Foraging ecology of ptarmigan at Meade River, Alaska. *Wilson Bull.* **92**, 341–351.

19

Tundra Grazing Systems and Climatic Change

R. L. Jefferies, J. Svoboda, G. Henry, M. Raillard, and R. Ruess

I. Introduction

A short growing season, relatively few degree-days above 0° C, and restricted availability of nitrogen, phosphorus, and water all combine to limit biological production in tundra ecosystems (Bliss, 1986). Yet some tundra ecosystems are relatively productive and can sustain large populations of mammalian and avian grazers such as muskoxen, caribou, lemmings, and snow geese. The impact of herbivory on these tundra grazing systems may be considerable, particularly because herbivores overlap in their use of different vegetation types. Although herbivores in many ecosystems consume only about 10% of net aboveground primary production (NAPP) (Crawley, 1983; Pimentel, 1988), lemmings and snow geese can consume between 50 and 90% (Schultz,

1968; Batzli *et al.*, 1980; Cargill and Jefferies, 1984). Thus, in some tundra grazing systems vertebrate herbivores, assisted by their gut decomposers, replace freeliving decomposers as the primary consumers and reducers of NAPP.

In the first part of this chapter, we examine and compare the characteristics of different northern grazing systems and the types of plant–animal interactions involved in maintaining these systems; we relate these features to plant production, nutrient cycling, and the forage requirements of herbivores. In the second part, we discuss the likely effects on arctic grazing systems of predicted climatic changes associated with increased amounts of radiatively active gases in the atmosphere.

We have had to be extremely selective in the material presented in this short review. For example, such topics as the paleoecology of tundra grazing systems (cf. Ritchie, 1983) and predicted sea level changes (cf. de Q. Robin, 1986) are not discussed, and the reader is referred to the citations. In addition, we have not discussed the effects of herbivory by arctic hares. Although this species is widespread, the animals do not occur at high enough densities to have a major effect on vegetation other than at the local level.

II. Tundra Grazing Systems

In the following descriptions, we emphasize the growth habits of forage species, their nutritional status, and the role of herbivores in modulating nutrient cycles (Table I).

A. Muskoxen–Graminoid–*Salix* Interactions: Pump-and-Idling Model

Largely as a result of hunting, muskoxen (*Ovibos moschatus*) in historical times were restricted to high-arctic oases and coastal plains, where plant productivity was locally high (Banfield, 1974). Because animals have been translocated and the species' range has expanded, muskoxen now occur in other areas as well (Klein, 1988). At the southern extremity of their present range, their diet is frequently dominated by willows (*Salix* spp.), but farther north graminoid species become increasingly important (Gray, 1973; Hubert, 1974; Parker and Ross, 1976; Wilkinson *et al.*, 1976; Klein, 1986). Average annual aboveground production of all *Salix* species drops from about 80 g m^{-2} in the Low Arctic to 10 g m^{-2} at 80°N (Klein, 1986).

Early in the season when soil temperatures are low, aboveground growth of willows and sedges is largely driven by a nutrient "pump," which moves nitrogen and phosphorus from storage organs to new leaves, shoots, and twigs (Chapin and Shaver, 1989). In summer muskoxen feed extensively in sedge meadows, where plant production exceeds the amount of forage consumed (Klein and Bay, 1990). The muskoxen feed about 50% of the time, and their daily food consumption is approximately 1.7–2.2% of their weight (Hubert, 1977). During summer, they supplement their dietary sodium by using mineral licks (Thing *et al.*, 1987). The foraging activities of muskoxen appear to be moderately coupled to sustaining plant production (e.g., removing plant litter, adding of feces and urine, maintaining juvenile plant growth) (Henry *et al.*, 1986; Henry and Svoboda, 1989).

The muskox reproductive cycle is strongly linked to forage availability (White and Fancy, 1986). In most areas, offspring are born several weeks before the period of plant growth. Milk production by cows is sustained during early lactation by body reserves (White and Fancy, 1986), and cows may lose weight during the first six weeks of lactation (Hubert, 1977). Ovulation and conception depend heavily on body condition before the rut, which occurs at the end of summer when females are regaining weight lost during gestation and lactation.

Asynchronous changes in metabolic rate and growth occur between both sexes from the end of summer through early winter; these changes are associated with calving, lactation, rutting activity, recovery from the rut, and the decline in plant productivity. Although males maintain high metabolic rates until after the rut, the rate in both sexes falls to a seasonal low from January to March. During this "idling" period for both plants and animals, muskoxen feed only 25% of the time (White and Fancy, 1986; Raillard and Svoboda, 1990). The animals feed mainly on dormant plant tissues (the basal portion of many leaves of "hayed-off" grasses and sedges remains green throughout the winter), but they also browse on leafless willow twigs. In northeast Greenland, the quality of the winter diet of muskoxen as measured by the protein to fiber ratio is about one-fourth that of the summer diet (Thing *et al.*, 1987). (The morphological adaptations allowing muskoxen to use hayed-off grasses and sedges more efficiently than caribou or reindeer have been discussed by Klein [1986]). In females, a high metabolic rate is reestablished just before calving.

Muskoxen are more poorly adapted than caribou for long distance travel through deep snow, and in winter they use valley slopes where the snow is shallow (Raillard and Svoboda, 1990). The unavailability of low-quality forage (hayed-off graminoids) in years with deep snow results in increased mortality of individuals and females failing to reproduce. Thus density-independent and -dependent processes interact to determine population sizes. At present, populations on Banks and Victoria islands and along the mainland coast of arctic Canada, especially near Queen Maud Gulf, are increasing by as much as 16% per year (Case *et al.*, 1989; McLean *et al.*, 1989). The reasons for the increase are not clear, although they may be related to favorable weather and the relaxation of hunting pressure since the early part of the twentieth century. If these rates of population increase continue, muskoxen are likely to have a significant impact on vegetation.

B. Caribou–Plant Interactions: Selective-Exploitation-and-Idling Model

The various subspecies of *Rangifer tarandus* occupy habitats from the taiga to the high-arctic islands in North America and Eurasia. Latitudinal differences

Table I Nutritional Characteristics of Tundra Grazing Systems

	% of NAPP[a] consumed in grazing system	Graminoid and lichen forage species	Regrowth capacity of primary forage species after defoliation	Nutrient feedback responses between herbivore and forage species[b]	Herbivore use of shrub/herb vegetation
Frequent recurrent grazing					
Herbivore-driven nutrient-flow model					
Lesser snow geese	80–90 (Salt marshes)	*Puccinellia phryganodes* *Carex subspathacea*	Capacity for regrowth high within growing season; continuous leaf production	+ + +	+/+
Oscillatory model					
Microtines (peak numbers)	≤50[c] (Sedge meadows)	*Arctophila fulva* *Dupontia fisheri* *Carex aquatilis/stans* *Eriophorum angustifolium* *Eriophorum vaginatum*	Capacity for regrowth within season increasingly limited from *A. fulva* → *E. vaginatum* unless there is high nutrient input; growth response often delayed to next season	+ +	+ –/+ + +
Intermittent grazing					
Pump and Idle model					
Muskoxen	3–50[c] (Sedge meadow)	*Carex aquatilis/stans* *Eriophorum angustifolium* *Eriophorum vaginatum*		+	+ +/+ +

394

Episodic grazing
Selective exploitation and idling model

| Caribou | 3–50[c] (Sedge meadow) | Carex aquatilis/stans Eriophorum angustifolium Eriophorum vaginatum | | + – | + + +/+ + |
| | | Lichen species | Very limited capacity for regrowth; recovery up to 15 years | – – | |

[a]NAPP = Net aboveground primary production.
[b]Feedbacks leading to increased NAPP.
[c]Modified from Bliss (1986).

395

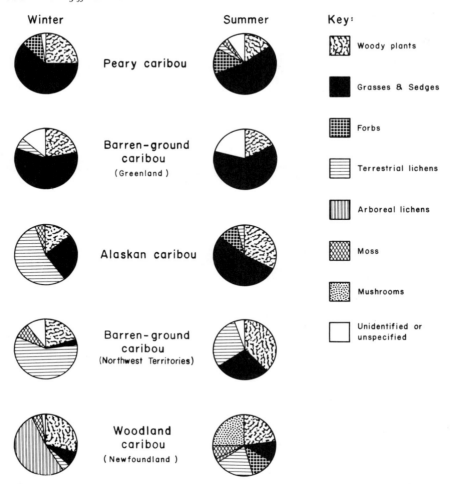

Figure 1   Latitudinal differences in caribou diets from the boreal forests to the High Arctic. [Modified from Leader-Williams (1988) with permission from Cambridge University Press.]

in caribou diets during summer and winter indicate that the animals forage selectively (Fig. 1). Populations of Peary caribou (*R. t. pearyi*) and of the Svalbard reindeer (*R. t. platyrhynchus*) and the Greenland reindeer (*R. t. groenlandicus*) from the Canadian and Soviet High Arctic and Greenland, respectively, feed on grasses and sedges, woody plants, forbs, and lichens (Michurin and Vakhtina, 1968; Parker and Ross, 1976; Miller *et al.*, 1982; Thomas and Edmunds, 1983). Seasonal migration routes are short, and the animals can survive winter without lichens as a dominant part of their diet. At lower latitudes, caribou feeding ecology is characterized by dependence on lichens. Barren-ground (*R. t. groenlandicus*) and Alaskan (*R. t. granti*) caribou migrate distances of up to 1000 km from the tundra to the taiga, where they forage for terrestrial lichens (Kelsall, 1968; Skoog, 1968; Skuncke, 1968). Values of in vitro digestibilities of plants in winter in the rumen fluids of Peary caribou are high for lichens (*Thamnolia vermicularis*, 62%; *Cetraria nivalis, C. cucullata,* and *C. islandica,* 69%) but low for some vascular plants (*Dryas integrifolia,* 22%; *Saxifraga oppositifolia,* 11%) (Thomas and Kroeger, 1980).

In adapting to this lichen–graminoid–woody plant food niche, caribou and reindeer have evolved efficient means for digging through snow (Leader-Williams, 1988). The animals have very broad hooves and are able to dig as deep as 60 cm as long as the snow is not hard. Caribou pawing partially separates vascular plant stems from moss. Some of the moss consumption in winter and early spring appears to be a direct consequence of this behavior (White, 1983; Klein, 1986), although Peary caribou and Svalbard reindeer apparently do select mosses and can digest mosses more efficiently than their southern counterparts (Thomas and Kroeger, 1980; Trudell *et al.*, 1980; Mathiesen *et al.*, 1984).

Caribou have narrow mouths and are selective feeders, unlike muskoxen, which are bulk feeders. Rates of forage intake in summer vary in relation to the growth habit of plants and their abundance (Trudell and White, 1981; Kuropat, 1984). Trudell and White found the mean intake rates of vegetation to be as follows: forbs, 13 g min^{-1}; willows and *Eriophorum vaginatum*, 6 g min^{-1}; and *Carex* spp., *Betula nana*, and lichens, 3–4 g min^{-1}. Assimilated food is deposited as fat (up to 30% of body weight) or is used for milk production (Ringberg, 1979; Leader-Williams, 1988). Caribou are less able than muskoxen to process food high in fiber, such as dried sedges and grasses. Selective feeding that produces only small changes in the quality or amount of food ingested may have "a greatly enhanced or multiplier effect on animal production" (White, 1983). Selective grazing by caribou in summer produces increased digestibility of the diet and an increase in daily weight gain compared with animals that cannot feed selectively. As White (1983) points out, animal performance is strongly influenced by temporal and spatial grazing patterns, selection of plants high in nutrients, and avoidance of plants rich in secondary compounds.

The recovery of tundra vegetation after exploitation by large numbers of animals is relatively slow (Fig. 2). For example, it took 15–20 years for complete restoration of a badly depleted winter lichen pasture (Chernov, 1985). Trampling by reindeer at high densities may severely damage lichens (Pegau, 1970). Twenty-two years following a crash in the reindeer population on an island in the Bering Sea, lichens had recovered to only 10% of the standing crop of lichens on an adjacent island with no history of grazing (Klein, 1987). Our own observations indicate that in summer, sedge meadows are badly damaged after some 40,000 caribou have foraged in them. Precisely when animals return to a particular site is not known; the return time may be on the order of years, as for willow ptarmigan (Andreev, 1988), which would allow the vegetation to recover.

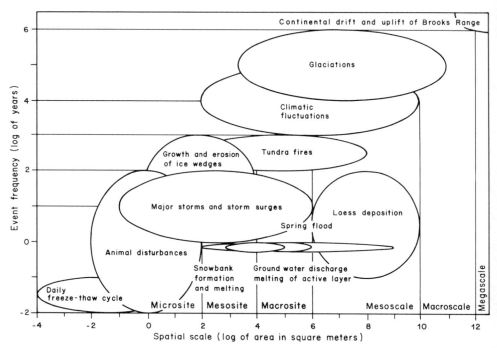

Figure 2   Spatial and temporal scales of natural disturbances on the North Slope, Alaska. [Modified from Walker and Walker (1991) with permission from Blackwell Scientific Publications.]

As in muskoxen, the metabolic rate of caribou falls in winter ("idling"), associated with a reduced food intake following the rut (White, 1983; Klein, 1986; Leader-Williams, 1988). Terrestrial or arboreal lichens on which caribou feed heavily in winter are relatively poor in nitrogen but rich in energy sources such as polysaccharides and lipids (Chapin and Shaver, 1988, 1989). The animals can sustain a negative nitrogen balance during winter by drawing on body protein reserves, although selection of green vascular plants beneath the snow and recycling of ammonia through the saliva offsets nitrogen loss. For caribou, as for muskoxen, carrying capacity may be largely determined by the availability of winter forage and by conditions on the winter range (Palmer, 1934; Klein, 1968; White, 1983). The inaccessibility of forage because of deep snow precedes massive animal die-offs, especially where emigration is not possible (Scheffer, 1951; Vibe, 1967; Klein, 1968; Reimers, 1977). Critically poor nutrition in severe winters also leads to poor fetal development and a decrease in birth weight in spring (White, 1983).

Caribou therefore change their migration routes to exploit seasonal differences in the availability of selected forage plants. The boom period of summer feeding is followed by reduced metabolic rates in winter and conservation of energy and nutrient resources (idling). Because both the frequency and intensity of grazing at a given site are episodic, the plant–herbivore interaction is essentially decoupled from the processes that sustain plant production; nutrient feedback loops between the animals and the vegetation are weak or absent.

C. Arctic Rodent–Plant Interactions: Oscillatory Model

The three groups of rodents in the Arctic are lemmings (*Dicrostonyx, Lemmus*), voles (*Microtus*), and ground squirrels (*Spermophilus*). Microtine rodents show cyclic fluctuations in population density, particularly in northern latitudes (Batzli, 1981; Laine and Henttonen, 1983), whereas populations of ground squirrels are relatively stable from year to year (Batzli, 1981). Theoretical and empirical evidence indicates the strong effects of nutritional factors on spatial and temporal variation in the sizes of microtine populations (Batzli, 1983; Laine and Henttonen, 1983). In contrast, stability of ground squirrel populations has been attributed to the territorial defense of burrows for breeding and hibernating (Carl, 1971; Green, 1977).

Among lemmings and voles, the proportional contribution of some plant species to the diet is greater than their proportional contribution to available plant biomass (Batzli and Jung, 1980); the dietary preferences of *Lemmus*, *Microtus*, and *Dicrostonyx* also differ. For example, in northern Alaska, the brown lemming (*L. sibiricus*) forages on monocotyledons and mosses, the collared lemming (*D. groenlandicus*) takes the leaves of dicotyledonous shrubs, and the tundra vole (*M. oeconomus*) feeds on both monocotyledons and dicotyledons (Batzli and Jung, 1980). Arctic rodents in other regions show similar preferences (Kalela *et al.*, 1961; Koshkina, 1961; Tast, 1974; Fuller *et al.*, 1975). Microtine rodents are active throughout the year, although in the winter food availability decreases. During winter Norwegian and brown lemmings eat the green stem bases of many monocotyledonous plants (Koshkina, 1961; Batzli and Pitelka, 1983). Collared lemmings eat a higher proportion of buds, twigs, and bark of shrubs (Batzli, 1983), and tundra voles in northern Finland feed heavily on rhizomes of *Eriophorum angustifolium* (Tast, 1972). Ground squirrels prefer herbaceous dicotyledons, particularly legumes and plants with a high water content (Batzli and Sobaski, 1980; McLean, 1981); in winter, the animals hibernate.

As Batzli (1983) points out, not only do diets of arctic microtine rodents vary with space and time, but the quality of the diet also varies. The digestibility of dicotyledons, monocotyledons, and mosses, based on the caloric value of each, is approximately 70%, 50%, and 25%, respectively, depending on the rodents' digestion efficiencies (Batzli and Cole, 1979; Batzli, 1983), although for individual species, digestion efficiencies maybe much lower (e.g., 35% for monocotyledons in brown lemmings). Secondary compounds generally increase from mosses to monocotyledons to forbs to deciduous shrubs and, finally, to evergreen shrubs (Jung *et al.*, 1979).

Until recently, there has been no satisfactory evidence that the average density of arctic rodents is directly related to the quality of available forage.

Schultz (1968) reported an increase in reproductive output of lemmings after tundra vegetation was fertilized, but population numbers did not increase the following summer because predators responded to "the pantry effect." Batzli (1983) and Batzli and Henttonen (1990), however, have provided strong circumstantial evidence that populations reach the highest densities in habitats with the highest-quality food, and it is only in such habitats that dramatic oscillations of high and low densities occur. The relative abundance of tundra and singing (*Microtus abreviatus*) voles in different habitats correlates strongly with the absolute abundance of a few forage plants (*Eriophorum angustifolium* and *Carex* spp. for tundra voles, and *Vaccinium uliginosum* and total forbs for singing voles; Batzli, pers. comm.). Diet preference may also be based on negative qualities of plant species—for example, extracts of *Carex* sp. reduce fat and enlarge the kidneys in weaning lemmings (Jung and Batzli, 1981)— rather than positive qualities of selected forage species. A nine-year study in Finland showed some synchrony between high population numbers and peak flowering years (Tast and Kalela, 1971; see also Laine and Henttonen, 1983). Synchronous changes in density and body size of all microtine populations were linked to cycles of plant growth and reproduction.

Two explanations for this phenomenon have been proposed; the second is an extension of the first. Tikhomirov (1959) proposed that the underlying cause is a kind of predator–prey interaction. At high densities, grazers severely restrict the growth of vegetation, thereby causing resource limitation, and populations of animals decline rapidly. Tast and Kalela (1971) extended the hypothesis by suggesting that favorable climatic conditions initially create a pulse of generative and vegetative growth that is exploited by microtines. Low reproductive effort among plants follows because they require time to replenish energy and nutrient reserves, and this low reproductive effort is associated with low numbers of microtines. Sudden declines may be associated with increased concentrations of inducible plant secondary compounds in plants as well as with low forage availability (Haukioja, 1981; Batzli, 1983). Over time the vegetation recovers, thereby providing resources for increasing numbers of microtines. In a detailed study, Oksanen and Ericson (1987) showed that the apparent flowering cycles in tundra and taiga populations of herbaceous plants were mainly a consequence of fluctuations in microtine populations, rather than the result of favorable weather patterns.

In conclusion, general levels of microtine herbivore populations may be set by food availability. Small herbivores with high reproductive potential are able to take advantage of a single favorable season so that their numbers increase temporarily beyond the system's capacity to support sudden population surges, with catastrophic consequences.

D. Lesser Snow Geese–Coastal Vegetation Interactions: Herbivore-Driven Nutrient-Flow Model

Large numbers of geese breed in the coastal areas of the Arctic and Subarctic in early summer, then migrate southward in early fall. One common, strongly colonial species is the lesser snow goose; current estimates indicate that more than two million of these birds inhabit the eastern Canadian Arctic alone (Kerbes, 1975; Boyd *et al.*, 1982). In these colonies the densities of nests may exceed 2,000 km^{-2} (Kerbes *et al.*, 1990). Goslings grow from 80 g at hatching to 1500 g in less than seven weeks, while the adults are regaining the loss of body weight ($\leq 40\%$) associated with egg laying and incubation (Cooke *et al.*, 1982).

In some habitats, such rapid gains in body weight are achieved because the geese modulate nitrogen flow (and that of other nutrients) in a system where plant production is nitrogen-limited (Jefferies, 1988a,b). The geese produce droppings approximately once every four minutes while feeding, effectively accelerating the nitrogen cycle by providing a readily available nitrogen source for plant growth (Bazely and Jefferies, 1985). Grazing during the growing season increases net aboveground primary production of coastal salt marshes by 40–100%, depending on the year (Jefferies, 1988a,b). Forage quality, as measured by nitrogen content, is also higher than that of ungrazed swards because the intense cropping gives rise to a higher percentage of young, nutrient-rich leaves (Cargill and Jefferies, 1984). The intense grazing of these lawns results in the consumption of about 90% of NAPP and produces little accumulation of plant litter. Cyanobacteria colonize the surface of bare sediments among the graminoid shoots and fix atmospheric nitrogen (Bazely and Jefferies, 1989a). This process replaces much of the nitrogen that is incorporated into the geese and exported from the site when they fly south.

These feedback responses between the herbivore and the forage plants sustain the production and nutritional quality of the vegetation (Jefferies, 1988a,b). The intense grazing is essential to maintain the grazing lawn. Erection of exclosures quickly alters plant species composition and decreases the frequency of forage species (Bazely and Jefferies, 1986).

Snow goose populations have traditionally been thought to be limited by density-dependent factors operating on the wintering grounds and migration routes. In both North America and in Europe, however, goose populations are rapidly increasing, probably as a result of changing agricultural and conservation practices in both wintering and spring staging areas.

In early spring, immediately after the snow melts but before plant growth begins, snow geese grub for roots and rhizomes of graminoid plants, creating patches where little vegetation remains (Jefferies, 1988a,b). Clonal regrowth starts from the periphery of small patches and proceeds inward. Where grubbing is extensive because so many birds are present, however, patch size increases each season, and the vegetation fails to recover. The soil organic layer dries out and erodes, exposing the underlying glacial gravels and marine clays. Large tracts of salt marsh have thus disappeared and become mud flats (Kerbes *et al.*, 1990). At more inland sites peat barrens have developed, covered only sparsely with bryophytes and grasses. The increasing numbers of geese and a succession of late spring snowmelts continue to have a profound effect on the availability of summer grazing pasture.

As coastal grazing lawns disappear, birds are forced to feed in inland sedge meadows after the hatching period. The tight nutrient coupling between the herbivore and its forage species begins to break down. Birds forage in small groups or family units; they are no longer strictly colonial and disperse within the region. Goslings web-tagged at hatching have been recaptured 60 km from the nest site six weeks later (Kerbes *et al.*, 1990). (Neither the adults, which are molting, nor the goslings are able to fly during this phase of their life cycle.) Graminoid plants of the sedge meadow, especially those with a cespitose growth habit, fail to grow rapidly in response to the overall effects of grazing (Frey, 1987). Recently, the foraging pattern of the geese has come to resemble that of caribou in summer: exploitation of selected species over a large area.

E. Invertebrate Herbivory

Insects are the dominant group of invertebrate herbivores at tundra sites, although in general the density, standing crop, and number of invertebrate herbivorous species is low (Haukioja, 1981). The severity of the climate and low vegetation diversity probably account for the low number of species. The systematics and ecology of arctic arthropods have been reviewed thoroughly by Danke (1981).

Lepidopterans and herbivorous hymenopterans, which are the most important herbivorous insect orders, are widely distributed throughout the Arctic, although large annual fluctuations in numbers occur, particularly at higher latitudes (Haukioja, 1981). Nitrogen appears to be the most limiting nutrient in the diet of these invertebrates. For example, in one study (Haukioja and Niemalä, 1974) seasonal decreases in nitrogen and in the nitrogen:phenolics ratio in birch leaves made it increasingly difficult for two invertebrates to obtain this element. (Phenolics inhibit the protease trypsin.) Tundra invertebrate herbivores are highly selective in their choice of food plants. When plant life-forms were ranked according to the feeding responses of four lepidopteran larvae, deciduous shrubs were preferred by all four species (MacLean & Jensen, 1985). Graminoids and evergreens ranked lowest. Compared with evergreen shrubs, deciduous shrubs, especially willow, support a large number of insect species. The results are consistent with the hypothesis of Bryant and Kuropat (1980) that weakly defended (palatable) plants are found where habitat quality and growth potential are high. Heavily protected, low-preference species, such as evergreens from infertile sites, are slow-growing, and much of the nutrient capital of the plant is allocated to leaves, which turn over slowly (see Chap. 18). Potentially, leaves are particularly sensitive to herbivory, so carbon compounds are heavily committed to defense (in secondary compounds and leaf structure) (Chapin, 1980).

Within soils, nematodes are abundant; numbers range from $0.1 \times 10^6 \text{ m}^{-2}$ to $10 \times 10^6 \text{ m}^{-2}$ at different tundra sites (MacLean, 1981). Within-site variation frequently exceeds between-site variation, although numbers are low in water-logged habitats. Few of the species appear to be plant feeding; the majority feed on bacteria and fungi. Since these are generally small organisms, nematode biomass is frequently less than 1 g dry wt m^{-2} (MacLean, 1981).

III. Comparisons among Tundra Grazing Systems

The foraging behavior of vertebrate herbivores can be related to the growth habits of the forage species and their capacity for regrowth following defoliation (see Table I). In *Puccinellia phryganodes* and *Carex subspathacea*, which are grazed by geese, new leaves are produced continuously, and old leaves die throughout the growing season (Kotanen and Jefferies, 1987; Bazely and Jefferies, 1989b). Given adequate nutrient input (fecal nitrogen), the plants are able to grow rapidly after defoliation. *Dupontia fisheri* (Mattheis *et al.*, 1976; Frey, 1987) and *Arctophila fulva* also produce leaves throughout the growing season. This growth habit contrasts with those of *Carex aquatilis* or *C. stans* and the *Eriophorum* species (Table I), in which tillers and leaf primordia are formed the previous season and develop rapidly as a new season begins (Henry, 1987). Although juvenile shoots may develop after defoliation of adult shoots, repeated shoot clipping restricts further aboveground growth within the season, unless readily available nutrients are added (Chapin and Shaver, 1985; Archer and Tieszen, 1986). Animals such as caribou, and to a lesser extent muskoxen, are forced to forage over a wide area to find adequate food. Given the animals' dietary preferences, the growth system of the plant shoot and the supply of available nutrients dictate the foraging patterns of the herbivores. The characteristically slow growth of lichens presents an extreme example of the long plant recovery time needed before further foraging bouts are possible. Yet the shortage of available forage within the season at one site can be offset by the abundance of sedge and lichen communities at similar sites.

Despite the variety of growth responses in forage plants, all these plant–animal interactions share important characteristics. Vertebrate herbivores consume a range of forage species in different habitats according to the plants' availability and nutritional quality, including the presence or absence of secondary compounds. The types of vegetation most heavily used as forage are communities dominated by graminoid and willow species, which occur over many latitudinal and climatic zones. In spring, however, the rapidly growing reproductive shoots of forbs rich in soluble carbohydrates, nitrogen, and phosphorus (e.g., *Pedicularis* spp.) also become a food source, and the nutrient input to the animal may be out of proportion to the amount of biomass eaten. One observes a scaling in both numbers and body size of some animal species (e.g., caribou, reindeer, lesser snow geese) at increasingly higher latitudes associated with a decrease in plant production and the availability of suitable habitat (Miller *et al.*, 1976). Within the Arctic, the amplitude of changes in microtine numbers is also tempered at higher latitudes. The shorter growing season allows fewer reproductive cycles and makes it more difficult for populations to proliferate within a season in response to favorable food production (Klein, 1986).

It is not fully established whether density-dependent factors, such as the availability of enough good-quality forage in winter, restrict population sizes of the different herbivores that remain in the Arctic. Strong circumstantial evidence does indicate that mortality increases in late winter or early spring. Adverse and unpredictable weather events may affect herbivore population numbers and plant production out of proportion to their frequency. Deep and prolonged snow cover appears to control populations of muskoxen and caribou (Parker, 1978). Even where the snow is relatively shallow, groundfast ice (Miller *et al.*, 1982) or a hard crust of snow (Raillard and Svoboda, 1990) may prevent foraging for most plants.

Except for sites where herbivores accelerate the decomposition cycle, all arctic grazing systems are "low-input" systems. Nutrient input and turnover are slow, and nutrients are efficiently recycled within plants (Chap. 13, 15, 16). Spatial and temporal shifts in resource use allow herbivores to sustain their own production with a minimum of nutritional input. By speeding nutrient turnover when they occur at high densities, lesser snow geese (Jefferies, 1988a) and lemmings (Schultz, 1968) can change a grazing system from low-input to high-input. The beneficial effects of biotic disturbances—increasing nutrient fluxes and the growth of forage plants—highlight how strongly these systems depend on positive feedback processes (see Table I). For this feedback to work, the herbivores' reproductive cycles must be in phase with the growth responses of forage plants to increased nutrient availability. Otherwise, population numbers are likely to drop rapidly, as among lemmings (Schultz, 1968).

IV. Tundra Grazing Systems and Climatic Change

The predicted rising of temperature poleward is likely to have profound effects on polar and boreal ecosystems. Each species is likely to respond differently. Thus there is little reason that present assemblages of species should remain intact and simply migrate northward as the climate shifts (Davis, 1986). Reorganization of communities is inevitable (Chap. 16). Table II (see also Chap. 2 and 3) summarizes predicted climatic conditions together with likely responses of vegetation and herbivores.

One prediction is that the growing season will lengthen from about 60–100 days at present to 100–130 days, depending on locality. Associated with this change is the likelihood that snow cover may persist longer in the spring than it does now because of increased snowfall and cloud cover, reducing insolation. These weather patterns may be expected in arctic coastal lowlands, where stratus clouds are common in spring. On the other hand, the snow-free season is likely to extend further into autumn, when both day length and photosynthetically active radiation (PAR) are significantly reduced from midsummer.

Table II Effects of Predicted Climatic Change on Plant–Animal Interactions in Tundra Grazing Systems

Climate	Vegetation	Herbivores
Temperature rise: Winter 5–7° C Summer 3–4° C	Northward movement of biogeographical zones, including tree line	Caribou: change in migration routes; difficulty in obtaining food in winter because of snow and increased ground icing; delayed arrival at calving grounds; losses during migration because of higher rivers; higher food availability in at least some habitats; possibly further range restriction for Peary caribou.
Cloud cover: higher. Insolation: lower.	Vegetational instability: successional changes	Rodents: likely population expansion, movement to new habitats; an increase in wet tundra likely to lead to population expansion of brown lemmings and tundra voles; more dry tundra likely to induce expansion of collared lemmings, singing voles, and red-backed voles, which show less propensity for cycling than the other two; predator populations therefore less likely to fluctuate in dry conditions.
Precipitation: little change or increase mainly in winter–spring only.	Dry summers: reduction of wet and mesic habitats, expansion of dry habitats.	
Snow cover: deeper, especially snowbanks.	Deepening of active layer.	
Snowmelt: delayed.	Downcutting of drainage channels, erosion of sediments.	
Albedo: high before snowmelt, loss of radiant energy.	Increased decomposition. Increased CO_2 liberation.	
Ice crusts: increase.	Decreased CH_4 production.	
Freeze–thaw: more cycles.	Release of nutrients via mineralization.	
Growing season: late start but extending into fall.	Positive response in the Low Arctic; little change in the High Arctic.	Muskoxen: adverse effects of deep snow and ground ice; otherwise, least likely to be affected of all herbivores.
Degree-days: fewer in summer, shift from the seasonal solar peak to the postsolstice, declining-radiation regime.	Increase in NAPP by ~100% in the short term, but possibly reduction in NAPP in wet habitats (drought effect) and areas where late snowfall persists.	Lesser snow geese: increased foraging in freshwater habitats associated with destruction of coastal salt marshes at sites immediately south of the spring snow line at the time of migration.
		All species: instability in population numbers.

The number of freeze–thaw cycles is also predicted to increase. In subarctic regions at present, the number of such cycles per year is between 15 and 25, whereas farther north the number decreases. With runoff occurring more frequently because of the increase in freeze–thaw cycles, downgrading of drainage systems is likely, particularly since the soil active layer may deepen because frozen sediments thaw. Oxidation rates of peats will also increase if surface peat layers dry out in the summer.

Higher temperatures, a longer growing season, increased decomposition rates, deepening of the active layer, and lower water tables may double present values of net primary production (cf. Chap. 7). Whether such an increase can be sustained beyond two or three decades depends on the capacity of the system to maintain the elevated levels of nutrients that are likely to result. Jonasson et al. (1986) reported that the carbon–nutrient balance of plants from high latitudes is affected by warm summers. Photosynthesis and carbohydrate production increase so that nutrient uptake, rather than carbon fixation, limits plant growth (Chap. 15 and 16). Excess carbohydrates not required for growth are converted into secondary metabolites (see Chap. 18). In their study, nutrient concentrations decreased (because of growth dilution), and phenols increased in four out of five species examined. If such shifts occur in response to climatic change, they are likely to reduce forge quality.

In riparian habitats, subarctic tall willows, such as *Salix pulchra, S. latifolia*, and *S. caprea*, will replace dwarf willows over much of the continental tundra. A rise in temperature associated with increased moisture may result in the displacement of *Carex stans–Eriophorum vaginatum* meadow communities by *Carex aquatilis–Eriophorum angustifolium* communities.

Increased snow cover is likely to alter the vegetation and affect the survival and reproductive success of animals. Plant communities beneath deep, persistent snow beds lasting late into the growing season are generally devoid of lichens; hence, any increase in snow depth followed by late melting may decrease lichen abundance (Chapin et al., 1990). Vegetation beneath deep snow is less likely to be found and eaten by caribou than thinly covered vegetation. The accumulation of snow in different areas may restrict animal movement, thus raising the energetic costs of feeding and locomotion. The presence of deep snow on the caribou calving grounds in spring will cause high calf mortality, not only because energetic costs are higher, but also because the reproductive shoots of sedges and herbs (which usually develop rapidly after snowmelt; Chapin et al., 1990) will fail to develop at this time. Such herbaceous shoots, which are high in nutritive value, are an important source of forage for calves. In contrast, subnivean herbivores such as lemmings and voles may gain access to forage as snow cover increases.

Continental populations of caribou and muskoxen will probably accommodate to these and other changes in the vegetation. Inevitably, shifts in the floristic composition of their diets will be coupled with longer periods of the year when the animals' metabolic rates are high. Migration routes and time spent in different areas are also likely to change, especially because some areas of tundra may become forested, and sea ice may not form between islands. A major factor that may limit nutrient and energy intake in warmer, drier summers is increased insect harassment (White *et al.*, 1975; Fancy, 1983; White and Fancy, 1986). This last point reminds us that we rarely know which density-dependent processes are responsible for regulating population numbers of arctic herbivores.

During the initial period of climatic change, population numbers of all herbivores will probably fluctuate more in association with the increasing instability of grazing systems. In the extreme north, changes in vegetational composition and primary production may initially be less dramatic, so that populations of muskoxen and caribou could accommodate shifts in environmental conditions. Triggered by the icing of vegetation or the failure of sea ice to form (Kischinskii, 1971, 1984), however, population crashes could occur in these regions. It is not so much average climatic conditions that affect population numbers, but rather the extremes prevailing in any one year. Given the biological inertia within ecological systems in response to environmental change, the adverse effects of weather patterns on any one population could be substantial.

Indirect effects of climatic change on tundra grazing systems are likely to be far-reaching and unpredictable (see Table II). For example, the threshold effects of grubbing by geese on the integrity of coastal ecosystems after late snowmelt were unforeseen. Similar phenomena may be expected in the future—hence the need for long-term field studies. If spring snowmelt continues to be late, as some models predict, salt marsh vegetation south of the snow line will be destroyed by grubbing, and the geese will move in summer to the coastal freshwater marshes and sedge meadows in ever-increasing numbers. Increased mobility of birds during the posthatching period, the breakdown of tightly coupled nutrient feedback processes, and the establishment of summer grazing patterns similar to those of caribou are likely. Also likely are major changes in the distributions and population sizes of other plants and animals linked to these marine trophic food chains (e.g., algae, shorebirds, seals, polar bears).

In conclusion, strong circumstantial evidence indicates that arctic herbivores and their forage species have shown considerable resilience in response to climatic and other changes since the last glaciation. With future climatic change likely to be an order of magnitude faster than before, however, we predict marked oscillations in herbivore numbers and the exploitation of different forage species. Nevertheless, although existing grazing systems are likely to change, they are unlikely to lose their character entirely or to disappear.

V. Summary

Tundra ecosystems sustain large populations of mammalian and avian herbivores. The plant–animal interactions characterizing the different grazing systems can be related to the growth habits of the forage species, their response to nutrients, and the herbivore's own role in modulating nutrient cycles. In ecosystems where graminoid forage species produce leaves continuously throughout the growing season, plant regrowth is rapid in response to defoliation and input of fecal nutrients. Feedback processes between the herbivore and the vegetation are well developed. In contrast, in systems based on forage species that do not produce leaves continuously, the growth response to defoliation and addition of nutrients is either absent or deferred, and feedback processes are weak. The shoot growth patterns of these selected plants dictate the foraging patterns of herbivores such as muskoxen, caribou, microtine voles, and snow geese.

Modifications of these grazing systems in response to climatic warming are likely. Reorganization of the species composition of plant communities appears inevitable, and any increase in the depth of snow cover is likely to lead to increased mortality of caribou and muskoxen. At present, a threshold effect linked to snow cover and high densities of snow geese has destroyed some coastal areas and changed the foraging pattern of the geese. Such effects, which are not easily predictable, may be expected to become more prevalent as the climate changes.

Acknowledgments

We wish to thank George Batzli, Terry Chapin, Ellen Chu, and David Klein, who made valuable suggestions for improving of earlier drafts of this manuscript. Donald (Skip) Walker generously allowed us to use an unpublished diagram. Catherine Siu kindly typed the manuscript and dealt efficiently with the inevitable changes.

References

Andreev, A. (1988). The ten-year cycle of the willow grouse of Lower Kolyma. *Oecologia* **76**, 261–267.

Archer, S. R., and Tieszen, L. L. (1986). Plant response to defoliation: Hierarchical considerations. *In* "Grazing Research at Northern Latitudes" (O. Gudmundsson, ed.), pp. 45–59. Plenum, New York.

Banfield, A. W. T. (1974). "The Mammals of Canada." National Museums of Canada, Ottawa.

Batzli, G. O. (1981). Populations and energetics of small mammals in the tundra ecosystem. *In* "Tundra Ecosystems: A Comparative Analysis" (L. C. Bliss, O. W. Heal, and J. J. Moore, eds.), pp. 377–396. Cambridge Univ. Press, Cambridge.

Batzli, G. O. (1983). Responses of arctic rodent populations to nutritional factors. *Oikos* **40**, 396–406.

Batzli, G. O., and Cole, F. C. (1979). Nutritional ecology of microtine rodents: Digestibility of forage. *J. Mammal.* **60**, 740–750.

Batzli, G. O., and Henttonen, H. (1990). Demography and resource use by microtine rodents near Toolik Lake, Alaska, U.S.A. *Arct. Alp. Res.* **22**, 51–64.

Batzli, G. O., and Jung, H. G. (1980). Nutritional ecology of microtine rodents: Resource utilization near Atkasook, Alaska. *Arct. Alp. Res.* **12**, 483–499.

Batzli, G. O., and Pitelka, F. A. (1983). Nutritional ecology of microtine rodents: Food habits of lemmings near Barrow, Alaska. *J. Mammal.* **64**, 648–655.

Batzli, G. O., and Sobaski, S. T. (1980). Distribution, abundance, and foraging patterns of ground squirrels near Atkasook, Alaska. *Arct. Alp. Res.* **12**, 501–510.

Batzli, G. O., White, R. G., MacLean, S. F., Jr., Pitelka, F. A., and Collier, B. D. (1980). The herbivore-based trophic system. *In* "An Arctic Ecosystem: The coastal tundra at Barrow, Alaska" (J. Brown, P. C. Miller, L. L Tieszen, and F. L. Bunnell, eds.), pp. 378–381. Dowden, Hutchinson & Ross, Stroudsburg, Pennsylvania.

Bazely, D. R., and Jefferies, R. L. (1985). Goose faeces: A source of nitrogen for plant growth in a grazed salt marsh. *J. Appl. Ecol.* **22**, 693–703.

Bazely, D. R., and Jefferies, R. L. (1986). Changes in the composition and standing crop of salt marsh communities in response to removal of a grazer. *J. Ecol.* **74**, 693–706.

Bazely, D. R., and Jefferies, R. L. (1989a). Lesser snow geese and the nitrogen economy of a grazed salt marsh. *J. Ecol.* **77**, 24–34.

Bazely, D. R., and Jefferies, R. L. (1989b). Leaf and shoot demography of an arctic stoloniferous grass, *Puccinellia phryganodes*, in response to grazing. *J. Ecol.* **77**, 811–822.

Bliss, L. C. (1986). Arctic ecosystems: Their structure, function, and herbivore carrying capacity. *In* "Grazing Research at Northern Latitudes" (O. Gudmundsson, ed.), pp. 5–26. Plenum, New York.

Bryant, J. P., and Kuropat, P. J. (1980). Selection of winter forage by subarctic browsing vertebrates: The role of plant chemistry. *Annu. Rev. Ecol. Syst.* **11**, 261–285.

Boyd, H., Smith, G. E. J., and Cooch, F. G. (1982). The lesser snow geese of the eastern Canadian Arctic. *Can. Wild. Serv. Occas. Pap.* **46**.

Cargill, S. M., and Jefferies, R. L. (1984). The effects of grazing by lesser snow geese on the vegetation of a sub-arctic salt marsh. *J. Appl. Ecol.* **21**, 669–686.

Carl, E. A. (1971). Population control in arctic ground squirrels. *Ecology* **52**, 395–413.

Case, R., Gunn, A., and Jackson, F. (1989). Status and management of muskoxen in the Northwest Territories. *In* "Proceedings of the Second International Muskox Symposium, Saskatoon" (P. F. Flood, ed.), pp. A12–A16. National Research Council, Ottawa, Canada.

Chapin, F. S., III. (1980). The mineral nutrition of wild plants. *Annu. Rev. Ecol. Syst.* **11**, 233–260.

Chapin, F. S., III, and Shaver, G. R. (1985). Individualistic growth response of tundra plant species to environmental manipulations in the field. *Ecology* **66**, 564–576.

Chapin, F. S., III, and Shaver, G. R. (1988). Differences in carbon and nutrient fractions among arctic growth forms. *Oecologia* **77**, 506–514.

Chapin, F. S., III, and Shaver, G. R. (1989). Differences in growth and nutrient use among arctic plant growth forms. *Funct. Ecol.* **3**, 73–80.

Chapin, F. S., III, *et al.* (1990). "Structure and Function of Terrestrial, Freshwater, and Estuarine Ecosystems." Arctic Research Consortium of the U.S. Workshop on Arctic System Science: A Plan for Action. Boulder, Colorado.

Chernov, Y. (1985). "The Living Tundra." Cambridge Univ. Press, Cambridge.

Cooke, F., Abraham, K. F., Davies, J. C., Findlay, C. S., Healey, R. F., Sadura, A., and Segin, R. J. (1982). "The La Pérouse Bay Snow Goose Project—A 13-Year Report." Department of Biology, Queen's University, Kingston, Ontario.

Crawley, M. J. (1983). "Herbivory: The Dynamics of Animal–Plant Interactions." Blackwell, Oxford.

Danke, H. V. (1981). "Arctic Arthropods." Entomological Society of Canada, Ottawa.

Davis, M. B. (1986). Climatic instability, time lags, and community disequilibrium. *In* "Community Ecology" (J. Diamond and T. J. Case, eds.), pp. 269–284. Harper and Row, New York.

de Q. G. Robin (1986). Changing the sea level. *In* "The Greenhouse Effect, Climatic Change, and Ecosystems." (B. Bolin, B. R. Döös, J. Jager, and R. A. Warwick, eds.), pp. 323–359. Wiley, New York.

Fancy, S. G. (1983). Movements and activity budgets of caribou near oil-drilling sites in the Sagavanirktok River flood plain, Alaska. *Arctic* **36**, 193–197.

Frey, I. R. (1987). Ecological studies on wild geese. M. S. thesis, University of Marburg, Germany.

Fuller, W. A., Martell, A. M., Smith, R. F. C., and Speller, S. W. (1975). High arctic lemmings, *Dicrostonyx groenlandicus*: I. Natural history observations. *Can. Field Nat.* **89**, 223–233.

Gray, D. R. (1973). Social organizations and behaviour of muskoxen (*Ovibos moschatus*) on Bathurst Island, N. W. T. Ph.D. diss., University of Edmonton, Alberta.

Green, J. E. (1977). Population regulation and annual cycles of activity and dispersal in the arctic ground squirrel. M. S. thesis, University of British Columbia, Vancouver.

Haukioja, E. (1981). Invertebrate herbivory at tundra sites. *In* "Tundra Ecosystems" (L. C. Bliss, O. W. Heal, and J. J. Moore, eds.), pp. 547–556. Cambridge Univ. Press, Cambridge.

Haukioja, E., and Niemalä, P. (1974). Growth and energy requirements of the larvae of *Dineura virididorsata* (Retz) (Hym., Tenthredinidae) and *Oporinia autumnata* (Bkh.) (Lep., Geometridae) feeding on birch. *Ann. Zool. Fenn.* **11**, 207–211.

Henry, G. H. R. (1987). "Ecology of Sedge Meadow Communities of a Polar Desert Oasis: Alexandra Fiord, Ellesmere Island, Canada." Ph.D. diss., University of Toronto, Ontario.

Henry, G. H. R., and Svoboda, J. (1989). Comparison of grazed and non-grazed high-arctic sedge meadows. *In* "Proceedings of the Second International Muskox Symposium, Saskatoon" (P. F. Flood, ed.), p. A47, National Research Council, Ottawa, Canada.

Henry, G. H. R., Freedman, B., and Svoboda, J. (1986). Survey of vegetated areas and muskox populations in east-central Ellesmere Island. *Arctic* **39**, 78–81.

Hubert, B. A. (1974). The social organization of antelope in relation to their ecology. *Behaviour* **58**, 215–267.

Hubert, B. A. (1977). Estimated productivity of muskox on Truelove Lowland. *In* "Truelove Lowland, Devon Island, Canada: A High Arctic Ecosystem" (L. C. Bliss, ed.), pp. 467–492. Univ. of Alberta Press, Edmonton.

Jefferies, R. L. (1988a). Vegetational mosaics, plant–animal interactions, and resources for plant growth. *In* "Plant Evolutionary Biology" (L. Gottlieb and S. K. Jain, eds.), pp. 341–369. Chapman and Hall, London.

Jefferies, R. L. (1988b). Pattern and process in arctic coastal vegetation in response to foraging by lesser snow geese. *In* "Plant Form and Vegetation Structure, Adaptation, Plasticity, and Relation to Herbivory" (M. J. A. Werger, P. J. M. van der Aart, H. J. During, and J. T. A. Verhoeven), pp. 281–300. SPB. Academic, The Hague.

Jonasson, S., Bryant, J. P., Chapin, F. S., III, and Anderson, M. (1986). Plant phenols and nutrients in relation to variations in climate and rodent grazing. *Am. Nat.* **128**, 394–408.

Jung, H. G., and Batzli, G. O. (1981). Nutritional ecology of microtine rodents: Effects of plant extracts on the growth of arctic microtines. *J. Mammal.* **62**, 286–292.

Jung, H. G., Batzli, G. O., and Seigler, D. S. (1979). Patterns in the phytochemistry of arctic plants. *Biochem. Syst. Ecol.* **7**, 203–209.

Kalela, O., Koponen, T., Lind, E. A., Skaren, U., and Tast, J. (1961). Seasonal change of habitat in the Norwegian lemming *Lemmus lemmus* (L). *Ann. Acad. Sci. Fenn. A, IV Biol.* **55**, 1–73.

Kelsall, J. P. (1968). The migratory barren-ground caribou of Canada. *Can. Wildl. Serv. Monogr.* **3**.

Kerbes, R. H. (1975). The nesting populations of lesser snow geese in the eastern Canadian Arctic: A photographic inventory of June 1973. *Can. Wildl. Serv. Rep. Ser.* **35**.

Kerbes, R. H., Kotanen, P. M., and Jefferies, R. L. (1990). Destruction of wetland habitats by lesser snow geese: a keystone species on the west coast of Hudson Bay. *J. Appl. Ecol.* **27**, 242–258.

Kischinskii, A. A. (1971). The modern state of the wild reindeer population on the Novosibirsky Islands (in Russian). *Zool. Zh.* **50**, 117–125.

Kischinskii, A. A. (1984). Insular populations of wild reindeer in the eastern sector of the Soviet Arctic and methods for their rational exploitation. *In* "Wild Reindeer of the Soviet Union" (E. E. Syroechkovskii, ed.), pp. 158–162. Dep. of the Interior, Washington, DC.

Klein, D. R. (1968). The introduction, increase, and crash of reindeer on St. Matthew Island. *J. Wild. Manage.* **32**, 350–367.

Klein, D. R. (1986). Latitudinal variation in foraging strategies. *In* "Grazing Research at Northern Latitudes" (O. Gudmundsson, ed.), pp. 237–246. Plenum, New York.

Klein, D. R. (1987). Vegetation recovery patterns following overgrazing by reindeer on St. Matthew Island. *J. Range Manage.* **40**, 336–338.

Klein, D. R. (1988). The establishment of muskox populations by translocation. *In* "Translocation of Wild Animals" (L. Nielsen and R. D. Brown, eds.), pp. 298–317. Wisconsin Humane Society, Milwaukee, and Kleberg Wildlife Research Institute, Kingsville, Texas.

Klein, D. R., and Bay, C. (1990). Foraging dynamics of muskoxen in Peary Land, northern Greenland. *Holarct. Ecol.* **13**, 269–280.

Koshkina, T. V. (1961). New data on the nutritional habits of the Norwegian lemming. *Byull. Mosk. Ova. Ispyt. Prir. Otd. Biol.* **66**, 15–32.

Kotanen, P., and Jefferies, R. L. (1987). The leaf and shoot demography of grazed and ungrazed plants of *Carex subspathacea. J. Ecol.* **75**, 961–975.

Kuropat, P. J. (1984). Foraging behavior of caribou on a calving ground in north western Alaska. M. S. thesis, University of Alaska, Fairbanks.

Laine, K., and Henttonen, H. (1983). Plant production and population cycles of microtine rodents in Finish Lapland. *Oikos* **40**, 407–418.

Leader-Williams, N. (1988). "Reindeer and South Georgia: The Ecology of an Introduced Population." Cambridge Univ. Press, Cambridge.

McLean, B. D., Jingfors, K., and Case, R. (1989). Distribution and abundance of muskoxen and caribou on Banks Island, July 1985. *In* "Proceedings of the Second International Muskox Symposium, Saskatoon" (P. F. Flood, ed.), p. A45, National Research Council, Ottawa.

McLean, I. G. (1981). Social ecology of the Arctic ground squirrel. Ph.D. diss., University of Alberta, Edmonton.

MacLean, S. F. (1981). Introduction: Invertebrates. *In* "Tundra Ecosystems: A Comparative Analysis" (L. C. Bliss, O. W. Heal, and J. J. Moore, eds.), pp. 509–516. Cambridge University Press, Cambridge.

MacLean, S. F., and Jensen, T. S. (1985). Food selection by insect herbivores in Alaskan arctic tundra: The role of plant life-form. *Oikos* **44**, 211–221.

Mathieson, S. D., Orpin, C. G., and Blix, A. S. (1984). Rumen microbial adaptation to fiber digestion in Svalbard reindeer. *Can. J. Anim. Sci.* **64**, (Suppl.), 261–262.

Mattheis, P. J., Tieszen, L. L., and Lewis, M. C. (1976). Responses of *Dupontia fisheri* to simulated lemming grazing in an Alaskan arctic tundra. *Ann. Bot.* **40**, 179–197.

Michurin, L. N., and Vakhtina, T. V. (1968). On winter feeding of wild reindeer (*Rangifer tarandus* L.) in arctic tundras of Taimyr (in Russian). *Zool. Zh.* **47**, 477–479.

Miller, F. L., Russell, R. H., and Gunn, A. (1976). Peary caribou and muskoxen on western Queen Elizabeth Island, N. W. T., 1972–74. *Can. Wildl. Serv. Rep. Ser.* **40**.

Miller, F. L., Edmunds, E. J., and Gunn, A. (1982). Foraging behaviour of Peary caribou in response to springtime snow and ice conditions. *Can. Wildl. Serv. Occas. Pap.* **48**.

Oksanen, L., and Ericson, L. (1987). Dynamics of tundra and taiga populations of herbaceous plants in relation to the Tihomirov–Fretwell and Kalela–Tast hypotheses. *Oikos* **50**, 381–388.

Palmer, L. J. (1934). Raising reindeer in Alaska. *U.S. Dep. Agric. Misc. Publ.* **207**, 1–41.

Parker, G. R. (1978). The diets of muskoxen and Peary caribou on some islands in the Canadian High Arctic. *Can.. Wildl. Serv. Occas. Pap.* **35**.

Parker, G. R., and Ross, R. K. (1976). Summer habitat use by muskoxen (*Ovibos moschatus*) and Peary caribou (*Rangifer tarandus pearyi*) in the Canadian High Arctic. *Polarforschung* **46**, 12–25.

Pegau, R. E. (1970). Effect of reindeer trampling and grazing on lichens. *J. Range Manage.* **23**, 95–97.

Pimentel, D. (1988). Herbivore population feeding pressure on plant hosts: Feedback evolution and host conservation. *Oikos* **53**, 289–302.

Raillard, M., and Svoboda, J. (1990). Muskox winter feeding strategies at Sverdrup Pass, 79°N, Ellesmere Island. *Muskox* **37**, 86–92.

Reimers, E. (1977). Population dynamics in two subpopulations of reindeer in Svalbard. *Arct. Alp. Res.* **9**, 369–381.

Ringberg, T. (1979). The Spitzbergen reindeer: Winter-dormant ungulate? *Acta Physiol. Scand.* **105**, 268–273.

Ritchie, J. C. (1983). "Past and Present Vegetation of the Far Northwest of Canada." University of Toronto Press, Toronto.

Scheffer, V. B. (1961). The rise and fall of a reindeer herd. *Sci. Mon.* **73**, 356–362.

Schultz, A. M. (1968). A study of an ecosystem: The arctic tundra. *In* "The Ecosystem Concept in Natural Resource Management" (V. M. Dyne, ed.), pp. 77–93. Academic Press, New York.

Skoog, R. O. (1968). Ecology of caribou (*Rangifer tarandus granti*) in Alaska. Ph.D. diss., University of California, Berkeley.

Skuncke, F. (1969). Reindeer ecology and management in Sweden. *Biol. Pap. Univ. Alaska* **8**, 1–82.

Tast, J. (1972). Annual variations in the weights of wintering root voles, *Microtus oeconomus*, in relation to their food conditions. *Ann. Zool. Fenn.* **9**, 116–119.

Tast, J. (1974). The food and feeding habits of the root vole, *Microtus oeconomus*, in Finnish Lapland. *Aquilo Ser. Zool.* **15**, 25–32.

Tast, J., and Kalela, O. (1971). Comparisons between rodent cycles and plant production in Finnish Lapland. *Ann. Acad. Sci. Fenn. A IV Biol.* **186**, 1–14.

Thing, H., Klein, D. R., Jingfors, K., and Holt, S. (1987). Ecology of muskoxen in Jameson Land, northeast Greenland. *Holarct. Ecol.* **10**, 95–103.

Thomas, D. C., and Edmunds, J. (1983). Rumen contents and habitat selection of Peary caribou in winter, Canadian Arctic Archipelago. *Arct. Alp. Res.* **15**, 97–105.

Thomas, D. C., and Kroeger, P. (1980). In vitro digestibilities of plants in rumen fluids of Peary caribou. *Artic* **33**, 757–767.

Tihomirov, B. A. (1959). "Relationship of the Animal World and the Plant Cover of the Tundra." Tundra Biology Institute, Komarov Academy of Sciences of the USSR.

Trudell, J., White, R. G., Jacobsen, E., Staaland, H., Ekern, K., Kildemo, K., and Gaare, E. (1980). Comparison of some factors affecting the in vitro digestibility estimate of reindeer forages. *In* "Proceedings of the Second International Reindeer/Caribou Symposium, Roros, Norway" (E. Reïmers, E. Gaare, and S. Skjenneberg, eds.), pp. 262–273. Directoratet for vilt og ferskvannsfisk Trondheim.

Trudell, J., and White, R. G. (1981). The effect of forage structure and availability on food intake, biting rate, and daily eating time of reindeer. *J. Appl. Ecol.* **18**, 63–81.

Vibe, C. (1967). Arctic animals in relation to climatic fluctuations. *Medd. Gronl.* **170**, 1–227.

Walker, D. A., and Walker, M. D. (1991). History and pattern of disturbance in Alaskan arctic terrestrial ecosystems: A hierarchical approach to analyzing landscape change. *J. Appl. Ecol.* **28** (in press).

White, R. G. (1983). Foraging patterns and their multiplier effects on productivity of northern ungulates. *Oikos* **40**, 377–384.

White, R. G., Thomson, B. R., Skogland, T., Person, S. J., Russell, D. E., Holleman, D. F., and Luick, J. R. (1975). Ecology of caribou at Prudhoe Bay, Alaska. *Biol. Pap. Univ. Alaska Spec. Rep.* **2**, 151–201.

White, R. G., and Fancy, S. G. (1986). Nutrition and energetics of indigenous northern ungulates. *In* "Grazing Research at Northern Latitudes" (O. Gudmundsson, ed.), pp. 259–270. Plenum, New York.

Wilkinson, P. F., Shank, C. C., and Penner, D. F. (1976). Muskox–caribou summer range relations on Banks Island, N.W.T. *J. Wildl. Manage.* **40**, 151–162.

20

Modeling the Response of Arctic Plants to Changing Climate

James F. Reynolds and Paul W. Leadley

I. Introduction

Although the exact effects of elevated carbon dioxide on global climate are unknown, consensus is growing among atmospheric modelers that global climate will change nonuniformly over the earth's surface during the next 50 years, with particularly dramatic impacts in polar regions (Chen and Drake, 1986; Chap. 2). Since it is impossible to experimentally determine the outcome of these climate changes, they essentially constitute an uncontrolled, Earth-scale "experiment" (Baes *et al.*, 1977; Ramanathan, 1988). The challenge facing arctic ecologists is how to knowledgeably predict the outcome of this experiment (e.g., increased temperatures, higher precipitation, changing season length, etc.) in the absence of long-term databases and at different levels of biological organization (e.g., plants, populations, ecosystems, and landscapes).

Investigating and predicting the response of biological systems to climate change is greatly complicated by the problem of scale. Living systems encompass 14 orders of magnitude of spatial and temporal scales (Osmond et al., 1980). For example, atmospheric CO_2 enters the biosphere at the level of the plant leaf, where cellular and physiological processes are reasonably well understood and the immediate (short-term) effects of elevated CO_2 concentration can be predicted with fairly high confidence. But the direct effects of CO_2 on cellular and leaf processes lead to changes in longer-term allocation patterns, growth, and other properties at the level of the whole plant (Strain, 1987). In turn, these effects feed back on lower hierarchical levels, for example, altering leaf photosynthetic capacity, and they are translated to higher hierarchical levels, for example, influencing ecosystem carbon uptake (Hilbert *et al.*, 1987), interactions among plants (Bazzaz *et al.*, 1985), and interactions among organisms of different trophic levels (Committee on Global Change, 1988). Additional complexity arises from species-specific variability in responses to CO_2 and from interactions with other environmental variables, mainly nutrients, temperature, and water (Strain and Cure, 1985).

Preliminary field studies on arctic plant response to nutrients (Chapin and Shaver, 1985a), elevated CO_2 (Hilbert *et al.*, 1987; Tissue and Oechel, 1987), and temperature (Chapin, 1987) provide insights into the nature of this complexity. For example, Figure 1 illustrates how arctic plants might respond to simultaneous increases in temperature and atmospheric CO_2. Numerous observations, including those of Shaver *et al.* (1986) for the arctic sedge *Eriophorum vaginatum*, have shown that plants tend to adjust their root:shoot ratio so as to maintain an internal carbon–nitrogen balance favorable for growth. The suite of potential interactions are illustrated using Davidson's (1969) model, which states that a functional balance exists between the size and activity of the shoot (which supplies carbohydrates) and the size and activity of the root (which supplies water and essential nutrients) (see Fig. 1). For example, if a plant invests newly produced photosynthate into shoot mass at the expense of root mass, the investment will favor carbon procurement, but relatively less root mass means lower nutrient and water uptake. Investment in root mass over shoot mass, however, will have the opposite effect. It is the integration of this whole-plant function that ultimately determines the ability of plants to compensate for imbalances and changes in the availability of resources in the environment (Chapin *et al.*, 1987).

It will be impossible to design more than a small fraction of the many different combinations of experiments needed to test plant response to environmental change. Even at relatively small spatial scales (e.g., one square meter), it is very difficult to experimentally determine the effects of elevated CO_2 and temperature on arctic systems because we cannot observe long-term changes that may lead to a new homeostasis (Oechel and Strain, 1985). But we can supplement limited empirical knowledge with the relatively inexpensive development of mathematical and computer simulation models. Modeling, along with experimentation, plays an essential role in our ability to predict plant and ecosystem responses to climate change (Dahlman, 1985; Reynolds and Acock, 1985). It is a powerful tool that can help extend our

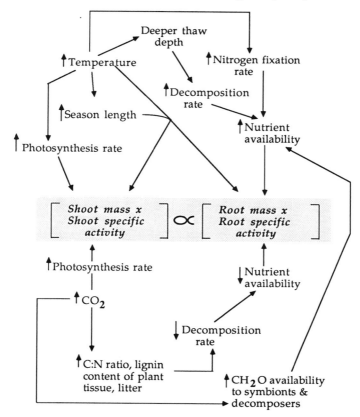

Figure 1 Some potential effects of increased temperature (from Chapin, 1983) and elevated carbon dioxide (from Oechel and Strain, 1985) on arctic vegetation. Effects are shown as modifiers of the specific activities and masses of shoots and roots (which are proportional) and, ultimately, on the functional balance of plants.

empirical knowledge on arctic systems by simulating the complex combinations of environmental–biotic interactions that are otherwise impossible to address.

In this chapter, we consider the modeling of whole-plant growth in arctic ecosystems. Although we emphasize whole plants, we also refer to other levels of organization (e.g., ecosystems and landscapes) to place this work in context. First, we review the concept of scale and define the types of models that result from considering scale. Second, we classify existing whole-plant arctic models based on these definitions and describe specific models used to predict whole-plant response to climate change. Third, we critique these models, addressing the types of models used, how comprehensive they are (and how this may affect their predictions), their underlying assumptions, and

Table I   Hierarchical View of Ecological Systems

Hierarchical level	Spatial scale[a] (m²)	Temporal scale	System[b]	
			Plant	Ecosystem
.				
.				
.				
Landscape	10^6	Decades–centuries	L + 3	L + 2
Toposequence or soil catena	10^2	Years–decades	L + 2	L + 1
Patch ecosystem	10^0	Months–years	L + 1	**L**
Plant	10^{-1}	Weeks–months	**L**	L − 1
Organs	10^{-2}	Hours–weeks	L − 1	L − 2
Cells	10^{-4}	Seconds–hours	L − 2	L − 3
.				
.				
.				

[a]Examples of "typical" values from Osmond *et al.* (1980) and Woodmansee (1988).
[b]**L** is the level at which one is interested in the system.

model validation. We examine the effect of changing season length on plant response to illustrate some of the problems inherent in developing such models. Finally, we present a strategy for developing arctic models in terms of hierarchical considerations and discuss the importance of appropriate scale in making long-term predictions of arctic ecosystem behavior.

II. Scales and Types of Models

Several fundamental properties of hierarchical systems are relevant to our discussion (O'Neill, 1988). First, the system of interest (i.e., level *L* in Table I) is itself a component of some higher-level system (*L* + 1). This higher level will influence (e.g., limit or bound, control, etc.) the behavior of *L* and will be the context in which the behavior of *L* is ultimately interpreted. Second, *L* can be subdivided into components of the next lower level (*L* − 1). These components serve as state variables in models of *L* and are studied to explain the mechanisms operating at *L*. We can generally ignore levels higher than *L* + 1 and lower than *L* − 1 in trying to understand *L* = 0 behavior: The former are too large and slow and, hence, appear constant at *L*, whereas the latter are too small and fast and appear as background static or variability that is averaged or filtered out at *L* (O'Neill, 1988).

We can now define several types of models arising from these scale considerations. Descriptions at any particular level in the hierarchy in Table I may be related or connected to the next higher level in a mechanistic fashion. That is, it is from *L* − 1 components that we understand or explain phenomena at L (Thornley, 1980). We refer to a model of a system's behavior at any chosen level in a hierarchy (e.g., organ, whole plant, ecosystem, etc.) as mechanistic when it is described by the interactions of lower-level components (for example, modeling whole-plant dynamics in terms of descriptions of leaves, stems, and roots).

In contrast, empirical and phenomenological models describe L system behavior without decomposing it into lower-level subsystems. Empirical models are based on statistical formulations (polynomial regression, for example) that correlate a set of observed variables to some response variable of interest. The form of a phenomenological model is derived from a general understanding of the process or system of interest, and thus the model equation is biologically meaningful and describes some logical or expected relationship between variables. An example of a phenomenological model is the logistic curve of whole-plant growth, $dW/dt = rW(W_{max} - W)$, where W is dry weight, r is the growth rate, and W_{max} is the maximum W attainable. This model does not seek to "plunge beneath the surface" in terms of detail but, instead, describes the process or system property directly, at the level of interest (Spanner, 1964; Thornley, 1976). The logistic model could, in turn, be used as a component in a mechanistic model of $L + 1$ phenomena, e.g., plant community dynamics. Mechanistic whole-plant growth models (such as those described in Section III) are typically composed of numerous empirical and phenomenological equations that describe $L - 1$ growth phenomena (Thornley, 1980).

III. Arctic Plant Growth Models

A. Review of Existing Models

A number of whole-plant growth models for vascular and nonvascular plants have been developed for arctic ecosystems, most of which were constructed during and immediately after the International Biological Programme (IBP) (Miller *et al.*, 1975, 1979; Wielgolaski, 1975). We examined 21 models with regard to the minimal set of processes and variables that Reynolds and Acock (1985) proposed as critical to predict plant growth under changing climatic conditions (Table II). We also rated the relative comprehensiveness of each model in terms of how many growth processes it considered (e.g., photosynthesis, allocation, nutrient uptake, etc.) and how many components it included as state variables (e.g., leaves, stems, carbon–nitrogen storage, roots, etc.). Although much variation exists, we classified arctic plant growth models into four general types, based on increasing mechanism and comprehensiveness: (1) budget, (2) flux, (3) semimechanistic, and (4) mechanistic (see Fig. 2).

Table II characterizes representative examples of each type. Budget models are highly empirical, with no attention to mechanism or comprehensiveness. Plants are treated as "black boxes": plant biomass is lumped into a single

Table II Representative Examples of Budget, Flux, Semimechanistic, and Mechanistic Models[a]

Variables & processes	Budget — O'Neill (1976)	Flux — Miller et al. (1972, 1976), Tiezsen et al. (1980)	Semi-mechanistic — NECS[b] Miller et al. (1983)	Semi-mechanistic — Leadley and Reynolds (1991a)	Mechanistic — ARTUS[c] Miller et al. (1985)	Mechanistic — Leadley and Reynolds (1991b)
Plant structure	Lumped					
Shoot		*		*	*	*
Leaves		*	*		*	*
Stems		*	*		*	*
Canopy architecture[d]					*	
Roots		*		*	*	*
Root architecture					*	
Photosynthesis	Fixed %					
Conductance[d]		*				
Leaf age						*
Leaf nitrogen[d]						
Temperature[d]		*	*	*	*	*
Irradiance[d]		*	*	*	*	*
CO_2[d]		*	*	*		*
CH_2O inhibition[d]			*	*		*
Conductance						
CO_2[d]						
Humidity[d]						
Leaf water potential[d]						
Nutrient uptake		*		*	*	*
Energy budget		*				
Allocation and growth						
Substrate availability[d]		*	*	*	*	*
Water status[d]				*	*	
Temperature[d]			*	*	*	*
Nutrient status[d]			*	*	*	*

Plant age[d]

Leaf expansion[d]

C:N storage[d]   Fixed %

Litterfall[d]

C:N of litter

Senescence[d]

Population dynamics

  Tiller numbers

  Competition

[a] An asterisk indicates that the process was included in the model.

[b] Northern Ecosystems Carbon Simulator

[c] ARctic TUndra Simulator

[d] From Reynolds and Acock (1985).

[e] Tiller number was included in the model but was fixed.

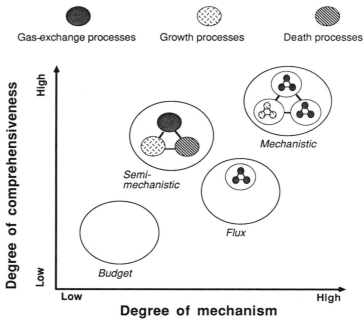

Figure 2   Types of arctic plant growth models based on relative degrees of comprehensiveness and mechanism. Comprehensiveness refers to the number of processes considered (e.g., gas exchange, growth, and death), whereas mechanism refers to the level of detail at which each of these processes is treated. Circles represent hierarchical levels as discussed in the text. The largest circle represents the whole plant, and the smallest the lowest level at which the model would be considered mechanistic.

box, and growth is described as the difference between inputs and outputs (usually fixed values). Semimechanistic models have low to medium comprehensiveness: plant growth tends to be described with a minimum number of $L - 1$ components and with heavy emphasis on phenomenological modeling. Mechanistic models have the highest overall mechanism and comprehensiveness and, consequently, are usually very detailed. Flux models are special cases of semimechanistic models that contain a high degree of mechanism and comprehensiveness in terms of gas exchange per se but tend to have low overall comprehensiveness in terms of other plant growth processes. Flux models have been typically used to study the relationship between single-leaf and canopy photosynthesis, water loss, or both; effects of species composition on canopy photosynthesis; and effects of canopy architecture on light interception and plant microclimate (Miller and Tieszen, 1972; Miller *et al.*, 1976; Tieszen *et al.*, 1980). Because of the emphasis on modeling at the IBP wet meadow sites (especially Barrow, Alaska), most of the models were developed for wet tundra ecosystems. Shrub-dominated tundra, polar deserts, and polar semideserts are poorly represented by models.

Table III summarizes the spatial and temporal scales used in the models. Only one attempt has been made to extrapolate findings from this small scale to a larger area: Miller *et al.* (1983) assumed that an "average" square meter was representative of arctic and boreal ecosystem types and parameterized the model for plants of polar deserts, wet sedge, tussock, low shrub, peatlands, and various northern boreal forests (see Table III). In general, the length of time models simulate is related to the type of model: the lower the degree of comprehensiveness and mechanism, the longer the simulation (see Table III). Budget and semimechanistic whole-plant models have been used to simulate decade- to century-long processes. Mechanistic models have been used mainly to simulate growth dynamics for relatively short periods, typically a single season. Flux models, which emphasize detailed gas-exchange processes, have the smallest time step (one hour) of all models and have been used exclusively for single-season simulations.

B. Predicting Plant Response to Climate Change: Case Studies

Several groups applied their models to predict possible responses of arctic plants to climate change. Miller *et al.* (1983), using a semimechanistic model with a monthly time step, attempted to predict plant response out to 50 years (Table IV). Other researchers, using models that were more comprehensive and mechanistic restricted their predictions to a single growing season (see Table IV). Although some trends appear from these results, they must be viewed cautiously because each study incorporates very different models, climate scenarios, and indices of plant response.

Net primary productivity (NPP) for vascular plants is predicted to increase by 3–40% with a 4° C increase in temperature (Table IV). Overall, vascular plants are predicted to respond more positively (as measured by NPP and photosynthesis) than mosses to such increases in temperature. Stoner *et al.* (1978a) predicted that a large increase in temperature (+11.1° C), accompanied by large increases in radiation and decreases in relative humidity, would lead to substantial decreases in seasonal net CO_2 exchange for both vascular plants and mosses (from 13 to 24%). In a simulation of wet meadow tundra conducted for conditions in Norway, Sandhaug *et al.* (1975) predicted that increased temperatures over a 10–12-day period in the spring would increase gross photosynthesis by almost the same amount as equivalent temperature increases over a 40-day period in midseason (see Table IV). Miller *et al.* (1984) predicted that increased precipitation (50% increase per year) would have little direct effect (<1%) on vascular plants or mosses in tussock tundra (Table IV). A doubling of atmospheric CO_2 concentration accompanied by a 4° C increase in air temperature over a 50-year period was predicted by Miller *et al.* (1983) to increase NPP in tussock tundra and wet arctic meadows by about 3 to 4%. Thus, nearly all simulations predict that with increased temperature, CO_2 concentrations, or both on the order expected by the middle of the next century, vascular plant productivity will increase, but moss productivity will show little effect.

Table III Comparison of Arctic and Selected Northern Ecosystem Vegetation Models

Vegetation type[a]	Model type[b]	Spatial scale of simulation		Temporal scale of simulation		Inclusion in model?			Location[c]	Reference
		Min.	Max.	Time step in model	Length of simulation	CO_2	Temperature	Nutrients		
Arctic ecosystem models[d]										
Wet meadow	Semi-mechanistic	m²	m²	Day	1 year	No	Yes	No	N,H,M	Bunnell & Scoullar (1975, 1981)
Wet meadow	Semi-mechanistic	m²	m²	Day	4 years	No	Yes	No	H	Sandhaug et al. (1975)
Wet meadow?	Budget	m²	m²	Year	100+ years	No	No	No	D	O'Neill (1976)
Several	Budget	m²	m²	Week	1 year	No	No	No	M[e]	Jones & Gore (1981)
All major	Semi-mechanistic	m²	Arctic	Month	50 years	Yes	Yes	Yes	M	Miller et al. (1983)
Bog	Budget	m²	m²	Year	5–55 years	No	No	No	M	Gore & Olson (1967)
Tussock tundra	Mechanistic	Ramet	m²	Day	1 year	No	Yes	Yes	N	Miller et al. (1984)
Arctic vascular plant models										
Wet meadow	Flux	m²	m²	Hour	1 year	Yes	Yes	No	N	Miller & Tieszen (1972); Miller et al. (1976); Tieszen et al. (1980)
Dupontia fisheri[f]	Mechanistic	m²	m²	Day	1 year	No	Yes	Yes	N	Miller et al. (1978a)
Dupontia fisheri	Mechanistic	Ramet	m²	Day	20 years	No	Yes	No	N	Lawrence et al. (1978)

Vegetation type[a]	Model type[b]	Spatial scale of simulation Min.	Spatial scale of simulation Max.	Temporal scale of simulation Time step in model	Temporal scale of simulation Length of simulation	Inclusion in model? CO$_2$	Inclusion in model? Temperature	Inclusion in model? Nutrients	Inclusion in model? Water status	Location[c]	Reference
Wet meadow	Flux	m²	m²	Hour	1 year	Yes	Yes	No	No	N	Stoner et al. (1978a)
Calluna bog	Semi-mechanistic	m²	m²	Day	1 year	No	Yes	No	No	M	Grace & Woolhouse (1974)
Dupontia fisheri	Mechanistic	Ramet	m²	Day	1 year	Yes?	Yes	Yes	Yes	N	Stoner et al. (1978b)
Eriophorum vaginatum	Semi-mechanistic	m²	m²	Year	50 years	Yes	Yes	Yes	Yes	N	Leadley and Reynolds (1991a)
Eriophorum vaginatum	Mechanistic	Ramet	m²	Day	1 year	Yes	Yes	Yes	Yes	N	Leadley and Reynolds (1991b)
Arctic nonvascular plant models											
Lichen	Flux	m²	m²	Hour	2–3 days	No	Yes	No	Yes	N	Lange et al. (1977), Lechowicz (1981)
Moss	Mechanistic	m²	m²	Day	2 years	No	Yes	No	Yes	M	Hayward & Clymo (1983)
Moss	Flux	m²	m²	Hour	1 year	Yes	Yes	No	Yes	N	Stoner et al. (1978a)
Moss	Flux	m²	m²	Hour	1 year	No	Yes	No	Yes	N	Miller et al. (1978b)
Moss	Mechanistic	m²	m²	Day	1 year	No	Yes	No	Yes	N	Miller et al. (1984)
Moss	Flux	m²	m²	Hour?	1 year	No	No	No	Yes	N	Harley et al. (1989)

[a]Vegetation classification after Bliss (1981).
[b]See text and Figure 2 for explanation of model types.
[c]Key: N = North Slope of Alaska, USA; H = Hardangervidda, Norway; M = Moor House, UK; D = Devon Island, Canada.
[d]Vascular plant component of ecosystem model.
[e]Validated for Moor House only.
[f]D. fisheri is a common grass in wet meadow tundra in northern Alaska.

423

Table IV Some Case Scenarios of Predictions for Plant Response to Climate Change in the Arctic[a]

Vegetation	Standard conditions	Time	Temp °C	CO_2	Ppt	Other	Vascular plants				Mosses			Reference
							NPP	Pg	Pn	R	NPP	Pg	Pn	
Tussock tundra	Average conditions for Eagle Creek, Alaska	1 year	+4				+40				-1			ARTUS: Miller et al. (1984)
Tussock tundra	-12° C (mean) 36° C (amplitude) Eagle Creek, Alaska	50 years	+4	× 2	+50%	Nutrient min.: +100%	+7				+1			NECS: Miller et al. (1983)
							+2.9 yr^{-1}	+3.1 yr^{-1}		+3.1 yr^{-1}				
Wet meadow	-12° C (mean) 36° C (amplitude) Barrow, Alaska		+4	× 2			-1				+1			
							+3.8 yr^{-1}	+4.2 yr^{-1}		+4.5 yr^{-1}				
Wet meadow	Temp: 4.5° C Rad: 361 $cal\ cm^{-2}\ d^{-1}$ Hum: 6.1 $g\ m^{-3}$ SWP: -1 bar Barrow, Alaska	1 year	+11.1 (hot season)			Rad: × 1.75 Hum: × 1.4			-13 to -16 above				-21 to -24	Stoner et al. (1978a)
Wet meadow	1972 season at Hardangervidda, Norway	1 year	+2[b]					+22				+1		Sandhaug et al. (1975)
Wet meadow	1971 season at Barrow, Alaska	1 year	+2[c]			Season length: +10 days		+24	+25 above			+0		Miller et al. (1976); Tieszen, et al. (1980)
	July 15-19, 1971, at Barrow, Alaska	1 day	Air: +5 Soil: +4						+2 above					
Dupontia fisheri	90-day season at Barrow, Alaska	1 year				Season length: +10 days	+7	+13	+13					Stoner et al. (1978b)

[a] Key: above = aboveground plant CO_2 exchange only; Hum = humidity; Min = mineralization; NPP = net primary productivity; Pg = gross plant CO_2 exchange; Pn = net plant CO_2 exchange; Ppt = precipitation; Rad = radiation; and SWP = soil water potential. Measures of Pn and Pg do not include soil respiration.

[b] Changed over a 40-day period in summer

[c] Changed over a 10-to-12-day period in spring

IV. Critique of Models

The predictions of models, such as those given in Table IV, are often cited as supporting evidence for the potential responses of arctic ecosystems to climate change (e.g., Chapin, 1984; Chapin and Shaver, 1985b; Oechel and Strain, 1985; Chapin, 1987; Committee on Global Change, 1988). It is therefore important to have a sense of the relative strengths and limitations of a model whose predictions are being cited. What are its assumptions? Is it able to respond to the particular climate change scenario imposed? What conditions has it been validated for? Was it developed at the appropriate level of mechanism and comprehensiveness for addressing climate change? In this section, we address some of these issues.

A. Model Assumptions

One question concerns whether a model has the capacity to respond to the suite of environmental variables that are expected to shift with climate change in the Arctic, specifically, increased temperature, season length, atmospheric CO_2 concentration, and nutrient availability (Chapin, 1984). Many environmental variables limit growth of arctic plants, but low temperature, a short growing season, low nutrient availability, and poor soil aeration (and low water availability in polar deserts, polar semideserts, mountains, and on ridges) appear to be most important (Chapin and Shaver, 1985b). Experiments where air temperature, nutrients, or CO_2 concentrations have been manipulated indicate that increased nutrients substantially increase productivity (Chapin and Shaver, 1985a; Shaver and Chapin, 1986; Kummerow *et al.*, 1987); increased air temperature increases overall productivity, although individual species responses may be positive or negative (Shaver and Chapin, 1986); and long-term exposure to elevated CO_2 has little effect on productivity (Billings *et al.*, 1982; Tissue and Oechel, 1987). Soil aeration may be an important limiting factor, but neither the mode of action nor probable effects of changes in soil aeration are well understood (Billings *et al.*, 1982; Chapin and Shaver, 1985b).

Given the importance of nutrients in arctic ecosystems (Chap. 13, 15, 16), it is surprising that so few (<1/3) models have considered nutrients (see Table III). All models (excluding the budget types) included temperature responses, whereas most did not include CO_2 concentration as a variable. No model addressed soil aeration, but it is difficult to assess the significance of this omission since so little is known about it. Few of the vascular plant models included water status (most were developed for mesic ecosystems), but all models for nonvascular plants did (these plants are poikilohydic and occupy the surface layer, which may become dehydrated, even in mesic systems). Including a response to an environmental variable in a model does not, of course, guarantee that the model will respond properly: the form of inclusion must be evaluated. Such evaluation can be difficult because assumptions are often not explicit, and it is difficult to know which assumptions lead to particular model behaviors. To more clearly assess the arctic models reviewed here, we will discuss the processes of net carbon fixation, nutrient uptake, allocation and growth, tissue losses, and phenology in the context of requirements we think important for modeling plant response to long-term climate change. (These requirements may exceed the original purpose of some of the models reviewed.)

1. Net Carbon Fixation

All of the models in Table III include net carbon fixation. In fact, in almost all cases, this is the process considered in most detail. By definition, such a focus is expected in flux models. But the budget model of O'Neill (1976) included no plant processes other than net energy capture and tissue losses, and the semimechanistic model of Miller *et al.* (1983) devoted more than half of the published computer code to net carbon fixation. Most likely, this emphasis stems from our relatively high understanding of photosynthetic and respiratory processes and the concomitant modeling expertise for these processes. Is this, however, an appropriate emphasis in arctic plants? There is good evidence that many arctic plants are not carbon limited (Warren-Wilson, 1966). High nonstructural carbohydrate concentrations (Chapin *et al.*, 1986) and lack of a long-term stimulatory effect of elevated CO_2 on photosynthesis (Tissue and Oechel, 1987) support this idea. End-product inhibition of photosynthesis was included in only two models (Miller *et al.*, 1983, 1984), even though it may play an important role in regulating photosynthesis in arctic plants. In contrast to many nutrient-limited systems, increased nutrients do not appear to increase photosynthetic rates in the Arctic and may in some cases decrease them (Bigger and Oechel, 1982).

Photosynthesis is an important process, but it may not be the most important process controlling arctic plant growth (Chap. 8, 15, 16). Accordingly, we suggest that further efforts in this area may not be as critical to modeling ecosystem response to climate change as the processes of nutrient uptake, allocation, tissue loss, and phenology.

2. Nutrient Uptake

Long-term (more than one year) predictions of arctic models that do not have nutrient controls should be viewed with considerable skepticism. Most of the long-term simulations attempted (i.e., Gore and Olson, 1967; Sandhaug *et al.*, 1975; O'Neill, 1976; Lawrence *et al.*, 1978), fall into this category; exceptions include Miller *et al.* (1983) and Leadley and Reynolds (1991a) (see Table III). Certainly, less emphasis has been placed on nutrient uptake than on modeling carbon fixation. Nutrient uptake was described by empirical functions in Miller *et al.* (1983, 1984) and by Michaelis–Menten uptake kinetics in Miller *et al.* (1978a), Stoner *et al.* (1978b), and Leadley and Reynolds (1991a). Because nutrient relations are so critical to understanding the long-term response of arctic plant growth, much more emphasis must be placed here.

3. Allocation and Growth

Schemes for modeling whole-plant carbon and nitrogen allocation either did not exist (e.g., resources assumed to be equally available to all plant parts) or were limited to empirical functions; for example, Bunnell and Scoullar (1975, 1981) used an allocation scheme based on an organ's proximity to the carbon source. When growth was treated explicitly, it was usually defined as the conversion of total nonstructural carbohydrates (TNC) into structural carbon fractions. In some cases the fractionation of carbon was quite detailed (e.g., Stoner *et al.*, 1978b), but usually it was restricted to TNC and structural components (e.g., Miller *et al.*, 1983). In cases where carbon and mineral nutrients were considered limiting, growth was usually set as the minimum of the maximum growth rate, available carbon, or available mineral nutrient (e.g., Miller *et al.*, 1984). From Figure 1, it is clear that complex interactions exist and further work on growth and allocation is important.

4. Tissue Losses

Mechanisms controlling the loss of plant organs and whole ramets are poorly understood. We believe that correctly modeling these processes is one of the keys to predicting the long-term response of arctic vegetation: it is the balance between production and losses that determines the standing crop in these perennial-dominated ecosystems. Several examples of models of tissue loss and their relationship to observed responses illustrate both the variety of modeling approaches and the problems arising from these approaches.

In their long-term simulations of northern ecosystems, Miller *et al.* (1983) assumed plant death was equal to growth. This model could not, therefore, simulate increased overwintering biomass, and the long-term predictions of this model were compromised by restricting long-term response to increased production. In a more mechanistic approach, Bunnell and Scoullar (1975, 1981) assumed that organs senesced when respiration exceeded carbon allocation to plant parts. This assumption may seem reasonable, but many arctic plants have high nonstructural carbohydrate concentrations, even in older tissues (Shaver and Billings, 1976), and senescence may be much more tightly coupled to aging and nutrient status. Likewise, the attempt by Lawrence *et al.* (1978) to model ramet mortality based on short-term carbon balance may not reflect controlling mechanisms. A more comprehensive approach to senescence was taken by Miller *et al.* (1984), who assumed that leaf senescence was a function of leaf nitrogen reserves, temperature, and leaf turnover rates. The model assumed a fixed number of ramets, however, and would behave incorrectly since arctic plants usually respond to altered environment (e.g., response to nutrient addition: Shaver *et al.*, 1986; and response to elevated CO_2: Tissue and Oechel, 1987) by increasing ramet number rather than ramet size.

Satisfactory models may remain elusive since the mechanisms controlling ramet and organ death are so poorly understood and so infrequently studied. Yet efforts to correctly model long-term plant production will be moot without this information. We suggest that considerable emphasis should be placed on experimental studies and model development for organ senescence and population dynamics.

5. Phenology

The general phenological pattern of growth has been described for many arctic vegetation types, but the relationships between phenology and environment are not well understood (see Chap. 9). Temperature controls when growth, flowering, and senescence begin (Larigauderie and Kummerow, 1990), but there may be an interaction with photoperiod. For those models with a resolution of a year, consideration of phenology is not relevant. But for those models with a time step of less than a year (most of the models developed so far) phenology may play an important role in controlling model behavior. Phenology is temperature-dependent, so phenological stages that are time-dependent (i.e., temperature-invariant), such as those in Miller *et al.* (1983) and Lawrence *et al.* (1978), will be inappropriate for temperature-change scenarios. Modeling approaches that allow for both time- and temperature-dependent controls on phenology, such as that of Miller *et al.* (1984), are more likely to respond properly.

B. Model Validation

Model validation provides a critical check on the appropriateness of underlying model assumptions, so models should be as extensively validated as possible. One of the most acute problems with the models we reviewed was the lack of validation data sets including changes in temperature, CO_2, or nutrients. Other serious problems included no validation (e.g., Miller *et al.*, 1983) or poor fits to validation data (e.g., Miller *et al.*, 1984). The model of Miller *et al.* (1984) was extensively validated, but predicted shoot productivity was only 41% of measured productivity, and predicted root length was off by an order of magnitude. Another problem, and one not easily remedied, is the paucity of long-term validation data sets in the Arctic. Thus, predicting the effects of long-term climate change will necessarily be based on validations with data sets of short-term response to these manipulations.

V. The Paradox of Model Complexity

The theory of hierarchies (de Wit, 1970; Allen and Starr, 1982; Salthe, 1985; Landsberg, 1986; O'Neill *et al.*, 1986) suggests that if we wish to simulate long-term, large-scale phenomena, including short-term, small-scale detail will hinder rather than help; in other words, it is unnecessary and counterproductive to look more than one level down in the search for "mechanistic" explanations of a system's behavior. Many ecological models have attempted to explicitly include processes and structures at several hierarchical levels. Although these models can be successful in simulating specific short-term processes, they are not necessarily well suited for long-term simulations (Allen and Starr, 1982). These models become difficult to maintain and verify, tend to be unstable, are not amenable to modification for new situations, and ultimately "lose" their mechanism as noise (Rexstad and Innis, 1985; Landsberg, 1986; Loehle, 1987; O'Neill, 1988).

At the other extreme, empirical models have a limited capacity for extrapolation and, therefore, are useful for describing what we know now but not for extending our knowledge of short-term response to long-term predictions. The advantages of these models are that they are simple to build; it is often easier to achieve stable long-term simulations, and the interpretation of long-term dynamics is usually more straightforward than with complex, mechanistic models. The strong inverse correlation between model complexity and the length of simulation attempted (see Table III) supports these theoretical principles. Thus we have arrived at a fundamental paradox: both theory and experience suggest that comprehensive, mechanistic models are unlikely to be useful for long-term simulations and that simple, empirical models are unlikely to be useful for making predictions outside the range of data on which they are based.

Given these scaling problems and the lack of long-term validation data sets, it is tempting to turn to short-term predictions made by highly mechanistic models validated with short-term data sets. Indeed, the most often cited predictions of response to climate change are those from the flux model of Miller *et al.* (1976). There are, however, serious problems associated with assuming that short-term responses are representative of long-term response, as illustrated by the following example.

Miller *et al.* (1976) predicted the effects of changes in season length on annual aboveground net CO_2 exchange in *Dupontia fisheri* canopies using a flux model (see Section III and Table II; Fig. 3, line A, shows the model's predictions). Stoner *et al.* (1978b) incorporated the photosynthetic response surface generated from Miller *et al.*'s model into a mechanistic whole-plant growth model for *D. fisheri* (see Table III) and predicted annual net CO_2 exchange for aboveground parts only and for the whole-plant. Stoner *et al.*'s predicted effects for net aboveground CO_2 exchange (Fig. 3: lines B and C) were significantly lower than those of Miller *et al.* This outcome was due to Stoner *et al.*'s inclusion of a number of $L - 1$ factors, not treated in the flux model, that significantly affect whole-plant growth. These factors include feedbacks of nutrient limitation on gas exchange and the timing of leaf growth. Whole-plant net CO_2 exchange (Fig. 3: line C) exhibited the most damped response to changing season length, because changes in the loss of CO_2 through root respiration partially compensated for changes in aboveground CO_2 flux. Thus a model's degree of comprehensiveness may significantly affect its predictions.

In both studies, changes in season length were modeled using single-year simulations with step increases or decreases from a 90-day standard. But season length is likely to change gradually in response to changes in global

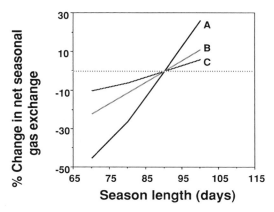

Figure 3   Predicted effect of changing season length on the percent change in annual net CO_2 exchange in *Dupontia fischeri* by models differing in their comprehensiveness. Responses were calculated as the percent change from a 90-day season length. A, Miller *et al.* (1976) flux model; B, Stoner *et al.* (1978b) (above ground only); C, Stoner *et al.* (1978b) (above- and below-ground).

climate. If so, how might the predictions differ in the above models? Do these responses reflect the sustainable response of vegetation to changes in season length? To investigate this, we used a semimechanistic model developed for *E. vaginatum* (Leadley and Reynolds, 1991a; Tables II, III). Also using a standard season length of 90 days, we ran our model with single-year step increases as well as with gradual increases or decreases in season length over a 40-year period (see Fig. 4a). Figure 4b summarizes the relative impacts of these two scenarios on seasonal gas exchange in *E. vaginatum*. The outcomes differ substantially depending on the approach. For a step change in season length from 90 to 110 days, the model predicts that annual net gas exchange will increase 19%, but the increase is only 1% if the season is lengthened gradually.

The underlying reasons for such differences become more apparent on closer examination of the model's temporal dynamics. We ran simulations with season length set at 90 days until year 10, where it was either increased to 110 days thereafter (a step increase), or one half day per year was added (a gradual increase) until year 40 (Fig. 4c). The high initial increase (+19%) in the step simulation is not sustained for long: in about 5 years the model reaches a steady state for a fixed seasonal length of 110, and the predicted percent change is down to 1%, equivalent to the predicted change in annual net gas exchange when the gradual change in season length eventually reaches 110 (see Fig. 4c). This behavior reflects the existence of several strong nutrient limitations on growth and subsequent feedback on carbon uptake (e.g., those shown in Fig. 1), and the model gradually reaches a new

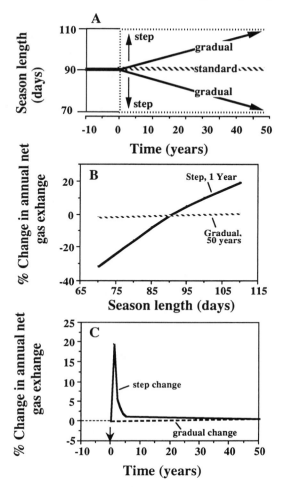

Figure 4   (A) Patterns of change in season length from a standard of 90 days used in simulations. All simulations were run for 10 years before changes in season length were started (year 0). (B) Predicted effects of changing season length on the percent change in annual whole-plant net productivity in *E. vaginatum*. Dashed line represents response after 50 years to a gradual increase or decrease in season length. Solid line represents response after 1 yr to a step increase or decrease in season length. (C) Time course of changes predicted by the model for step versus gradual change in season length from 90 to 110 days.

equilibrium point. These results are consistent with Oechel and Strain's (1985) argument that short-term responses are not necessarily indicative of long-term responses because feedbacks (e.g., nutrient limitations) act on longer time scales.

VI. A Strategy for Future Modeling

Some modelers are optimistic that long-term simulations using comprehensive, highly mechanistic models are possible because of recent advances in our understanding of ecosystem function, modeling expertise, and computer technology (Onstad, 1989). Although we do not rule out the possibility that such complex models could be used for long-term simulations in the Arctic—the limited success of the model of Lawrence *et al.* (1978) for *D. fischeri* is an example—theoretical and practical experience would argue otherwise. Ideally, one would like to build a relatively simple model that captures the essential mechanisms controlling plant response to altered climate. The model developed by Miller *et al.* (1983) is a good example. (Perhaps problems with some of their assumptions and the lack of validation have kept the predictions of this model from being more widely cited.) Yet although this approach has considerable merit, it cannot address questions about how fine-scale processes might affect outcomes.

We agree with Bazzaz and Sipe (1987) that prediction across hierarchical levels (either upward or downward) "requires a knowledge of process at each level and the scales of interaction and regulations between levels." Hence, we think the best modeling strategy is to capitalize on the advantages of each of type of model shown in Figure 2. Each model can be used at the scale for which it is best suited, and the interaction of the models allows information from different hierarchical levels to contribute to understanding the system's behavior. The exact nature of this interaction between models at different hierarchical levels is the subject of much of the current effort in ecosystem modeling.

One approach to coupling hierarchically related models is to use simplification and aggregation schemes to incorporate detail from models with greater mechanism into models with lesser mechanism. In fact, Stoner *et al.* (1978b) took this approach. They ran the highly mechanistic photosynthesis model of Miller *et al.* (1976) over a wide range of environmental conditions and incorporated the resultant photosynthetic response surface into a more comprehensive, but less mechanistic, plant growth model.

Another approach is to apply the information generated by models of differing mechanism indirectly; that is, knowledge generated from a model at one level can be used in the context of the entire set of knowledge to build or alter a model at another level. Interaction between hierarchical levels can be facilitated by a common framework, such as those advocated by Overton (1977) and Reynolds *et al.* (1989). The following example shows how a model with a short time step (one day) can be used to test hypotheses about within-year processes and how this knowledge can then be used in building a model with a longer time step (one year).

Global circulation models predict that arctic temperatures may increase 8–14° C during midwinter and 2–5° C during midsummer with a doubling of the atmospheric CO_2 concentration (Chen and Drake, 1986; Chap. 2). The

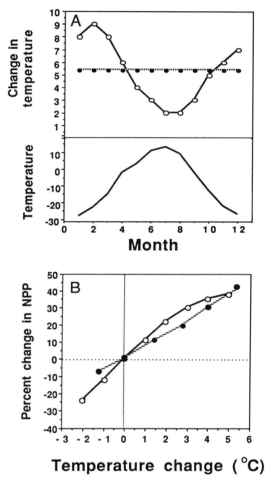

Figure 5   (A) Examples of the temperature curves used for climate change scenarios. The upper panel shows examples of uniform and nonuniform distributions of temperature increase used for simulations in (B). Both distributions represent a mean annual increase of 5.4° C. The lower panel is the current temperature regime. (B) Predicted changes in net primary productivity for *E. vaginatum* when daily ambient temperature is increased uniformly versus nonuniformly during the growing season. ●, uniform; ○, nonuniform.

semimechanistic model (one-year time step) of Leadley and Reynolds (1991a; see Tables II, III) is, however, restricted to mean annual temperature. This presents an interesting dilemma: it is not clear a priori whether mean annual temperature, which does not include any information about seasonal patterns, will be an adequate characterization of this temperature change. We used a mechanistic model with a one-day time step (Leadley and Reynolds, 1991b; see

Tables II, III) to determine if the use of mean annual temperature would cause errors in the yearly (semimechanistic) model. The mechanistic model was run for a summer with a typical temperature regime found at Toolik Lake in northern Alaska (Fig. 5a: bottom), and then rerun with two scenarios: (1) daily temperature was increased uniformly over the season, and (2) daily temperature was nonuniformly increased during the summer. Figure 5a (top) shows an example of uniform and nonuniform increases, each reflecting an annual mean increase of 5.4° C. Figure 5b shows the effects of the two temperature scenarios on predicted net primary productivity in *E. vaginatum*. Very little difference resulted in response to uniform versus nonuniform temperature changes with the daily model over the range of annual mean temperature increases likely to occur during climate warming. This suggests that annual mean temperature change may be an acceptable simplification to include in semimechanistic models for longer-term simulations.

VII. Summary

The need to extrapolate from arctic plant response to short-term manipulations to long-term predictions of response to climate change will probably require a fundamentally different approach than the development of single models at specific levels of hierarchy. We should attempt to identify models developed for vascular and nonvascular arctic tundra plants at several levels of mechanism. But only a few existing models can respond to the set of environmental variables that are expected to be altered by changes in global climate. This constraint, and problems with assumptions and validation, suggests that we are not yet in a position to predict the effects of global climate change on arctic vegetation. Nevertheless, we hope we can adapt the best parts of previous modeling efforts into a unified hierarchy of models to produce long-term simulations that are both tractable and soundly based on mechanism.

Acknowledgments

This chapter is a contribution of the U.S. Department of Energy program, "Response, Resistance, Resilience to, and Recovery from Disturbance in Arctic Ecosystems," grant #DE-FG03-84ER60250. We would like to thank David Hilbert, Anne Larigauderie, and John Tenhunen for comments on the manuscript.

References

Allen, T. F. H., and Starr, T. B. (1982). "Hierarchy: Perspectives for Ecological Complexity." Univ. of Chicago Press, Chicago.
Baes, C. F., Jr., Goeller, H. E., Olson, J. S., and Rotty, R. M. (1977). Carbon dioxide and climate: The uncontrolled experiment. *Sci. Am.* **65**, 87–97.

Bazzaz, F. A., Garbutt, K., and Williams, W. E. (1985). *In* "Direct Effects of Increasing Carbon Dioxide on Vegetation" (B. R. Strain and J. D. Cure, eds.), pp. 155–170. DOE/ER-0238. Nat'l. Tech. Info. Serv., Springfield, Virginia.

Bazzaz, F. A., and Sipe, T. W. (1987). *In* "Potentials and Limitations of Ecosystem Analysis" (E.-D. Schulze and H. Z. Zwölfer, eds.), pp. 203–227. Springer-Verlag, Berlin.

Billings, W. D., Luken, J. O., Mortensen, D. A., and Peterson, K. M. (1982). Arctic tundra: A source or sink for atmospheric carbon dioxide in a changing environment? *Oecologia* **53**, 7–11.

Bigger, C. M., and Oechel, W. C. (1982). Nutrient effect on maximum photosynthesis in arctic plants. *Holarct. Ecol.* **5**, 158–163.

Bliss, L. C. (1981). North American and Scandinavian tundras and polar deserts. *In* "Tundra Ecosystems: A Comparative Analysis" (L. C. Bliss, O. W. Heal, and J. J. Moore, eds.), pp. 5–34. Cambridge Univ. Press, Cambridge.

Bunnell, F. L., and Scoullar, K. A. (1975). ABISKO II: A computer simulation model of carbon flux in tundra ecosystems. *In* "Structure and Function of Tundra Ecosystems" (T. Rosswall and O. W. Heal, eds.), pp. 425–448. Ecological Bulletins (Stockholm), Vol. 20.

Bunnell, F. L., and Scoullar, K. A. (1981). Between-site comparisons of carbon flux in tundra using simulation models. *In* "Tundra Ecosystems: A Comparative Analysis" (L. C. Bliss, O. W. Heal, and J. J. Moore, eds.), pp. 685–715. Cambridge Univ. Press, Cambridge.

Chapin, F. S., III. (1983). Direct and indirect effects of temperature on arctic plants. *Polar Biol.* **2**, 47–52.

Chapin, F. S., III. (1984). The impact of increased air temperature on tundra plant communities. *In* "The Potential Effects of Carbon Dioxide–Induced Climatic Changes in Alaska" (J. J. McBeath, ed.), pp. 143–148. School of Agriculture and Land Resources Management Misc. Publ. 83-1, University of Alaska, Fairbanks.

Chapin, F. S., III. (1987). Environmental controls over growth of tundra plants. *Ecol. Bull. (Copenhagen)* **38**, 69–76.

Chapin, F. S., III, and Shaver, G. R. (1985a). Individualistic growth response of tundra plant species to environmental manipulations in the field. *Ecology* **66**, 564–576.

Chapin, F. S., III, and Shaver, G. R. (1985b). Arctic. *In* "Physiological Ecology of North American Plant Communities" (B. F. Chabot and H. A. Mooney, eds.), pp. 16–40. Chapman and Hall, New York.

Chapin, F. S., III, Shaver, G. R., and Kedrowski, R. A. (1986). Environmental controls over carbon, nitrogen, and phosphorus fractions in *Eriophorum vaginatum* in Alaskan tussock tundra. *J. Ecology* **74**, 167–195.

Chapin, F. S., III, Bloom, A. J., Field, C. B., and Waring, R. H. (1987). Plant responses to multiple environmental factors. *Bioscience* **37**, 49–57.

Chen, L. T. A., and Drake, E. T. (1986). Carbon dioxide increase in the atmosphere and oceans and possible effects on climate. *Annu. Rev. Earth Planet. Sci.* **14**, 201–235.

Committee on Global Change. (1988). "Toward an Understanding of Global Change." National Academy Press, Washington, DC.

Dahlman, R. C. (1985). Modeling needs for predicting responses to CO_2 enrichment: Plants, communities, and ecosystems. *Ecol. Modell.* **29**, 77–106.

Davidson, R. L. (1969). Effects of root/leaf temperature differentials on root/shoot ratios in some pasture grasses and clover. *Ann. Bot.* **33**, 561–569.

de Wit, C. T. (1970). Dynamic concepts in biology. *In* "Predictions and Measurement of Photosynthetic Productivity" (I. Setlik, ed.), pp. 17–23. PUDOC, Wageningen, Netherlands.

Gore, A. J. P., and Olson, J. S. (1967). Preliminary models for accumulation of organic matter in an *Eriophorum/Calluna* ecosystem. *Aquilo Ser. Bot.* **6**, 297–313.

Grace, J., and Woolhouse, H. W. (1978). A physiological and mathematical study of growth and productivity of a *Calluna–Sphagnum* community. IV. A model of growing *Calluna*. *J. Appl. Ecol.* **11**, 281–295.

Harley, P. C., Tenhunen, J. D., Murray, K. J., and Beyers, J. (1989). Irradiance and temperature effects on photosynthesis of tussock tundra *Sphagnum* mosses from the foothills of the Philip Smith Mountains, Alaska. *Oecologia* **79**, 251–259.

Hayward, P. M., and Clymo, R. S. (1983). The growth of *Sphagnum*: Experiments on, and simulation of, some effects of light flux and water-table depth. *J. Ecol.* **71**, 845–863.

Hilbert, D. W., Prudhomme, T. I., and Oechel, W. C. (1987). Response of tussock tundra to elevated carbon dioxide regimes: Analysis of ecosystem carbon dioxide flux through nonlinear modeling. *Oecologia* **72**, 466–472.

Jones, H. E., and Gore, A. J. P. (1981). A simulation approach to primary production. *In* "Tundra Ecosystems: A Comparative Analysis" (L. C. Bliss, O. W. Heal, and J. J. Moore, eds.), pp. 241–256. Cambridge Univ. Press, Cambridge.

Kummerow, J., Mills, J. N., Ellis, B. A., Hastings, S. J., and Kummerow, A. (1987). Downslope fertilizer movement in arctic tussock tundra. *Holarct. Ecol.* **10**, 312–319.

Landsberg, J. J. (1986). Experimental approaches to the study of the effects of nutrients and water in carbon assimilation by trees. *Tree Physiol.* **2**, 423–425.

Lange, O. L., Gieger, I. L., and Schulze, E.-D. (1977). Ecophysiological investigations in lichens of the Negev desert. V. A model to simulate net photosynthesis and respiration of *Ramalina maciformis*. *Oecologia* **28**, 247–259.

Larigauderie, A., and Kummerow, J. (1990). The sensitivity of phenological events to changes in nutrient availability for several plant growth forms in the arctic. *Holarct. Ecol.* **114**, 38–44.

Lawrence, B. A., Lewis, M. C., and Miller, P. C. (1978). A simulation model of population processes of arctic tundra graminoids. *In* "Vegetation and Production Ecology of an Alaskan Arctic Tundra" (L. L. Tieszen, ed.), pp. 599–619. Springer-Verlag, New York.

Leadley, P. W., and Reynolds, J. F. (1991a). Long-term response of an arctic sedge (*Eriophorum vaginatum*) to climate change: A simulation study. Unpublished manuscript.

Leadley, P. W., and Reynolds, J. F. (1991b). Short-term response of an arctic sedge (*Eriophorum vaginatum*) to climate change: A simulation study. Unpublished manuscript.

Lechowicz, M. J. (1981). The effects of climatic pattern on lichen productivity: *Cetraria cucullata* in the arctic tundra of Northern Alaska. *Oecologia* **50**, 210–216.

Loehle, C. (1987). Applying artificial intelligence techniques to ecological modeling. *Ecol. Modell.* **38**, 191–212.

Miller, P. C., and Tieszen, L. L. (1972). A preliminary model of processes affecting primary production in the arctic tundra. *Arct. Alp. Res.* **4**, 1–18.

Miller, P. C., Collier, B. D., and Bunnell, F. L. (1975). Development of ecosystem modeling in the tundra biome. *In* "Systems Analysis and Simulation in Ecology" (B. C. Patten, ed.), Vol. III, pp. 95–115. Academic Press, New York.

Miller, P. C., Stoner, W. A., and Tieszen, L. L. (1976). A model of stand photosynthesis for the wet meadow tundra at Barrow, Alaska. *Ecology* **57**, 411–430.

Miller, P. C., Stoner, W. A., Tieszen, L. L., Allessio, M., McCown, B., Chapin, F. S., III, and Shaver, G. R. (1978a). A model of carbohydrate, nitrogen, and phosphorus allocation and growth in tundra production. *In* "Vegetation and Production Ecology of an Alaskan Arctic Tundra" (L. L. Tieszen, ed.), pp. 577–598. Springer-Verlag, New York.

Miller, P. C., Oechel, W. C., Stoner, W. A., and Sveinbjörnsson, B. (1978b). Simulation of CO_2 uptake and water relations of four arctic bryophytes at Point Barrow, Alaska. *Photosynthetica* **12**, 7–20.

Miller, P. C., Billings, W. D., and Oechel, W. C. (1979). A modeling approach to understanding plant adaptation to low temperatures. *In* "Comparative Mechanisms of Cold Adaptation" (L. S. Underwood, L. L. Tieszen, A. B. Callahan, and G. E. Folk, eds.), pp. 181–214. Academic Press, New York.

Miller, P. C., Kendall, R., and Oechel, W. C. (1983). Simulating carbon accumulation in northern ecosystems. *Simulation* **40**, 119–131.

Miller, P. C., Miller, P. M., Blake-Jacobson, M., Chapin, F. S., III, Everett, K. R., Hilbert, D. W., Kummerow, J., Linkins, A. E., Marion, G. M., Oechel, W. C., Roberts, S. W., and Stuart, L. (1984). Plant–soil processes in *Eriophorum vaginatum* tussock tundra in Alaska: A systems modeling approach. *Ecol. Monogr.* **54**, 361–405.

Oechel, W. C., and Strain, B. R. (1985). Native species response to increased atmospheric CO_2 concentration. *In* "Direct Effects of Increasing Carbon Dioxide on Vegetation" (B. R. Strain and J. D. Cure, eds.), pp. 117–154. DOE/ER-0238. Natl. Tech. Info. Serv., Springfield, Virginia.

O'Neill, R. V. (1976). Ecosystem persistence and heterotrophic regulation. *Ecology* **57**, 1244–1253.

O'Neill, R. V. (1988). Hierarchy theory and global change. *In* "Scales and Global Change" (T. Rosswall, R. G. Woodmansee, and P. G. Risser, eds.), pp. 29–45. Wiley, New York.

O'Neill, R. V., DeAngelis, D. L., Waide, J. B., and Allen, T. F. H., (1986). "A Hierarchical Concept of Ecosystems." Princeton Univ. Press, Princeton, New Jersey.

Onstad, D. W. (1989). Population-dynamics theory: The roles of analytical, simulation, and super-computer models. *Ecol. Modell.* **43**, 111–124.

Osmond, C. B., Björkman, O., and Anderson, D. J. (1980). "Physiological Processes in Plant Ecology." Springer-Verlag, Berlin.

Overton, W. S. (1977). A strategy of model construction. *In* "Ecosystem Modeling in Theory and Practice" (C. A. S. Hall and J. W. Day, eds.), pp. 49–73. Wiley, New York.

Ramanathan, V. (1988). The greenhouse theory of climate change: A test by an inadvertent global experiment. *Science* **240**, 293–299.

Reynolds, J. F., and Acock, B. (1985). Predicting the response of plants to increasing carbon dioxide: A critique of plant growth models. *Ecol. Modell.* **29**, 107–129.

Reynolds, J. F., Acock, B., Dougherty, R. L., and Tenhunen, J. D. (1989). A modular structure for plant growth simulation models. *In* "Biomass Production by Fast-Growing Trees" (J. S. Pereira and J. J. Landsberg, eds.), pp. 123–134. Kluwer Academic, Boston.

Rexstad, E., and Innis, G. S. (1985). Model simplification—three applications. *Ecol. Modell.* **27**, 1–13.

Sandhaug, A., Kjelvik, S., and Wielgolaski, F. E. (1975). A mathematical simulation model for terrestrial tundra ecosystems. *In* "Fennoscandian Tundra Ecosystems" (F. E. Wielgolaski, ed.), Part 2, Animals and Systems Analysis, pp. 251–266. Springer-Verlag, Berlin.

Shaver, G. R., and Billings, W. D. (1976). Carbohydrate accumulation in tundra graminoid plants as a function of season and tissue age. *Flora* **165**, 247–267.

Shaver, G. R., and Chapin, F. S., III. (1986). Effect of fertilizer on production and biomass of tussock tundra, Alaska, U.S.A. *Arct. Alp. Res.* **18**, 261–268.

Shaver, G. R., Chapin, F. S., III, and Gartner, B. L. (1986). Factors limiting seasonal growth and peak biomass accumulation in *Eriophorum vaginatum* in Alaskan tussock tundra. *J. Ecol.* **74**, 257–278.

Spanner, D. C. (1964). "Introduction to Thermodynamics." Academic Press, New York.

Stoner, W. A., Miller, P. C., and Oechel, W. C. (1978a). Simulation of the effect of the tundra vascular plant canopy on the productivity of four plant species. *In* "Vegetation and Production Ecology of an Alaskan Arctic Tundra" (L. L. Tieszen, ed.), pp. 371–387. Springer-Verlag, New York.

Stoner, W. A., Miller, P. C., and Tieszen, L. L. (1978b). A model of plant growth and phosphorus allocation for *Dupontia fisheri* in coastal, wet-meadow tundra. *In* "Vegetation and Production Ecology of an Alaskan Arctic Tundra" (L. L. Tieszen, ed.), pp. 559–576. Springer-Verlag, New York.

Strain, B. R. (1987). Direct effects of increasing atmospheric CO_2 on plants and ecosystems. *Trends Ecol. & Evol.* **2**(1), 18–21.

Strain, B. R., and J. D. Cure, eds. (1985). "Direct Effects of Increasing Carbon Dioxide on Vegetation." DOE/ER-0238. Nat'l. Tech. Info. Serv., Springfield, Virginia.

Thornley, J. H. M. (1976). "Mathematical Models in Plant Physiology." Academic Press, New York.

Thornley, J. H. M. (1980). Research strategy in the plant sciences. *Plant, Cell Environ.* **3**, 233–236.

Tieszen, L. L., Miller, P. C., and Oechel, W. C. (1980). Photosynthesis. *In* "An Arctic Ecosystem: The Coastal Tundra at Barrow, Alaska" (J. Brown, P. C. Miller, L. L. Tieszen, and F. L. Bunnell, eds.), pp. 102–139. Dowden, Hutchinson & Ross, Stroudsburg, PA.

Tissue, D. T., and Oechel, W. C. (1987). Response of *Eriophorum vaginatum* to elevated CO_2 and temperature in the Alaskan tussock tundra. *Ecology* **68**, 401–410.

438   *James F. Reynolds and Paul W. Leadley*

Warren-Wilson, J. (1966). An analysis of plant growth and its control in arctic environments. *Ann. Bot.* **30**, 383–402.

Wielgolaski, F. E. (1975). Functioning of Fennoscandian tundra ecosystems. *In* "Fennoscandian Tundra Ecosystems" (F. E. Wielgolaski, ed.), Part 2, Animals and Systems Analysis, pp. 300–326. Springer-Verlag, Berlin.

Woodmansee, R. G. (1988). Ecosystem processes and global change. *In* "Scales and Global Change" (T. Rosswall, R. G. Woodmansee, and P. G. Risser, eds.), pp. 11–27. Wiley, New York.

Summary

21

Arctic Plant Physiological Ecology in an Ecosystem Context

F. S. Chapin III, R. L. Jefferies, J. F. Reynolds, G. R. Shaver, and J. Svoboda

I. Ecophysiology of Individual Processes

Ecophysiological studies of plants provide a strong basis for predicting changes in species distributions that might result from climatic warming. Arctic plants are extremely tolerant of low temperature. They have a low temperature optimum for photosynthesis and are relatively insensitive to a wide range of temperatures around this optimum (Chap. 8). Work in the Soviet Union, however, demonstrates that this and other physiological traits vary considerably among arctic plants; this variation is related to both their phytogeographic origin and present habitat requirements. The high respiratory potential of arctic plants enables them to maintain at low temperatures relative growth rates comparable to those of temperate plants growing under much warmer conditions (Chap. 8, 9). Effective photosynthesis at low temperature is achieved by maintaining large quantities of chloroplasts and mitochondria (Chap. 8), entailing a substantial nitrogen cost (Chapin and Shaver, 1985; Korner, 1989). The high respiratory potential results in a high maintenance respiration (Chap. 8) and reduced growth (Tieszen, 1978) under warmer conditions, perhaps explaining the restriction of most arctic plants to relatively cold environments. Likewise, uptake of water (Chap. 12) and nutrients (Chap. 15) occurs readily at low temperatures. In contrast, nitrogen fixation in the tundra is a highly temperature-sensitive process, largely restricted to favorable microenvironments, such as the warm surface of bare soil and moss mats (Chap. 14).

Plants of the arctic tundra are basically shade plants, with low light compensation and saturation points (Chap. 8). Again, however, arctic species vary considerably in their light requirements, so that the expected increase in arctic cloudiness (Chap 2) may cause differential photosynthetic responses among species. Some vascular species have low net photosynthetic rates during the day but maintain substantial rates during the arctic "night" (Chap. 8), whereas more-light-requiring species photosynthesize only during the day. Mosses, which generally grow beneath a vascular plant canopy, are particularly effective at using low light, and their growth is strongly reduced when the vascular plant canopy is removed (Chap. 10).

Arctic plants differ from alpine plants in requiring a long photoperiod to prevent senescence and maintain growth (Chap. 5,9). This requirement for a long photoperiod may restrict the growing season of some arctic species in a warmer climate (Chap. 9). The ability to exploit a longer growing season in the event of climatic change may depend on the substantial, but largely unquantified, genetic variability found among local populations and in the buried seed pool (Chap. 17).

The major adaptations of plants to a short growing season relate to life-history traits. Arctic plants initiate inflorescences one or more seasons before flowers are produced to allow sufficient time for development (Chap. 9). The short, unpredictable growing season largely excludes annuals from the Arctic (Chap. 5, 17). Most tundra plants have well-developed stores of carbohydrates (Chap. 9) and nutrients (Chap. 16), which appear necessary to support initial growth in a short growing season (Billings and Mooney, 1968). The carbohydrate stores, however, are comparable to levels found in ecologically similar temperate plants and may reflect storage to avert catastrophe rather than a reserve to support spring growth, at least in those species that have overwintering leaves (Chap. 9). To support spring growth, nutrient stores are generally depleted more severely than carbohydrate stores because soils are still frozen at the time of greatest nutrient demand for growth (Chap. 16). In contrast, conditions are more favorable for photosynthesis at this time of year than in other seasons (Chap. 8). Mosses and lichens are particularly effective at exploiting the short growing season because they grow opportunistically whenever conditions are favorable (Chap. 10).

Because of the effective adaptation of arctic plants to low temperature, low light, and a short growing season, water and nutrients are important as potential limiters of tundra productivity. Plant growth is nutrient-limited in virtually all arctic environments (Chap. 14, 16). Plants minimize this limitation by effective nutrient recirculation from senescing leaves (Chap. 16), by maintaining substantial capacity for nutrient uptake (through both a high physiological potential and a large root biomass; Chap. 15), and by effective absorption of amino acids, a form of nitrogen that may be less effectively exploited by temperate species (Chap. 15). As with photosynthesis and growth, however, major differences exist among tundra species in nutrient uptake and use. Evergreen shrubs retain nutrients in leaves longer than do deciduous plants (Chap. 16), but they have a lower potential to absorb nutrients from soil (Chap. 15). Mosses are a particularly effective filter for nutrients released from the vascular plant canopy and from snowmelt because they live at the ground surface (Chap. 10) and have a high physiological capacity for absorbing nutrients (Chap. 14). Hence, changes in nutrient availability resulting from climate change may alter the species composition of the tundra because of differential capacities for absorbing and retaining nutrients.

Although water stress may be relatively unimportant in wet low-arctic communities, it may be extremely important in limiting plant growth in dry environments, which become progressively more abundant with increasing latitude (Chap. 4). The mechanisms by which tundra plants minimize water stress in dry environments are basically the same as those in temperate plants: osmotic adjustment to maintain turgor at low water potential and reduction in transpiration rate through stomatal closure (Chap. 12). Because tundra species differ in their potential to withstand water stress, the development of drier soils in a warmer climate (Chap. 3) may lead to substantial changes in species composition of different arctic habitats (Chap. 12).

Studies of the biotic environment make up the major new research areas in arctic physiological ecology over the past decade. Competition is important in the Arctic, particularly in the Low Arctic, and plants differ in their competitive ability, depending on the environment in which they grow (Chap. 6, 17). Those species that are competitively inferior in low-nutrient, late successional environments have a higher reproductive allocation (Chap. 17) and are more likely to colonize recently disturbed habitats (Chap. 6) or to become prominent in the buried seed pool (Chap. 17). Although most dominant tundra plants are long-lived perennials, seed reproduction is just as important to their lifetime success as in their temperate counterparts (Chap. 17). Tundra plants also vary substantially in their capacity to deter herbivores or to regrow following defoliation. Evergreen shrubs are strongly protected by secondary metabolites but are sensitive to defoliation, whereas deciduous shrubs, and especially graminoids, are less well protected chemically but capable of compensatory growth in response to herbivory (Chap. 18, 19).

Perhaps the greatest new insights coming from recent work on the physiological ecology of arctic plants relate to the diversity of physiological mechanisms by which arctic plants maintain themselves in an extreme environment. This diversity may provide the basis for the patterning of vegetation in the Arctic at various spatial scales (Chap. 4), although at present we are unable to predict the pattern convincingly from physiological processes alone.

II. Physiological Ecology in an Ecosystem Context

As a heritage of the International Biological Programme, substantial progress has been made in understanding the physiological ecology of arctic plants in an ecosystem context (Wielgolaski, 1975; Bliss, 1977; Brown *et al.*, 1980; Bliss *et al.*, 1981; Miller *et al.*, 1984). Yet, as this book has clearly indicated, we often know least about those physiological processes that are most critical in determining controls over ecosystem function. This is to be expected because, until recently, the main motivation for arctic plant physiological ecology has been to understand how plants cope with the environment, not what is the role of plants in controlling ecosystem processes.

Photosynthetic research has a strong history in the Arctic, much of it aimed at estimating carbon gain by entire plant canopies and thus gross annual production (e.g., Tieszen, 1978). This research set the tone for attempts to link physiological and ecosystem processes. More recently, work on photosynthesis and respiration has been related to patterns of plant growth and maintenance (Chap. 8) and to net plant and ecosystem carbon balance (Chap. 7). Little is currently known about the maintenance of stems and roots, making it difficult to construct realistic whole-plant carbon budgets. Only relatively recently have researchers asked whether carbon balance is the most important factor directly limiting the growth of tundra plants. The growth of arctic plants often shows less response to CO_2 or light than to nutrients (Chap. 7, 9). On the other hand, seasonal growth patterns may be constrained by the high carbon demands of maintenance respiration (Chap. 8). The nature of the links between photosynthesis and growth remains an important area for future research, particularly for mosses and lichens, which are ecologically important in the Arctic but not well studied (Chap. 10).

Plant growth is difficult to investigate in the tundra because of the clonal structure and large belowground biomass of many arctic plants. Although we are learning more about patterns of root phenology in relation to plant growth, patterns of root production and turnover are known only qualitatively. Similarly, patterns of shoot and stem growth have only recently been linked to estimates of production and turnover per unit area of ground (Chap. 9, 16). Generally, the natality and mortality of plant organs are less well documented than growth (Chap. 9). Demographic approaches provide a promising tool to address birth and death rates of plant parts (Chap. 17), but these estimates must be expressed per unit area of ground to provide a useful link to plant carbon balance.

Much of the past work on nutrient uptake has emphasized physiological adaptation, and only recently has attention been given to the significance of the various forms of nitrogen present in the soil and the relative importance of different ecological factors that might control nutrient uptake in the field (Chap. 13, 15). Nothing is now known about the function of mycorrhizae as a nutrient source or carbon sink under field conditions. Nutrient limitation of growth is widespread in the Arctic, indicating that an understanding of plant nutrient balance is an important link not only to nutrient cycling but also to carbon flux in the tundra (Chap. 7, 9, 16). Yet little is known about nutrient acquisition and loss by whole plants or per unit area of ground (Chap. 16). The information available suggests that efficient internal nutrient recirculation within plants is a key factor minimizing demand for nutrients from the soil. Although the nutrient relations of mosses are thought to be quite different from those of vascular plants, these relations have received little attention in the Arctic (Chap. 15). In the High Arctic, nitrogen fixation by cyanobacteria may be a key factor in the nitrogen budget of the tundra (Chap. 14).

Most past work on water relations sought to determine the degree of water stress in arctic plants. Although water stress appears to be a relatively minor factor in wet and moist tundra (Chap. 12), transpiration may play a critical role in determining the water and energy balance of the tundra surface, patterns of soil moisture, and the extent of methane transport from the soil to the atmosphere. Thus, in an ecosystem context, the water relations of plants may have a much more important role than previously appreciated. The water relations of mosses are particularly critical, because, when dry, mosses represent an effective barrier to both heat and water fluxes and therefore strongly influence soil heat and water budgets (Chap. 10). These ecosystem aspects of plant water relations are largely unexplored. In the High Arctic, water directly limits plant growth. In the Low Arctic, water indirectly limits growth because it determines the rates of decomposition and nutrient flow to the root surface.

Biotic interactions also regulate productivity, nutrient cycling, and species composition of tundra habitats. Recent research puts life-history traits such as seed production, germination, and establishment in the context of individual longevity and successional dynamics in the tundra (Chap. 6, 17), but little is known about longevity and turnover rates of individuals in such environments. In the tundra, as in most ecosystems, the links among physiological characteristics, competitive success, and distributions along environmental gradients are tenuous. Exploring these links is an important area for future research. In many ecosystems, herbivory is important in regulating productivity and nutrient cycling, both through direct effects and through disturbance (Chap. 19). Secondary metabolites in plants, which strongly influence patterns of herbivory (Chap. 18), are also important in an ecosystem context because of their effect on the quality of organic matter and rates of decomposition (Chap. 13, 19).

By influencing the quality of organic matter, soil temperature, and moisture regimes, plants affect feedback processes in tundra ecosystems (Chap. 3, 13). Yet these feedback processes are little understood and have not been effectively tied to plant ecophysiological traits. This domain remains the biggest challenge in linking the physiological ecology of arctic plants to ecosystem processes.

III. Physiological Ecology and Climate Change

A. Direct Effects

The expected climate change in the Arctic is likely to be more pronounced and rapid than anything arctic vegetation has experienced (Chap. 2, 5). Some aspects of the change are much less certain than others, and, unfortunately, those climatic characteristics that are least understood are likely to have the greatest effect on the response of arctic vegetation and ecosystems (Table I). The scenario we describe is inevitably simplistic and should be viewed as an initial hypothesis to be refined or disproven by future research. It particularly applies to low-arctic tundra.

Atmospheric concentrations of CO_2 and other greenhouse gases are rising now and will continue to rise in the forseeable future (see Table I; Fig. 1; Chap. 2). The normal photosynthetic response to CO_2 is highly constrained in the present tundra environment because most tundra plants are strongly nutrient-limited (Chap. 7). The vegetation will likely respond fully to increased CO_2 only if ecosystem feedbacks reduce the nutrient limitation of plant growth through increases in nutrient availability (Chap. 13, 14).

Air temperatures will probably increase, but the direct effect of higher air temperature on photosynthesis and growth may be relatively minor because photosynthesis in arctic plants has a broad temperature-response curve (Chap. 8), and growth is often constrained more by nutrients than by carbohydrates or air temperatures (Chap. 7, 9, but see Chap. 8). On the other hand, soil temperature, which will probably also increase (see Table I), will enhance decomposition both directly and indirectly by increasing thaw depth, drainage, and aeration (Chap. 3). Greater nutrient availability and higher root temperature will increase plant growth, making plants more responsive to an increase in CO_2. The extent of soil warming will be strongly affected by the plant-mediated feedback processes discussed below.

The growing season will probably lengthen somewhat in response to increased temperature, although more snowfall may offset this effect. Plant response to this change is uncertain. If plant growth is constrained by nutrient availability in spring and by photoperiod in autumn (Chap. 5, 9), there may be little added growth during a longer season (Chap. 20). The vegetation response to a longer snow-free season may be governed more strongly by microbial responses to a longer growing season than by direct responses of vegetation. Any increases in decomposition and mineralization rates are likely to cause corresponding increases in plant growth (Chap. 13). The extent of plant response would affect the leakiness of tundra ecosystems to nutrient loss.

Cloudiness and precipitation may increase, but the magnitude, and even the direction, of change in precipitation over the entire Arctic is uncertain (see Table I). If, as predicted, most of the increase in precipitation occurs as winter snow, this will offset the tendency toward early snowmelt and will have a minimal effect on summer soil moisture. Increased temperature will enhance evapotranspiration, but the extent of soil drying may range from substantial in

Table I Certainty of Climatic Change and Its Direct Effects on Vegetation and the Anticipated Impact of These Changes on the Function of Tundra Ecosystems

	Certainty of change	Impact of change
Change in environment		
↑ CO_2	+ + +	+
↑ Air temperature	+ + +	+
↑ Season length	+	+ + +
↑ Vapor pressure deficit	+ + +	+ +
↑ Cloudiness	+	+
↑ Precipitation	+	+ + +
↑ Soil temperature	+ +	+ + +
↑ Thaw depth	+ +	+ + +
↑ Drainage	+ +	+ + +
↓ Soil moisture	+	+ + +
↑ Nutrient availability	+	+ + +
↑ Fire	+	+ + +
Direct effects on vegetation		
↑ Photosynthesis	+ +	+ +
↑ Nutrient uptake	+	+ + +
↑ N fixation	+ +	+ +
↑ Transpiration	+ +	+ +
↑ Methane transport	+	+ + +
↑ Production	+	+ + +
↑ Nutrient status	+	+ + +
Indirect effects on ecosystems		
↑ Litter quality	+	+ + +
↓ Nutrient immobilization	+	+ + +
↑ Decomposition	+	+ + +
↑ Herbivory	+	+ + +
↑ Methane oxidation	+	+ + +
↑ Fire frequency	+	+ + +
Feedbacks to globe		
↓ Albedo	+	+
↑ CO_2 release	+	+ + +
↓ Methane release	+	+ + +

upland tundra to minimal in wet tundra (Chap. 3, but see Chap. 10). In polar desert or dry heath, a drier soil may reduce plant growth because of direct drought effects (Chap. 12), whereas in moist tussock tundra, improved drainage caused by deeper thawing of the soil may substantially enhance decomposition and nutrient availability to plants. In these two systems, changes in soil moisture may be the most important effects of climate change. As the surface soil and mosses dry, thermal conductance and soil warming lessen, tending to offset the drying effects. As mentioned above, soil moisture in waterlogged sites may change little, but if soils do become better drained, decomposition (and, consequently, plant growth) will increase (Chap. 7, 13). Clearly, we must improve our predictions of changes in soil moisture and their effects on other physical parameters, decomposition processes, and vegetation.

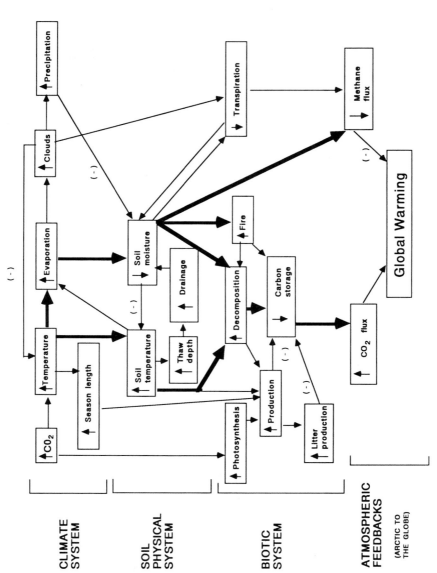

Figure 1 Major cause-and-effect linkages among climate, soil, and biota in the Arctic and their feedbacks to global climate. Thickness of arrows indicates expected strength of effect.

B. Plant-Mediated Feedbacks

If the predominant effects of climate change in the Boreal Zone and Low Arctic are warmer and drier soils and more frequent fires (Chap. 3, 18), the vegetation will probably shift toward a greater abundance of shrubs, less *Sphagnum*, and perhaps, in the long run, colonization by trees (Chap. 5, 18). In the High Arctic we can expect increased plant cover and invasion by low shrubs (Chap. 4). This altered vegetation would respond differently from the present tundra vegetation to climate change, increasing the complexity of our predictions. The vegetation would be more responsive to CO_2 (Chap. 7) and air temperature (Chap. 9). Soil temperature, on the other hand, may not increase because of greater shading by the plant canopy and possible invasion by *Sphagnum* (Chap. 3, 10). Present patterns of herbivory in different areas of the Arctic would change, and in some areas the dominant herbivore may be replaced. For example, caribou may replace muskoxen in some areas, whereas moose may replace caribou if the boreal forest extends northward (Chap. 18, 19). Because of enhanced litter quality, decomposition would increase, resulting in greater productivity and a shift in the composition of vegetation toward shrubs (Chap. 13, 18). More evapotranspiration would dry the soil further, increasing decomposition and productivity (Chap. 12). Quite different vegetation scenarios are also possible. Past climatic warming from glacial periods to interglacials caused *Sphagnum* to increase, leading to peat accumulation and presumably lower nutrient availability (Chap. 10). Now, however, we are moving from an interglacial to an even warmer climate. Clearly, predictions based on projected changes in vegetation are even more tenuous than the predictions of direct climatic effects on the present tundra community. To improve these predictions, we must improve our understanding of the biotic interactions and feedback processes in tundra ecosystems.

C. Feedback Processes Affecting Climate

Although the immediate responses of tundra plants to impending climate change are important in their own right, these responses may also provoke further changes in global climate. If decomposition increases more than productivity, CO_2 flux to the atmosphere will increase substantially and could enhance global warming. Similarly, warming of wet, poorly drained soils might increase methanogenesis, or drying of soils might enhance methane oxidation (thereby reducing methane emissions), in turn either increasing or decreasing the atmospheric greenhouse effect. Increased shrubbiness would reduce the albedo of winter snow and enhance atmospheric warming; this effect would be particularly pronounced if trees invaded the present tundra zone.

IV. Conclusion

Studies of the physiological ecology of arctic plants have provided, and will continue to provide, new insights into the mechanisms by which plants adapt to and tolerate a severe physical environment. But understanding the physiological basis of the plant-mediated feedback processes controlling the functioning of ecosystems demands very different types of studies. These involve the examination of ecological processes occurring over longer time scales than those traditionally studied. We need to go beyond the short-term kinetics of resource acquisition and examine the constraints on plant and microbial carbon and nutrient budgets. Moreover, the processes that control ecosystem functioning may differ from those mediating plant activity. For example, the water relations of mosses, which govern soil heat and water flux, may be more important to understanding the water budgets of tundra ecosystems than are mechanisms of osmotic adjustment in vascular plants. Finally, we need to place considerable emphasis on feedbacks between plants and soil and between herbivores and plants if we are to predict the impacts of climatic change.

Integrated studies of the mechanisms regulating biogeochemical cycling and biological productivity in the Arctic are now urgently needed. Climatic changes, which may have already begun, are expected to be most pronounced in the Arctic. Many native human populations in the Arctic depend heavily on local biological resources and would be affected by a change in climate. Furthermore, the sensitivity of arctic ecosystems to human activities associated with resource extraction and exploitation will undoubtedly change with changes in vegetation and the thermal balance of the tundra. Thus climatic change could strongly alter the interactions between human populations and their arctic environment.

The response of arctic ecosystems to global climate change could in turn feed back substantially on global climate. Large carbon pools in tundra soil could augment atmospheric CO_2 and climatic warming if decomposition is stimulated more than primary production. The tundra is also an important contributor to the global methane budget, so environmental changes affecting methane flux could have an important global effect. An improved understanding of physiological controls over biotic processes and of the feedbacks among these processes is key to predicting the changing role of the Arctic in the global system.

References

Billings, W. D., and Mooney, H. A. (1968). The ecology of arctic and alpine plants. *Biol. Rev. Camb. Philos. Soc.* **43**, 481–529.

Bliss, L. C., ed. (1977). "Truelove Lowland, Devon Island, Canada: A High Arctic Ecosystem." Univ. of Alberta Press, Edmonton.

Bliss, L. C., Heal, O. W., and Moore, J. J., eds. (1981). "Tundra Ecosystems: A Comparative Analysis." Cambridge Univ. Press, Cambridge.

Brown, J., Miller, P. C., Tieszen, L. L., and Bunnell, F. L., eds. (1980). "An Arctic Ecosystem: The Coastal Tundra at Barrow, Alaska." Dowden, Hutchinson & Ross, Stroudsburg, Pennsylvania.

Chapin, F. S., III, and Shaver, G. R. (1985). Arctic. *In* "Physiological Ecology of North American Plant Communities" (B. F. Chabot and H. A. Mooney, eds.), pp. 16–40. Chapman and Hall, London.

Körner, C. (1989). The nutritional status of plants from high altitudes: A worldwide comparison. *Oecologia* **81**, 379–391.

Miller, P. C., Miller, P. M., Blake-Jacobson, M., Chapin, F. S., III, Everett, K. R., Hilbert, D. W., Kummerow, J., Linkins, A. E., Marion, G. M., Oechel, W. C., Roberts, S. W., and Stuart, L. (1984). Plant–soil processes in *Eriophorum vaginatum* tussock tundra in Alaska: A systems modeling approach. *Ecol. Monogr.* **54**, 361–405.

Tieszen, L. L., ed. (1978). "Vegetation and Production Ecology of an Alaskan Arctic Tundra." Springer-Verlag, New York.

Wielgolaski, F. E., ed. (1975). "Fennoscandian Tundra Ecosystems." Part 1, Plants and Microorganisms. Springer-Verlag, Berlin.

Index

Physiological Ecology
A Series of Monographs, Texts, and Treatises

Continued from page ii